Progress in the Chemistry of Organic Natural Products

The volumes of this classic series, now referred to simply as "Zechmeister" after its founder, Laszlo Zechmeister, have appeared under the Springer Imprint ever since the series' inauguration in 1938. It is therefore not really surprising to find out that the list of contributing authors, who were awarded a Nobel Prize, is quite long: Kurt Alder, Derek H.R. Barton, George Wells Beadle, Dorothy Crowfoot-Hodgkin, Otto Diels, Hans von Euler-Chelpin, Paul Karrer, Luis Federico Leloir, Linus Pauling, Vladimir Prelog, with Walter Norman Haworth and Adolf F.J. Butenandt serving as members of the editorial board.

The volumes contain contributions on various topics related to the origin, distribution, chemistry, synthesis, biochemistry, function or use of various classes of naturally-occurring substances ranging from small molecules to biopolymers.

Each contribution is written by a recognized authority in the field and provides a comprehensive and up-to-date review of the topic in question. Addressed to biologists, technologists, and chemists alike, the series can be used by the expert as a source of information and literature citations and by the non-expert as a means of orientation in a rapidly developing discipline.

All contributions are listed in PubMed.

A. Douglas Kinghorn · Heinz Falk ·
Simon Gibbons · Yoshinori Asakawa · Ji-Kai Liu ·
Verena M. Dirsch
Editors

Natural Product Diversity of Australian Rainforests

Volume 128

By Anthony Richard Carroll

Springer

Editors
A. Douglas Kinghorn
College of Pharmacy
Ohio State University
Columbus, OH, USA

Heinz Falk
Institute of Organic Chemistry
Johannes Kepler University Linz
Linz, Austria

Simon Gibbons
Natural and Medical Sciences Research Center
University of Nizwa
Nizwa, Oman

Yoshinori Asakawa
Faculty of Pharmaceutical Sciences
Tokushima Bunri University
Tokushima, Japan

Ji-Kai Liu
School of Pharmaceutical Sciences
South Central University for Nationalities
Wuhan, China

Verena M. Dirsch
Department of Pharmaceutical Sciences
University of Vienna
Vienna, Austria

ISSN 2191-7043 ISSN 2192-4309 (electronic)
Progress in the Chemistry of Organic Natural Products
ISBN 978-3-031-90473-8 ISBN 978-3-031-90474-5 (eBook)
https://doi.org/10.1007/978-3-031-90474-5

This Springer imprint is published by the registered company Springer Nature Switzerland AG
The registered company address is: Gewerbestrasse 11, 6330 Cham, Switzerland

About This Book

The book reviews the natural product chemistry of Australian rainforest plants, with a particular focus on their fascinating phylogenetic relationships. It discusses the unique flora of Australia, which is of Gondwanan origin, and how this has led to a diverse range of natural products. The text explores the chemical constituents of various plant groups, including those of gymnosperms, basal angiosperms, magnoliids, monocots, and eudicots, with an emphasis on the Australian rainforest species. Key points covered include the historical context of Australian flora, its high level of endemism, and the impact of past climatic changes on rainforest distribution. The book also details the specific phytochemicals found in these plants (about 2400), such as alkaloids, lignans, and terpenoids, and their biological activities. Overall, this comprehensive volume highlights the importance of Australian rainforests as a rich source of unique natural products and emphasizes the need for further research in this area, in particular given that only approximately 10% of these rainforest species have been investigated so far.

Contents

Contributor

Anthony Richard Carroll School of Environment and Science, Griffith University, Southport, QLD, Australia

Australian Rainforests—A Living Museum of Natural Product Chemical Diversity

Anthony Richard Carroll

Contents

1 Introduction

The current biogeographic distribution of plants is linked to their evolution in response to climatic, edaphic and biotic factors, tectonic plate movements and long-distance dispersal facilitated by wind, ocean currents, birds and mammals [1]. After the breakup of Pangaea about 200 million years ago (mya), land plant evolution

A. R. Carroll (✉)
School of Environment and Science, Griffith University, Gold Coast Campus, Southport, QLD, Australia
e-mail: a.carroll@griffith.edu.au

A. D. Kinghorn, H. Falk, S. Gibbons, Y. Asakawa, J.-K. Liu, V. M. Dirsch (eds.),
Natural Product Diversity of Australian Rainforests, Progress in the Chemistry of Organic Natural Products 128, https://doi.org/10.1007/978-3-031-90474-5_1

occurred independently on the two supercontinents Laurasia and Gondwana and this has resulted in uniquely distinct floras in the Northern and Southern Hemispheres even after the breakup of these supercontinents [2]. An example of this disjuncture is represented by the lineages of Spermatophytes (vascular seed producing plants). Australia's flora is distinctly of Gondwanan origin and their associated natural products (NP) reflects this. This contribution explores the NPs chemistry of Australian rainforest spermatophytes from a phylogenetic perspective.

The spermatophytes include the gymnosperms and angiosperms. Gymnosperms are vascular plants that have exposed seeds and no flowers while angiosperms have flowers and seeds that are fully enclosed within a fruit.

2 Phylogeny and Taxonomy of Gymnosperms and Angiosperms

The natural product chemistry of plants is linked intrinsically to their phylogeny. To understand the chemical diversity of seed plants requires some knowledge of taxonomic relationships between species.

2.1 Gymnosperm Phylogeny

The gymnosperms have an ancient lineage dating back to the mid-Devonian (385 mya). They diverged from ancestral angiosperms during the early Carboniferous (350 mya) and radiated extensively by the end of the Paleozoic (252 mya) to become the dominant land plants throughout the Mesozoic (252–65 mya) [3]. The 1035 extant species of gymnosperms are divided across three classes, eight orders, 13 families and 83 genera [3, 4]. Most of the genera (66) are monotypic or contain less than five species and only three genera (*Cycas*, *Pinus*, and *Podocarpus*) have upwards of a hundred or more species. Broadly, these classifications include the cycads (Cycadales), ginkgo (Ginkgoales), and a class, Pinopsida, that places the Gnetophytes (Gnetidales) sister to the conifers, Pinales, with the Cupressales and Araucariales forming two additional orders [4]. Climatic and tectonic changes over geological time have resulted in waves of extinction and radiation of gymnosperm species. A significant proportion (39%) of the world's forests (mostly in the northern hemisphere) are dominated by gymnosperms from the families Pinaceae, Cupressaceae, and Taxaceae. By contrast, in the southern hemisphere, biogeographical distribution patterns of the Cupressaceae subfamily Callitroideae, and the families Cycadaceae, Zamiaceae, Araucariaceae and Podocarpaceae can be explained mainly by vicariance associated with the breakup of Gondwana followed by dispersal into Asia or North America. The Gondwanan origins of many Gymnosperm lineages is compelling with 31 genera found in Australia, New Zealand, New Caledonia, and New Guinea and a further eight found in South America and seven in Southern Africa [3, 4].

2.2 Angiosperm Phylogeny

Angiosperm diversification led to a decline in gymnosperm dominance by the early Cenozoic (65 mya). Angiosperms are now the dominant land plants with over 300,000 species, occurring in all ecological niches on land and dominating tropical forests, savannah, heathland, and arid habitats. The temporal and spatial origins of angiosperms is hotly debated by paleobotanists. Molecular clocks inferred from DNA sequencing of extant species in conjunction with fossil records, suggest a Permian to Jurassic origin (300–200 mya), potentially from Pteridosperms (seed ferns) with massive diversification to generate most extant plant families occurring in the early Cretaceous (145 mya) [5]. Angiosperm geographic origins remain unresolved. The earliest fossil records have been found in northern parts of Gondwana, but the scarcity of fossils makes it difficult to track their true spatial origins. The combined evidence suggests that angiosperms originated in a warm, wet tropical region of the world and an initial wave of morphological diversification also occurred in the tropics. Subsequent waves of diversification and speciation occurred in the Cenozoic (from 65.5 mya) in temperate and arid regions making these regions repositories of younger, species rich, lineages. The presence of angiosperms in areas of the world with tropical climates since the early Cretaceous (145 mya), has resulted in these regions accumulating diverse species over time. Higher family/genus diversity and older lineages are found in regions with a history of tropical climates compared to temperate/arid regions, but speciation and diversification rates are higher in the temperate/arid regions. This suggests that the ancient tropical rainforests are refugia of more ancient lineages that have undergone multiple extinction events [5].

A worldwide consortium of plant taxonomists has developed a phylogenetic tree to classify the flowering plants based on evolutionary evidence provided by morphology and genomics [6]. The latest update recognizes 416 families of angiosperms grouped into 70 orders and over 300,000 species [7]. From an evolutionary perspective, these orders are further grouped based on morphological features that are considered either rudimentary or more advanced with genomic evidence to support or refute these physical attributes. The basal angiosperms (Amborellales, Nymphaeales, Austrobaileyales (ANA)), represented by three orders and seven families, are a small group comprising a few hundred species derived from ancestral angiosperms that diverged from the lineage that led to most other flowering plants. The Magnoliids also possess morphological traits that resemble those found in the earliest angiosperms and with anatomical similarities to gymnosperms. The Magnoliids are represented by approximately 10,000 species in five orders and 19 families. The Monocotyledons (Monocots) possess only one embryonic leaf. This clade includes the highly species rich orchids and grasses and comprises 11 orders, 77 families and 70,000 species. The Eudicotyledons (Eudicots) is the most diverse clade with over 200,000 species that typically possess two embryonic leaves. The 313 families in 45 orders are further subdivided into basal eudicots and core eudicots with the core eudicots further subdivided mostly into two major clades—the Superrosids and Superasterids (Fig. 1).

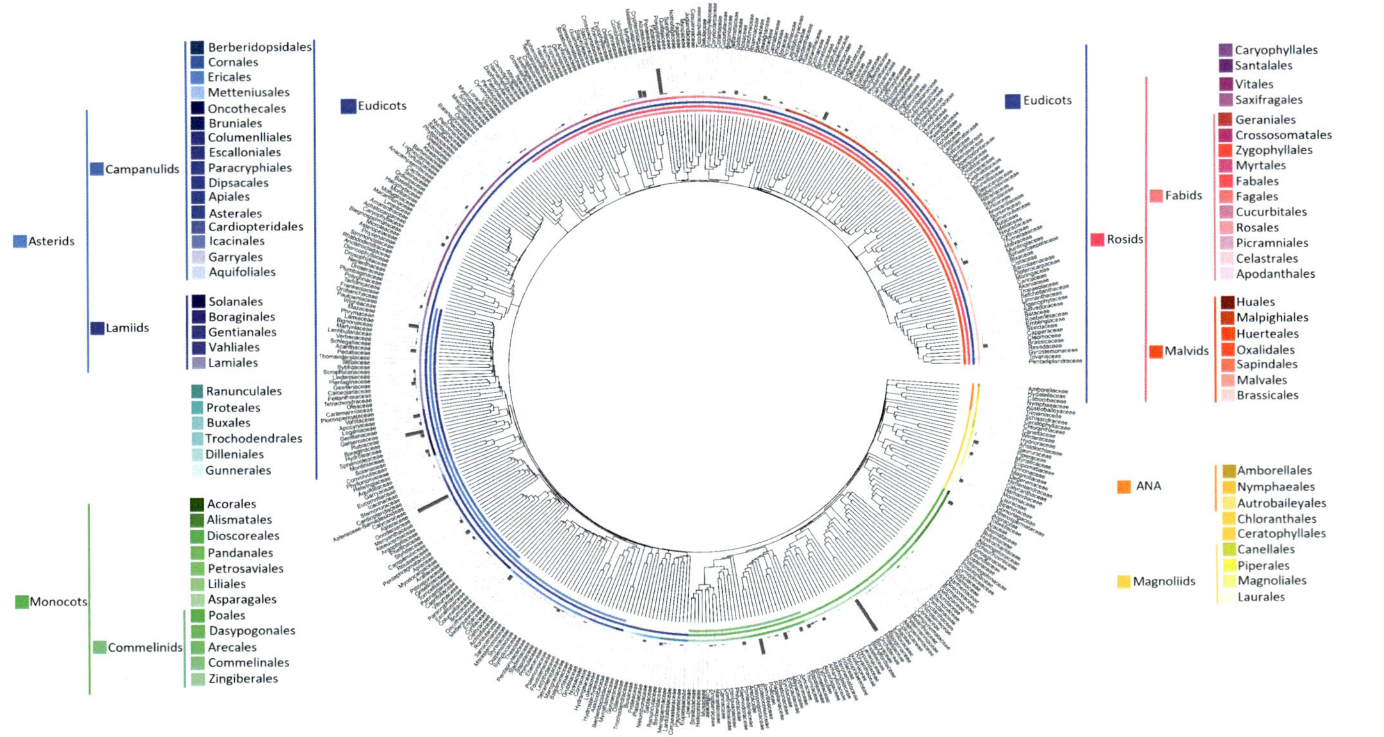

Fig. 1 Molecular phylogeny of the Angiosperms [7]. The bar chart track shows the proportion of species in each family. The Orchidaceae is the largest family with ~28,000 species

2.3 Phylogeny and Plant Natural Products

Over 200,000 natural products and their derivatives have been reported from gymnosperms and angiosperms [8]. These compounds can be subdivided into structural classes based either on their biosynthetic origins (terpenoids, lignans, flavonoids, isoflavonoids, phloroglucinols, polyketides, peptides, etc.) or by chemical features (alkaloids). A striking trend that appears when natural products, classified by structure class, are mapped with producer plant phylogeny, is the predictability of natural product distribution patterns [8]. Put simply, the biosynthetic enzymes required to make specific natural products is genetically encoded and this matches exquisitely with recent results obtained from molecular phylogenetic studies. Some natural products are polyphyletic (i.e., they are likely to be examples of convergent evolution where their biosynthesis has evolved independently in several unrelated taxa), while other natural product structure classes are monophyletic (i.e., the biosynthetic enzymes evolved once). The biosynthesis of pyrrolizidine alkaloids for example, appears to have evolved independently at least three times based on enzymatic analysis. In contrast, although the occurrence of benzyl isoquinoline alkaloids (BIA) is segregated into clusters of related families interspersed across disparate angiosperm phylogeny, the detection of key BIA biosynthetic enzymes across the diversity of angiosperms has shown them to be present but dormant across the non-BIA producing angiosperms and this suggests a monophyletic origin of BIAs prior to the divergence of the eudicots [9, 10]. In this context, one might expect there to be regional segregation of NP chemical diversity based on the unique floras that occur in different continents. This volume explores this from an Australasian perspective.

3 Australian Flora

3.1 History and Endemism

Australia is one of 17 countries considered to be megadiverse [11]. It is home to unique flora and fauna that in part owe their evolutionary lineages to the heritage of Australia as an eastern sector of the supercontinent Gondwana. The separation of Australia from Antarctica about 45 mya saw the rafting of plant and animal species with shared origins to those found on other southern continental land masses that formed Gondwana [12]. These include a wide range of flora and fauna in common across the southern hemisphere. Vertebrates include the ratite birds of South America, South Africa, New Zealand, Australia, and New Guinea, the parrots of Australasia and South America, the songbirds of Australia, New Guinea, and New Zealand and the monotremes and marsupials of Australia, New Guinea, and South America (Plate 1). Massive diversification and radiation within some of these faunas (for example the marsupials and songbirds) occurred almost exclusively in Australia in isolation.

Plate 1 Lemuroid ringtail possums (*Hemibelideus lemuroides*). A montane rainforest marsupial species from north Queensland. (Painting AR Carroll)

Some gymnosperm and angiosperm plant families also show a distinctive Gondwanan distribution with species from the Cycadaceae, Araucariaceae, Podocarpaceae, Nothofagaceae, Winteraceae, Atherospermataceae, Proteaceae, Myrtaceae, and Elaeocarpaceae occurring in both Australia and one or more of South America, South Africa, New Zealand, New Caledonia, and New Guinea. Even families with more widespread distributions comprise clades that are either Gondwanan or Laurasian [13]. The Rutaceae, Proteaceae, and Myrtaceae are families that have diversified and radiated almost exclusively in Australia. Vicariance is not the only explanation for the dominance of Gondwanan flora in Australia. Molecular phylogenetic analysis has increasingly suggested that long distance dispersal via transoceanic or land bridge routes has played an equally important role in shaping the extant flora of Australia, particularly in the tropics. Three main routes for this dispersal are from Africa (and Madagascar) from the west, South America from the east, and Asia from the north (Fig. 2). These long-distance dispersal events likely occurred prior to the Australian continental plate colliding with the Pacific plate about 20 mya [13]. Prior to this collision, diversification within key families, most notably within the Myrtaceae, Rutaceae, and Proteaceae, occurred in isolation for over 40 million years.

The Australian flora is highly endemic with over 80% of plant species found nowhere else [13]. Due to northern migration, the continent now occupies a latitude characterized by aridity with 75% of Australia classified as arid or semi-arid [14]. This was not always the case and when Australia separated from Antarctica 45 mya ago, the continent was largely covered in rainforest (mesic-adapted vegetation) and today some of Australia's rainforests can be considered the oldest in the world with

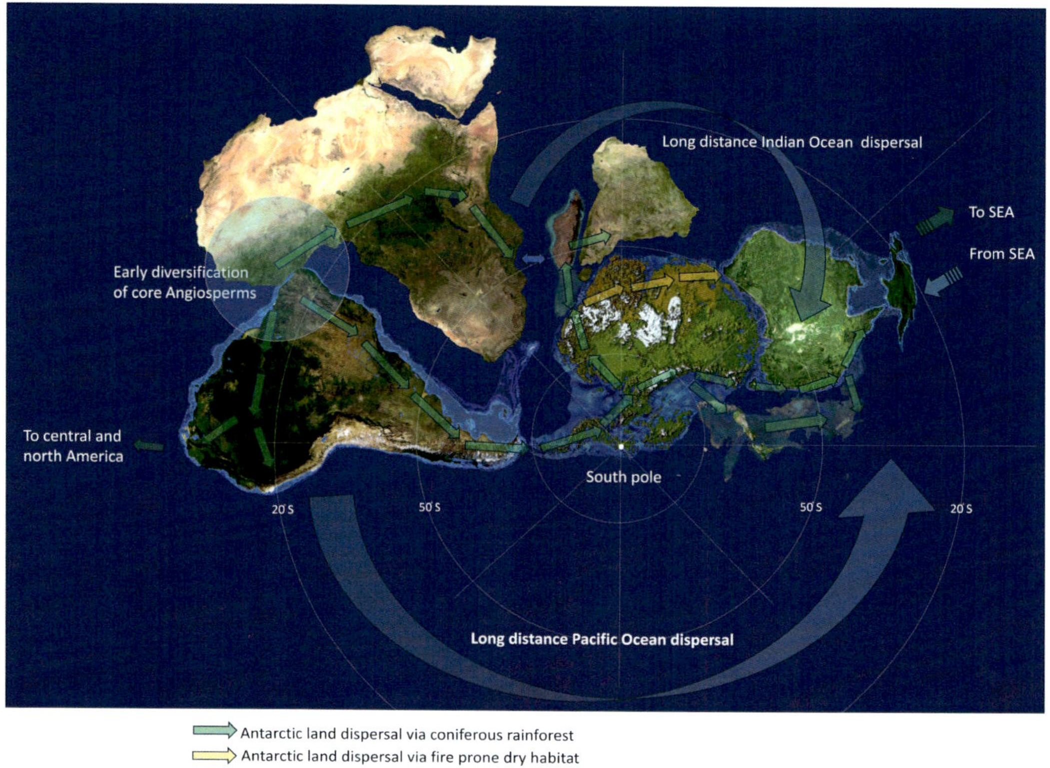

Fig. 2 Fragmented Gondwana 90 mya showing dispersal routes of ancient lineages of angiosperms into Australia from a region in northern Africa/South America where early diversification of the core angiosperms (Magnoliids, Monocots and Eudicots) potentially occurred. Some of these lineages then diversified in Australia before dispersing into other parts of the world via long-distance oceanic dispersal or land bridges

uninterrupted survival for over 180 million years [15]. Much of this diverse mesic-adapted flora most likely either originated or became established in Gondwana during the Cretaceous era (145–66 mya) and was widespread during the Tertiary period (66–2.6 mya) [16]. Mesic-adapted vegetation in Australia now includes the very wet ecosystems of rainforest, swamps, wet heath, and alpine vegetation surrounded by vegetation occupying regions that are seasonally wet including sclerophyllous forests and heaths. The changing climate of Australia led to contraction of mesic-adapted communities during the late Miocene (10.4–5 mya) and through adaptation to the drying climate some of these lineages evolved and speciated into the diverse dry adapted sclerophyllous vegetation that dominates much of Australia's landscape today. The ebb and flow of drier and wetter climates also corresponds with the ebb and flow of the extent of rainforests throughout Australia. During dry periods the rainforests contracted to refugia along perennial water courses and mountaintops and then expanded out under more favorable wetter climates. As such, the distribution of rainforest species across the Australian landscape is quite dispersed and disjunct with remnant pockets of rainforest species occurring in locations surrounded by large areas of sclerophyllous vegetation. To confound this, there is also evidence that Australia's first nations people dispersed some rainforest species to disparate locations, and this indicates that the current locations of these species is not the result of contraction of ancestral populations [17, 18]. The largest areas of the once widespread rainforest vegetation occur in disjunct remnant populations along the eastern seaboard of Australia. These refugia have been linked throughout geological history by climate bridges where favorable temperature and rainfall during interglacial periods facilitated expansion of these rainforest hotspots only to again contract as the climate became drier. Species that have limited capacity for seed and pollen dispersal remain restricted to these refugia, while other species with more adaptable seed and pollen dispersal traits and capacity to adapt to changing water and nutrient regimes, have expanded and speciated. The lack of phosphorus replenishment through glaciation and volcanism in most Australian soils has resulted in widespread nutrient depleted soils and to adaptations by many Australian flora to phosphorus depleted soils [19]. Compared with western and central Australia, the eastern coastline of Australia has been subjected to volcanic activity over the last 65 million years up until about 10,000 years ago and this has resulted in a mosaic of fertile basaltic sediments across the eastern escarpment of Australia. The Australian rainforests therefore occupy unique climatic, edaphic, and topographic niches that are refuges for once more widespread flora of the occupied landscapes that dominated areas of Gondwana. They also contain the ancestors of many of the sclerophyllous plant species that now dominate much of the modern Australian landscape. For example, the large rainforest tree *Stockwellia quadrifida*, found only on rainforested mountain ranges in northeastern Queensland, is an ancestor to the Eucalypts that now dominate the drier forests and woodlands across Australia [20]. As the Australian continent drifted north it encountered the Pacific plate thus providing a bridge across the Wallace line for dispersal of southern biota and invasion by Eurasian biota. Although this has influenced the flora of northern Australia to some extent, endemism is high across all Australian floral regions [13].

The first nations people of Australia have managed, manicured, and utilized the rainforest of Australia for over 60,000 years and, at the time of European colonization of Australia in 1788, rainforest covered 1% of the continent [21]. Indigenous fire management practices have in some locations restricted the encroachment of rainforests [22]. Over the last 240 years clearing for agriculture, mining, water reservoirs, timber extraction and urbanization has seen a dramatic contraction of rainforest. A patchwork of rainforest remnants now covers 36,000 km^2 of land (3% of Australia's native forest area), approximately half of pre-European levels [23]. The largest areas lost are the lowland tropical and subtropical rainforests of Queensland and New South Wales that contained the highest concentrations of species diversity. Australia is estimated to contain 18,700 angiosperms (7% of world species diversity) and 120 gymnosperms (~11.7% of world species diversity), of which 93% are endemic [24]. Of the 83 gymnosperm genera, 19 (23%) occur in Australia including several monotypic genera or genera with under five species (10 genera) that occur nowhere else. A further 12 genera occur in New Zealand, New Caledonia, and New Guinea [59]. Many of the Asian gymnosperms likely have a Gondwanan origin and have dispersed north via Australasia. The occurrence of the species poor genera suggests that they are likely to be relicts of more species-rich assemblages that have contracted into climatically favorable refugia. The phylogenetic relationships within the gymnosperms, further points to a Laurasian/Gondwanan split and the generically diverse Australian gymnosperms are a repository of past diversity and are of ancient antiquity. Australia's rainforests host 37 species of gymnosperms.

Within the flowering plants, southwest Western Australia is recognized as a biodiversity hotspot, with ~12,000 species most of which are found nowhere else [11]. Interestingly, this biodiversity is concentrated into highly species-rich genera because of massive speciation, but diversity at the family level is much more restricted. In contrast, the dispersed rainforests of eastern Australia, also a megadiversity hotspot, have high family level biodiversity but much lower species per genus level biodiversity. This matches with the rainforests being refugial pockets of once more widespread vegetation that has seen waves of extinction rather than speciation. Consequently, the Australian rainforests represent an important link to a past flora and, given that much of Australia's sclerophyllous vegetation own their origins to this ancient rainforest stock, it also provides a link to this now widespread and uniquely Australian flora.

The Daintree region in north Queensland is considered to contain the world's oldest rainforest since it has vegetated the region continuously for over 180 million years. The antiquity and uniqueness of Australia's rainforests has been recognized by the UNESCO World Heritage Commission in four locations: the Wet Tropics of Queensland World Heritage Site (containing the largest area of tropical rainforest in Australia), K'gari (Fraser Island) World Heritage Area, Queensland (the largest sand island in the world and home to the only tall rainforest growing on sand), the Gondwanan Rainforests of Australia World Heritage Site (the subtropical rainforests of South East Queensland and New South Wales), and the Tasmanian Wilderness World Heritage Area (home to the temperate rainforests of Tasmania) [25].

3.2 The Flora of Australian Rainforests

Webb et al. have classified Australia's rainforest into three ecofloristic regions; (A) subtropical and temperate humid evergreen rainforests, (B) tropical humid evergreen grading into highly season raingreen (monsoon) forests, and (C) subtropical moderately seasonal humid/subhumid raingreen forests [26]. Further subdivision of each region generates eight ecofloristic provinces based on the dominant floristic elements. This classification scheme aligns with the paleogeographic evidence that Australia's rainforests are mostly of Gondwanan origin. Region A covers the rainforest found along the coast and adjacent ranges of southeast Queensland down to southern Victoria and across Bass Strait into western Tasmania. Region B includes the rainforests of central and northeast Queensland and disjunct areas in the northwest Northern Territory and the Kimberley region of Western Australia. Region C mainly covers disjunct areas to the west of the Great Dividing Range from central NSW up to north Queensland (Plate 2 and Fig. 3).

There is no single source of data listing the rainforest plant taxa of Australia and so the number of Australian rainforest seed plant taxa is difficult to quantify. Based on data obtained from several sources such as the Australian Tropical rainforest plants database (eighth edition) [28], Rainforest plants of Australia (Rockhampton to Victoria) [29] and records of Tasmania rainforest species [30], the best estimate is approximately 3,800 species, subspecies, and varieties. These are divided into 37 gymnosperm and 3,767 angiosperm taxa, and all the gymnosperms are endemic. The angiosperms include 3,090 species that are either wholly endemic to Australia or have ranges extending into New Guinea and/or other parts of Malesia (Indonesia, the Malaysian peninsula, Borneo and the Philippines) or Oceania (including New Caledonia, Vanuatu, Solomon Islands, and Fiji). Another 383 species have extended ranges into Southeast Asia and beyond and 294 angiosperm species have naturalized into Australia's rainforests (Fig. 4). In the following section that describes the chemistry of Australian rainforest species, compounds from species collected in Malesia or Oceania but not studied in Australia are included. For the most part, those Australian rainforest species with ranges extending into continental Asia (northern China and India), only the chemistry obtained from plant material collected in Australia is discussed. Some of these species have been extensively studied outside of Australia and references to reviews of this chemistry are included where relevant.

The tropics contains the largest number of species and the most overlap with southeast Asian species in keeping with the theory that after the Australian tectonic plate collided with the Pacific plate, island hopping across the Wallace line facilitated bidirectional dispersal. Long-distance transoceanic seed dispersal was also facilitated by birds, fruit bats and ocean currents.

Rainforests further south in Australia have higher levels of endemism and species of Gondwanan origin. There is a distinct latitudinal zonation of species diversity that reflects different ancestral flora communities. Given that the land area covered by rainforest is so low, it is remarkable that the species richness in these areas is very high in comparison with the total seed plant diversity of Australia. Family

Plate 2 Rainforest ecosystems top row (gallery palm dominated rainforest, tropical lowland rainforest, gymnosperm dominated subtropical rainforest), middle row (temperate rainforest dominated by gymnosperms and ferns, littoral rainforest, low land tropical rainforest) bottom row (subtropical rainforest, rainforest with emergent Eucalypts, gymnosperm dominated temperate rainforest). (Photographs AR Carroll)

diversity is high in the Australian rainforests with taxa from 187 families and 942 genera being present. Australia's rainforests contain a disproportionate number of Amborellales, Nymphaeales, Austrobaileyales, and Magnoliid angiosperms and even within Eudicot families, there are higher proportions of species of ancient lineage. Of the 29 families of Amborellales, Nymphaeales, Austrobaileyales, and Magnoliids, 15 have species in Australian rainforest. Three families, Eupomatiaceae, Himantandraceae, and Austrobaileyaceae, are restricted to continental Australia (defined

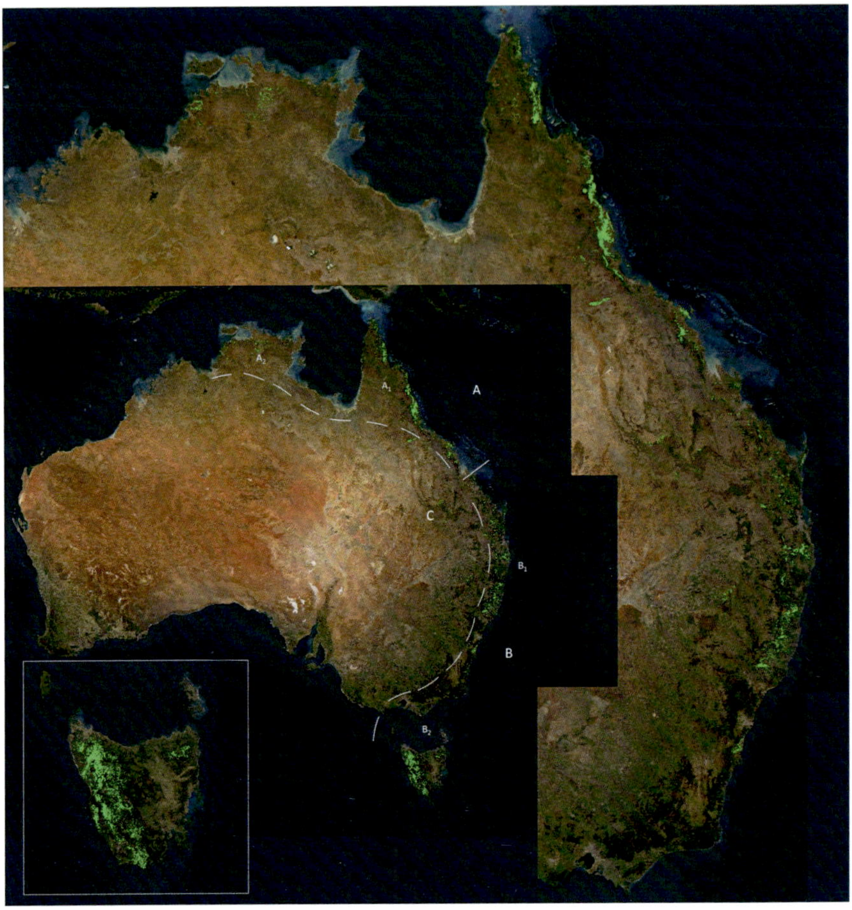

Fig. 3 Map of rainforest distribution in Australia. Information obtained from the Forests of Australia (2018) dataset [27]. Floristic regions A, B, and C from Webb et al. [26], represent distinct floral assemblages that reflect different climatic-edaphic-topographic conditions throughout their evolution. Region A contains the tropical rainforests of north Queensland (A_1) and the monsoon forests of Northern Territory and Western Australia (A_2). Region B contains the subtropical rainforests of north New South Wales and south Queensland (B_1) and the temperate rainforests of Victoria and Tasmania (B_2). Region C contains the dry rainforests and vine thickets west of the Great Dividing Range

here as Australia and New Guinea). Another family, Trimeniaceae, occurs only in continental Australia and Pacific islands. Two families, Atherospermataceae and Winteraceae, have disjunct distributions in continental Australia, New Caledonia, New Zealand, and the Americas. The Calycanthaceae, with only three genera, has

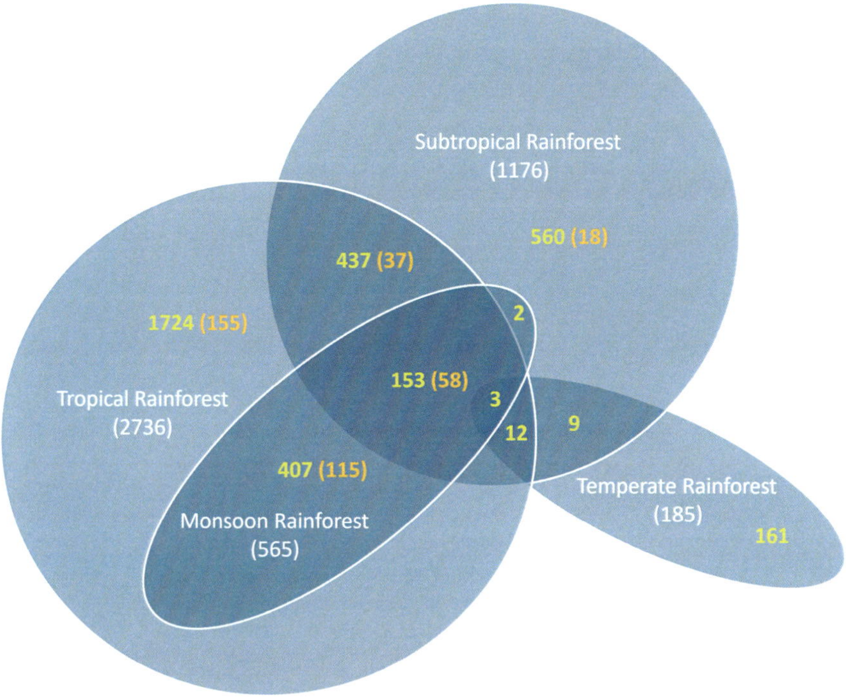

Fig. 4 Graph showing the number of angiosperm species in tropical (North Queensland), subtropical (Northern New South Wales/Southern Queensland), monsoon (Northern Territory/Northwestern Australia) and temperate rainforests (South New South Wales/Victoria/Tasmania). Numbers: white—total species in rainforests from floristic regions; yellow—total species within each isolated or overlapping block, orange—species also found in Asia

a monotypic genus, *Idiospermum*, that is restricted to the tropical lowland rainforests of North Queensland. *Idiospermum australiense* is the sole 90-million-year-old southern hemispheric remnant of this Gondwanan family from which significantly later migration and diversification led to the northern hemisphere distribution of the only other species in the family [31]. Australia's rainforest also contains high genus diversity of Fabaceae (54), Proteaceae (35), Myrtaceae (38), Rubiaceae (35), Rutaceae (31), and Orchidaceae (91).

4 Chemical Constituents of Australian Rainforest Spermatophytes

Constituents from approximately 10% of Australia's rainforest spermatophytes (439) have resulted in 2,385 natural products being reported (~2% of all plant NPs). This is notable since Australia's rainforests account for just 0.024% of the earth's land

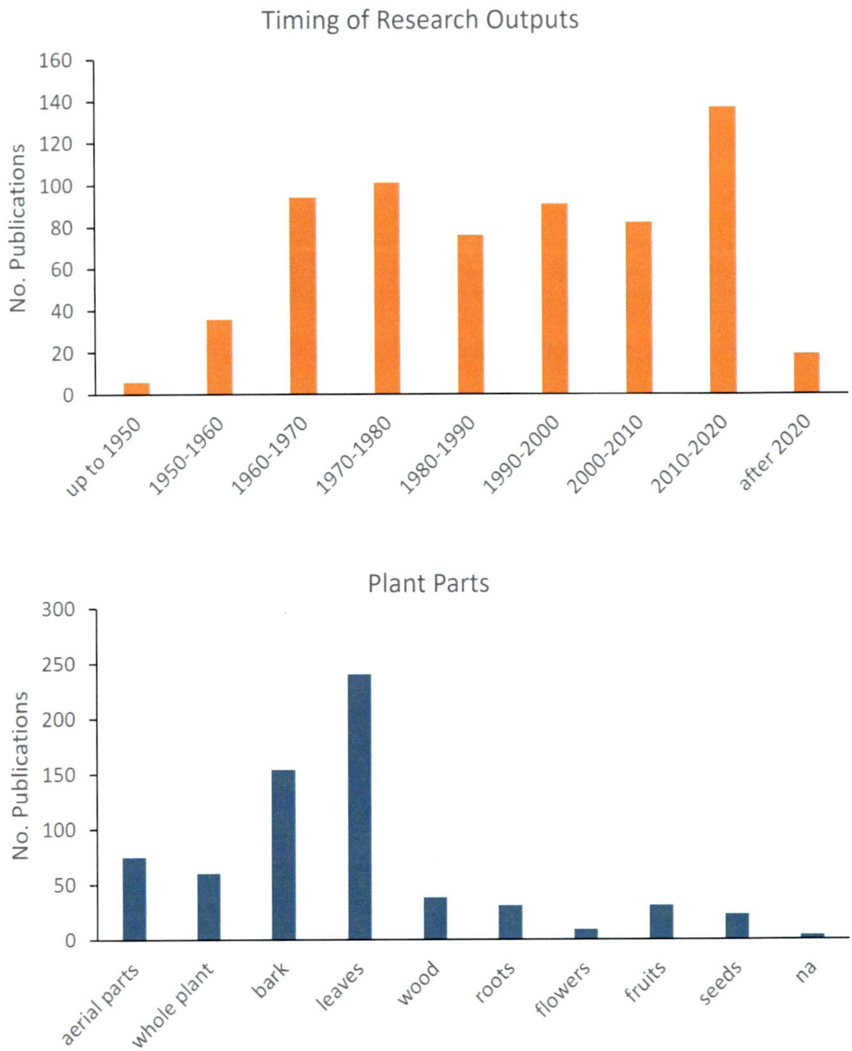

Fig. 5 Decadal frequency of publications reporting natural product from Australian rainforest plants and number of publications reporting on the natural product chemistry of various rainforest plant parts

area (149 million km^2) and ~0.18% of global spermatophyte diversity. In addition, through analysis of the large natural product database, DEREP-NP, [32] (containing ~113,000 plant NPs) that have been annotated herein with taxonomic information, 1723 NPs have been reported from 129 spermatophyte species, collected mostly in Asia, but also those occurring naturally in Australian rainforests [32].

Two major collections of Australian rainforest plant species have been undertaken to study their NP chemistry and medicinal properties. The first—the Australian

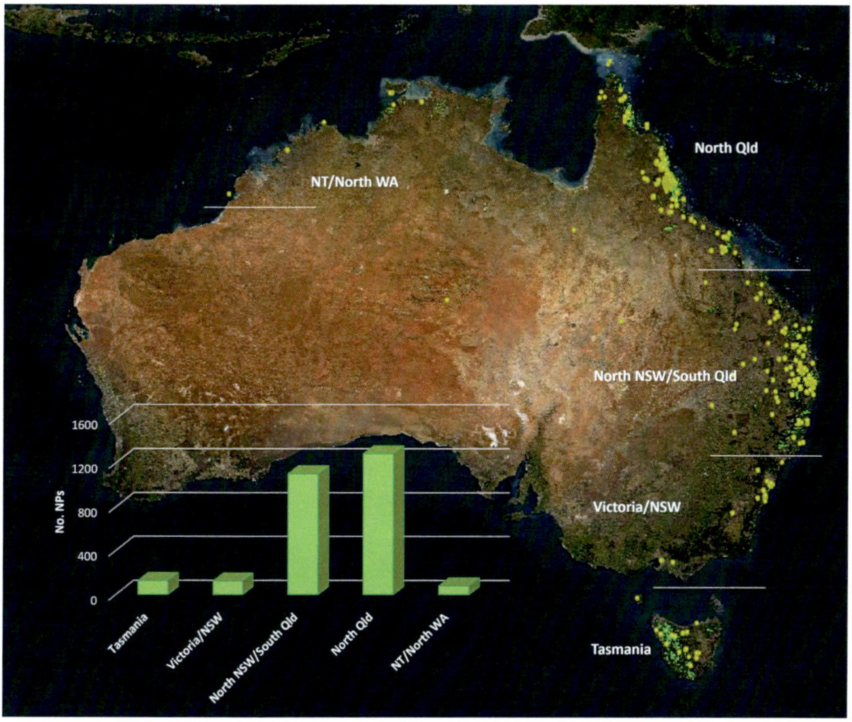

Fig. 6 Collection locations of rainforest plant species that have yielded natural products. The bar chart inset shows the number of natural products reported from rainforest plants collected in different floristic regions

Phytochemical Survey, was carried out by the Australian government research organization CSIR (now CSIRO) between 1940 and 1980 initially in response to the need to find alternative sources of essential medicines when supply routes to Australia were interrupted during the Second World War [33]. The survey quickly expanded after the cessation of fighting to become a major scientific enterprise involving botanists, pharmacologists, and organic chemists. Readers interested in the history of the Australian Phytochemical Survey are directed to the publication of Price et al. [33]. The survey used field-based chemical spot tests to identify potential alkaloid-producing species. Information summarized in the book "Plants for Medicine" shows that 602 Australian rainforest species were surveyed [34]. Between 1993 and 2007, collections of Queensland plants were undertaken as part of a collaborative drug discovery project funded by AstraZeneca (initially Astra) with Griffith University (GU) (Brisbane, Australia) [35]. Over 15 years, ~20,000 plant samples were collected by Queensland herbarium botanists, and this included ~10,000 samples from 1,750 Queensland rainforest species. Unsurprisingly, a substantial proportion of species previously collected during the Australian Phytochemical Survey were also present in the GU collection (450 species), but this was complemented with a high number of species not previously collected (1300 species) [36]. High-throughput screening of extracts obtained from these samples across a range of biological targets of interest to AZ led to many publications highlighting new chemistry and bioactivities. The project was complemented by a new strategy in the early 2,000's to also generate, and screen lead-like fraction libraries and the provision of purified NPs for biological evaluation [37, 38]. Many of the NPs isolated during this time had previously been reported from other plant species but are reported from specific Australian rainforest species for the first time here [39]. Since the conclusion of the AZ project in 2007, the plant samples continue to be a valuable resource (NatureBank) available to researchers from around the world to access under license for drug discovery [40]. Another major biodiscovery enterprise by Q-biotics (previously Ecobiotics) focuses specifically on North Queensland rainforest species and ecological criteria are used to target specific species for collection [41]. A further 81 endemic species not found in either the NatureBank or Australian Phytochemical Survey collections have also been reported to contain unique NPs. From the 1960's to 2010, the number of research studies reporting on constituents from Australian rainforest plants averaged between 8–10 papers per year, but this has increased over the decade from 2010 to 2020 to 14 per year. Leaf, bark, aerial parts, and whole plant (the latter two both generally obtained from understorey plant species) are the most studied plant materials. Little attention has been paid to reproductive tissues (flowers, fruits, and seeds) while roots (often difficult to access) have only been studied sporadically (Fig. 5). Rainforest species that have yielded natural products have been sampled across all habitat types with the most compounds being obtained from species in the tropics and subtropics (Fig. 6). Since only ~10% of Australia's rainforest species have been investigated, the paucity of studies on specialized plant parts highlights that there is still a lot of natural product chemistry to be discovered from Australia's rainforests.

The following section provides a summary of the natural products reported from Australian spermatophytes. Since there are ~2,400 unique natural products that have been documented, only some important examples are covered in detail below. For all compounds, their sources and references are summarized in the accompanying tables. Bioactivities of natural products isolated from rainforests species are described if they are unique or possess activities of relevance for drug discovery and development. The criteria used here follow those utilized elsewhere, but, in general, cytotoxicity weaker than 10 μM, microbial *MIC* values weaker than 64 μM, and enzymatic activities weaker than 10 μM, are not discussed [42].

4.1 Chemistry of Australian Rainforest Gymnosperms

Ninety-two natural products have been reported from Australian rainforest gymnosperms. The natural products were reported from cycads, and pines from the families Araucariaceae, Podocarpaceae, and Cupressaceae (Fig. 7).

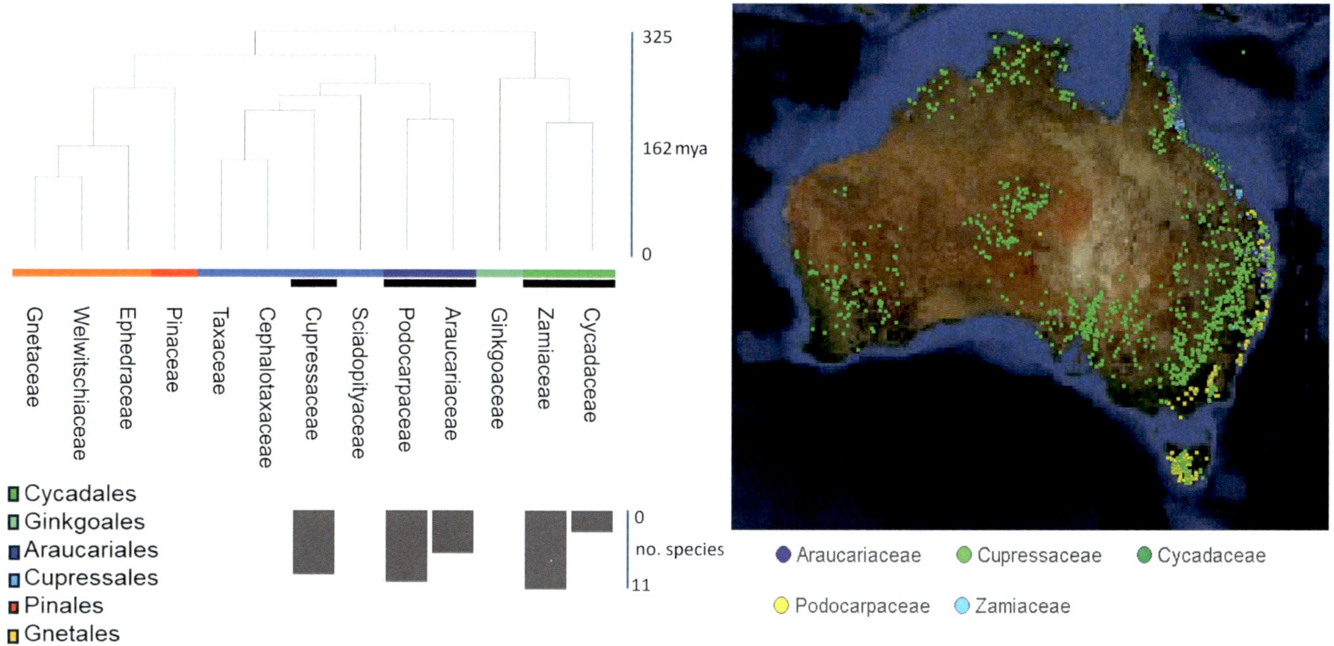

Fig. 7 Age calibrated phylogenetic tree of the Gymnosperm families color-coded by orders [3]. The black tabs under the tree represent families that have species in Australian rainforests. The bar chart depicts the number of species found in Australian rainforests in each family. The map of the distribution of herbarium voucher specimen records for Australian gymnosperm species that have been chemically investigated was obtained from the Atlas of Living Australia [43]

4.1.1 Cycadales

Only five species of cycads (Cycadaceae and Zamiaceae) have been investigated (Table 1, Fig. 8 and Plate 3) and these yielded the azoxyglycosides macrozamin (**1**) and cycasin (**2**) [44, 45]. These azoxyglycosides are highly toxic.

Table 1 Distribution of natural products isolated from the cycads

Species	Collection location[a]	Plant part[b]	Compound class	Compound(s)	Refs.
Cycadaceae					
Cycas media	SEQ	lf	Azoxyglycoside	**1**	[44]
Zamiaceae					
Bowenia serrulata	NQ	s	Azoxyglycoside	**1, 2**	[45]
Bowenia spectabilis	NQ	s	Azoxyglycoside	**1**	[45]
Lepidozamia hopei	NQ	s	Azoxyglycoside	**1**	[45]
Lepidozamia peroffskyana	NQ	s	Azoxyglycoside	**1**	[45]

[a] SEQ = Southeast Queensland, NQ = North Queensland, [b] lf = leaves, s = seeds

Plate 3 Selection of Australian rainforest Cycadales that have been studied. *Lepidozamia peroffskyana* (left and middle, inset—mature seeds) (Mt. Tamborine, Southeast Queensland), *Bowenia spectabilis* (right, Mission Beach, north Queensland). (Photographs AR Carroll)

Fig. 8 Chemical structures of compounds **1** and **2** isolated from cycads

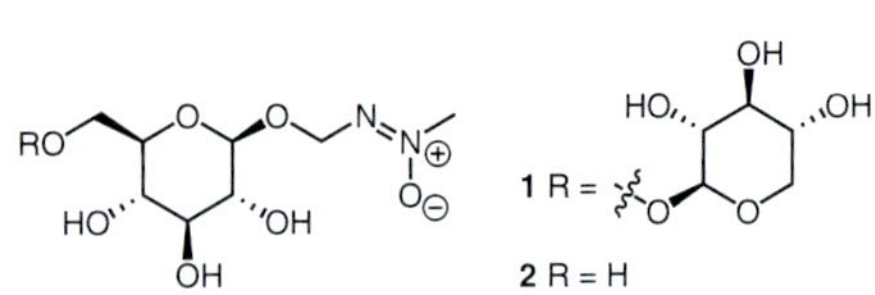

4.1.2 Pinopsida

Six species in the Araucariaceae, including the living fossil *Wollemia nobilis*, known only from fossils until a stand of mature trees was discovered in rainforest lining a canyon in the Blue Mountains west of Sydney in 1995, have been studied (Plate 4, Table 2) [46]. These species are the only Australian gymnosperm source of labdane diterpenoids (Fig. 9) (agatholic acid (**3**), agathic acid (**4**), agathic acid 16-methyl ester (**5**), 13*β*-hydroxylabda-8,14-dienoic acid (**6**), methyl 13*β*-hydroxylabda-8,14-dien-19-oate (**7**), 13*β*-hydroxylabda-7,14-dienoic acid (**8**), methyl 13*β*-hydroxylabda-7,14-dien-19-oate (**9**), 13*β*-hydroxylabda-8(17),14-dienoic acid (**10**), methyl 13*β*-hydroxylabda-8(17),14-dien-19-oate (**11**), levopimaric acid (**12**), communic acid (**13**), 7-hydroxy-labda-8(17),13(16),14-trien-19-yl-(*E*)-coumarate (**14**), 7-hydroxy-labda-8(17),13(16),14-trien-19-yl-7′-*O*-methyl-(*E*)-coumarate (**15**), 7-hydroxy-labda-8(17),13(16),14-trien-19-yl-(*Z*)-coumarate (**16**), 7-hydroxy-labda-8(17),13(16),14-trien-19-yl-7′-*O*-methyl-(*Z*)-coumarate (**17**), 7-oxocallitrisic acid (**18**), methyl communate (**19**), methyl isocupressate (**20**), 15-hydroxy-(8*E*),13-labdadien-19-oic acid (**21**), methyl acetyl-isocupressate (**22**), methyl 15-acetoxy-(8*E*),13-labdadien-19-oate (**23**), methyl 7-oxo-19-acetoxy-(8*E*),13-labdadien-15-oate (**24**), (8*E*)-labda-19-ol (**25**), 15-hydroxy-(8*E*),13-labdadien-19-al (**26**), 15-acetoxy-(8*E*),13-labdadien-19-ol (**27**), 15-acetoxy-(8*E*),13-labdadien-19-al (**28**), (8*E*),13-labdadien-15,19-diacetate (**29**), acetyl-isocupressic acid (**30**), isocupressic acid (**31**), sandaracopimaric acid (**32**), wollemol (**33**), and wollemolide (**34**)) [47–55].

Most gymnosperm bis-flavonoids (agathisflavone (**35**), 7″-*O*-methyl-agathisflavone (**36**), cupressuflavone (**37**), 7,7″-di-*O*-methylagathisflavone (**38**), 7-*O*-methylcupressuflavone (**39**), di-*O*-methylcupressuflavone (**40**), bilobetin (**41**), agathisflavone-4′,7,7″-trimethyl ether (**42**), cupressuflavone-4′,7,7″-trimethyl ether (**43**), cupressuflavone-4′,4″,7,7″-tetramethyl ether (**44**), hinokiflavone (**45**), 7,4-di-*O*-methylagathisflavone (**46**), kayaflavone (**47**), 7,7,4,4-tetra-*O*-methyl-amentoflavone (**48**), 7,4,7,4-tetra-*O*-methyl-robustaflavone (**49**)) are also reported from Araucariaceae species (Fig. 10) [51, 56–61].

Araucariaceae species have also yielded the β-amyrin derivative 2-*O*-acetyl-11-keto-boswellic acid (**50**) [51], cinnamate derivatives (phloretic acid (**51**), 4-*n*-butoxylphenylpropanetriol (**52**), 5-*p*-(*Z*)-coumaroylquinic acid (**53**), and 5-*p*-(*E*)-coumaroylquinic acid (**54**)) [51, 62], the shikimic acid derivative, quinic acid (**55**), carotenoid glycosides ((6*R*,9*S*)-3-oxo-α-ionol-9-*O*-*β*-D-glucopyranoside (**56**), and (6*S*,9*S*)-roseoside (**57**)) [62], a glycoside of pinene, namely, 6-*O*-acetyl-pina-2-ene-4,10-diol-10-*O*-D-glucopyranoside (**58**) [55], and the flavonoid 4′-*O*-methyl-scutellarein (**59**) (Fig. 11) [55]. The four species studied from the sister family Podocarpaceae have yielded isoflavones, irisolidone (**60**) and biochanin A (**61**) [63], an anthocyanin, cyanidin-3-rutinoside (**62**) [64], steroids (podecdysones A (**63**)

Plate 4 Assorted Australian rainforest gymnosperms. Top row *Araucaria cunninghamii* (left), *Agathis robusta* (right), bottom row left to right: *Podocarpus elatus*, *Atherotaxis cupressoides* (left) *A. biwillii* (right). (Photographs AR Carroll)

and B (**64**)) [65, 66], and isoprenylated phenylpropanoids such as dacrinial (**65**), zanthoxylol (**66**), and dacriniol (**67**) (Fig. 11) [67].

Table 2 Distribution of natural products isolated from gymnosperms (Araucariaceae, Podocarpaceae, Cupressaceae)

Species	Collection location[a,b]	Plant part(s)[c]	Compound class	Compound(s)	Refs.
Araucariaceae					
Agathis atropurpurea	NQ	b	Diterpenoid	**3**	[47]
Agathis microstachya	NQ	b	Diterpenoid	**4**, **5**	[48]
Agathis robusta	cult. (Italy) SEQ	lf	Bisflavonoid	**35–37**	[56]
	SEQ	res	Diterpenoid	**6**, **7**, **8–13**	[49, 50]
Araucaria bidwillii	cult. (India, Egypt) SEQ	lf	Bisflavonoid	**36**, **38–43**	[51, 57, 58]
	cult. (Egypt) SEQ	lf	Cinnamate	**51**	[51]
	cult. (Egypt) SEQ	lf	Diterpenoid	**14–18**, **50**	[51]
	cult. (Egypt) SEQ	lf	Triterpenoid	**50**	[51]
Araucaria cunninghamii	cult. (India) SEQ	lf	Bisflavonoid	**36**, **38–41**, **44–48**	[59]
	cult. (China) SEQ	lf	Carotenoid	**56**, **57**	[62]
	cult. (China) SEQ	lf	Cinnamate	**52–54**	[62]
	cult. (India) SEQ	lf	Diterpenoid	**19**	[52]
	cult. (China) SEQ	lf	Shikimate	**55**	[62]
Wollemia nobilis	cult. (Italy) NSW	tw/cone	Bisflavonoid	**36**, **40**, **42–44**, **46**, **48**, **49**	[60, 61]
	cult. (Italy) NSW	lf/cone	Diterpenoid	**4**, **19**, **30**, **31–34**	[53–55]
	cult. (Italy) NSW	cone	Monoterpenoid	**58**	[55]
	cult. (Italy) NSW	cone	Flavonoid	**59**	[55]
Podocarpaceae					

(continued)

Table 2 (continued)

Species	Collection location[a,b]	Plant part(s)[c]	Compound class	Compound(s)	Refs.
Prumnopitys amara	NQ	hw	Isoflavonoid	**60**, **61**	[63]
Podocarpus lawrencei	Tas	lf, cone	Anthocyanin	**62**	[64]
Podocarpus elatus	SEQ	b	Steroid	**63**, **64**	[65, 66]
Lagarostrobos franklinii	Tas	hw	Phenyl propane	**65–67**	[67]
Cupressaceae					
Athrotaxis cupressoides	Tas	lf	Diterpenoid	**80**	[71]
	Tas	lf, tw	Homoerythrane	**68–78**	[68]
Athrotaxis laxifolia	Tas	lf	Diterpenoid	**80**	[71]
	Tas	lf, tw	Homoerythrane	**68**, **69**, **72–75**, **79**	[69]
Athrotaxis selaginoides	Tas	lf	Diterpenoid	**80**	[71]
	Tas	lf, tw	Homoerythrane	**68**, **69**, **72–76**, **79**	[69]
	Tas	hw	Lignan	**81–83**	[72]
Callitris glaucophylla	NQ	lf, tw	Sesquiterpenoid	**84–87**	[73, 74]
Callitris intratropica	NQ	w	Lignan	**93**, **94**	[75]
	NQ	lf	Sesquiterpenoid	**88**, **89**	[75]
	cult. (S Africa) NQ	lf	Sesquiterpenoid	**90**, **91**	[76]
Callitris macleayana	NQ	lf	Bisflavonoid	**95**	[78]
	SEQ	lf	Lignan	**93**	[78]
	NQ	hw	Sesquiterpenoid	**92**	[77]

[a] SEQ = Southeast Queensland, NQ = North Queensland, NSW, New South Wales, Tas = Tasmania, cult. = cultivated, [b] natural location shown for those species studied from cultivated material, [c] lf = leaves, s = seeds, tw = twigs, res = resin, ap = aerial parts, rt = roots, fl = flowers, wp = whole plant, b = bark, hw = heartwood, w = wood

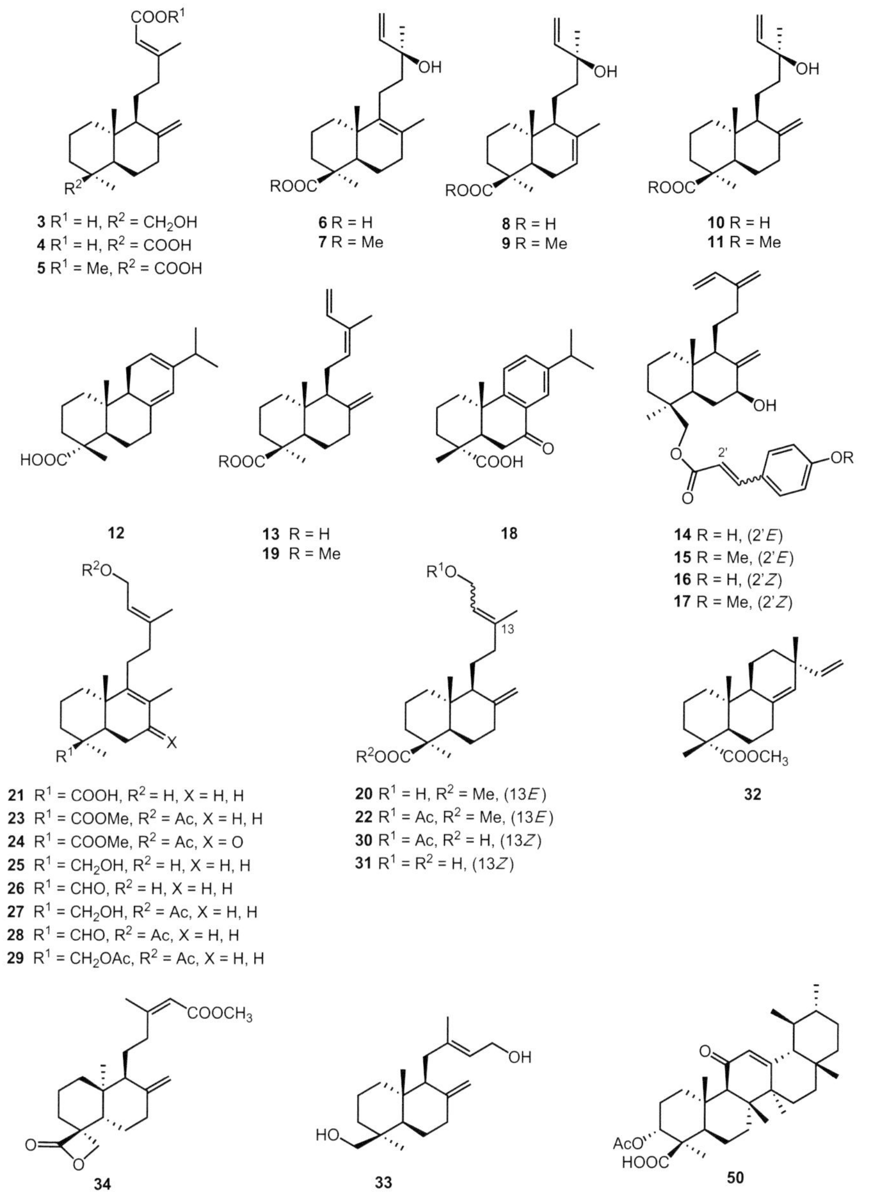

Fig. 9 Structures of terpenoids **3**–**33** and **50** reported from Australian rainforest gymnosperms

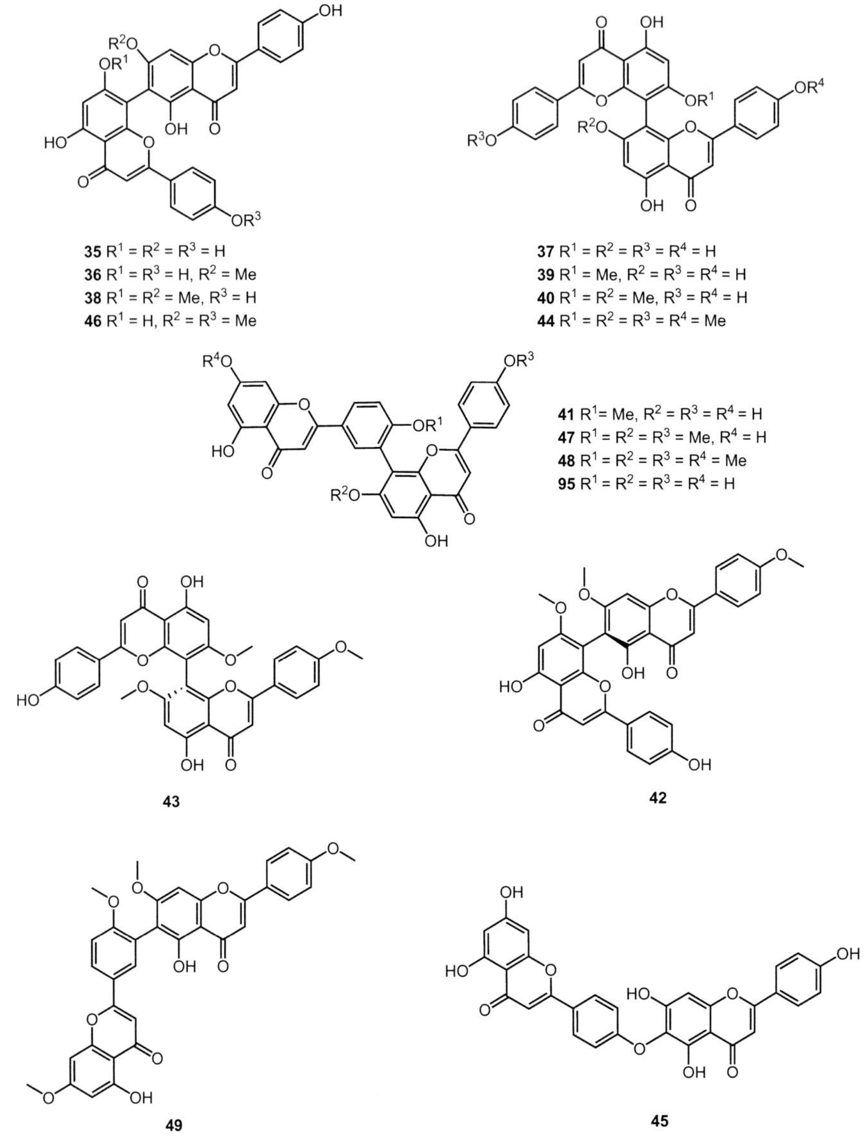

Fig. 10 Structure of bis-flavonoids **35**–**49** and **95** isolated from Australian rainforest gymnosperms

The three *Athrotaxis* species from the family Cupressaceae found in temperate rainforests in Tasmania have yielded a series of homoerythrinan alkaloids (Fig. 12). These include taxodine (**68**), 3-*epi*-schelhamnericine (**69**), *O*-methylathrocupressine (**70**), 2-*epi*-homoerythratine (**71**), homoerythratine (**72**), 2-hydroxytaxodine (**73**), 2-hydroxyisotaxodine (**74**), 2-*epi*-hydroxyisotaxodine (**75**), athrocupressine (**76**), 2-acetoxytaxodine (**77**), 2-acetoxyisotaxodine (**78**), and selaginoidine (**79**). These

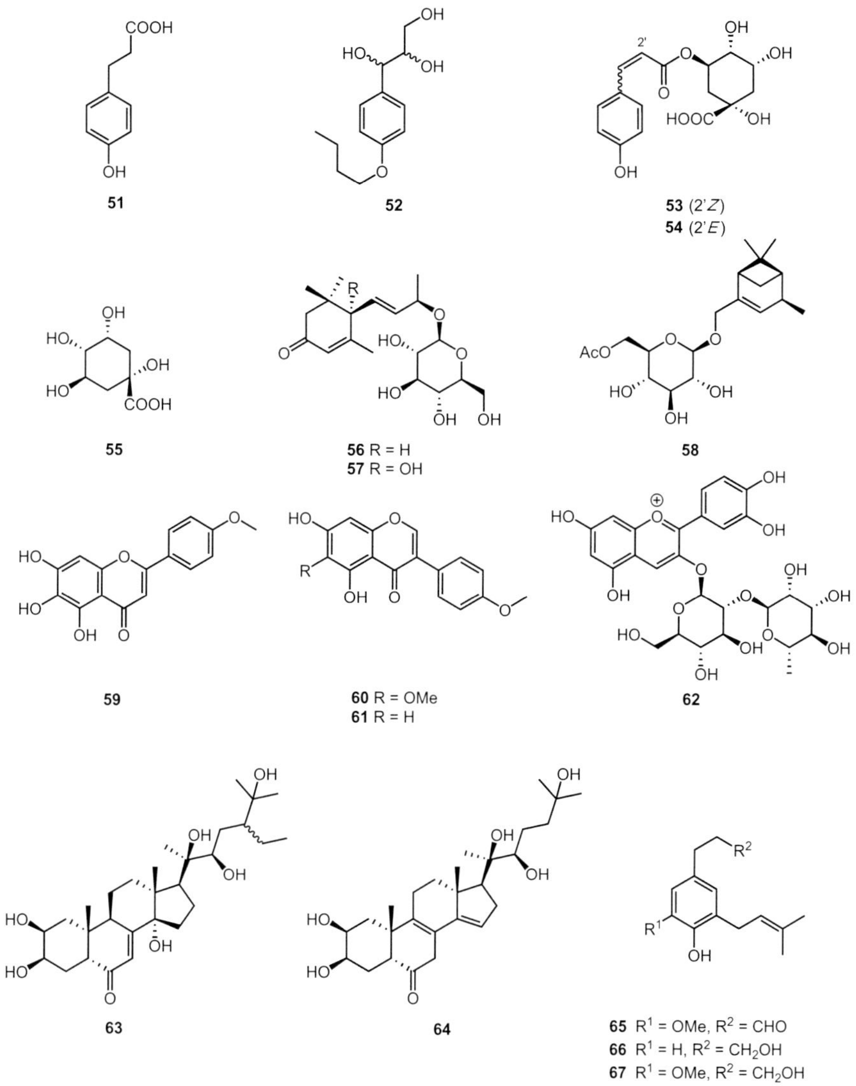

Fig. 11 Structures of compounds **51**–**67** isolated from Australian rainforest gymnosperms

alkaloids are related to those found in the monocot family Colchicaceae [68–70]. The *Athrotaxis* species also contain the diterpenoid 8*β*-hydroxyisopiniarene (**80**) [71], while *A. selaginoides* contains three decarboxylated lignans, hinokiresinol (**81**), agatharesinol (**82**) and athrotaxin (**83**) [72]. Three *Callitris* species (Cupressaceae) have yielded a series of sesquiterpenoids: ilicic acid methyl ester (**84**),

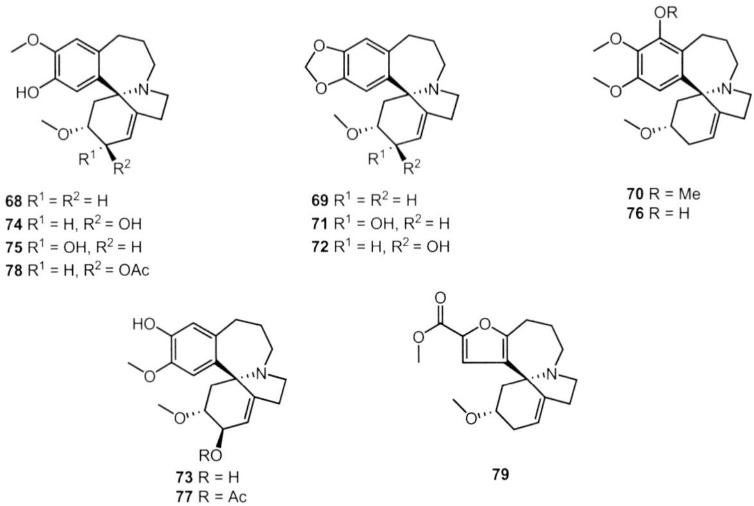

Fig. 12 Structures of alkaloids **68–79** reported from Australian rainforest gymnosperms

costic acid (**85**), dihydrocolumellarin (**86**), columellarin (**87**), callitrin (**88**), callitrisin (**89**), (−)-eudesma-1,4(15)-11-triene (**90**), (−)-α-guaioxide (**91**), and 3,10-dihydroxydielmentha-5,11-diene-4,9-dione (**92**) [73–77]. The lignans deoxypodophyllotoxin (**93**) and deoxypodorhizon (**94**) [75, 78], and a bis-flavone amentoflavone (**95**) were also reported [78]. The sesquiterpenoids (**84–87**) show termiticidal activity (Fig. 13) [73, 74].

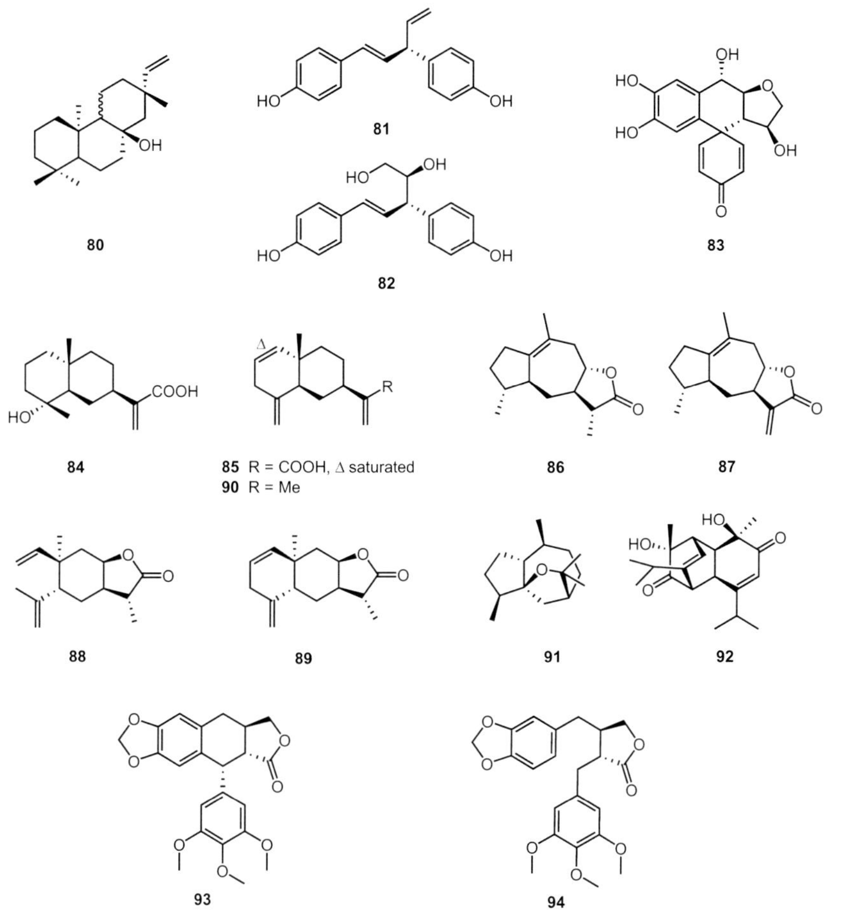

Fig. 13 Structures of terpenoids **80**, **84–92**, and lignans **81–83**, **93**, and **94** reported from Australian rainforest gymnosperms

4.2 Chemistry of Australian Rainforest Basal and Magnoliid Angiosperms

The angiosperms are the major flora of Australian rainforests, and the source of most natural products reported. All but 10 of the 49 orders that contain Australian rainforest species have had at least one species that has been investigated chemically (Fig. 14).

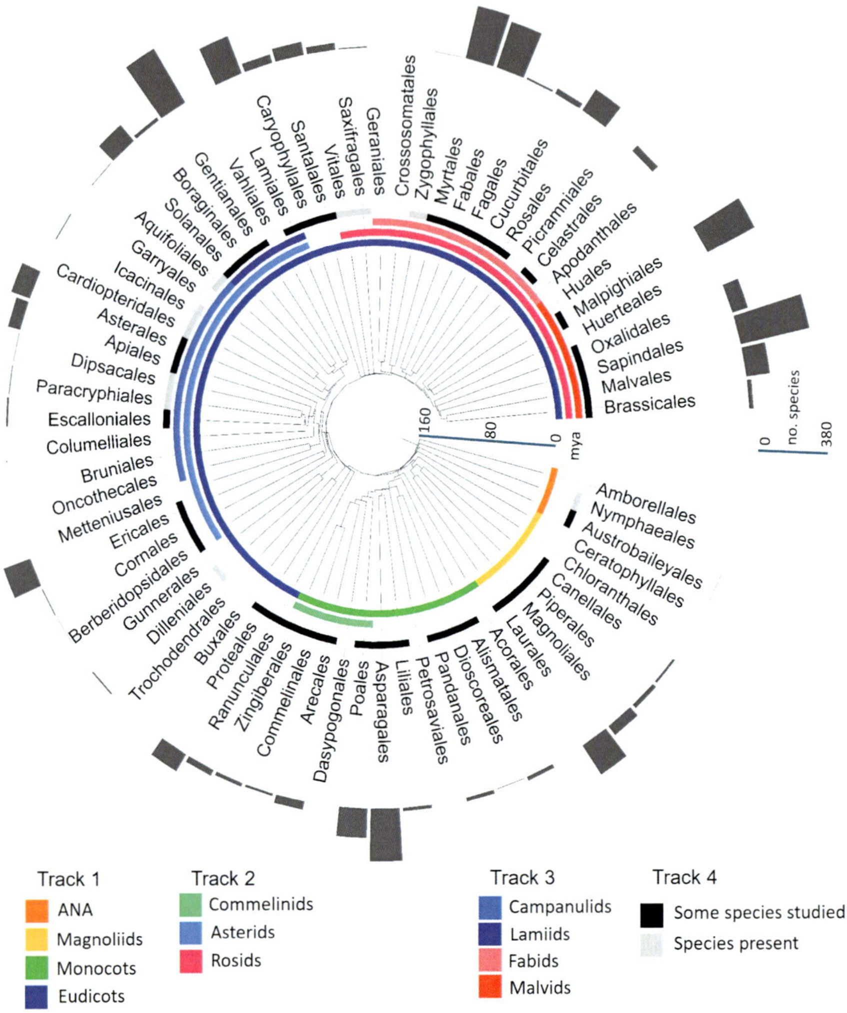

Fig. 14 Dated molecular phylogenetic tree of the orders of Angiosperms [7] highlighting clades and higher organizations. Orders with at least one Australian rainforest species present (grey) or species chemically studied (black, and species numbers (bar ring))

4.2.1 Basal Angiosperms

As noted earlier, the rainforests of Australia are home to a disproportionate number of species displaying anatomical features that either resemble those found in gymnosperms or the earliest fossil angiosperms [79]. These include species in the basal angiosperms (Amborellales, Nymphaeales, and Austrobaileyales), the Magnoliids, Monocots, and basal eudicots and basal lineages within specific eudicot families.

The Amborellales, Nymphaeales, and Austrobaileyales grade of flowering plants includes *Amborella trichopoda* (the sole species in the order Amborelales), found only in New Caledonia, the water lilies (Nymphaeales), with a cosmopolitan distribution, and the Austrobaileyales (an order consisting of three families of woody lianas, shrubs and trees) that are found in Australia, the Pacific islands, southeast Asia, and the Americas. Collectively, these basal angiosperms diverged from the core angiosperms (Mesangiospermae) prior to their diversification.

Austrobaileyales

Within the Austrobaileyales, the lianas *Austrobaileya scandens* ((Plate 5), sole species in the family Austrobaileyaceae) and *Trimenia moorei* (one of only eight species in the family Trimeniaceae) only occur in the tropical and subtropical rainforests of Australia, respectively.

Both species contain lignans (Table 3, Fig. 15), with aryl tetralin lactones (deoxypicropodophyllin (**96**), austrobailignan-1 (**97**), austrobailignan-2 (**98**), austrobailignan-9 (**99**), β-peltatin-A methyl ether (**100**), deoxypodophyllotoxin (**93**)), aryltetralins (austrobailignan-3 (**101**), austrobailignan-8 (**102**), and austrobailignan-4 (**103**)), and diaryl butanes (austrobailignan-5 (**104**) and austrobailignan-6 (**105**)), along with oxidized *seco*-lignans (7*S*,8*S*,8′*S*)-nicotlactone B (**106**), (8*R*,8′*S*)-marphenol K (**107**), and marphenol A (**108**) being reported from *A. scandens* [80, 81]. In addition, both *A. scandens* and *T. moorei* (reported under its synonym *Piptocalyx moorei*) each contain a tetrahydrofuran, with calopiptin (**109**)

Plate 5 *Austrobaileya scandens* vine and paintings of its flower and fruits. (Photograph and paintings AR Carroll)

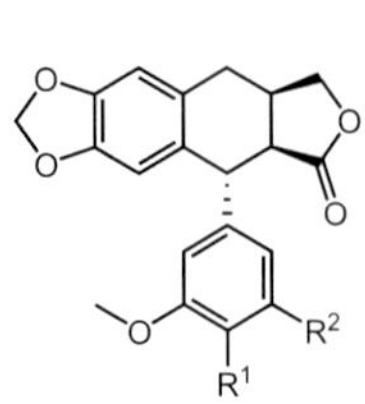

96 $R^1 = R^2 = OMe$
98 $R^1–R^2 = OCH_2O$

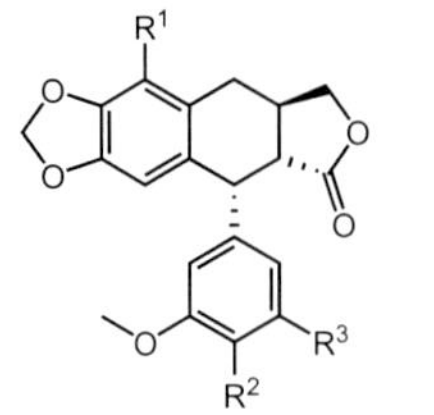

97 $R^1 = H$, $R^2–R^3 = OCH_2O$
99 $R^1 = OMe$, $R^2–R^3 = OCH_2O$
100 $R^1 = OMe$, $R^2 = R^3 = OMe$

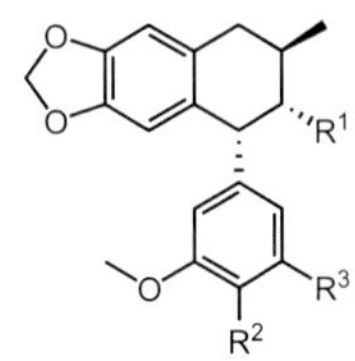

101 $R^1 = Me$, $R^2–R^3 = OCH_2O$
102 $R^1 = COOH$, $R^2–R^3 = OCH_2O$
103 $R^1 = Me$, $R^2 = R^3 = OMe$

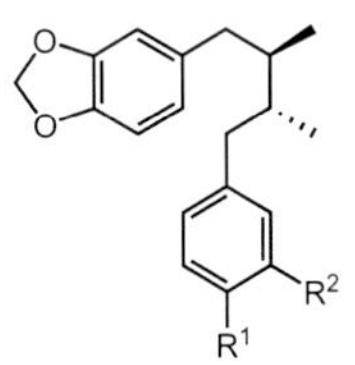

104 $R^1–R^2 = OCH_2O$
105 $R^1 = OH$, $R^2 = OMe$

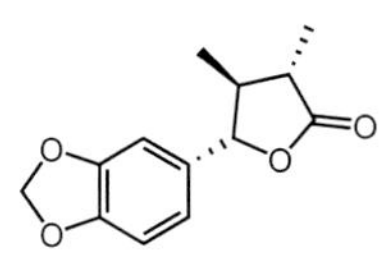

106

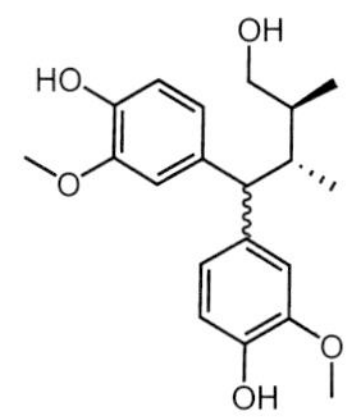

107

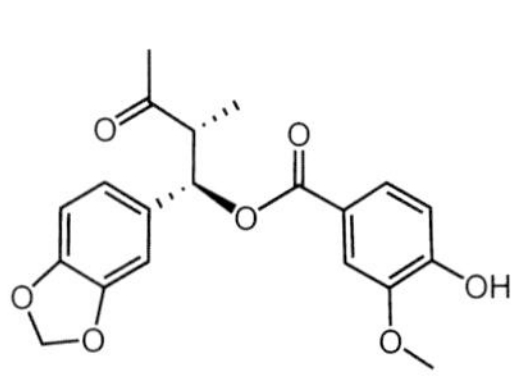

108

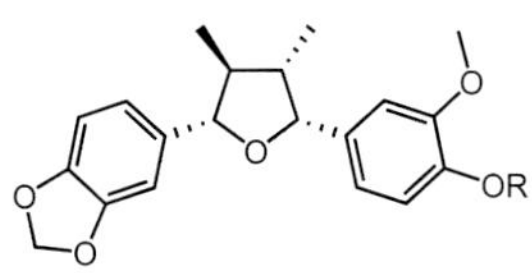

109 R = Me
110 R = H

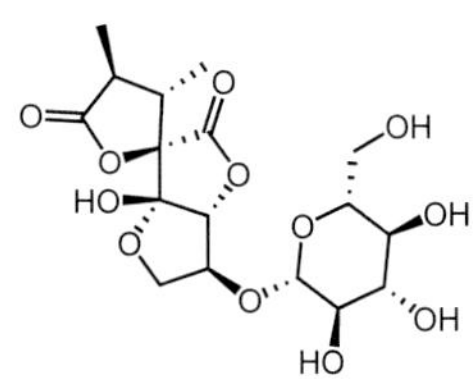

111

Fig. 15 Structures of compounds **96–111** reported from *A. scandens* and *T. moorei*

Table 3 Distribution of natural products isolated from the Austrobaileyales

Species	Collection location[a]	Plant part[b]	Compound class	Compound(s)	Refs.
Austrobaileyaceae					
Austrobaileya scandens	NQ	ap	Lignan	**96–108, 110**	[80, 81]
Trimeniaceae					
Trimenia moorei (*Piptcalyx moorei*)	NNSW	lf	Glycoside	**111**	[84]
	NNSW	lf	Lignan	**109**	[82]

[a] NQ = North Queensland, NNSW = Northern New South Wales, [b] lf = leaves, ap = aerial parts

being the methyl ether of austrobailignan-7 (**110**) [80, 82]. In keeping with the bioactivity reported for other aryltetralin lactones, compounds **97**–**99** show potent cytotoxicity towards PC3 cells, while marphenol (**108**) induces S cell cycle arrest in PC3 cells [81]. Calopiptin is antitrypanosomal towards *Trypanosoma cruzi* with an IC_{50} of 12.6 μM but is not antileishmanial nor cytotoxic [83]. *Trimenia moorei* also contains the glycoside piptoside (**111**) [84]. This constituent was originally linked to the intensely bitter taste of the leaves, but later studies did not corroborate this finding. Since the only investigation of *T. moorei* occurred in the early 1960s, revisiting its chemistry is likely to yield additional lignans with potential antiprotozoal activity and shed further light on the bitter constituents.

4.2.2 Magnoliids

The Australian rainforest Magnoliids have representatives from 12 families across all four orders: Canellales, Piperales, Magnoliales, and Laurales.

Canellales

The order Canellales has species distributed in South and Central America, New Zealand, New Caledonia, New Guinea and other parts of Malesia, and the Solomon Islands, with one species in Madagascar (Fig. 16). The Winteraceae has a Gondwanan origin, but diversification within the family occurred in the Paleocene after the breakup of Gondwana and the current species distributions (Fig. 16) are likely the result of long-distance dispersal events [85].

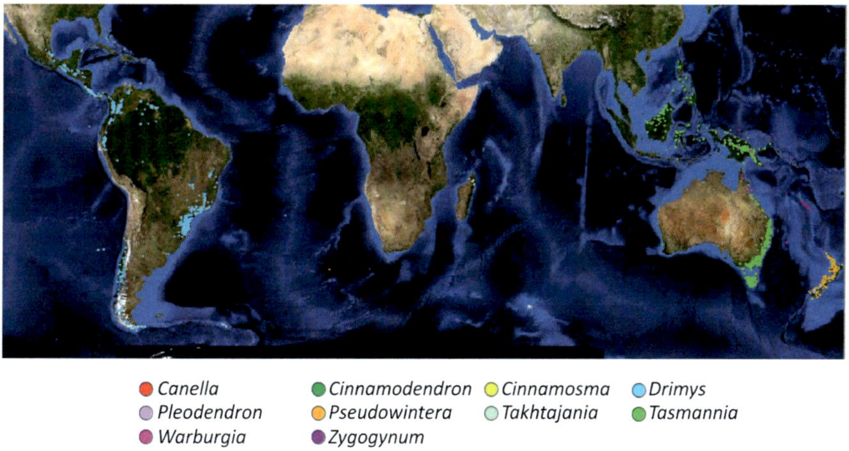

Fig. 16 Worldwide herbarium records of collection locations for all species in the Canellales adapted from supporting information provided in [5, 43]

Plate 6 Flowers and fruits of *Tasmannia* species. (Photographs, AR Carroll)

Table 4 Distribution of natural products isolated from the Winteraceae

Species	Collection location[a]	Plant part[b]	Compound class	Compound(s)	Refs.
Tasmannia lanceolata (*Drimys lanceolata*)	Tas	lf	Phenyl propane	**113–116**	[86]
	Tas	fr	Sesquiterpenoid	**112**, **117–137**	[86, 88]
Tasmannia xerophila	NSW	lf	Sesquiterpenoid	**138**, **139**	[90]
Zygogynum semecarpoides (*Bubbia semecarpoides*)	NQ	b	Aryl tetralone	**140**	[91]

[a] NQ = North Queensland, NSW = New South Wales, Tas = Tasmania, [b] lf = leaves, fr = fruits, b = bark

Two genera (*Tasmannia* and *Zygogynum* (syn. *Bubbia*)) are represented by 12 species in Australian rainforests (Plate 6). Drimane sesquiterpenoids and aryl propenes characterize the family, and *Tasmannia lanceolata* (from Tasmania) is no exception (Table 4, Fig. 17). This species is termed colloquially Tasmanian pepper, notably because it contains high concentrations of polygodial (**112**) in its fruits and leaves [86, 87]. The co-occurrence of myristicin (**113**), safrole (**114**) and other arylpropenes (eugenol (**115**), and methyl eugenol (**116**)) impart both a peppery and numbing sensation when the fruits and leaves are eaten as a condiment [86]. Over 20 drimanes have been reported from *T. lanceolata* (**112**, *epi*-polygodial (**117**), warburganal (**118**), 1β-acetoxy-9-deoxyisomuzigadial (**119**), changweikang aldehyde (**120**), polygonic acid (**121**), polypiperic acid (**122**), *epi*-polypiperic acid (**123**), changweikang acid A (**124**), 3a,4,5,6,7,7a-hexahydro-3a,7,7-trimethyl-1*H*-inden-3-carboxylic acid (**125**), cinnamolide (**126**), 7-ketoconfertifolin (**127**), dendocarbin A (**128**), dendocarbin L (**129**), dendocarbin M (**130**), ugandenial A (**131**), fuegin (**132**), 3β-acetoxy-7-ketoconfertifolin (**133**), 3-β-acetoxydrimenin (**134**), drimendiol (**135**), 1β-acetoxypolygodial (**136**), and 1β-acetoxyisodrimeninol (**137**)) [86, 88]. Some of these sesquiterpenoids possess potent antifungal and antibacterial activities [89]. The

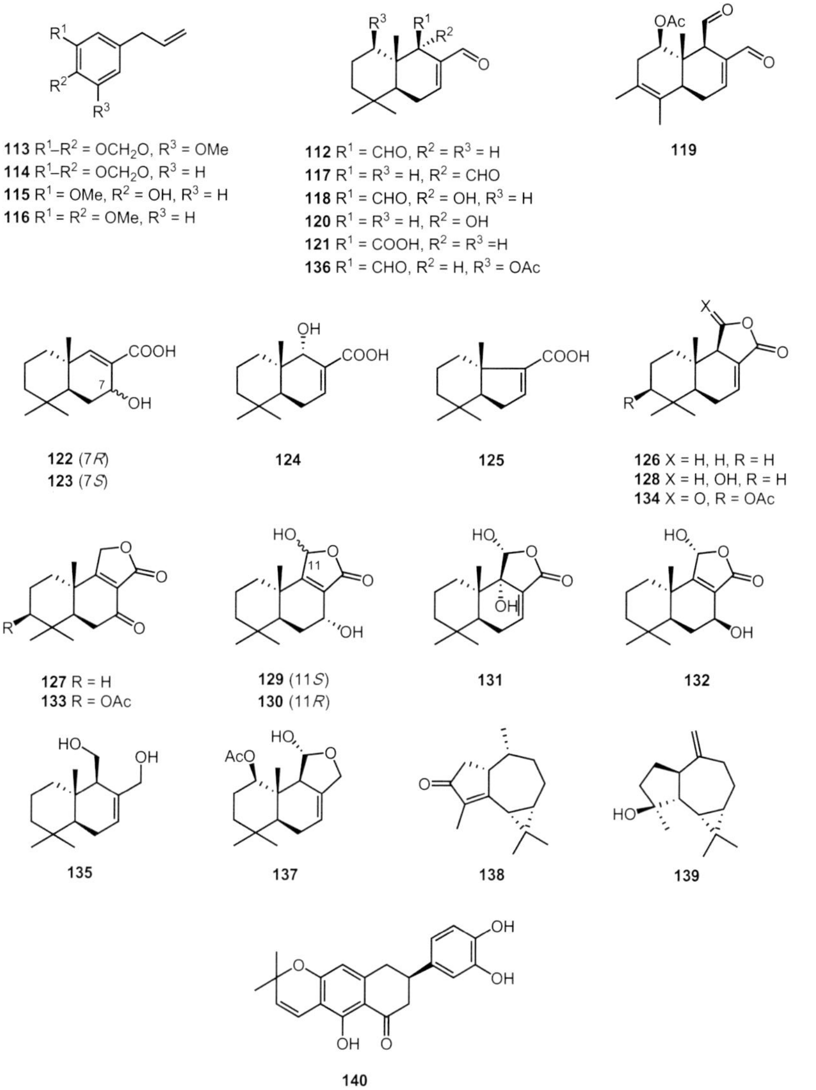

Fig. 17 Structures of phenylpropanes **113–116**, terpenoids **112–139**, and aryl tetralone **140** reported from Australian rainforest Winteraceae species

pungent properties of **112** and related dialdehydes are the result of agonism of vanilloid receptors [86]. The leaves of *T. xerophila* have yielded two sesquiterpenoids, spathulenol (**138**) and cyclocolorenone (**139**) [90].

A novel approach to identify compounds in complex mixtures using ^{1}H DOSY NMR was applied to identify these sesquiterpenoids (Fig. 18). This method used the intrinsic diffusion coefficient (D) of a molecule (a surrogate for molecular weights) to separate NMR resonances in the NMR tube. A model was developed to predict D for compounds based on physicochemical properties, and then, comparison of the experimental D and NMR derived structural features for diffusion grouped resonances with predicted D and calculated structural features in the DEREP-NP database has allowed accurate identification of compounds present in the database by simply acquiring a ^{1}H DOSY spectrum (a three-minute experiment). This methodology has advantages over MS-based metabolomics methods, because compounds with the same molecular formula can be differentiated based on structural features that are only identifiable by NMR analysis [90].

Although only two species of *Tasmannia* so far have been studied, Australian mainland rainforest species (*T. xerophila* and *T. stipitata*) were reported with even more pungent properties, and we have found both to contain high concentrations of polygodial [91]. The three Australian *Zygogynum* species were assigned previously to the genus *Bubbia*, but morphological and molecular evidence has since been used to ascribe them to *Zygogynum* (a genus found in New Caledonia, the Solomon Islands and Malesia) [92]. Aryltetralones such as zygolone A (**140**) appear to be a defining chemical marker for the genus *Zygogynum*, since they have not been reported from any other Winteraceae genera [93]. An initial investigation of Australian *Zygogynum semecarpoides* (syn. *Bubbia semecarpoides*) has shown that it contains zygolone A (**140**) and additional investigations of the other Australian species may shed a further light on their relationship with other species in the genus found elsewhere [91]. The aryl tetralones are likely to have a polyketide biogenesis.

Piperales

The Piperales is a large order with over 4,500 species that have a cosmopolitan distribution. The families Aristolochiaceae and Piperaceae have species in Australian rainforests [94]. A small number of species have been investigated chemically (Table 5).

Only one of the 13 Australian species in the Aristolochiaceae (*Aristolochia acuminate*; syn. *A. tagala*) has been investigated, and this was conducted on material collected in both Yunnan in southern China and Madagascar (Figs. 19 and 20) [95, 96]. Eight aristolactam-type alkaloids (aristolactam BII (**141**), aristolactam II (**142**), sauristolactam (**143**), aristolactam I (**144**), 7-methoxyaristolactam IV (**145**), aristolactam AII (**147**), aristolactam-*N*-*β*-D-glucoside (**147**), **and** cepharanone A *N*-*β*-D-glucoside (**148**)) and nine aristolochic acid derivatives (aristolochic acid 1 (**149**), aristolochic acid D (**150**), 3-hydroxy-4-methoxy-10-nitrophenanthrene-1-carboxylic acid methyl ester (**151**), ariskanin A (**152**), ariskanin D (**153**), ariskanin E (**154**), aristolochic acid C (**155**), ariskanin C (**156**), and ariskanin B (**157**)) were reported. The aristolochic acids, containing a 9-nitrophenanthrene unit, are unique to *Aristolochia* species [95, 96]. These natural products are important to the survival of birdwing

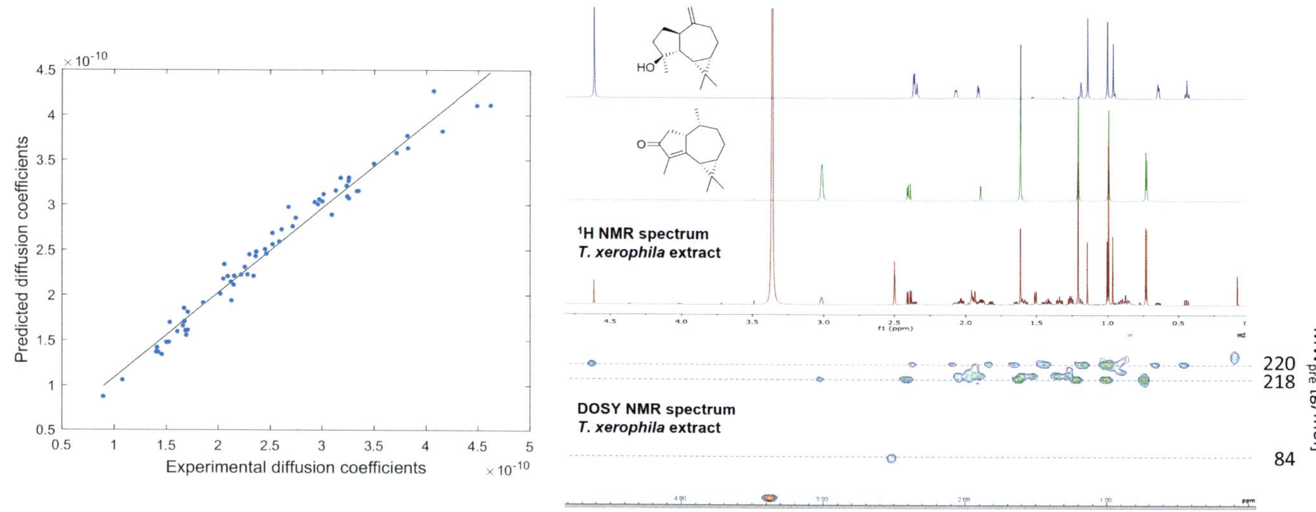

Fig. 18 Diffusion coefficient correlation curve for NPs (n = 63, $R^2 = 0.99$) based on a model that uses physicochemical properties of a molecule to predict its *D* and application of this method to identify sesquiterpenoids **88** and **89** in an extract from *T. xerophila* by ^{1}H DOSY NMR spectroscopy. The blue and green ^{1}H NMR spectra are projections obtained from the ^{1}H DOSY spectrum

Table 5 Distribution of natural products isolated from the Piperales

Species	Collection location[a]	Plant part[b]	Compound class	Compound(s)	Refs.
Aristolochiaceae					
Aristolochia acuminata (*Aristolochia tagala*)	South China, MAD	ap	Aristolactam	**141–157**	[95, 96]
Piperaceae					
Piper hederaceum var. *hederaceum*	NNSW	w	Amide	**158–165**	[99]
	NNSW	w	Phenyl propane	**166, 167**	[99]
Piper caninum	NQ	b	Monoterpenoid	**168, 169**	[100]
Piper interruptum	Thailand	st	Amide	**171**	[101]
	Thailand	st	Ester	**172**	[101]
	Thailand	st	Neolignan	**170**	[101]

[a] NQ = North Queensland, NNSW = Northern New South Wales, MAD = Madagascar, [b] ap = aerial parts, st = stems, b = bark, w = wood

butterfly larvae that feed on *Aristolochia* vines. The compounds provide feeding-deterrent properties that may protect the larvae from predators [97]. The birdwing butterflies, having the largest wingspan of any butterfly species in the world, evolved in Australia, but now have centers of diversity in New Guinea and parts of Malesia [98].

Only three of 14 Piperaceae species have been studied (Fig. 21). The wood of *Piper hederaceum* var. *hederaceum*, a vine from Northern NSW, contained a series of eight alkyl amides ($\Delta^{\alpha,\beta}$-dihydropiperine (**158**), *N*-isobutyl-(2*E*,4*E*)-octadienamide (**159**), 3,4-methylenedioxycinnamoylpiperidide (**160**), *N*-isobutyl-(2*E*,4*E*)-decadienamide (**161**), fagaramide (**162**), piperlongumine (**163**), piperine (**164**), and chavicine (**165**)) possessing either piperidine or isobutylamine groups. These moieties are linked to the hot peppery taste of pepper species used as condiments [99]. Two phenyl propenes, hydroxyisodillapiole (**166**) and dillapiole (**167**), were also reported [99]. *Piper caninum* from North Queensland contained two cinnamate esters of borneol, (+)-bornyl *p*-coumarate (**168**) and (+)-bornyl *p*-caffeate (**169**) [100]. A third species, *P. interruptum*, investigated only from plant material collected in Thailand, contains the neolignan eupomatenoid 1 (**170**) (originally isolated from *Eupomatia laurina*—see later), the alkyl isobutylamide pipercallosine (**171**), and the bisepoxide, crotepoxide (**172**) [101].

Magnoliales

The Magnoliales originated in Gondwana and, based on fossil records from Brazil, were actively diversifying in the middle third of the Cretaceous [102]. Four of the six families, Eupomatiaceae, Annonaceae, Himantandraceae, and Myristicaceae in

Fig. 19 Herbarium location records for *Aristolochia accuminata* [5, 43]. Flowers (top right) and fruits (left middle) of *Aristolochia praevenosa* (food plant of the endangered Richmond birdwing butterfly *Ornithoptera richmondia*). Male Cairns birdwing butterfly (*Ornithoptera euphorion*) that feeds on *A. acuminata* (bottom left), *Piper mestonii* (bottom middle and right). (Photographs AR Carroll)

the Magnoliales are represented in Australian rainforests. Two families, Eupomatiaceae and Himantandraceae, are restricted to continental Australia, and each is represented by a single genus with three *Eupomatia* (Eupomatiaceae) and two *Galbulimima* (Himantandraceae) species, respectively. Both genera have a species with an extended range into New Guinea (Fig. 22). The constituents reported from both of these families are unusual with alkaloids and lignans predominating (Tables 6 and 7). Two other families, Myristicaceae and Annonaceae, that are far more widespread and genus/species rich, are also present in Australia's rainforests, but none of the four Myristicaceae species and only six of 45 Annonaceae species have been investigated (Table 8).

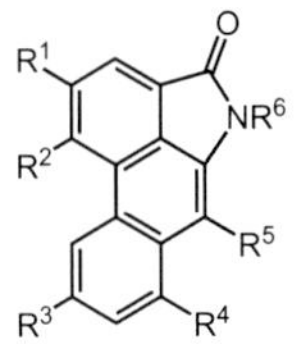

141 $R^1 = R^2 = OMe$, $R^3 = R^4 = R^5 = R^6 = H$
142 $R^1–R^2 = OCH_2O$, $R^3 = R^4 = R^5 = R^6 = H$
143 $R^1 = OH$, $R^2 = OMe$, $R^3 = R^4 = R^5 = H$, $R^6 = Me$
144 $R^1–R^2 = OCH_2O$, $R^3 = R^5 = R^6 = H$, $R^4 = OMe$
145 $R^1–R^2 = OCH_2O$, $R^3 = R^4 = R^5 = OMe$, $R^6 = H$
146 $R^1 = OH$, $R^2 = OMe$, $R^3 = R^4 = R^5 = R^6 = H$
147 $R^1–R^2 = OCH_2O$, $R^3 = R^5 = H$, $R^4 = OMe$, $R^6 = \beta$-D-glc
148 $R^1–R^2 = OCH_2O$, $R^3 = R^4 = R^5 = H$, $R^6 = \beta$-D-glc

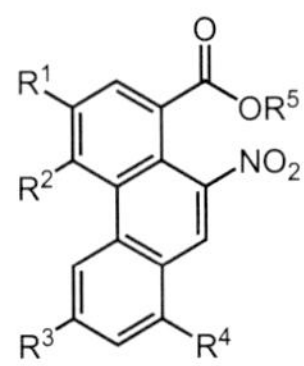

149 $R^1–R^2 = OCH_2O$, $R^3 = OH$, $R^4 = OMe$, $R^5 = H$
150 $R^1–R^2 = OCH_2O$, $R^3 = R^5 = H$, $R^4 = OMe$
151 $R^1 = OH$, $R^2 = OMe$, $R^3 = R^4 = H$, $R^5 = Me$
152 $R^1 = R^2 = OMe$, $R^3 = R^4 = H$, $R^5 = Me$
153 $R^1 = OH$, $R^2 = OMe$, $R^3 = H$, $R^4 = OMe$, $R^5 = Me$
154 $R^1 = R^2 = OMe$, $R^3 = H$, $R^4 = OMe$, $R^5 = Me$
155 $R^1–R^2 = OCH_2O$, $R^3 = OH$, $R^4 = R^5 = H$
156 $R^1 = R^2 = OMe$, $R^3 = OH$, $R^4 = H$, $R^5 = Me$
157 $R^1 = R^3 = OH$, $R^2 = OMe$, $R^4 = H$, $R^5 = Me$

Fig. 20 Structures of alkaloids **141–157** isolated from *Aristolochia acuminata*

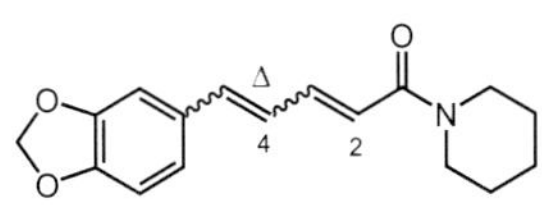

158 (2*E*),Δ saturated
164 (2*E*,4*E*)
165 (2*Z*,4*Z*)

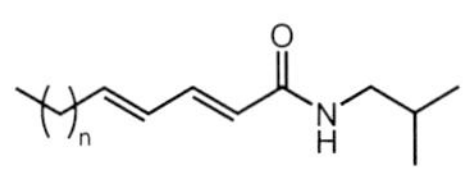

159 n = 2
161 n = 4

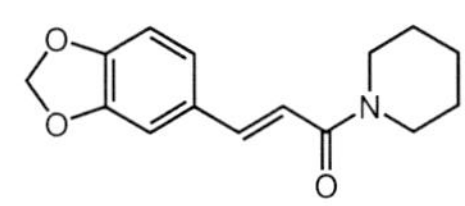

160

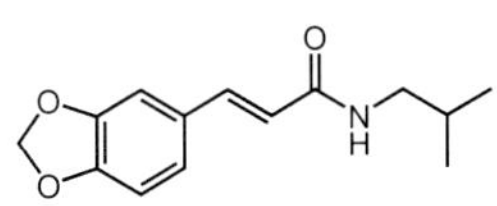

162

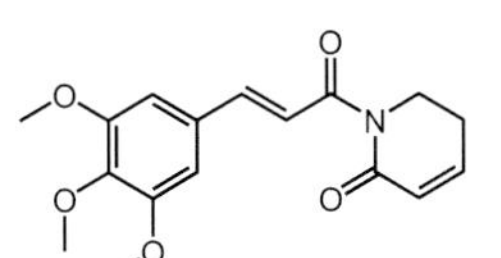

163

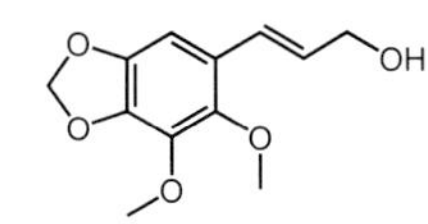

166

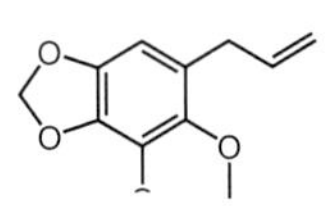

167

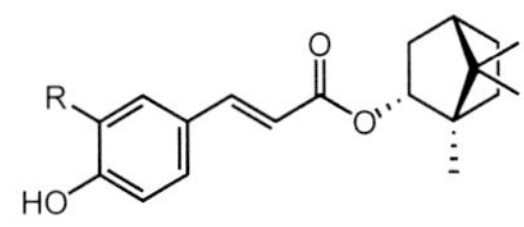

168 R = H
169 R = OH

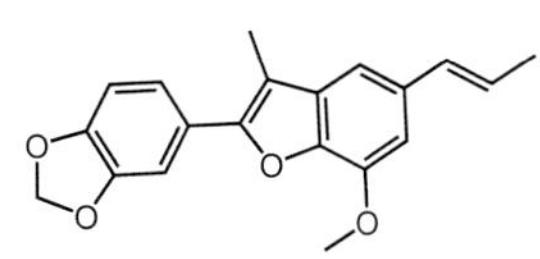

170

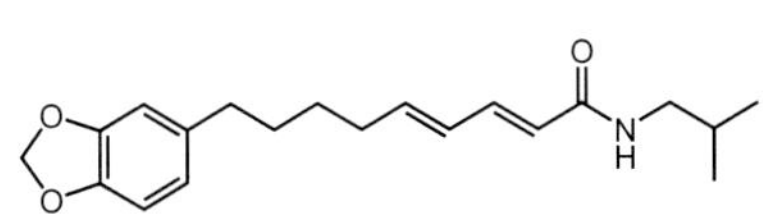

171

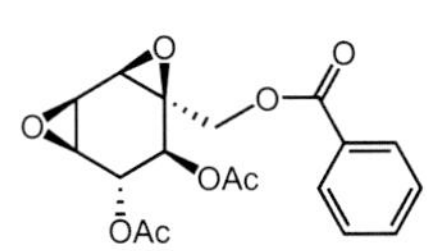

172

Fig. 21 Structures of compounds **158–172** isolated from Piperaceae species

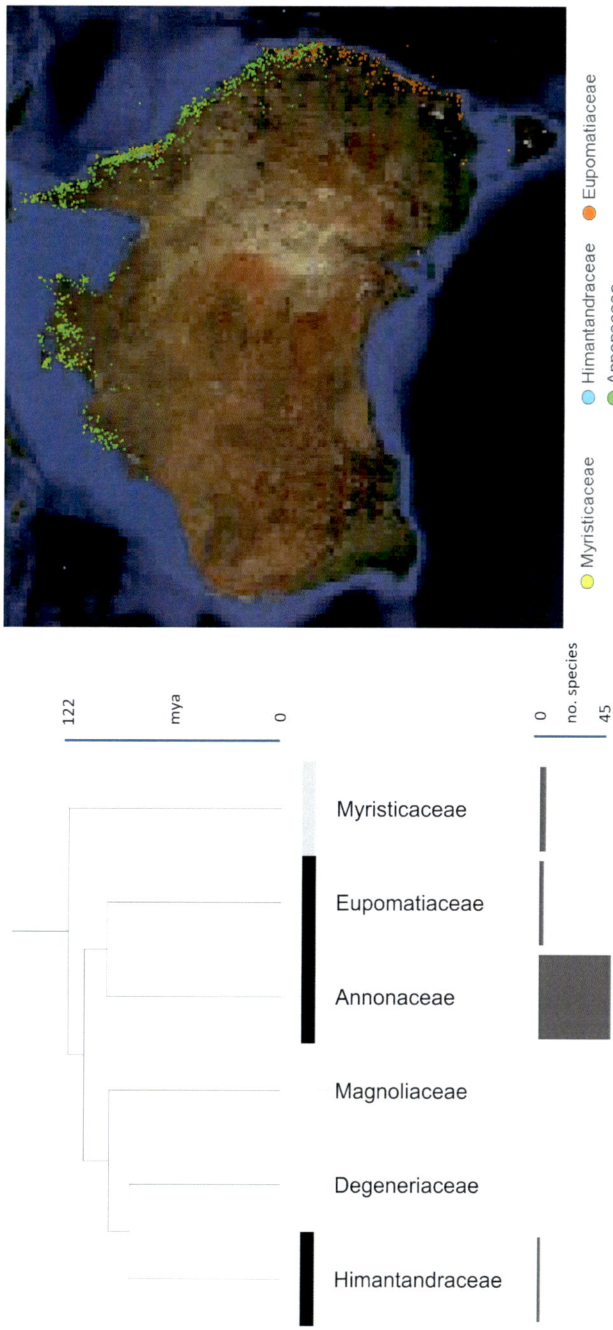

Fig. 22 Dated molecular phylogeny of the Magnoliales [7] highlighting families with species found (grey) and studied (black) from Australian rainforests. The total species present in each Australian rainforest family are depicted in the bar chart. The map shows the collection location of herbarium records for all Australian rainforest Magnoliales species [5, 43]

Table 6 Distribution of natural products isolated from the Eupomatiaceae

Species	Collection location[a]	Plant part(s)[b]	Compound class	Compound(s)	Refs.
Eupomatia laurina	PNG	b	Aporphine	**187–189**	[119]
	PNG	lf	Flavonoid	**248**	[139]
		fl	Ester	**173**	[103]
	PNG, NQ, NSW	b, fl	Lignan	**215–224**	[105, 123–125]
	PNG, NQ	b, r	Naphthyridine	**176–180**	[104–109]
	PNG, NQ	b, lf	Neolignan	**170, 193–204, 209–213**	[105, 119–121]
Eupomatia bennettii	SEQ	ap	Aporphine	**187**	[117]
		fl	Ester	**174, 175**	[103]
	SEQ	ap, tub	Lignan	**218, 220, 237–244**	[118, 135]
	SEQ	ap	Naphthyridine	**180–183**	[106]
	SEQ	ap, tub	Neolignan	**170, 193–196, 205–208, 214**	[106, 118]

[a] SEQ = Southeast Queensland, NQ = North Queensland, NSW = New South Wales, PNG = Papua New Guinea, [b] lf = leaves, ap = aerial parts, r = roots, tub = tubers, b = bark, fl = flowers

Eupomatiaceae

Eupomatia species have primitive floral traits including the absence of petals and sepals with staminodes (infertile stamens) arranged in a Fibonacci whorl taking their place (Plates 7 and 8). The fertile stamens are on the outside of this whorl. The combined stamen/staminode floral structure is called an androecium. The flowers also have a carpel that is fused to form a flat-topped disk with no styles. The flowers are pollinated by weevils, *Elleschodes* sp., that are attracted by an intense and fruity ester scent in the morning [103]. These weevils are the exclusive pollinators of *Eupomatia* species and have not been recorded in any other plants. The female parts of the flower are fertile first and masses of weevils consume parts of the staminodes and mate during this period. In the late afternoon, the flower closes and the fertile stamens then open to release pollen that covers the departing weevils. The whole androecium detaches from the flower during the night and the weevils then go on to pollinate other flowers the following day. The detached androecium is a brood substrate for the larvae of the weevils.

The scent of *E. laurina* is almost exclusively methyl (2*S*)-methyl butanoate (**173**) while that of *E. bennettii* is more of a mixture with differences in composition occurring between the male and female phases of flowering. The major components are methyl (2*R*)-2-acetoxy-4-methyl pentanoate (**174**) (78%) and (2*R*)-2-hydroxy-4-methyl pentanoate (**175**) (16%) [103]. A similar mutualistic relationship occurs in the New Caledonian Winteraceae species *Zygogynum* (syn.

Exospermum) *stipitatum* and its weevil pollinator *Palontus exospermi*. The anatomically unusual characteristics of *Eupomatia* species are matched by their chemical constituents. Collections of plant material of *E. laurina* from North Queensland, New Guinea, and Sydney and of *E. bennettii* from Southern Queensland have yielded 59 natural products including polyaromatic alkaloids of unusual structure (Fig. 23), diverse neolignans (Fig. 24) and lignans (Figs. 25 and 26) with unique ring systems. The first alkaloid reported was eupolauridine (**176**), an indeno[1,2,3-*ij*][2,7]naphthyridine without precedent [104, 105]. Other unusual alkaloids were later reported, including 3-methoxyeupolauridine (**177**), two benzo[*h*]pyrrolo[4,3,2-*de*]quinolin-4(5*H*)-ones (eupolauramine (**178**) and hydroxyeupolauramine, (**179**)),

Plate 7 *Eupomatia laurina* flowers, whole plants, *Elleschodes* sp. weevils pollinating the flowers and flower buds. (Photographs AR Caroll)

Plate 8 *Eupomatia bennettii* flowers, whole plants, *Elleschodes* sp. weevils pollinating the flowers and flower buds. (Photographs AR Carroll)

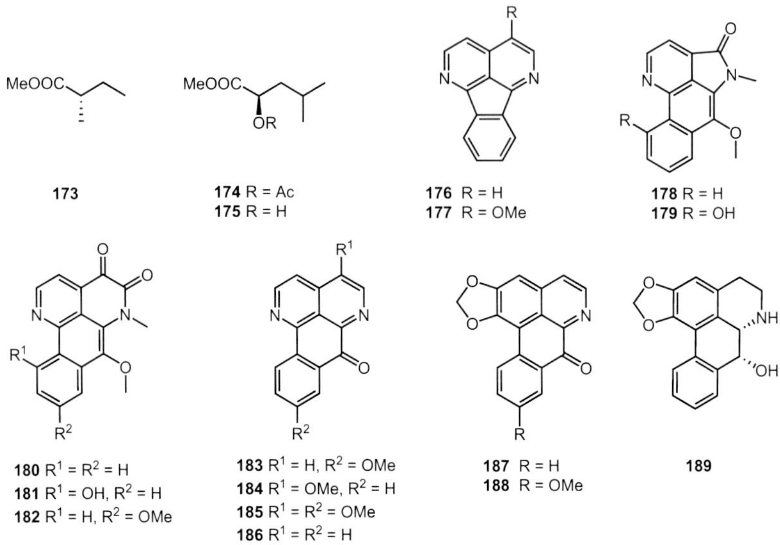

Fig. 23 Structures of volatile floral chemicals **173–175** and alkaloids **176–189** isolated from *Eupomatia* species

three 4*H*-naphtho[1,2,3-*ij*][2,7]naphthyridine-4,5(6*H*)-diones (imbilines 1–3 (**180–182**)), and four 7*H*-naphtho[1,2,3-*ij*][2,7]naphthyridin-7-ones (eupomatidines 1–3 and sampangine (**183–186**)) [104–109].

The co-occurrence of liriodenine (**187**) and methoxyliriodenine (**188**), common oxo-aporphine alkaloids, and of norushinsunine (**189**), reported from many Magnoliid species, gave a clue to the biosynthetic origins of the *Eupomatia* alkaloids (Scheme 1). Oxidative cleavage of catechols termed extradiol oxidation has been shown to be mediated by extradiol catechol dioxygenases in plants and microbes [110]. These dioxygenases cleave the carbon–carbon bond adjacent to the phenolic hydroxy group in catechols generating a glyoxylate and ketone. Decarboxylation of the glyoxylate by a glyoxylate decarboxylase probably generates an aldehyde that then condenses with ammonia to produce the naphthyridine ring found in the eupomatidines, or oxidative decarboxylation generates cleistopholine (**190**) found in Annonaceae species. Oxidation of the A ring of the naphthyridine first leads to a hydroxylated sampangine intermediate and then further oxidation generates the imbiline structures. Oxidative decarboxylation of imbilines (experimentally verified through reaction of imbiline 1 with hydrogen peroxide under basic conditions [106]) then generates the eupolauramine structures. A side reaction of eupomatidine proposed to involve a UV light-induced extrusion of carbon monoxide, generates eupolauridine [111]. Azafluoranthene alkaloids isolated from Menispermaceae species and aza-fluorenes such as onychine (**191**) from Annonaceae species have been proposed to be generated by an analogous reaction [112]. The *Eupomatia* alkaloids along with their analogues have since been reported from a limited

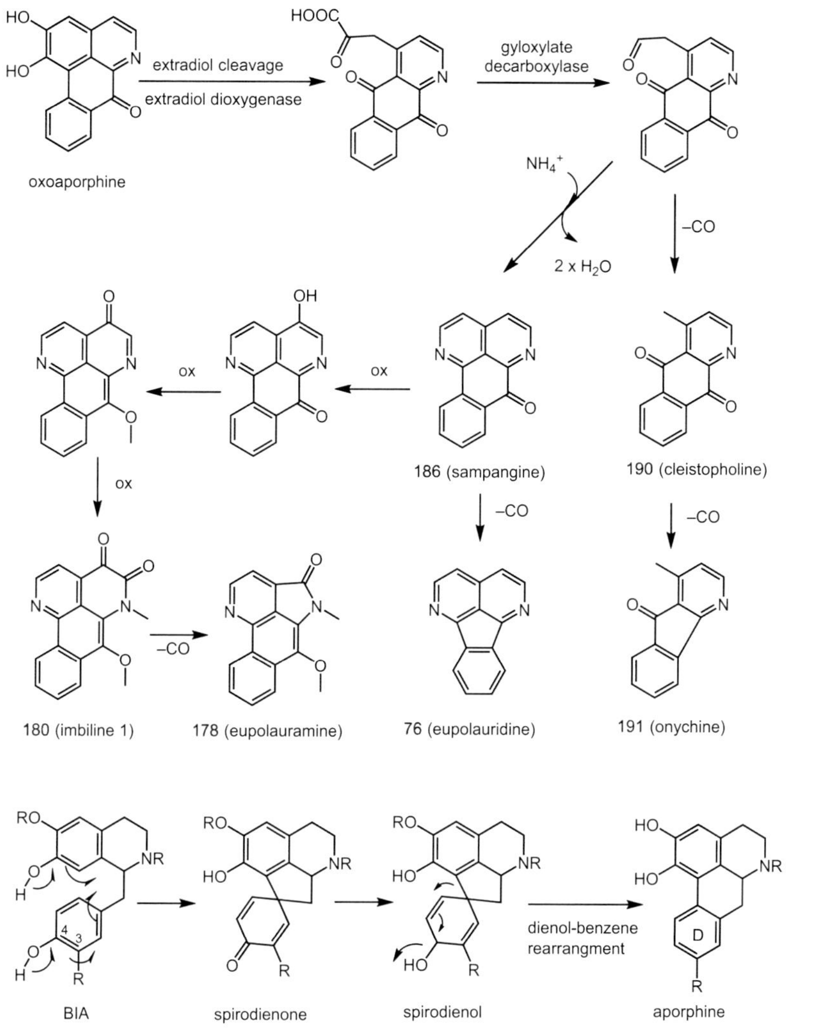

Scheme 1 Proposed biogenesis of the *Eupomatia* alkaloids

number of Annonaceae species [113]. This is unsurprising given the close taxonomic relationship between the Eupomatiaceae and Annonaceae. *Eupomatia* was once included in the Annonaceae. Interestingly, most of the alkaloids that contain either naphthyridines or extradiol-modified aporphines are found in the basal grade of the Annonaceae (*Anaxagorea*, *Cananga*, *Cleistopholis*, *Ambivia*, and *Porcella*), suggesting that this biosynthetic attribute has an ancient origin that has been abandoned or became dormant in subsequent diversification and speciation within the Annonaceae [114].

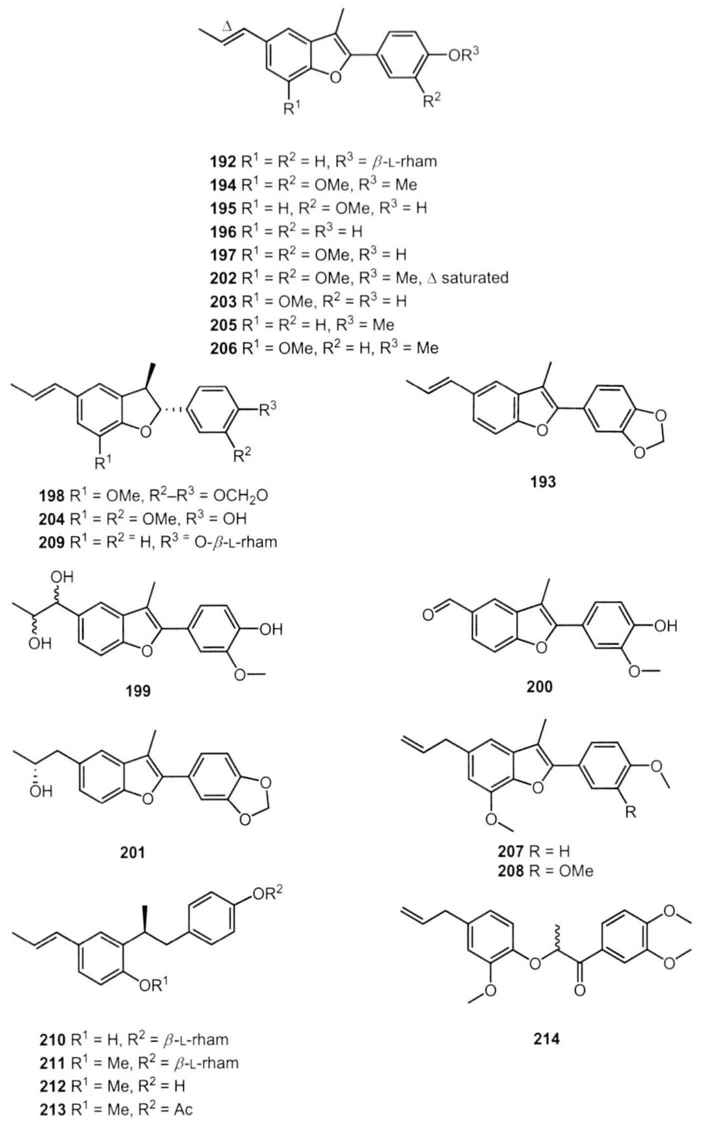

Fig. 24 Structures of neolignans **192–214** isolated from *Eupomatia* species

The *Eupomatia* alkaloids (mainly of the sampangine class) show cytotoxic, antiplasmodial and antifungal activities [115–117]. An interesting side note regarding the aporphine alkaloids is that they are made by oxidative diphenolic coupling of benzyl isoquinoline alkaloids and so those with only a *para*-phenol in the benzyl moiety generate the aporphine skeleton through *ortho-para* coupling, to first generate

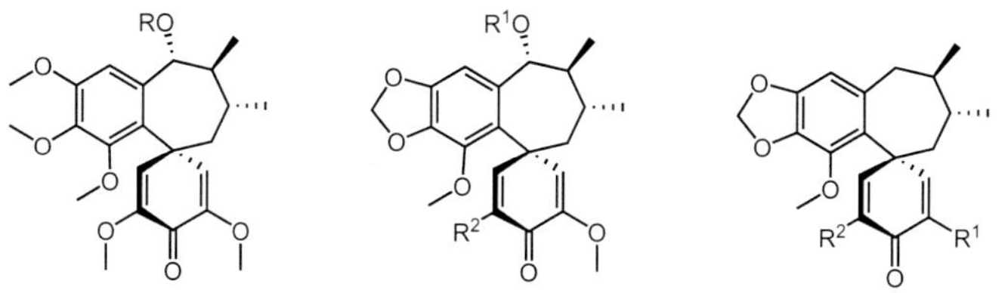

215 R = Ac
216 R = H
217 R = COPh

218 R^1 = COPh, R^2 = OMe
219 R^1 = COPh, R^2 = H
220 R^1 = Ac, R^2 = H
221 R^1 = Ac, R^2 = OMe

222 R^1 = OMe, R^2 = H
223 R^1 = H, R^2 = OMe

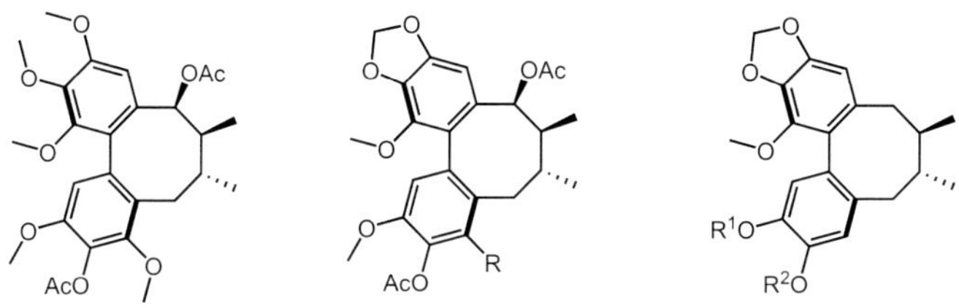

225

226 R = OMe
227 R = H

228 R^1 = H, R^2 = Me
229 R^1 = Me, R^2 = H

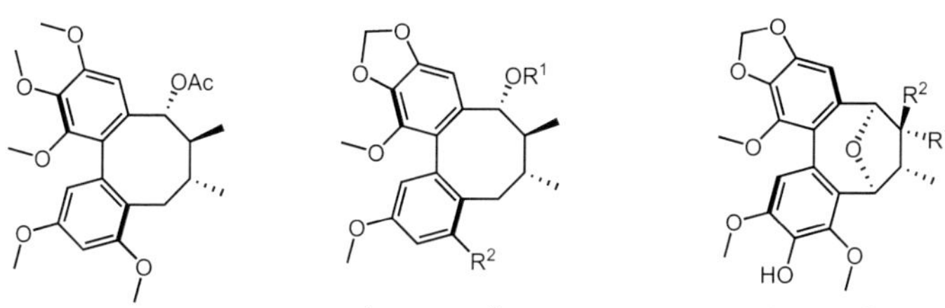

230

231 R^1 = COPh, R^2 = OMe
232 R^1 = Ac, R^2 = OMe
233 R^1 = COPh, R^2 = H
234 R^1 = Ac, R^2 = OMe

235 R^1 = Me, R^2 = H
236 R^1 = H, R^2 = Me

Fig. 25 Structures of *Eupomatia*-derived lignans **215–223** and their dienone-phenol **225–229** or dienol-benzene **230–234** rearrangement products and related *Gymnotheca*-derived dibenzocyclooctenes **235** and **236**

the spirodienone, with the aporphine product generated by ketone reduction and dienol-benzene rearrangement leading to an aporphine with no oxygen substituents in the D-ring [112]. Those benzyl isoquinoline alkaloids with a 3,4-dioxybenzyl moiety can directly produce a D-ring dioxygenated aporphine via *ortho-para* coupling with radical oxidation of the hydroxy group attached to C-3 in the benzyl moiety or a mono-oxygenated D-ring aporphine via radical oxidation of the hydroxy attached to C-4, subsequent *ortho-para* coupling to generate the spirodienone, and then reduction and dienol-benzene rearrangement.

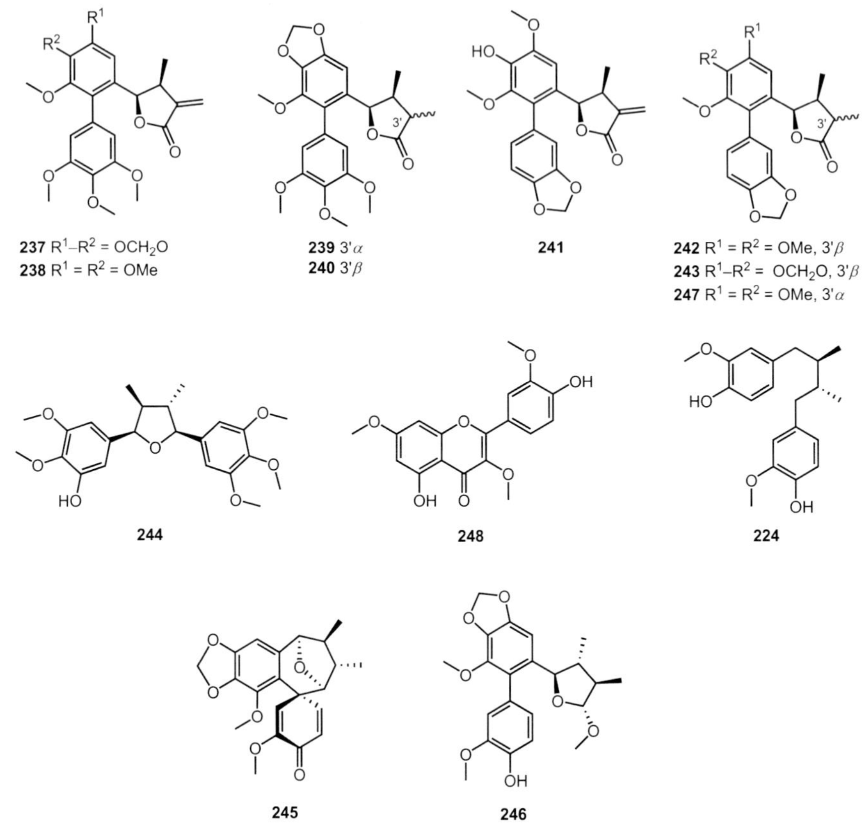

Fig. 26 Structures of additional lignans **224**, **237–244**, and **247** and a flavonoid **248** obtained from *Eupomatia* species and related lignans **245** and **246** isolated from *Gymnotheca* spp.

A diverse group of lignans and neolignans also has been found in *Eupomatia* species. The neolignans (Fig. 24) exclusively result from diphenolic oxidative coupling of either *p*-hydroxyphenylpropenes or 3,4-dihydroxyphenylpropenes (Scheme 2). Three classes have been reported. The most abundant class is represented by the eupomatenoids (2-phenyl-3-methyl-5-propylbenzofurans), with 19 variants reported (eupomatenoids 1–18 and eupomatene B (**170**, **192–209**)) [105, 106, 118–121]. These compounds result from a radical coupling between the aromatic carbon *ortho* to the phenol and *para* to the side chain in one phenylpropane, with C-2 of the propane side chain in a second phenyl propene (C-3:C-8 linkage) followed by a nucleophilic attack of the quinone methide by the *ortho*-phenol to generate the dihydrobenzofuran. Four additional compounds (eupomatenes A (**210**), C (**211**), D (**212**), and E (**213**)) also contain the C-3:C-2′ linkage but lack the furan elaboration [121]. Bennettinone (**214**) is an *O*-4:*C*-8 coupled neolignan and its structure was confirmed by a biomimetic synthesis using a diphenolic oxidative coupling reaction with silver oxide as the radical oxidant [118].

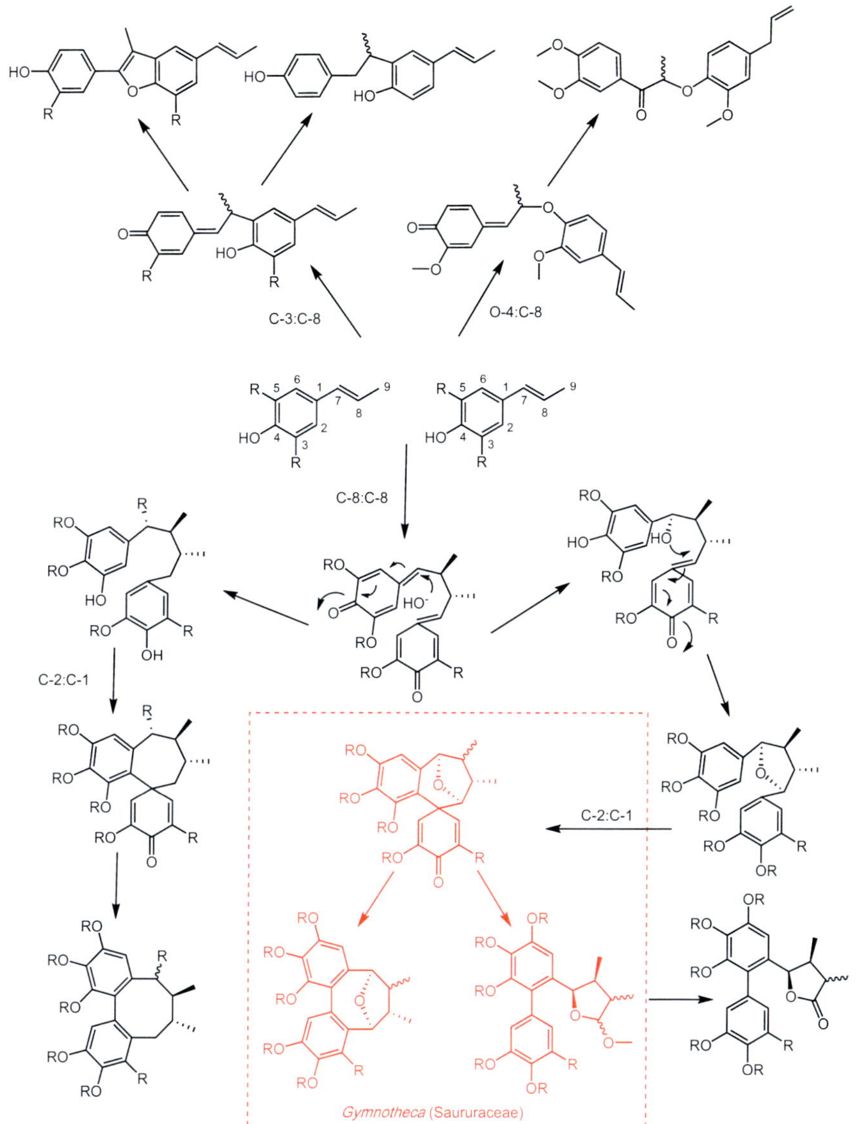

Scheme 2 Proposed biogenesis of the *Eupomatia* neolignans and lignans

The eupodienone lignans (Fig. 25) (C-8:C-8 coupled) found in *Eupomatia* species were, until 2012, without precedent. They are most closely related to the dibenzocyclooctanes found exclusively in the Schisandraceae (Austrobaileyales) [122]. They differ from the dibenzocyclooctanes since they contain a seven-membered ring that links one of the aromatic rings of the precursor diaryl butane to a cyclohexadienone moiety. A diphenolic oxidation coupling reaction like that used to generate aporphine alkaloids via a spiro-cyclohexadienone is a likely pathway. Nine congeners have been reported from *E. laurina* and *E. bennettii* (eupodienones 1–9 (**215**–**223**)) [105, 106, 123–125]. The diarylbutane (−)-dihydroguaiaretic acid (**224**) was also reported from *E. laurina* flowers [125]. Since 2012, several eupodienone analogues have been reported from *Gymnotheca chinensis* and *G. involucrata* (Saururaceae), Magnoliid species in the Piperales [126–134]. The eupodienones undergo a dienone-phenol rearrangement under acidic conditions to generate dibenzocyclooctanes via migration of the alkyl carbon attached to the dienone eupodienone 1 (**215**) yields (a*R*,5*S*,6*S*,7*R*)-5,10-diacetoxy-1,2,3,9,11-pentamethoxy-6,7-dimethyl-5,6,7,8-tetrahydrodibenzo[*a,c*]cyclooctene (**225**), **218** yields (a*R*,5*R*,6*S*,7*R*)-5,10-diacetoxy-1,9,11-trimethoxy-6,7-dimethyl-2,3-methylenedioxy-5,6,7,8-tetrahydrodibenzo[*a,c*]cyclooctene (**226**), **220** yields (a*R*,5*R*,6*S*,7*R*)-5,10-diacetoxy-1,11-dimethoxy-6,7-dimethyl-2,3-methylenedioxy-5,6,7,8-tetrahydrodibenzo[*a,c*]ccooctene (**227**), **222** yields (a*R*,6*R*,7*R*)-3,12-dimethoxy-6,7-dimethyl-10,11-methylenedioxy-5,6,7,8-tetrahydrodibenzo[*a,c*]cycloocten-2-ol (**228**), and finally **223** yields (a*R*,6*R*,7*R*)-2,12-dimethoxy-6,7-dimethyl-10,11-methylenedioxy-5,6,7,8-tetrahydrodibenzo[*a,c*]cycloocten-3-ol (**229**)) [124]. This leads to an "unnatural" oxygenation pattern in the newly formed aromatic ring compared to that found in all Schisandraceae dibenzocyclooctanes. Alkyl migration in both the dienone-phenol and dienol-benzene rearrangements (a reaction carried out on the dienols generated from sodium borohydride reduction of the eupodienones 1 (**215**), 4 (**218**), 5 (**219**), 6 (**220**), and 7 (**221**) yielding dibenzocyclooctenes (**230**–**234**) respectively) in the eupodienone series, differs from the expected aryl migration that had previously been observed in cyclohexadienone-containing alkaloids [112]. *Gymnotheca chinensis* and *G. involucrata* also yielded two "unnatural" dibenzocyclooctanes (gymnothelignans L (**235**) and M (**236**)) [126].

Eupomatia bennettii also produces a series of *seco*-lignans (Fig. 26), eupomatilones 1–7 (**237**–**243**), as major components of both the aerial parts and tubers [135]. The eupomatilones are oxygenated biphenyls substituted by a tetrahydrofuranone. The *ortho-ortho* substitution of one of the phenyl groups leads to restricted rotation about the biphenyl bond, leading to atropisomerism. In the four compounds with a symmetrical 2,3,4-trimethoxyphenyl substituent, resonances for two *meta*-coupled aromatic protons and three aromatic methoxy proton resonances are observed. In the other three compounds with asymmetric 3,4-dioxyphenyl rings, doubling of all resonances in the ^{1}H NMR spectra occurs because chirality is imparted by atropisomerism about the biphenyl bond, thereby generating diastereoisomers (Fig. 27). If the energy barrier for restricted rotation is large, it is conceivable that the diastereomers of eupomatilone 5 (**241**) should be separable. Acquiring ^{1}H NMR data at variable temperatures through successive incremental heating of both eupomatilone 1 (**237**) and eupomatilone 5 (**241**) from 298 K to above 375 K allowed coalescence temperatures to be determined for each resonance and this permitted a free energy of activation to be calculated. This showed that the energy of activation at room temperature is 60 $kJmol^{-1}$, an energy barrier lower than that needed for chromatographic separation of atropisomers but high enough to observe atropisomers by NMR spectroscopy [135, 136]. The eupomatilones are most likely derived from a eupodienone precursor possessing a furano bridge between C-6 and C-9.

Interestingly, *E. bennettii* also yielded a tetrahydrofuran lignan, eupobennettin (**244**), which has the same relative and absolute configurations as calopiptin (**109**) and austrobailignan-7 (**110**) discussed earlier [118]. Diphenolic coupling (C-2:C-1) of the 4″-demethyl precursor of eupobennettin would generate a eupodienone with the correct configuration at C-6, C-7, and C-8 (Scheme 2). Hydroxylation at C-6 leading to ring opening of the seven-membered ring and aromatization of the dienone then leads to the eupomatilone system after oxidation of the hemiketal to an ester. *Gymnotheca chinensis* and *G. involucrata* have furnished this tetrahydrofurnanoeupodienone intermediate gymnothelignan U (**245**) and also contain ketal congeners (such as gymnothelignan P (**246**)) of the eupomatilone series, providing direct evidence for this biosynthesis proposal [129, 131, 135]. The eupomatilone structures have been confirmed through total synthesis and this also necessitated the correction of the configuration at C-3′ in eupomatilone-6 to **247** [137, 138]. The ketal congeners of the eupomatilones have been screened for cytotoxic effects but were mostly weak or inactive towards several cancer cell lines [130]. Likewise, the *Gymnotheca* eupodienones were also not cytotoxic [129]. A flavone 3,7,3′-trimethylquercitin (**248**) was also reported from the bark of *E. laurina* [139]. The two *Eupomatia* species studied to date are super-producers of chemical diversity with structural classes in common with those found in the basal angiosperms, the Piperales and other Magnoliids. A third species, *E. barbata*, was only formally described in 2002 and its chemistry would be interesting to investigate.

Annonaceae

The Australian Annonaceae are dominated by species in the tribes Miliuseae and Uvariaeae that are likely to have migrated to Australia from their centers of diversity in New Guinea [140]. Five other Australian species, *Cananga odorata* (Plate 9), *Artabotrys carnosipetalus*, *Goniothalamus australis*, *Xylopia maccreae*, and *X. monosperma* are from more basal lineages (Fig. 28). Annonaceae are major producers of BIAs (Table 7, Fig. 29) and this class is found in most of the Australian species that have been investigated with the aporphine alkaloid subclass being the most highly represented.

Aporphine alkaloids have been reported from *Huberantha nitidissima* (**187**, **189**, (−)-ushinsunine (**249**)) (collections from Australia and New Caledonia) [141, 142], *Goniothalamus australis* ((−)-anonaine (**250**), asimilobine (**251**)) [143], *Mitrephora diversifolia* (oxostaphanine (**252**), thailandine (**253**)) [39, 144], *Pseuduvaria villosa* (**187**), and *Polyalthia xanthocarpa* (*N*-nornuciferine (**254**)) [39]. Two of these alkaloids, **249** and **250**, have weak antiplasmodial activity [143]. An unusual C-3:C-10 linked and brightly colored bisaporphine alkaloid, beccapolinium hydroxide (**255**), originally reported from *Polyalthia cauliflora* var. *beccarii*, was also found in *M.*

Table 7 Distribution of natural products isolated from the Annonaceae

Species	Collection location[a]	Plant part(s)[b]	Compound class	Compound(s)	Refs.
Huberantha nitidissima	NQ, NC	b, wp	Aporphine	**187**, **189**, **249**	[141, 142]
	NC	wp	BIA	**256**, **257**	[142]
	NC	wp	Bis BIA	**258–264**	[142]
	NC	wp	Protoberberine	**265**	[142]
Pseuduvaria villosa	NQ	b	Aporphine	**189**	[39]
Goniothalamus australis	NQ	ap	Aporphine	**249**, **250**, **268**, **269**	[143]
	NQ	ap	Aristolactam	**143**, **146**, **266**, **267**	[143]
	NQ	ap	Lactone	**273**, **274**	[143]
Mitrephora diversifolia	CY	b, r	Aporphine	**251–253**, **271**, **272**	[39, 144]
	CY	b	Bis aporphine	**255**	[39]
Polyalthia xanthocarpa	NQ	lf	Aporphine	**254**	[39]
Meiogyne cylindrocarpa	MAL	b	Butenolide	**275–281**	[146]
	MAL	b	Sesquiterpenoid	**282**, **283**	[147]

[a] NQ = North Queensland, CY = Cape York, NC = New Caledonia, MAL = Malaysia [b] lf = leaves, ap = aerial parts, r = roots, wp = whole plant, b = bark

diversifolia as part of the AZGU project [39]. Benzyl isoquinoline alkaloids (reticuline (**256**), protosinomenine (**257**)), bis-BIAs (dauricine (**258**), daurisoline (**259**), isodaurisoline (**260**), 7-*O*-methyllindoldhamine (**261**), 7′-*O*-methyllindoldhamine (**262**), dauriciline (**263**), lindoldhamine (**264**)), and the protoberberine, stepholidine (**265**), were also reported from New Caledonian *H. nitidissima* [142]. Aporphine lactams (**143**, **146**, enrerocarpam II (**266**), and caldensine (**267**)) have been found in *G. australis* [143]. These compounds are either the same or closely related to alkaloids almost exclusively reported in rainforest vines from the genus *Aristolochia* (Aristolochiaceae). Unlike the aristolactams that lack *N*-methylation, **143** and **267** are *N*-methylated and this imparts more potent antiplasmodial activities when compared to those of their of non-methylated congeners [143]. Two unusual quinolones, goniothalines A and B (**268** and **269**), from *G. australis*, at first glance look to be unrelated to the aporphine co-isolates. However, they are likely derived from an aporphine related to annolatine (**270**) (reported from *Annona montana*) via two extradiol cleavage reactions [145]. First, B-ring extradiol cleavage, decarboxylation, and decarbonylation leads to a cleistopholine derivative and second, an extradiol cleavage in the D ring and decarboxylation generates goniothaline. Two additional extradiol-cleaved onychine derivatives (5,8-dihydroxy-6-methoxyonychine (**271**) and 5-hydroxy-6-methoxyonychine (**272**)) were reported from a second collection of *M. diversifolia* from Cape York [144]. Moderate activity towards a chloroquine-resistant strain of *Plasmodium falciparum* was reported for compound **272**.

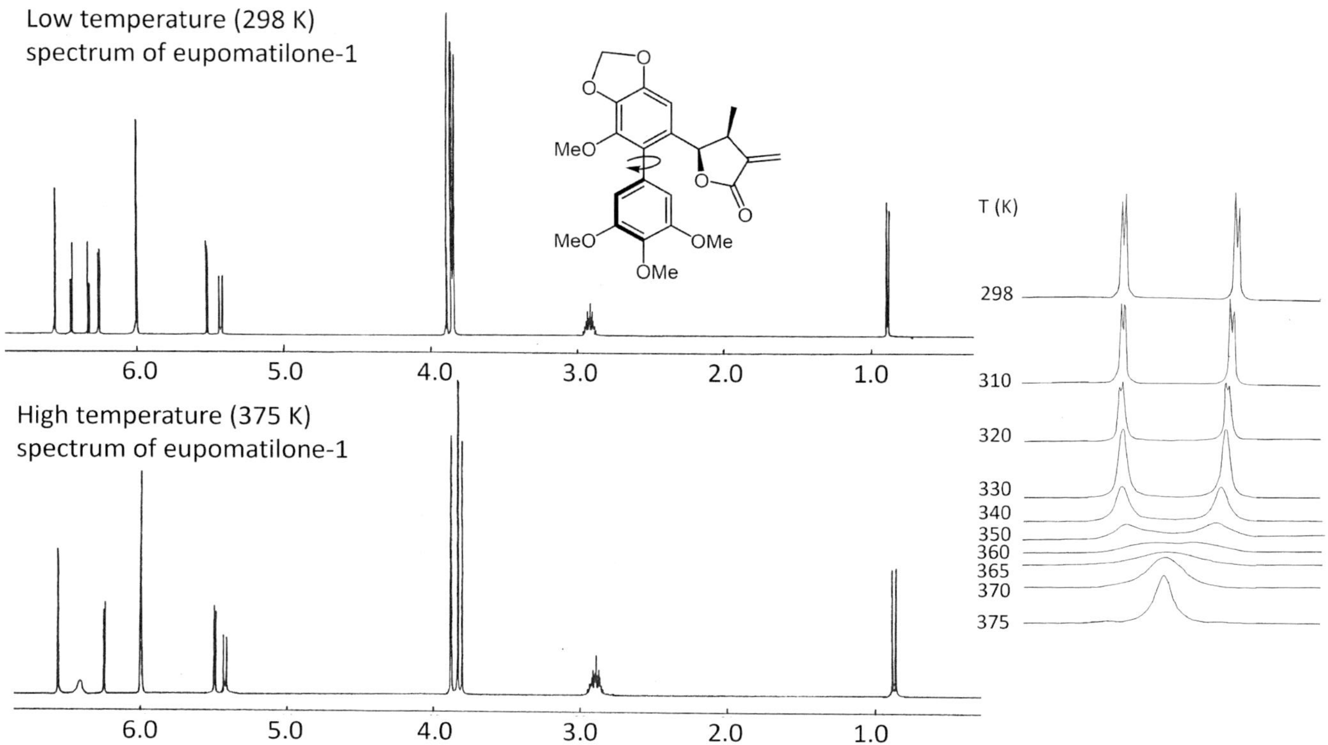

Fig. 27 Variable temperature analysis of eupomatilone 1 (**237**) by ^{1}H NMR spectroscopy. Doubled resonances observed at 298 K coalescing at 375 K. (inset shows the sequence of coalescence as temperature is increased (AR Carroll, Ph.D. thesis, 1989, [136])

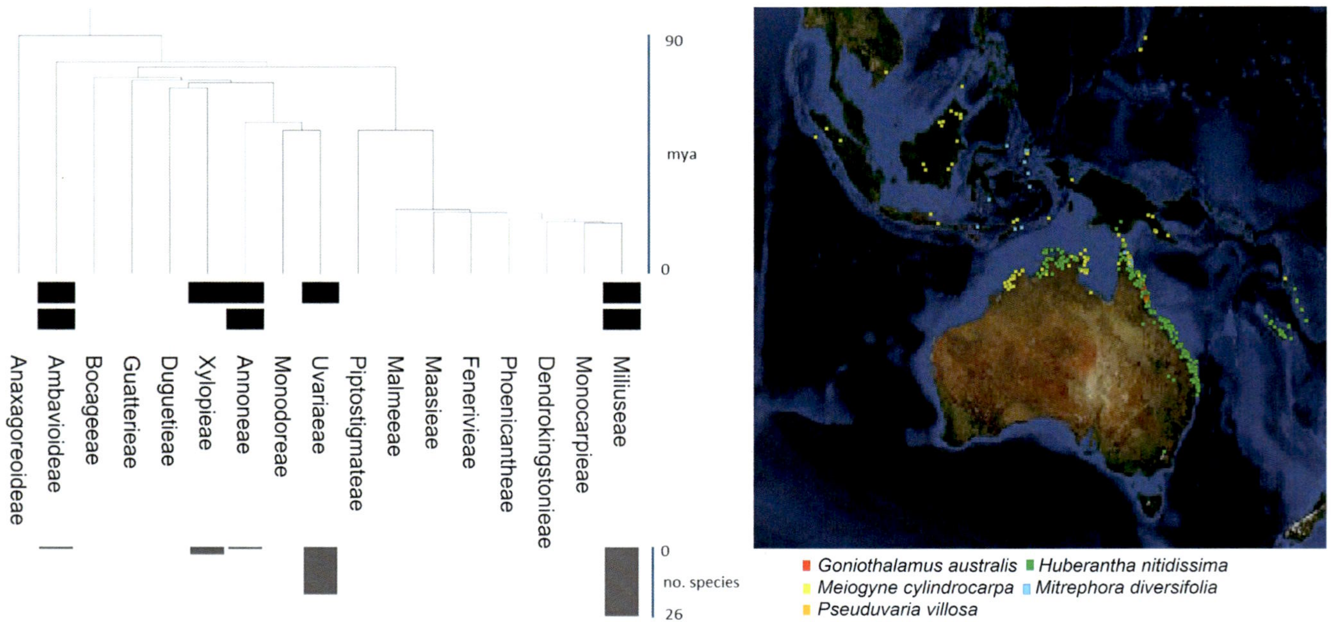

Fig. 28 Dated molecular phylogeny of the tribes from the Annonaceae [140] showing those that have species present (grey) and studied (black) and the number of species occurring in Australian rainforest; Map showing the distribution of species studied based on herbarium records [5, 43]

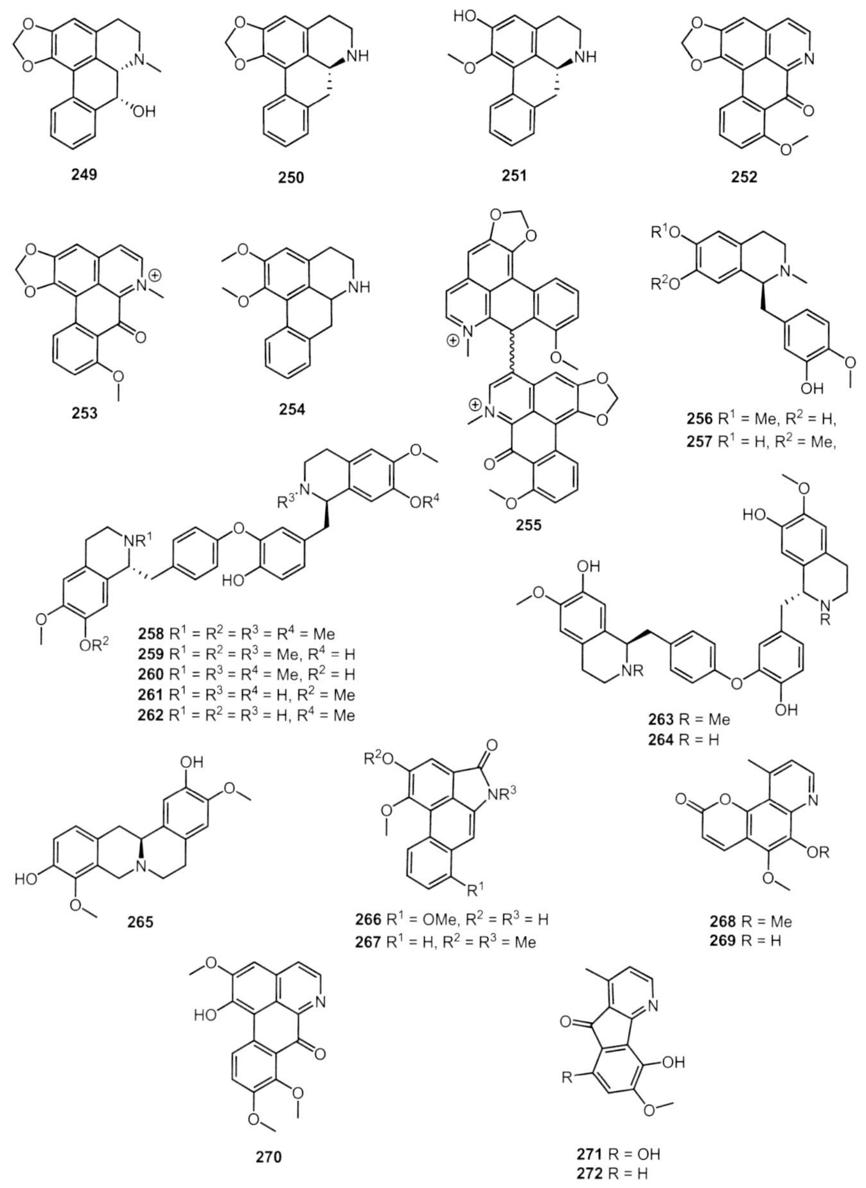

Fig. 29 Alkaloids **249–272** reported for Australian rainforest Annonaceae species

Plate 9 Flowers of Australian rainforest Annonaceae: *Pseuduvaria froggattii* (top left), *Meiogyne stenopetala* (top right) *Cananga odorata* (bottom left) and green spotted triangle butterfly (*Graphium agamemnon*) that feeds on *Goniothalamus*, *Mitrephora*, *Pseuduvaria* and *Polyalthia* species (bottom right). (Photographs and painting AR Carroll)

Additional Annonaceae natural products reported from rainforest species (Fig. 30) were two aryl lactones altholactone (**273**) and (+)-goniofufrone (**274**) from *G. australis* [143], seven alkynyl butanolides (sapranthins A–G **275**–**281**) [146] and diastereomeric bis-sesquiterpenoids (meiogynin A (**282**) and 1-*epi*-meiogynin A (**283**)) from *Meiogyne cylindrocarpa* (an Australian native but this investigation was on plant material collected in Malaysia) [147]. Two of the alkynyl butanolides were weakly cytotoxic.

Himantandraceae

Galbulimima baccata (Fig. 31) is a tall tree with disjunct populations in the rainforests of southeast and Northeast Queensland while *Galbulimima belgraveana* is found in New Guinea. The hallucinogenic properties of *G. belgraveana* have led to its use in religious ceremonies and its other properties have seen it also used traditionally to

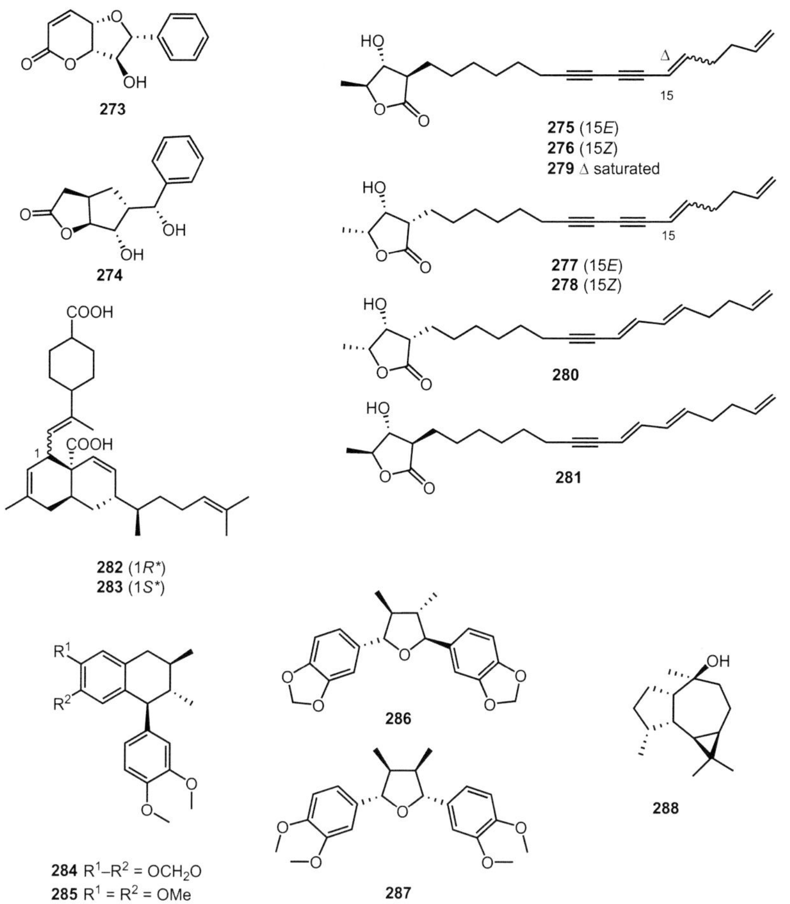

Fig. 30 Structures of compounds **273**–**283** reported for Australian Rainforest Annonaceae species and lignans **284**–**287** and a sesquiterpenoid **288** reported from *Galbulimima* spp.

treat abdominal pain and induce sleep [148]. It has been noted that eating pigeons that have consumed *G. belgraveana* fruits can lead to hallucinogenic experiences [28].

Chemical and pharmaceutical investigations of these species were a highlight of the Australian Phytochemical Survey. Both species contain aryltetralins (galcatin (**284**) and galbulin (**285**) and 1,4-diaryl-2,3-dimethylfurans lignans (galbacin (**286**), galgravin (**287**)) [149], the sesquiterpenoid viridiflorol (**288**) (Fig. 32) [150], and a remarkable diversity of unusual pseudoalkaloids (Table 8). The two 1,4-diaryl-2,3-dimethylfurans are both symmetrical thus differing in their configurations about the

Fig. 31 *Galbulimima baccata* and its disjunct populations in the rainforests of Southeast and North-east Queensland and *G. belgraveana* distribution in PNG and Indonesia (painting of *G. baccata* AR Carroll)

furan unit when compared to the related furan lignans reported from *Austrobaileya*, *Trimenia*, and *Eupomatia* [149].

The alkaloids (Figs. 32 and 33) form 11 structurally related subclasses and include 40 congeners [39, 148, 150–157]. They are unlike any other alkaloids reported from plants but bare a very distant similarity to alkaloids reported from zoanthid corals (such as zoanthamine A (**289**)) that are made by symbiotic dinoflagellates [158]. Many of the *Galbulimima* alkaloids are three dimensionally complex. The himandridine class, containing 20 congeners, as exemplified by himandrine (**290**), himandridine (**291**), himbosine (**292**), GB 1–GB 12 (**293–304**), GB 27 (**305**), GB 28 (**306**), GB 14 (**307**), GB 25 (**308**), and GB 24 (**309**), have up to 11 stereogenic centers and six contiguous rings. From a drug discovery perspective, himbacine (**310**) a molecule containing a piperidine that is ethylene bridged to a decalin annulated to a δ-lactone, originally showed antispasmodic and potent cardioselective M2/M4 muscarinic antagonistic activity [159]. This led scientists at Schering-Plough USA, to use **310** as a lead to develop novel treatments for Alzheimer's disease [159]. Unfortunately, these efforts failed but the synthetic analogues produced in the lead-optimization program were later included in a high-throughput screening library, and one of the analogues (**311**) showed potent protease-activated receptor-1 (PAR-1) activity [160]. This ultimately resulted in the himbacine analogue, voxapar (**312**), being developed into the first-in-class PAR-1 drug to treat patients with acute coronary syndrome and secondary prevention of cardiovascular events in high-risk patients [160].

The *Galbulimima* alkaloids have been grouped into four classes: Class I includes eight himbacine analogues (**310**, himbeline (**313**), GB 15 (**314**), himandravine (**315**),

Table 8 Distribution of natural products isolated from the Himantandraceae

Species	Collection location[a]	Plant part(s)[b]	Compound class	Compound(s)	Refs.
Galbulimima baccata	NQ	b	Lignan	**284–287**	[149]
	NQ	b, r, lf	Pseudoalkaloid	**290–310**, **314**, **316–332**, **334**	[39, 148, 151–157]
	NQ	lf	Sesquiterpenoid	**288**	[150]
Galbulimima belgraveana	PNG	b	Pseudoalkaloid	**290**, **310**, **313**, **315**, **316**, **318**	[151]

[a] NQ = North Queensland, PNG = Papua New Guinea, [b] lf = leaves, r = roots, b = bark

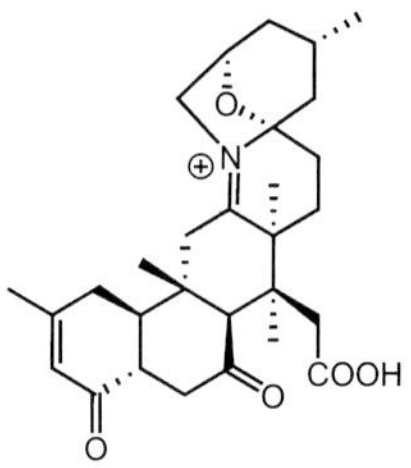

289 (zoanthamine A)

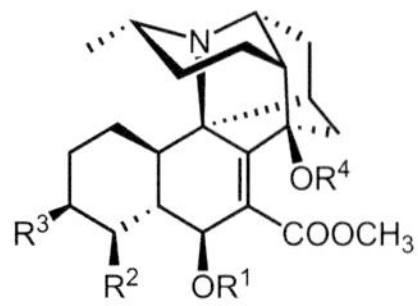

290 R^1 = COPh, R^2 = OMe, R^3 = R^4 = H
291 R^1 = COPh, R^2 = OMe, R^3 = OH, R^4 = H
292 R^1 = R^4 = Ac, R^2 = R^3 = OAc
293 R^1 = Ac, R^2 = OCOPh, R^3 = OAc, R^4 = H
294 R^1 = R^4 = Ac, R^2 = R^3 = OAc
295 R^1 = H, R^2 = OH, R^3 = OAc, R^4 = Ac
296 R^1 = H, R^2 = OH, R^3 = OCOPh, R^4 = Ac
297 R^1 = H, R^2 = OH, R^3 = OH, R^4 = Ac
298 R^1 = COPh, R^2 = OMe, R^3 = OAc, R^4 = H
299 R^1 = COPh, R^2 = OMe, R^3 = OH, R^4 = Ac
300 R^1 = R^3 = R^4 = H, R^2 = OMe
301 R^1 = Ac, R^2 = OMe, R^3 = R^4 = H
302 R^1 = R^4 = Ac, R^2 = OMe, R^3 = H
303 R^1 = R^3 = H, R^2 = OMe, R^4 = Ac
304 R^1 = R^4 = Ac, R^2 = OAc, R^3 = H
305 R^1 = COPh, R^2 = OMe, R^3 = OAc, R^4 = Ac
306 R^1 = R^4 = H, R^2 = OMe, R^3 = OH
307 R^1 = Ac, R^2 = R^3 = R^4 = H
308 R^1 = R^2 = R^3 = R^4 = H

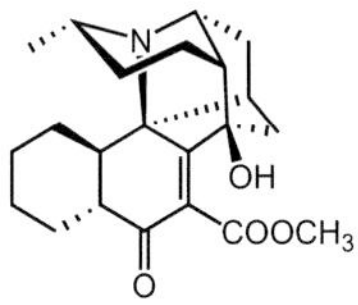

309

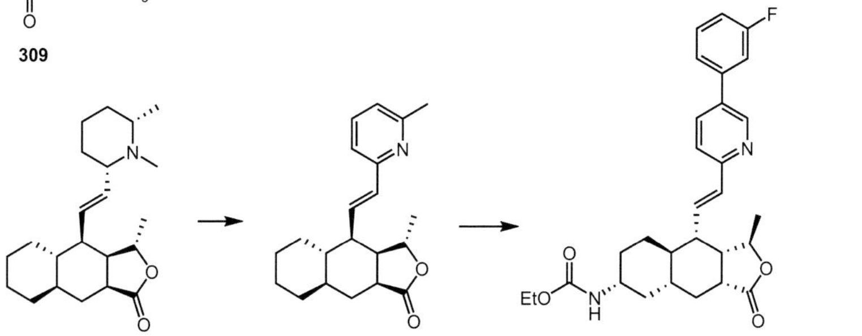

310 (himbacine) **311** **312** (vorapaxar)

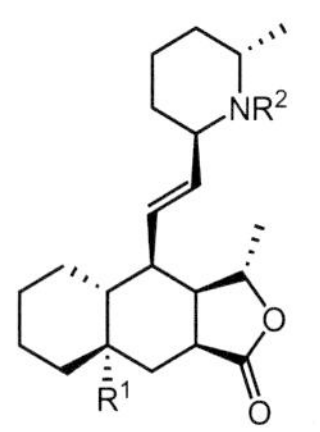

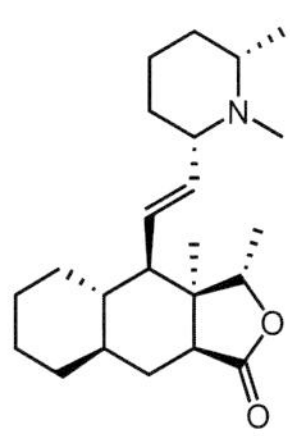

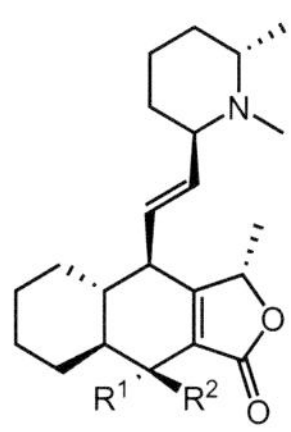

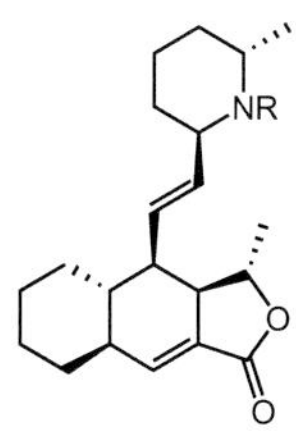

313 R^1 = R^2 = H
314 R^1 = OH, R^2 = Me

315

316 R^1 = OH, R^2 = H
317 R^1 = H, R^2 = OH

318 R = Me
319 R = H

Fig. 32 Structures of alkaloids **290–310** and **313–319** reported from *Galbulimima* spp. along with a related pseudoalkaloid zoanthamine A (**289**) and drug leads **311** and **312**

himgrine (**316**), GB18 (**317**), himgravine (**318**), GB 23 (**319**)), Class II, 20 himandrine analogues (**290–309**), Class III, six himbadine/himgaline analogues (himbadine (**320**), GB 13 (**321**), GB 22 (**322**), GB 30 (**323**), himgaline (**324**), GB 16 (**325**)) and Class IV contains six miscellaneous structures (GB 17 (**326**), GB 29 (**327**), GB 26 (**328**), GB 19 (**329**), GB 21 (**330**), and GB 20 (**331**)). The final structure in the series, GB 31 (**332**), is an oxidation product that lacks the piperidine moiety.

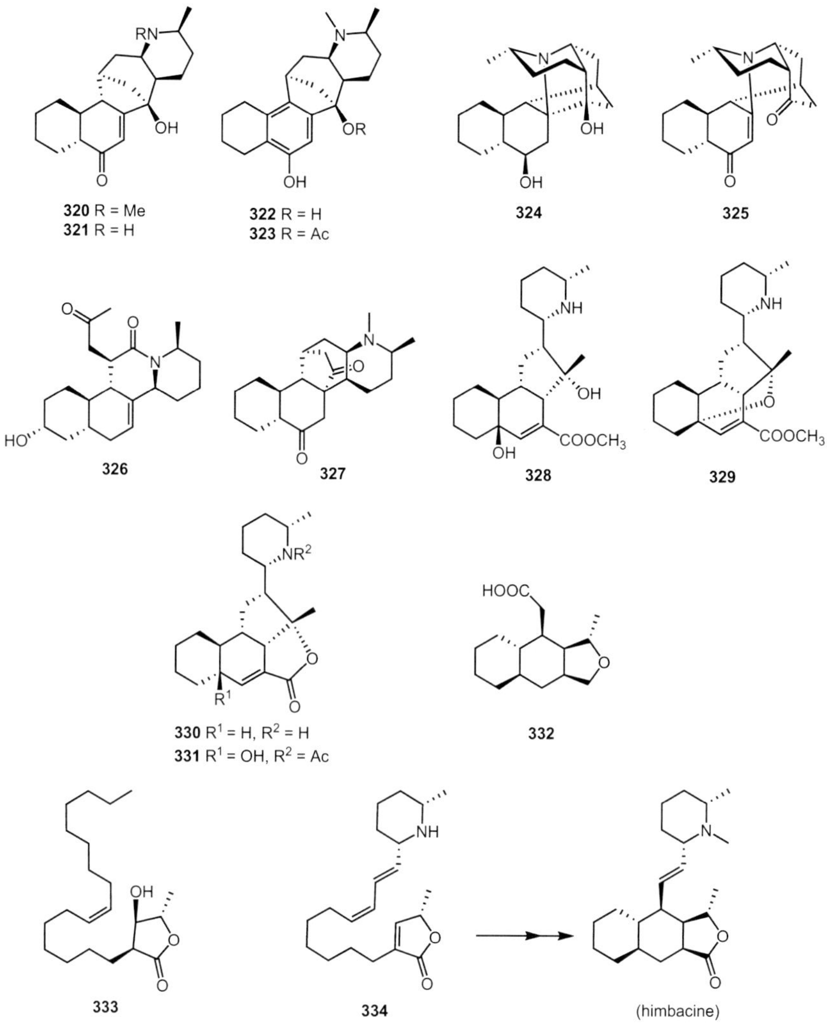

Fig. 33 Structures of the alkaloids **320–331**, and **334** and an oxidized alkaloid derivative **332** reported from *Galbulimima* spp. and the biosynthetically related butenolide **333** reported from *Zygoynum acsmithii*

Unlike other complex pseudoalkaloids reported from the angiosperm families Daphniphyllaceae, Ranunculaceae, and Escalloniaceae, all of which are derived from diterpenoids and triterpenoids, the origin of the *Galbulimima* alkaloids (like the *Zoanthus* alkaloids) appears to be a polyketide. A comprehensive biogenetic pathway was proposed initially by Bennett and later refined by Karuso and coworkers [148, 161]. At the core to this biogenesis is a linear 20-carbon polyketide substituted at C-3 by a carboxylic acid. Closely related C_{20} alkyl butanolides are almost exclusively found in Magnoliid species. The sapranthins (**275**–**281**) reported from *Meiogyne cylindrocarpa* (Annonaceae) [146], various C_{20} butanolides reported from Lauraceae species, and secobutanolides and butanolides from *Zygogynum acsmithii* (Winteraceae, New Caledonia) (e.g., **333**) [162] are tantalizingly like this elusive precursor. Even the alkyl end of the polyketide chain in the sapranthins is well disposed to undergo an amination reaction to generate the 2-methyl-piperidine. While isolating alkaloids for the AZGU project the present author and his associates found a linear precursor (**334**) in *Galbulimima belgraveana* bark from Papua New Guinea [39], adding further evidence for the proposed biogenesis.

Laurales

There are seven families in the Laurales and five have species that occur in Australian rainforests. The order has an ancient Gondwanan origin with most of the families being represented predominantly in the southern hemisphere [163]. The Australian rainforest species are spread across the Calycanthaceae (*Idiospermum australiense* (the basal species from the Calycanthaceae)), four species in the Hernandiaceae, 29 in the Monimiaceae, 130 in the Lauraceae, and 12 of the 17 in the Atherospermataceae (Fig. 34). Natural product chemical investigations have been undertaken on at least one species in each of these families.

Calycanthaceae

The Calycanthaceae sits in a basal relationship to all other Laurales [163]. The only Australian species in the family, *Idiospermum australiense* (Plate 10), is a large tree and the sole survivor and remnant of the family's ancestral Gondwanan origins in the upper Cretaceous. The other nine extant species are shrubs in two small genera (*Calycanthus* and *Chimonanthus*) that are much more recent migrants (mid-Miocene) to North America and eastern Asia [31]. *Idiospermum australiense* was originally described in the early 1900's and thought to be extinct after land clearing of the area where it was originally collected. The death of cattle on the Daintree River in north Queensland in the 1970's after consuming large unknown seeds eventually led to the plant's rediscovery. Two populations of *I. australiense* have now been found and are protected in national parks.

The fruits subsequently were investigated and shown to contain the toxic bis-quinoline (+)-calycanthine (**335**) and bis-indole (−)-chimonanthime (**336**) (both previously reported form other Calycanthaceae species), and a novel tris-indole (−)-idiospermuline (**337**) (Fig. 35 and Table 9) [164]. The absolute configuration of **337** was established by X-ray crystallography of its trimethiodide. Idiospermuline (**337**) is a μ*M*-potent cholinergic antagonist. These structures are unrelated to any others found in the Magnoliids but are closely related to alkaloids reported from the tribe Psychotrieae in the Rubiaceae [165]. Idiospermuline (**337**) possesses a different relative and absolute configuration at several of its chiral centers when compared to

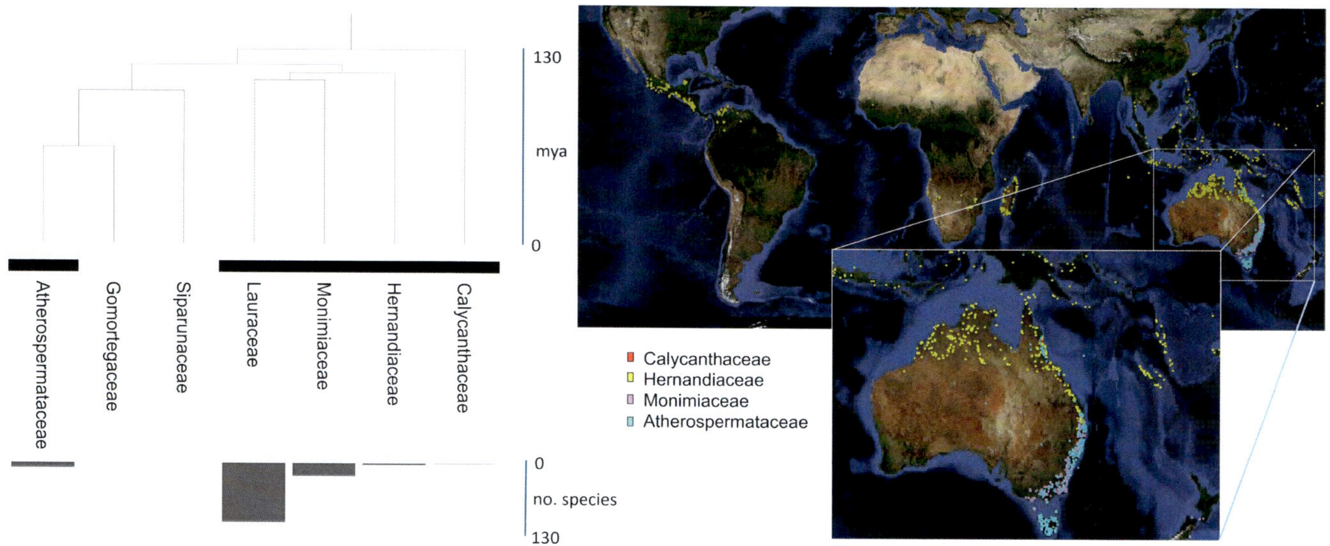

Fig. 34 Dated molecular phylogeny of the Laurales [7], showing those families that have species present (grey) and studied (black) and the number of species occurring in Australian rainforest. The map shows the worldwide distribution of chemically investigated Australian species in the families Hernandiaceae, Calycanthaceae, Monimiaceae and Atherospermataceae based on herbarium records [5, 43]

Table 9 Distribution of natural products isolated from the Calycanthaceae, Hernandiaceae, and Monimiaceae

Species	Collection location[a]	Plant part(s)[b]	Compound class	Compound(s)	Refs.
Calycanthaceae					
Idiospermum australiense	NQ	s	Indole	**335–337**	[164]
Hernandiaceae					
Gyrocarpus americanus subsp. *americanus*	CQ, NQ	b, wp	BIA	**343**	[39, 167]
	NQ, CQ	st, b	Bis BIA	**348**, **349**	[39, 167]
	NQ	st	Dibenzopyrrocoline	**345**	[39]
Gyrocarpus americanus subsp. *sphenopterus*	NQ	lf	BIA	**343**	[39]
Hernandia albiflora	NQ	lf	Aporphine	**342**	[39]
	NQ	tw, lf	Phenanthrene	**346**, **347**	[39, 168]
Hernandia bivalvis	SEQ	lf	Aporphine	**338**, **339**	[39, 166]
	SEQ	lf	BIA	**344**	[39]
Hernandia nymphaeifolia	NQ	lf	Aporphine	**340**, **341**	[39]
	VAN	b	Bis BIA	**350–354**	[169]
Monimiaceae					
Hedycarya angustifolia	Tas	wp	BIA	**355**	[170]
	Tas	wp	Monoterpenoid	**356**	[171]

[a] SEQ = Southeast Queensland, NQ = North Queensland, CQ = central Queensland, Tas = Tasmania, VAN = Vanuatu, [b] lf = leaves, tw = twigs, wp = whole plant, b = bark, st = stems, s = seeds

related Psychotrieae indoles and has a rare *N*-methylation pattern. This points to polyphyletic origins for this alkaloid class.

Hernandiaceae

Constituents from the four Hernandiaceae species found in Australian rainforests are all benzyl isoquinoline alkaloids (Fig. 36). They include five aporphines (laurotetanine (**338**), hernandine (**339**), hernangerine (**340**), *nor*-neolitsine (**341**), and *nor*-isodomesticine (**342**)) [39, 166], two BIAs (magnocurarine (**343**), orientaline (**344**)) [39, 167], one dibenzopyrrocoline (litcubinine (**345**)) [39], two ethyl aminophenanthrenes (thalicthuberine (**346**), *nor*-thalicthuberine (**347**)) [39, 168], two bisbenzylisoquinoline macrocycles ((−)-isotetrandrine (**348**), phaeanthine (**349**) [167] and five aporphine-benzylisquinoline dimers (6-northalicarpine (**350**), 6a,7-dehydrothalmelatine (**351**), hebridamine (**352**), vilaportine (**353**), thalicarpine 2′-*N*-oxide (**354**)) [169].

Plate 10 *Idiospermum australiense* flowers showing their color change from white to dark red as they mature. The flowers are borne on branches of this tall tree. (Photographs AR Carroll)

Fig. 35 Structures of the alkaloids **335–337** reported from *Idiospermum australiense* seeds

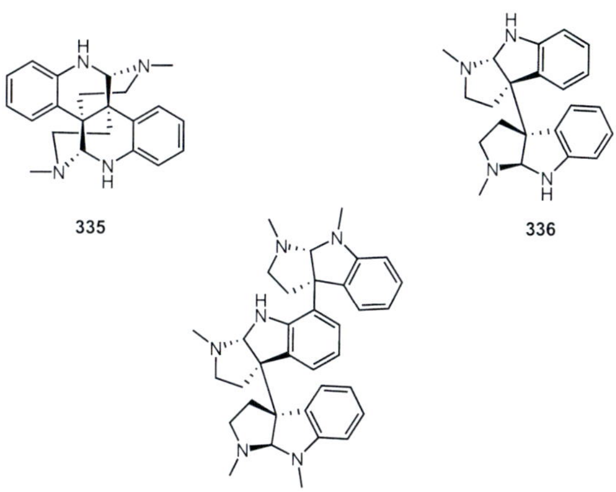

335

336

337

Monimiaceae

The Monimiaceae is sister to the Hernandiaceae. Only *Hedycarya angustifolia*, from King Island in the Bass Strait has been studied, and it yielded the benzyl isoquinoline alkaloid isoevanine (**355**) and a monoterpenoid (−)-5-*exo*-hydroxyborneol (**356**) (Fig. 36) [170, 171].

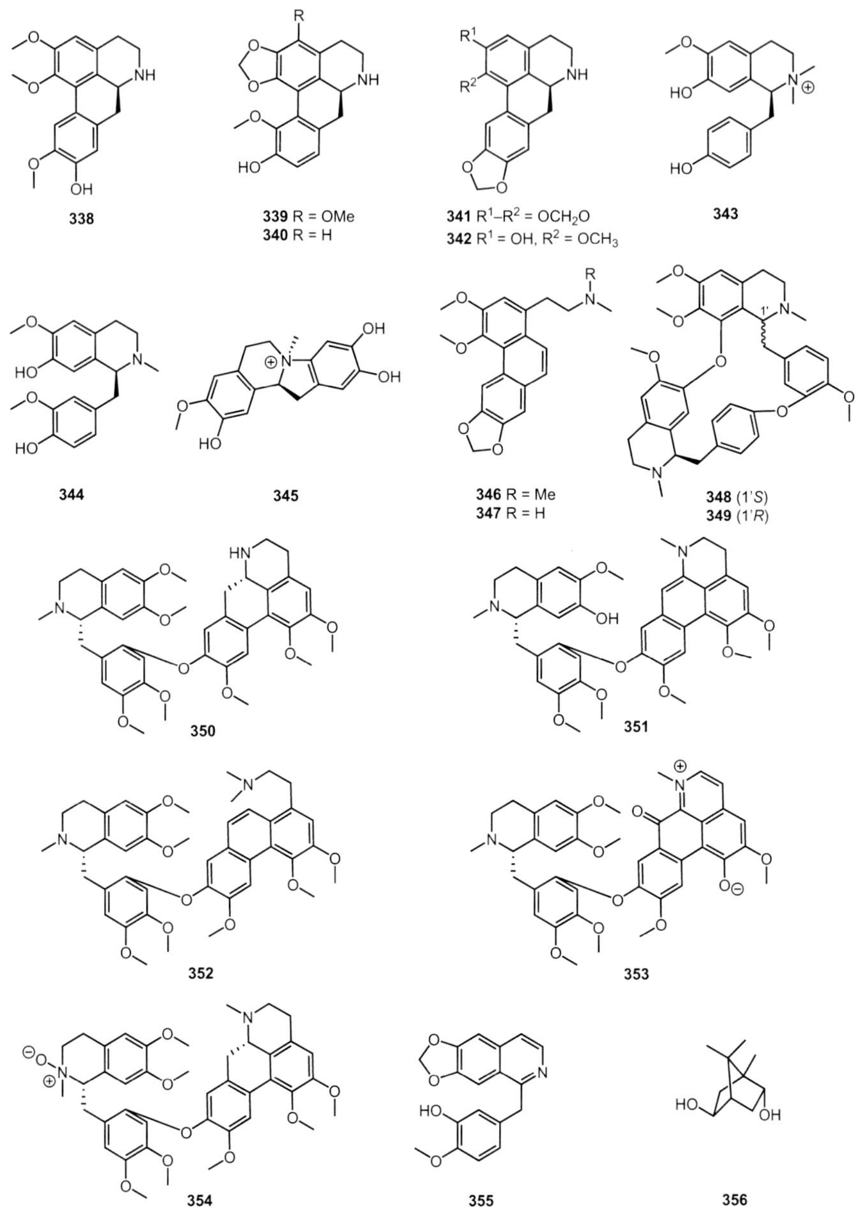

Fig. 36 Structure of the alkaloids **338**–**355** and the monoterpenoid **356** reported from Hernandiaceae and Monimiaceae species

Lauraceae

In Australian rainforests the family Lauraceae is dominated by species from the tribe Cryptocaryeae (81%, 105 species). This tribe is postulated to have originated in northern Africa, dispersed to South America and then on to Australia, approximately 90 mya [172]. Northern dispersal of Lauraceae taxa into Asia then occurred much more recently. This suggests a near basal position for the Cryptocaryeae within the Lauraceae (Fig. 37). The genus *Cassytha* is represented by three species in Australia's rainforest but its position in the phylogenetic tree has been difficult to place. The extant Cryptocaryeae are predominantly Southern Hemispheric. While *Cassytha filiformis* (Plate 11) a coastal vine has widespread distribution in Africa, central and south America, Japan and southeast Asia, all other *Cassytha* species are restricted to Australasia.

Thirty-five Australian rainforest Lauraceae species (Plate 11) have been investigated chemically (Table 10) and this has resulted in a diverse collection of alkaloids and polyketides being reported. The alkaloids (Figs. 38 and 39) reported from the Australian rainforest Lauraceae include aporphines (**338**, (6a*R*)-*nor*-mecambroline (**357**), (6a*R*)-roemerine (**358**), (6*S*,6a*R*)-roemerine-*N*-oxide (**359**), (6*R*,6a*R*)-roemerine-*N*-oxide (**360**), (6a*S*)-actinodaphnine (**361**), (6a*S*)-laurolitsine (**362**), (6a*S*)-boldine (**363**), laurelliptine **364**), isoboldine (**365**), *N*-methyllaurotetanine (**366**), boldine free base (**367**), laurolitsine (**368**), nantenine (**369**), domesticine (**370**), *nor*-domesticine (**371**), isocorydine (**372**), zenkerine (**373**), sparsiflorine (**374**), cassythine (**375**), (+)-ocoteine (**376**), cassythidine (**377**), *N*-methylactinodaphnine (**378**), and roemerine free base (**379**)) [39, 173–179], benzyl isoquinoline alkaloids (**256**, **343**, **344**, coclaurine (1*S*)-juziphine (**380**), (1*S*)-norjuziphine (**381**), oblongine (**382**), 1′-hydroxymagnocurarine (**383**), (**384**), norcoclaurine (**385**), phyllocryptonine (**386**)) [39, 178–180], ethylaminophenanthrene derivatives (cryptopleurospermine (**387**), atherosperminine (**388**)) [39, 174], the piperidine, pleurospermine (**389**) [181], phenanthroquinolizidine alkaloids (cryptopleuridine (**390**), cryptopleurine (**391**), antofine (**392**)) [39, 182, 183], morphinanes (sinoacutine (**393**), dihydro-sinomeninol (**394**), tetrahydrosalutaridinol (**395**)) [39, 176], pavine alkaloids (isocaryachine (**396**), crychine (**397**)) [39] and dibenzopyrrocoline alkaloids (**345**, cryptowoline (**398**), crytpowolinol (**399**), cryptaustoline (**400**), and cryptowolidine (**401**)) [39].

Higher alkaloid compound class diversity has been reported from Australian species in the Cryptocaryeae compared to Lauraceae and Cinnamomeae species since all classes described above are found in the Cryptocaryeae. The phenanthroquinolizidines **390–392** are vesicant molecules found in the bark and wood of *Cryptocarya pleurosperma*, *C. laevigata*, *C. cercophylla*, or *C. meisneriana*, and forestry workers have long avoided *C. pleurosperma* because of severe blistering that results from contact with the freshly cut timber [28]. This class of compound has also been reported from species in the Apocynaceae, Moraceae and Urticaceae suggesting polyphyletic biosynthetic origins [184]. The compounds are likely to originate from condensation of a 3,4-dioxyacetophenone with either pyrroline or dehydropiperidine

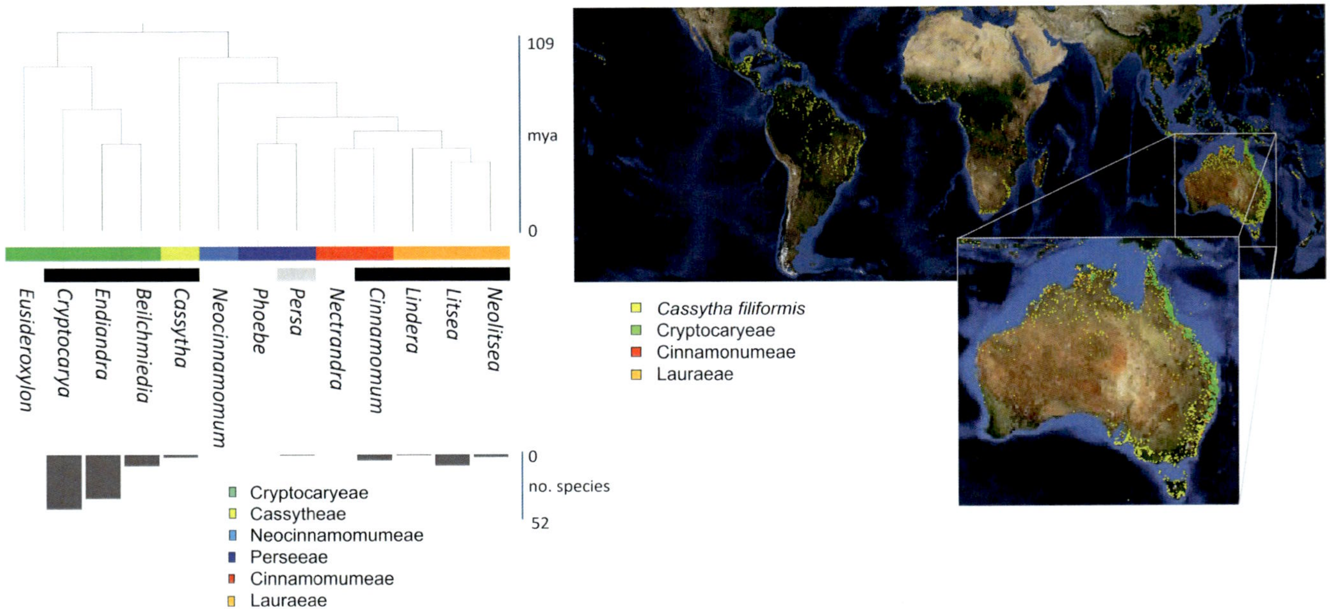

Fig. 37 Dated molecular phylogeny of the Lauraceae genera (color-coded by tribes [172]) showing those that have species present (grey) and studied (black) and the number of species occurring in Australian rainforest. The map shows the worldwide distribution of chemically investigated Australian species in the Lauraceae based on herbarium records [5, 43]

Plate 11 Lauraceae species *Cassytha filiformis* (left, flower and fruits inset) *Endiandra* species (top row flowers and fruit) *Cinnamomum* sp. (bottom middle and inset), *Cryptocarya laevigata* (middle right fruit and flowers), *Neolitsea dealbata* fruits (bottom right). (Photographs AR Carroll)

(derived from ornithine or lysine respectively) followed by a second condensation with an oxygenated acetophenone.

Eight aporphine alkaloids have been reported from Australian *Cassytha* species (**364**, **365**, **369–371**, **375–377**) and these compounds share close similarities to alkaloids reported from Australian *Neolitsea* species (**357–363**, **368**) (including a methylenedioxy group attached at C-6, C-7 of the B-ring) [175, 176, 179]. A small number of species in the Lauraceae tribe (*Litsea*, *Neolitsea*, and *Lindera*) and *Cinnamomum* species from the Cinnamomeae are likely to be more recent migrants to Australia from the north [172]. These Laurasian descendants all contain aporphine alkaloids closely related to those reported from Australian *Cryptocarya*, *Beilschmiedia*, and *Cassytha*. Interestingly, none of the *Endiandra* species studied contain alkaloids.

Linear and complex polycyclic polyketide acids (Fig. 40) substituted at one end by an aryl group (either phenyl or 3,4-methylenedioxyphenyl) are another class of compounds reported from Australian Cryptocaryeae species. The compounds are derived from Claisen condensations of a cinnamic acid with either four, five, six, seven, or nine malonates. These molecules include the endiandric acids A–D (**402–405**), 3″,4″-methylenedioxyendiandric acid A (**406**) from various *Endiandra* and *Beilschmiedia* species, cryptocaryoic acids A–C (**407–409**), cryptocaryone (**410**), 7′,8′-dihydroobolactone (**411**) and obolactone (**412**) from *Cryptocarya* species, and EBC-73 (**413**) from *Cinnamomum laubatii* [185–188]. The simplest structures are linear polyoxygenated acids or esters, but a striking feature of some of the other

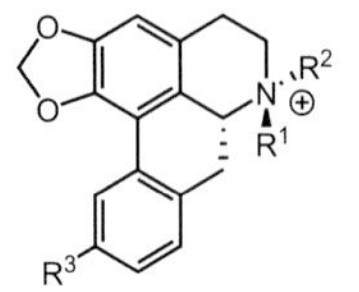

357 $R^1 = R^2 = H$, $R^3 = OH$
358 $R^1 = R^3 = H$, $R^2 = Me$
359 $R^1 = Me$, $R^2 = OH$, $R^3 = H$
360 $R^1 = OH$, $R^2 = Me$, $R^3 = H$

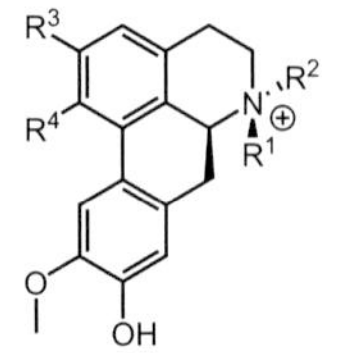

361 $R^1 = H$, $R^2 = Me$, $R^3–R^4 = OCH_2O$
362 $R^1 = R^2 = H$, $R^3 = OH$, $R^4 = OMe$
363 $R^1 = Me$, $R^2 = H$, $R^3 = OH$, $R^4 = OMe$

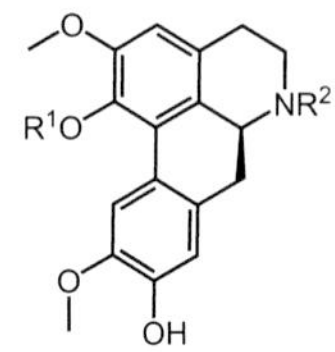

364 $R^1 = R^2 = H$
365 $R^1 = H$, $R^2 = Me$
366 $R^1 = R^2 = Me$

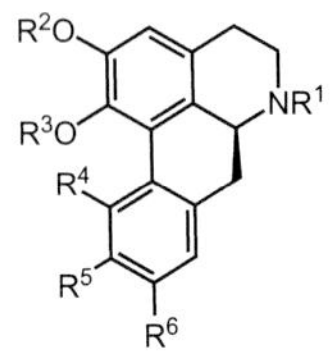

367 $R^1 = R^3 = Me$, $R^2 = R^4 = H$, $R^5 = OMe$, $R^6 = OH$
368 $R^1 = R^2 = R^4 = H$, $R^3 = Me$, $R^5 = OMe$, $R^6 = OH$
369 $R^1 = R^2 = R^3 = Me$, $R^4 = H$, $R^5–R^6 = OCH_2O$
370 $R^1 = R^2 = Me$, $R^3 = R^4 = H$, $R^5–R^6 = OCH_2O$
371 $R^1 = R^3 = R^4 = H$, $R^2 = Me$, $R^5–R^6 = OCH_2O$
372 $R^1 = R^2 = R^3 = Me$, $R^4 = OH$, $R^5 = OMe$, $R^6 = H$
373 $R^1 = R^3 = R^4 = R^6 = H$, $R^2 = Me$, $R^5 = OMe$
374 $R^1 = R^3 = R^4 = R^6 = H$, $R^2 = Me$, $R^5 = OH$

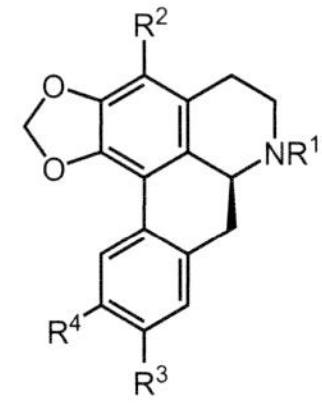

375 $R^1 = H$, $R^2 = R^4 = OMe$, $R^3 = OH$
376 $R^1 = Me$, $R^2 = R^3 = R^4 = OMe$
377 $R^1 = H$, $R^2 = OMe$, $R^3–R^4 = OCH_2O$
378 $R^1 = Me$, $R^2 = H$, $R^3 = OH$ $R^4 = OMe$

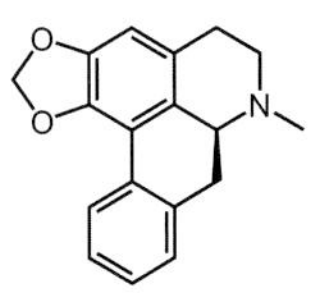

379

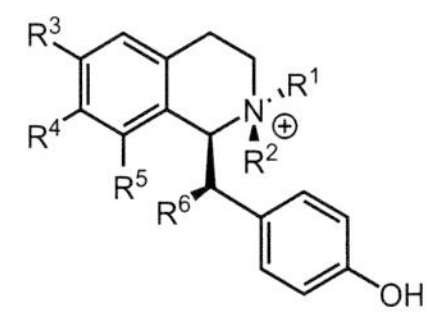

380 $R^1 = R^3 = R^6 = H$, $R^2 = Me$, $R^4 = OMe$, $R^5 = OH$
381 $R^1 = R^2 = R^3 = R^6 = H$, $R^4 = OMe$, $R^5 = OH$
382 $R^1 = R^2 = Me$, $R^3 = R^6 = H$, $R^4 = OMe$, $R^5 = OH$
383 $R^1 = R^2 = Me$, $R^3 = OMe$, $R^4 = R^6 = OH$, $R^5 = H$,

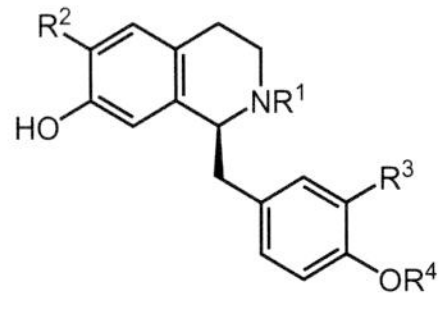

384 $R^1 = R^3 = R^4 = H$, $R^2 = OMe$
385 $R^1 = R^3 = R^4 = H$, $R^2 = OH$

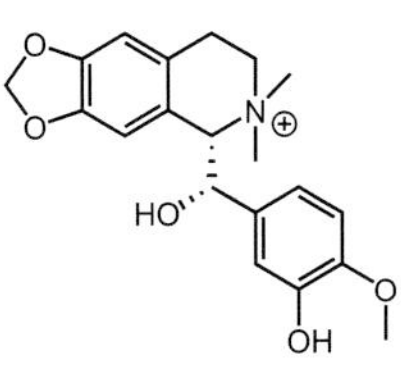

386

Fig. 38 Structures of the alkaloids **357**–**386** reported from Lauraceae species

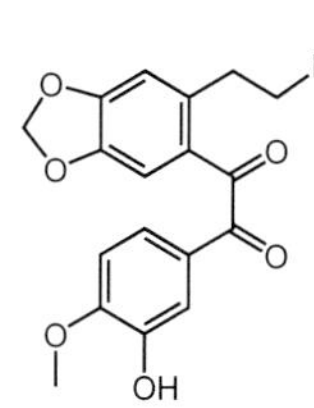

387

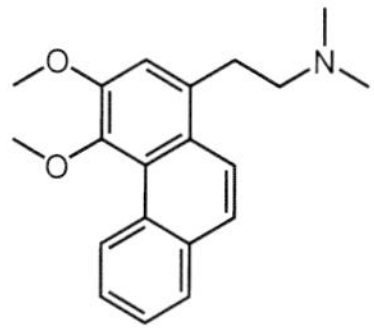

388

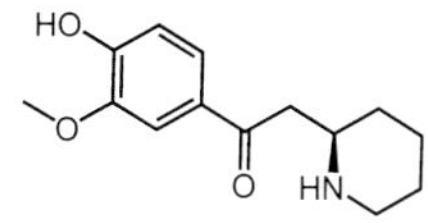

389

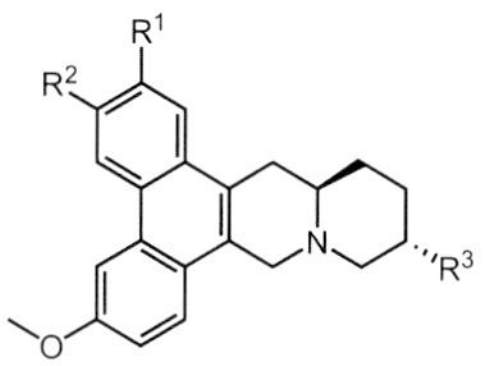

390 R^1–R^2 = OCH_2O, R^3 = OH
391 R^1 = R^2 = OMe, R^3 = H

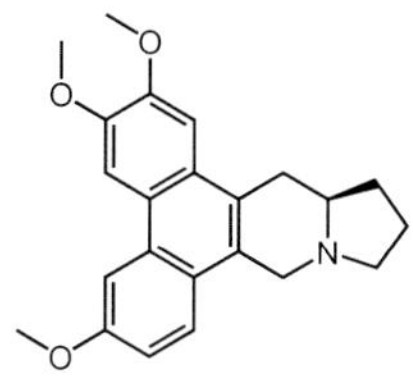

392

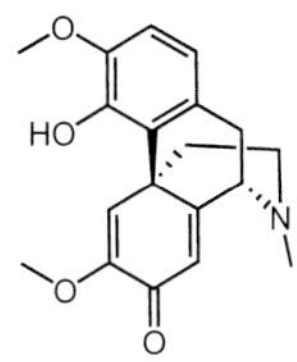

393

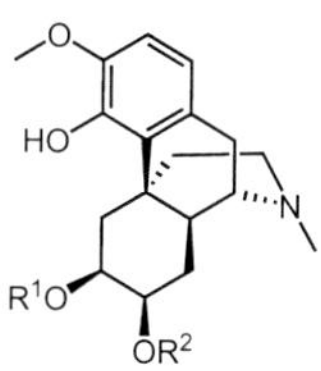

394 R^1 = H, R^2 = Me
395 R^1 = Me, R^2 = H

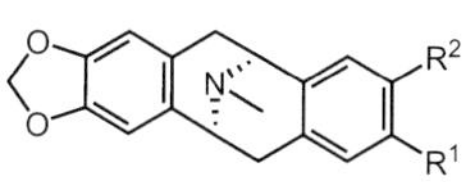

396 R^1 = OH, R^2 = OMe
397 R^1–R^2 = OCH_2O

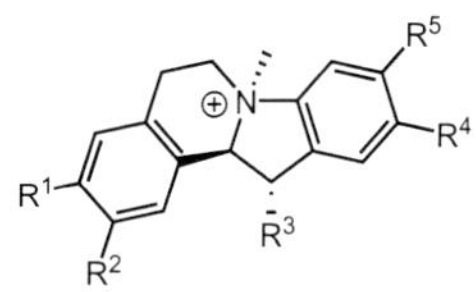

398 R^1 = OMe, R^2 = OH, R^3 = H, R^4–R^5 = OCH_2O
399 R^1 = OMe, R^2 = R^3 = OH, R^4–R^5 = OCH_2O
400 R^1 = R^4 = R^5 = OMe, R^2 = OH, R^3 = H
401 R^1 = OH, R^2 = OMe, R^3 = H, R^4–R^5 = OCH_2O

Fig. 39 Structures of the alkaloids **387–401** reported from Lauraceae species

molecules is their complex interconnected ring systems that include cyclobutanes linked to cyclopentane, cyclohexane or both. These molecules have been proposed to result from photochemical cyclizations of polyunsaturated linear acids [161]. The aryl tetralones from *Zygogynum* (Winteraceae) discussed earlier form another group that share a common phenyl polyketide precursor [92]. Related molecules have been reported from several *Beilschmiedia* species from Africa, East Asia, and South America and *E. kingiana* from central Malesia and some are reported to have weak to moderate anti-inflammatory activity [189]. The obolactones **411** and **412** show moderate antitrypanosomal activity against *T. brucei* and EBC-73 (**413**) is moderately cytotoxic [187, 188]. The terminal dihydro-α-pyran moiety of the obolactones and EBC-73 (**413**) is also shared with two additional small polyketides, massoia lactone (**414**) and 6-heptyl-5,6-dihydro-2*H*-pyran-2-one (**415**) reported from *Cryptocarya cocosoides* and the cytotoxic polyketides EBC-23–EBC-25 (**416–418**) and EBC-72 (**419**), EBC-75 (**420**), and EBC-76 (**421**) reported from *Cinnamomum laubatii* from north Queensland [187, 190, 191]. Polyketide EBC-73

Table 10 Distribution of natural products isolated from the Lauraceae

Species	Collection location[a]	Plant part(s)[b]	Compound class	Compound(s)	Refs.
Beilschmiedia brunnea	NQ	lf, r	Morphinane	**394**, **395**	[39]
Beilschmiedia elliptica	SEQ	b, lf	Aporphine	**338**, **364**, **365**, **373**, **374**	[39, 173]
	SEQ	b	BIA	**384**	[39]
Beilschmiedia obtusifolia	NQ	b	Phenyl polyketide acid	**402**	[185]
Beilschmiedia oligandra	NQ	lf	Phenyl polyketide acid	**402**, **406**	[185]
	NQ	lf	Meroterpenoid	**426**	[185]
Beilschmiedia tooram	NQ	lf	Phenyl polyketide acid	**403**, **404**	[185]
Beilschmiedia volckii	NQ	lf	Lignan	**430**	[185]
Endiandra anthropophagorum	NQ	r	Lignan	**223**, **427–429**	[193, 194]
Endiandra introrsa	NNSW	lf	Phenyl polyketide acid	**402–405**	[185]
Endiandra jonesii	SEQ	lf	Phenyl polyketide acid	**403**, **404**	[185]
Cryptocarya angulata	NQ	b, r	Aporphine	**372**, **378**, **379**	[39, 174]
	NQ	b	Phenanthrene	**388**	[174]
Cryptocarya cercophylla	NQ	b	Dibenzopyrrocoline	**392**, **399**	[39]
Cryptocarya cocosoides	NQ	lf	Pyrone	**414**, **415**	[191]
Cryptocarya densiflora	NQ	b	Pavine	**396**, **397**	[39]
Cryptocarya exfoliata	CY	b	BIA	**381**, **384**	[39]
Cryptocarya foveolata	SEQ	b	BIA	**256**	[178]
Cryptocarya laevigata	NQ	r	BIA	**386**	[39]
	SEQ	fl	Dibenzoisoquinoline	**391**	[39]
	NQ	b, r	Dibenzopyrrocoline	**398–401**	[39]
	NQ	fl	Piperidine	**389**	[39]
Cryptocarya macdonaldii	SEQ	b	BIA	**383**	[39]

(continued)

Table 10 (continued)

Species	Collection location[a]	Plant part(s)[b]	Compound class	Compound(s)	Refs.
Cryptocarya mackinnoniana	NQ	lf	Chalcone	**424**	[186]
	NQ	lf	Phenyl polyketide acid	**407–410**	[186]
Cryptocarya meisneriana	SEQ	b	Dibenzopyrrocoline	**392**, **399**	[39]
Cryptocarya murrayi	NQ	st	BIA	**343**, **382**, **383**	[39]
	NQ	st	Phenanthrene	**388**	[39]
Cryptocarya obovata	SEQ	lf	Coumarin	**425**	[188]
	SEQ	lf	Flavonoid	**422**, **423**	[188]
	SEQ	lf	Pyrone	**411**, **412**	[188]
Cryptocarya pleurosperma	NQ	b	BIA	**386**	[39]
	NQ	b	Dibenzoisoquinoline	**387**, **390**, **391**	[39, 182, 183]
	NQ	lf	Piperidine	**389**	[181]
Cryptocarya triplinervis	SEQ	b	Aporphine	**372**	[174]
	NQ	b	BIA	**384**	[39]
Cryptocarya vulgaris	CQ	wp	BIA	**383**	[39]
Cassytha filiformis	CY	wp	Aporphine	**375–377**	[175]
Cassytha pubescens	VIC	wp	Aporphine	**364**, **365**, **369–371**	[176]
	VIC	wp	Morphinane	**393**	[176]
Cinnamomum laubatii	NQ	b	BIA	**256**	[180]
	NQ	fr	Pyrone	**413**, **416–421**	[187, 190]
Cinnamomum oliveri	NQ	b	BIA	**384**	[39]
Litsea bindoniana	NQ	r	Aporphine	**368**	[39]

(continued)

Table 10 (continued)

Species	Collection location[a]	Plant part(s)[b]	Compound class	Compound(s)	Refs.
Litsea fawcettiana	NQ	st	Dibenzopyrrocoline	**400**	[39]
Litsea glutinosa	CY, CQ	lf, st	Aporphine	**365–368**, **378**	[39, 177]
Litsea leefeana	NQ	lf	Aporphine	**367**, **368**	[178]
	NQ, CQ	lf, st	BIA	**256**, **344**	[39, 178]
Neolitsea brassii	NQ	b	BIA	**256**	[39]
	NQ	b	Dibenzopyrrocoline	**345**	[39]
Neolitsea dealbata	NQ, SEQ	b	Aporphine	**357–363**, **368**	[39, 179]
	NQ, SEQ	b	BIA	**380**, **381**	[39, 179]
	NQ	b	Sesquiterpenoid	**431–433**	[196]
	NNSW, NQ	lf, b	Triterpenoid	**434–436**	[195, 196]
Lindera queenslandica	NQ	b	BIA	**385**	[39]

[a] SEQ = Southeast Queensland, NQ = North Queensland, CQ = central Queensland, CY = Cape York, Queensland, NNSW = Northern New South Wales, [b] lf = leaves, r = roots, fl = flowers, fr = fruits, wp = whole plant, b = bark, st = stems

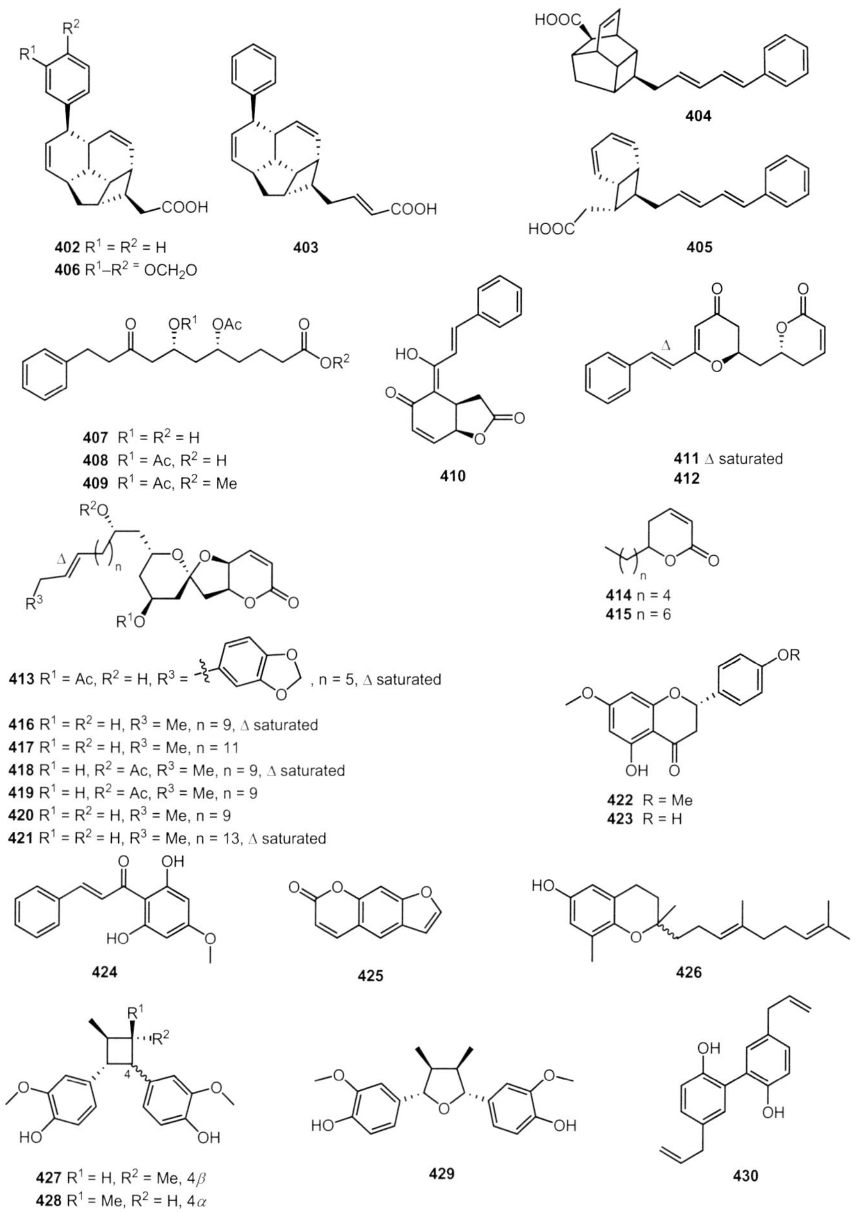

Fig. 40 Structures of the compounds **402–430** reported from Lauraceae species

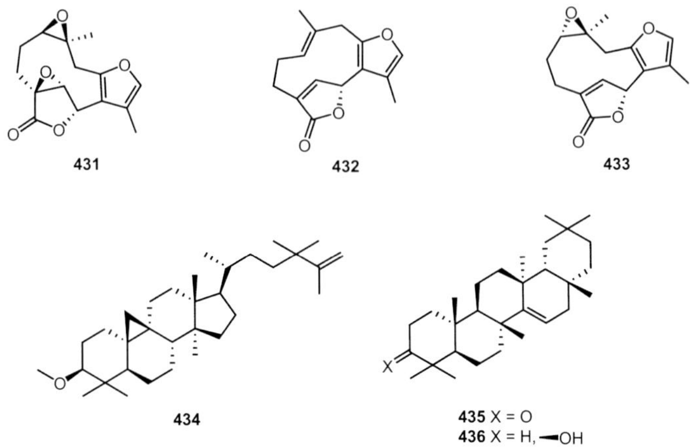

Fig. 41 Structures of compounds **431**–**436** reported from Lauraceae species

(**413**) has been synthesized stereo-selectively in nine linear steps [187]. These polyketides share a polymalonate biogenesis with the *Cryptocarya* compounds derived from either five or six malonates and the *Cinnamomum* compounds originating from 13 to 15 malonates. Other polyketide-derived α-pyrones are found predominately in Lauraceae, Annonaceae, Piperaceae, and Lamiaceae species outside of Australia [192]. This points to an ancient evolutionary origin for these structures. The phenyl-substituted polyketides also share features with flavonoids that are derived from successive condensations of cinnamic acid with three malonates. What sets these molecules apart is that they contain many more carbons than are used to produce flavonoids. Only three flavonoid/chalcones have been reported from Australian Lauraceae species, namely, 4′,7-di-*O*-methyl naringenin (**422**), 7-*O*-methyl naringenin (**423**), and 2′,6′-dihydroxy-4′-methoxychalcone (**424**) [186, 188].

The near basal position of the genera *Beilschmiedia*, *Cryptocarya*, and *Endiandra* in the Lauraceae suggests that the capacity to make phenyl polyketides originated early in ancestral Lauraceae and their absence in younger clades points to either a loss of biosynthetic capacity or dormancy. This is an example of chemical diversity linked to ancient lineages that are lost in younger clades. Australia's rainforests, which serve as a refuge for ancient Lauraceae lineages, can be considered a living museum, preserving a rich assortment of their ancient natural product chemical diversity.

A coumarin psoralen (**425**) has also been reported from *Cryptocarya*, while the tocopherol analogue oligandrol (**426**) was isolated from *Beilschmiedia oligandra* [185, 188]. Four lignans (**223**, endiandrin A (**427**), endiandrin B (**428**), and nectandrin B (**429**)) from *Endiandra anthropophagorum* include a cyclobutyl derivative **427** that possesses potent glucocorticoid receptor-binding affinity [193, 194]. The diarylfuran **429** and diarylbutane **223** co-isolates were much less active in a glucocorticoid receptor assay. The neolignan magnolol (**430**), previously reported from *Magnolia* species, was found in *Beilschmiedia volkii* [185]. Three sesquiterpenoid

Plate 12 Flowers of *Doryphora sassafras* and leaves of *Daphnandra apetala*. (Photographs AR Carroll)

furans, linderadine (**431**), linderalactone (**432**), and pseudoneolinderane (**433**), and the three triterpenoids cycloneolitsin (**434**), taraxerone (**435**), and taraxerol (**436**), have been reported from *Neolitsea dealbata* (Fig. 41) [195, 196].

Atherospermataceae

The Atherospermataceae (Plate 12) is a family that is wholly Gondwanan in distribution with most species (and genera) found in Australian rainforests (six *Daphnandra*, two *Doryphora*, two *Atherosperma*, and one *Dryadodaphne*), and with New Guinea (two *Dryadodaphne*), New Zealand (one *Laurelia*), New Caledonia (one *Nemuaron*), and Chile/Argentina (one *Laurelia* and one *Laureliopsis*) accounting for the rest. The natural product chemistry of species in this family has been investigated extensively (Table 11), and most of these studies are on Australian rainforest species. Alkaloids have been reported exclusively.

Benzyl isoquinoline alkaloids are the sole natural product class reported from Australian species (Figs. 42 and 43). The majority of these are macrocyclic bis-benzyl isoquinoline alkaloids (33 congeners, berbamine (**437**), **348**, fangchinoline (**438**), telobine (**439**), apateline (**440**), *N*-methylapateline (**441**), *N*-methylnorapateline (**442**), 1,2-dehydroapateline (**443**), 1,2-dehydrotelobine (**444**), johnsonine (**445**), *O*-methylrepandine (**446**), repandine (**447**), micranthine (**448**), *O*-methylmicranthine (**449**), *O,N*-dimethylmicranthine (**450**), repandinine (**451**), (−)-*nor*-tenuipine (**452**), tenuipine (**453**), de-*N*-methyltenuipine (**454**), dielsine (**455**), pseudorepanduline (**456**), repanduline (**457**), daphnoline (**458**), daphnandrine (**459**), oxycanthine (**460**), aromoline (**461**), *homo*-aromoline (**462**), (−)-nortenuipine-2-*β*-*N*-oxide (**463**), daphnine (**464**), dryadine (**465**), and dryadodaphnine (**466**)) in which the two isoquinoline moieties are ether linked and the two benzyl moieties are ether linked [39, 197–207]. Worldwide, about 30% of macrocyclic bis-BIAs are found in the Atherospermataceae

Table 11 Distribution of natural products isolated from the Atherospermataceae

Species	Collection location[a]	Plant part(s)[b]	Compound class	Compound(s)	Refs.
Atherosperma moschatum subsp. *integrifolium*	Tas	b	Aporphine	**187**, **372**, **471–473**	[197, 211, 212]
	Tas	b	Bis BIA	**348**, **437**	[197]
	Tas	b	Phenanthrene	**388**, **479**	[218]
Daphnandra sp.	NNSW	b	Bis BIA	**438**, **449**, **450**, **452**	[198]
Daphnandra apatela	SEQ	lf	BIA	**256**, **384**	[39]
	SEQ	b	Bis BIA	**439**, **440**, **443**, **444**	[39, 199]
	SEQ	lf	Phenanthrene	**346**, **480**	[39]
Daphnandra johnsonii	NSW	b	Bis BIA	**441**, **442**, **445–447**, **451**, **452**	[200]
Daphnandra micrantha	NNSW	b	Bis BIA	**448**, **458**, **459**	[201, 202]
Daphnandra repandula	NQ	fl	BIA	**344**	[39]
	NQ	b	Bis BIA	**441**, **446**, **447**, **451**, **452**, **455–457**, **460**, **461**, **463**, **464**	[203–205]
Daphnandra tenuipes	NNSW	st, b, lf	Bis BIA	**451**, **453**, **454**, **457**, **461**	[39, 202]
Doryphora aromatica	NQ	st, b	Aporphine	**249**, **372**	[39, 206]
	NQ	st, b	BIA	**256**, **344**, **384**	[39]
	NQ	b, lf	Bis BIA	**443**, **458**, **459**, **461**, **462**, **467–470**, **348**	[206, 207]
	NQ	lf	Isoquinoline	**484**	[207]
Doryphora sassafras	SEQ	lf, b, st	Aporphine	**187**, **249**, **338**, **372**, **474**	[39, 213–215]
	SEQ	b	Aristolactam	**475**	[214]
	SEQ	b,st, lf	BIA	**256**, **344**, **476–478**	[39, 213–215, 217]

(continued)

Table 11 (continued)

Species	Collection location[a]	Plant part(s)[b]	Compound class	Compound(s)	Refs.
	SEQ	b	Choline	**485**	[214]
	SEQ	lf,b	Isoquinoline	**481–483**	[213, 214]
Dryadodaphne trachyphloia	PNG	b	Aporphine	**187**, **338**, **473**	[216]
	PNG	b	Bis BIA	**465**, **466**	[216]

[a] SEQ = Southeast Queensland, NQ = North Queensland, NNSW = Northern New South Wales, NSW = New South Wales, Tas = Tasmania, PNG = Papua New Guinea, cult = cultivated, [b] lf = leaves, fl = flowers, b = bark, st = stems

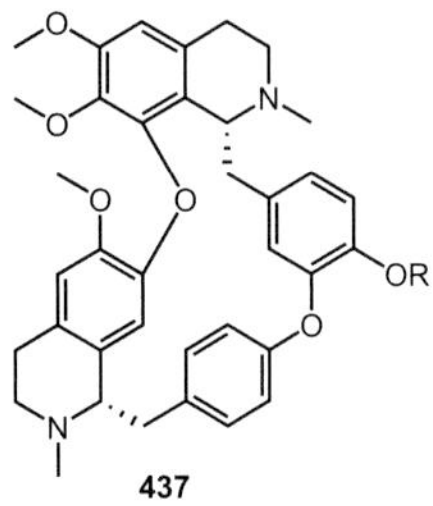

437

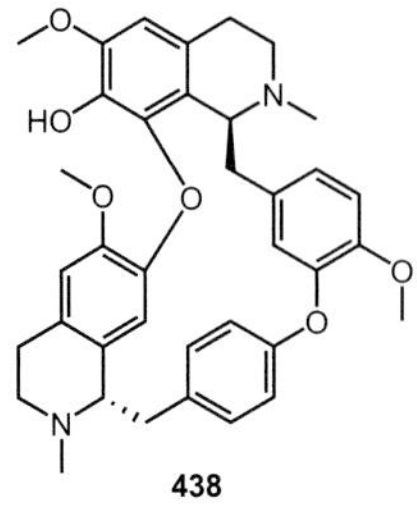

438

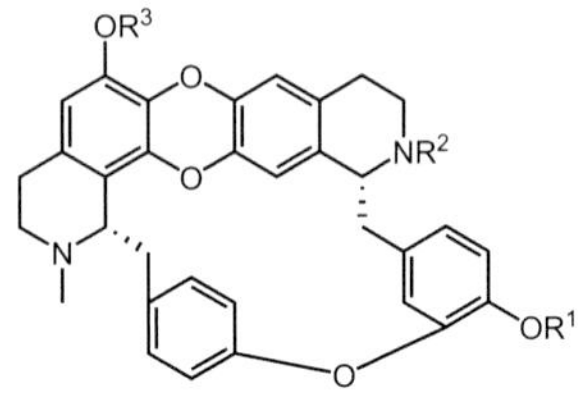

439 $R^1 = R^3 = Me$, $R^2 = H$
440 $R^1 = R^2 = H$, $R^3 = Me$
441 $R^1 = H$, $R^2 = R^3 = Me$
442 $R^1 = R^3 = H$, $R^2 = Me$

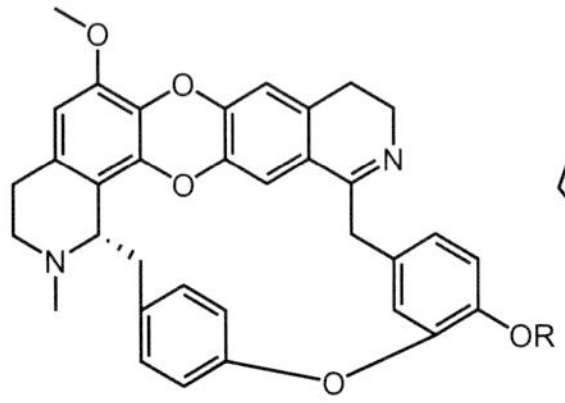

443 R = H
444 R = Me

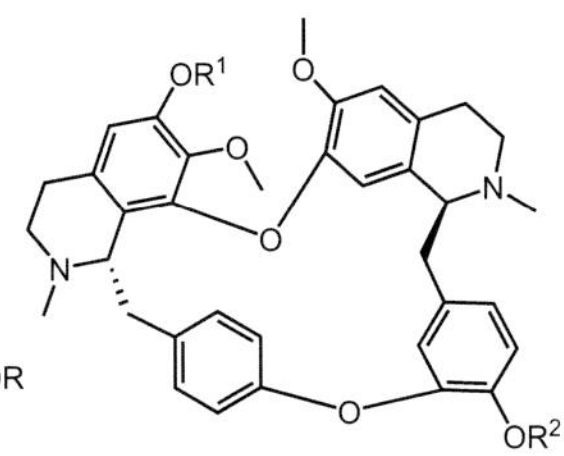

445 $R^1 = H$, $R^2 = Me$
446 $R^1 = R^2 = Me$
447 $R^1 = Me$, $R^2 = H$

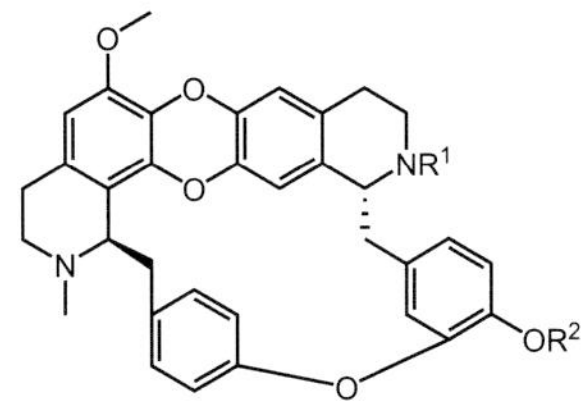

448 $R^1 = R^2 = H$
449 $R^1 = H$, $R^2 = Me$
450 $R^1 = R^2 = Me$

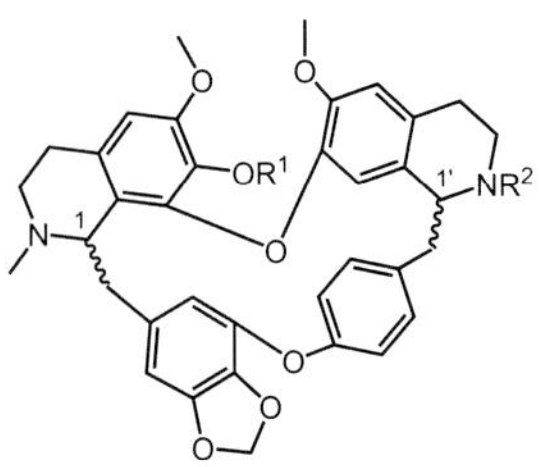

451 $R^1 = R^2 = Me$, (1*S*,1'*S*)
452 $R^1 = H$, $R^2 = Me$, (1*S*,1'*S*)
453 $R^1 = R^2 = Me$,
454 $R^1 = Me$, $R^2 = H$

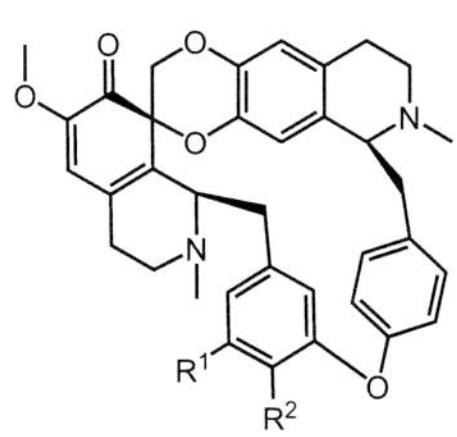

455 $R^1 = OMe$, $R^2 = OH$
456 $R^1 = H$, $R^2 = OMe$
457 R^1–$R^2 = OCH_2O$

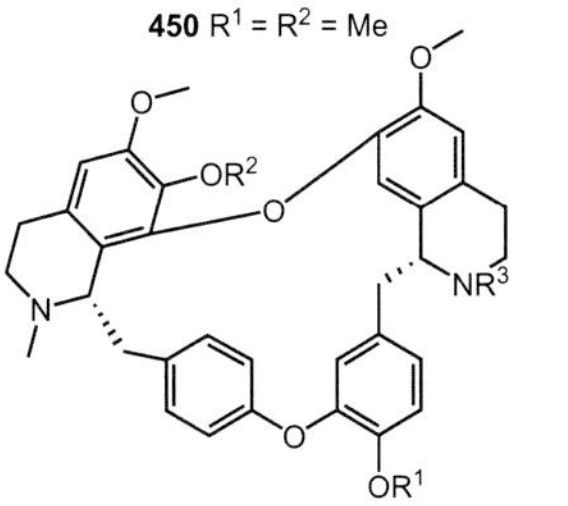

458 $R^1 = R^2 = R^3 = H$
459 $R^1 = Me$, $R^2 = R^3 = H$
460 $R^1 = H$, $R^2 = R^3 = Me$
461 $R^1 = R^2 = H$, $R^3 = Me$
462 $R^1 = R^3 = Me$, $R^2 = H$

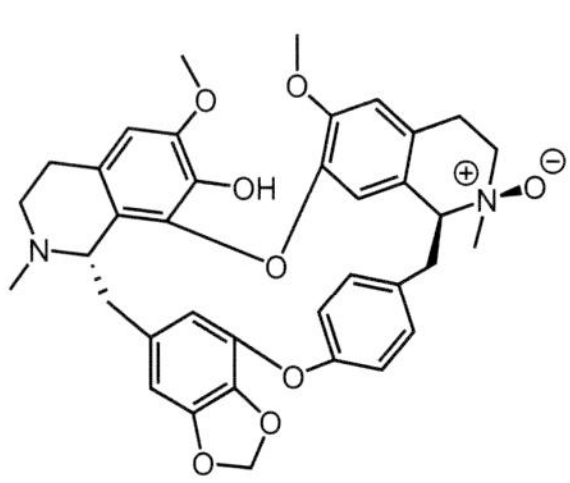

463

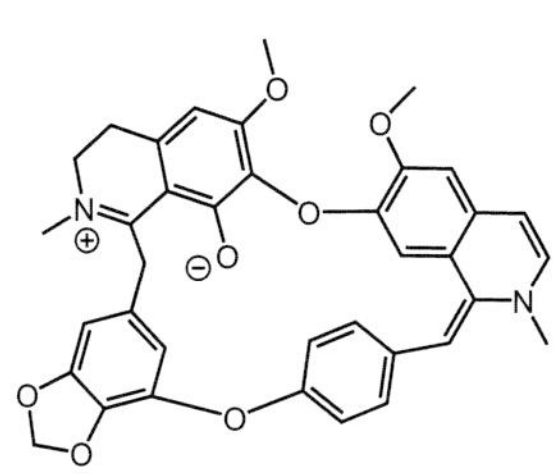

464

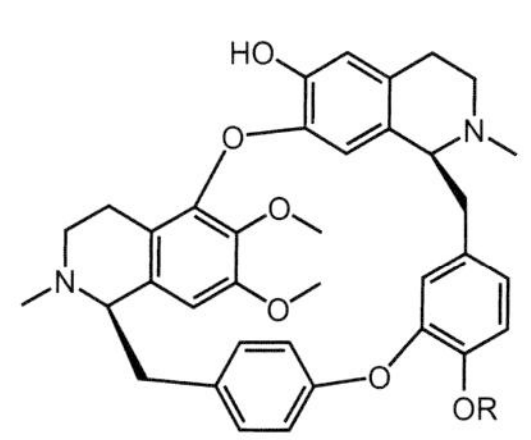

465 R = Me
466 R = H

Fig. 42 Structures of the alkaloids **437–466** reported from Atherospermataceae species

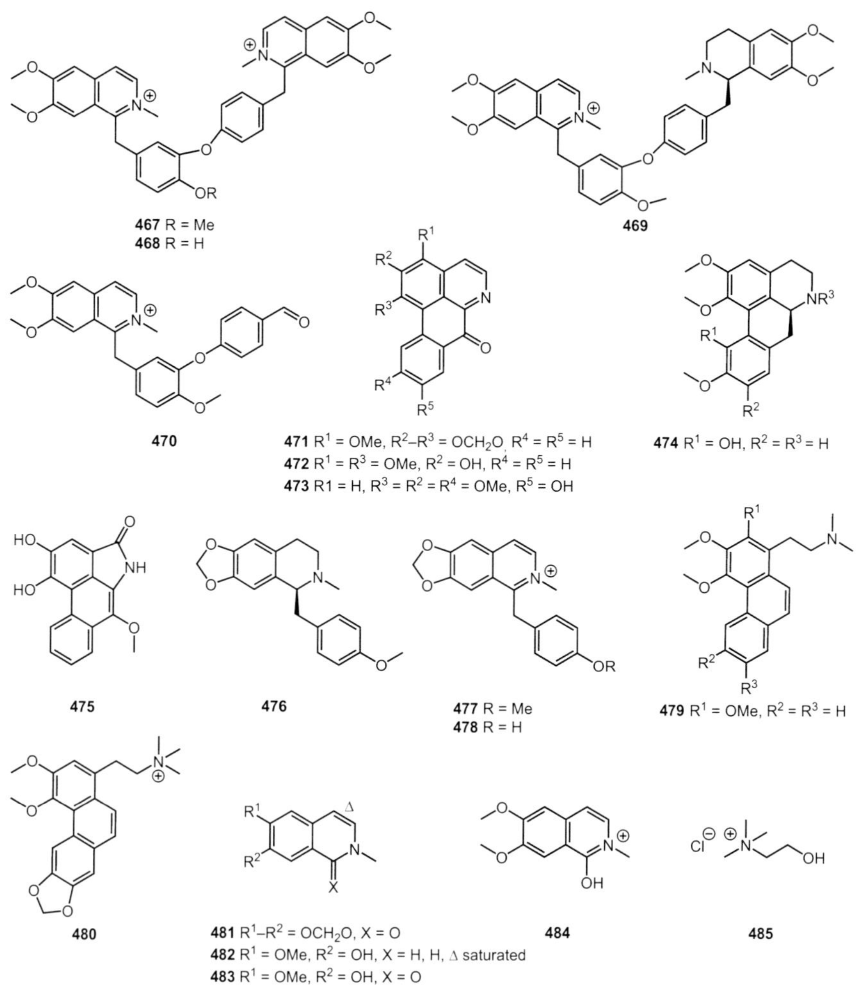

Fig. 43 Structures of the alkaloids **467–485** reported from Atherospermataceae species

with the remainder almost exclusively found in the Menispermaceae. These structures can be challenging to assign definitively, because both ether linkage points can be ambiguous with benzyl isoquinoline acting as either an oxygen donor or acceptor in the diphenolic oxidative coupling reactions. Extensive degradation chemistry or close examination of heteronuclear 2D NMR correlations and nuclear Overhauser effect correlations have been used to establish the oxygen linkages [208]. The relative and absolute configurations at C-1 and C-1′ further complicate such structural assignments. Several compounds have dibenzodioxin linkages. Even the compounds that have a single ether linkage between the two isoquinoline moieties have proved difficult to assign. Daphnine (**464**), for example, possesses a 7,7′ ether linkage unique among bis-BIAs and its structure was finally secured by X-ray crystallography after extensive spectroscopic and chemical studies failed to provide a definitive assignment [204]. *Daphnandra* species also contain unique variants of the macrocyclic bis-BIAs. Compounds **455–457** each contain an isoquinolone linked to a second isoquinoline via a spiro center at C-8 [202–204]. Many of the bis-BIAs were first isolated between 1950 and 1980 but little bioactivity was reported. Subsequently, micranthine and its *O*-methyl derivative were re-isolated from *Daphnandra apetala* during the AZGU collaboration in the 1990's as potent δ-opioid agonists and further testing of other macrocyclic bis-BIA analogues isolated from various *Daphnandra*, *Doryphora*, and Menispermaceae species resulted in five orders of magnitude differences in potencies, and *O*-methylmicranthine proved to be the most potent [39, 209]. This prompted an in-house medicinal chemistry program to develop synthetic analogues inspired by micranthine as potential treatments for chronic pain [210].

Four ether-linked acyclic bis-BIAs, phaeantharine (**467**), 12-de-*O*-methylphaeantharine (**468**), (*R*)-nomimantharine (**469**), and nominanthranal (**470**) have been isolated from *Doryphora aromatica* from north Queensland [207]. Two of these compounds (**467** and **468**) are moderate inhibitors of mycobacteria and MRSA bacterial strains [207]. Several aporphine alkaloids (**187**, **372**, **338**, atherospermidine (**471**), moschatoline (**472**), atheroline (**473**), and norisocorydine (**474**)) [39, 197, 206, 211–216], an aporphine lactam (doryflavine (**475**)) [214], benzyl isoquinoline alkaloids (**256**, **344**, **384**, doryafranine (**476**), 5-[(4-hydroxyphenyl)methyl]-6-methyl-1,3-dioxolo[4,5-*g*]isoquinolinium trifluoroacetate (**477**), and 5-[(4-methoxyphenyl)methyl]-6-methyl-1,3-dioxolo[4,5-*g*]isoquinolinium chloride (**478**)) [39, 213–215, 217], ethylamine phenanthrene alkaloids (**388**, **346**, methoxyatherosperminine (**479**), 1,2-dimethoxy-*N*,*N*,*N*-trimethylphenanthro[2,3-*d*][1,3]dioxole-4-ethanaminium trifluoroacetate (**480**)) [39, 218], isoquinoline alkaloids (doryanine (**481**), corypalline (**482**), doryphornine (**483**), and *N*-methyl-6,7-dimethoxyisoquinolone (**484**)) [207, 213, 214], and choline chloride (**485**) [214] have also been reported from Australian Atherospermataceae species.

4.3 Chemistry of Australian Rainforest Monocotyledon Angiosperms

Although 30 families of monocots are found in Australian rainforests and although the orchids represent one of the most genus- and species-rich families, very few chemical investigations have been undertaken on this group.

4.3.1 Dioscoreaceae and Stemonaceae

The native yam *Dioscorea transversa* (Dioscoreaceae) is the only one of three Australian rainforest species in the family Dioscoreaceae to be investigated, and it has yielded the two steroidal saponins 1β-hydroxyprotogracillin (**486**) and protogracillin (**487**) (Table 12 and Fig. 44) [219]. The Stemonaceae, a family of crawling and climbing herbs in the Order Pandanales, is of Asian origin. The Australian species *Stemona lucida*, from Cape York, have yielded a series of four dihydrophenanthrenes (stemanthrenes B-D (**488–490**), and stemanthrene A (**491**)), five 2-phenylbenzofurans (stemofuran G (**492**), stemofuran W (**493**), stilbostemin B (**494**), stilbostemin D (**495**), and stilbostemin F (**496**)), one dihydrostilbene dihydropinosylvin (**497**), and one *Stemona* alkaloid, stemofoline (**498**) (Fig. 44) [39, 220].

4.3.2 Liliales

The order Liliales is mostly Southern Hemispheric in distribution and contains ten families with five found in Australian rainforests. Species from only two families

Table 12 Distribution of natural products isolated from the Dioscoreaceae, Stemonaceae, Colchicaceae, and Smilacaceae

Species	Collection location[a]	Plant part(s)[b]	Compound class	Compound(s)	Refs.
Dioscoreaceae					
Dioscorea transversa	SEQ	wp	Steroid	**486, 487**	[219]
Stemonaceae					
Stemona lucida	CY	wp	*Stemona* alkaloid	**498**	[39]
	CY	tub	Stilbene	**488–497**	[220]
Colchicaceae					
Kuntheria pedunculata	NQ	wp/lf	Homoerythrane	**500–509**	[39, 222, 223]
Tripladenia cunninghamii	NNSW	wp	Colchicine	**499, 516, 517**	[224, 225]
	NNSW	wp	Homoaporphine	**510–514**	[224, 226]
	NNSW	wp	Homomorphane	**515**	[225]
Gloriosa superba	SEQ	r	Colchicine	**499, 518, 519**	[39]
Smilacaceae					
Smilax glyciphylla	NSW	wp	Chalcone	**520–522**	[227, 228]
	NSW	wp	Flavonoid	**523–527**	[228]

[a] SEQ = Southeast Queensland, NQ = North Queensland, CY = Cape York, NNSW = Northern New South Wales, NSW = New South Wales, [b] lf = leaf, tub = tuber, r = root, wp = whole plant

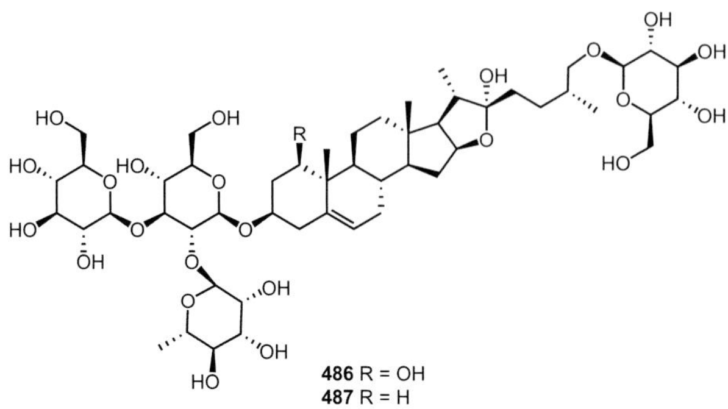

486 R = OH
487 R = H

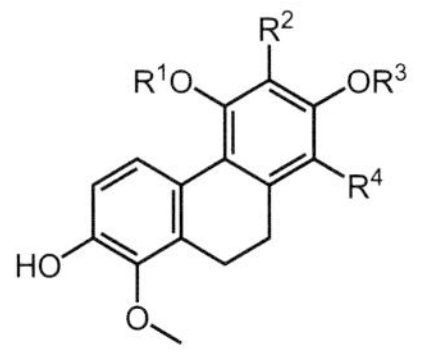

488 $R^1 = R^4 = H$, $R^2 = Me$, $R^3 = Me$
489 $R^1 = R^2 = R^4 = Me$, $R^3 = H$
490 $R^1 = R^4 = Me$, $R^2 = R^3 = H$
491 $R^1 = R^2 = Me$, $R^3 = R^4 = H$

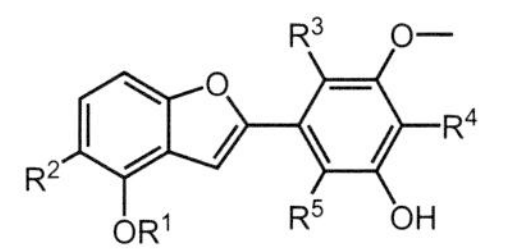

492 $R^1 = R^2 = R^3 = H$, $R^4 = R^5 = Me$
493 $R^1 = R^2 = Me$, $R^3 = R^4 = R^5 = H$
494 $R^1 = R^2 = R^3 = R^4 = R^5 = H$
495 $R^1 = R^2 = R^3 = R^4 = H$, $R^5 = Me$
496 $R^1 = R^2 = H$, $R^3 = R^4 = R^5 = Me$

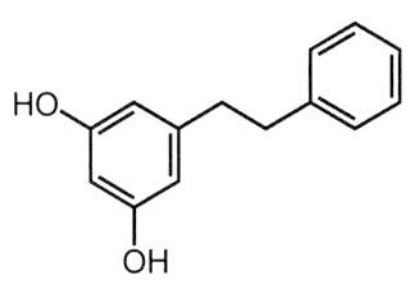

497

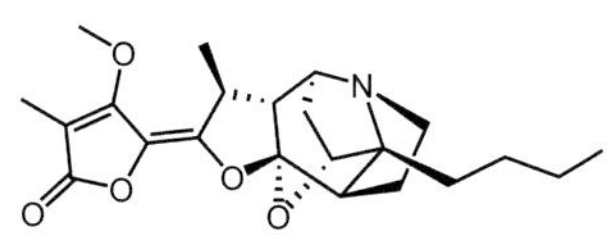

498

Fig. 44 Structures of the compounds **486–498** reported from Dioscoreaceae and Stemonaceae species

have been chemically investigated (Table 12). The ancestors of the order may have originated prior to the final breakup of Gondwana and the current vicariance across southern continents points to dispersal across the connected landmasses of Gondwana [221]. The family Colchicaceae has an Australian origin with the tribe Tripladenieae (containing exclusively Australian rainforest species) being a relic of the ancestral distribution of the family (Plate 13) [221]. The family is now mainly found in the Southern Hemisphere but has key species of medicinal importance in North America and Europe (Fig. 45).

The family is renowned for species that produce colchicine (**499**), a tubulin binder and a drug used to treat gout [69]. The Australian Tripladenieae genera *Tripladenia* and *Kuntheria* have yielded 21 homoerythrinan alkaloids (schelhammerine (**500**), schelhammericine (**501**), alkaloid A (**502**), alkaloid B (**503**), alkaloid G (**504**), alkaloid H (**505**), alkaloid E (**506**), schelhammeridine (**507**), alkaloid J (**508**), and alkaloid K (**509**)), homoaporphine alkaloids (kreysigine (**510**),

Plate 13 Top row left to right: fruits of *Dioscorea transversa* and flowers of *Tripladenia cunninghamii* and bottom row left to right: *Gloriosa superba* flowers and fruits of *Smilax australis*. (Photographs AR Carroll)

multifloramine (**511**), floramultine (**512**), kreysiginone (**513**), and dihydrokreysiginone (**514**), a homomorphinan alkaloid, kreysiginine (**515**) [39, 222–226], along with three colchicine analogues (**499**, deacetylcolchicine (**516**), and *N*-formyl-*N*-desacetylcolchicine (**517**)) (Fig. 46) [39, 224, 225]. Colchicine (**499**) and two other analogues, 2,10-didemethylcolchicine (**518**) and 2-demethylcolchicine (**519**), have also been found in *Gloriosa superba* [39]. These alkaloids are found throughout the Colchicaceae except the homoerythrinan type, which are restricted to Australian species. The homoerythrinan alkaloids are also key components of gymnosperms from Tasmania (*Athrotaxis* (Cupressaceae)) and China (*Cephalotaxus*, Taxaceae), and two angiosperm genera *Dysoxylum* (Meliaceae) and *Phelline* (Phellinaceae) pointing to their polyphyletic origins. None of the Australian Colchicaceae alkaloids has been evaluated for bioactivity.

The Smilacaceae is the second Liliales family studied from Australian rainforests (Plate 12). Only *Smilax glyciphylla* from the Sydney region has been investigated (Fig. 47). It yielded several glycosylated chalcones (glycyphyllins A–C (**520**–**522**) and flavonoids (catechin (**523**), kaempferol-3-*O*-β neohesperidoside (**524**),

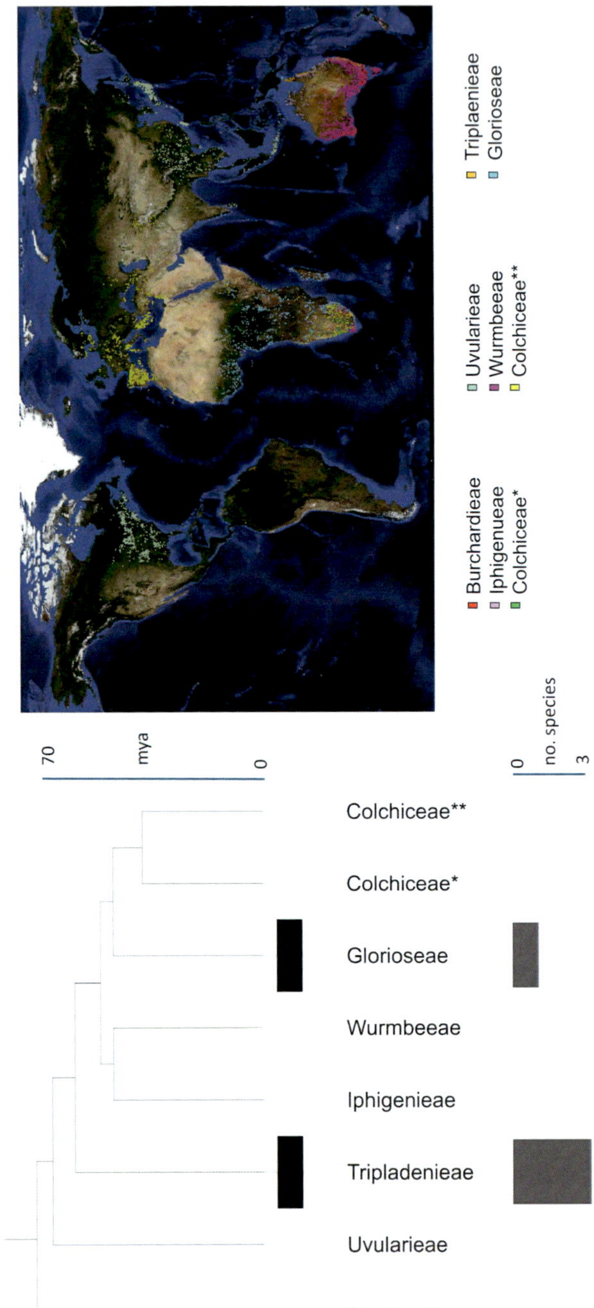

Fig. 45 Dated molecular phylogeny of the Colchicaceae tribes [221] showing those that have species present and studied (black) and the number of species occurring in Australian rainforest. The map shows the worldwide distribution of species in the Colchicaceae color coded by tribe based on herbarium records [5, 43]

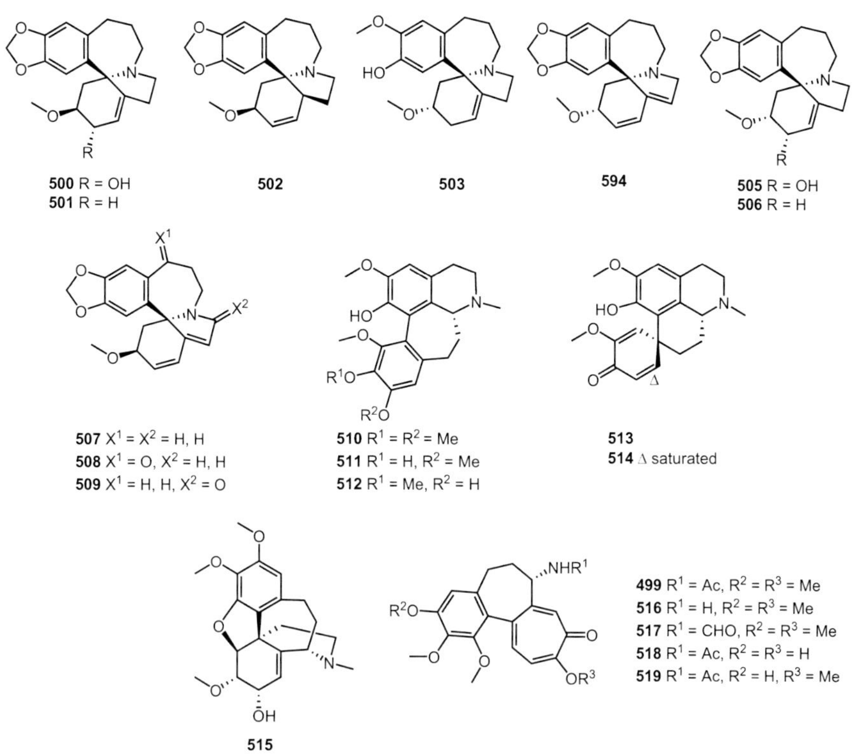

Fig. 46 Structures of the alkaloids **500–519** reported from Colchicaceae species

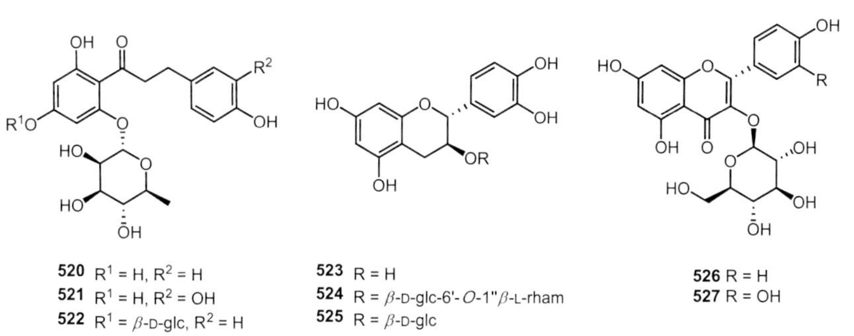

Fig. 47 Structures of compounds **520–527** reported from *Smilax glyciphylla*

(2*R*,3*R*)-dihydrokaempferol-3-*O*-*β*-D-glucopyranoside (**525**), kaempferol-3-*O*-*β*-D-glucopyranoside (**526**), and isoquercitrin (**527**)), of which three (**521**, **523**, **527**) showed antioxidant activity [227, 228].

4.3.3 Asparagales

The Order Asparagales has eight families represented in Australian rainforests (Fig. 48). Although Australia's rainforests house a very rich diversity of orchids (over 230 species in 90 genera) only one species has been chemically investigated. *Liparis nugentiae* contains the pyrrolizidine alkaloid hexahydro-1*H*-pyrrolizine-1-carboxylic acid (**528**) (Table 13, Fig. 49) [39].

The wholly Australian restricted monogeneric family Blandfordiaceae has four species in the genus *Blandfordia*. These plants are commonly referred to as Christmas bells (Fig. 48). Mostly found in heathlands, seeds obtained from the Tasmanian species *B. punicea* (that occurs on the edge of rainforest in damp heath) were studied in the U.K. and shown to contain a series of polyoxygenated steroids (punisterone (**529**), ecdysone (**530**), 20-hydroxyecdysone (**531**), 5*β*,20-dihydroxyecdysone (**532**), ponasterone C (**533**), pterosterone (**534**), and turkesterone (**535**)) (Fig. 49) [229].

The flax lilies *Dianella* (Asphodelaceae) often contain blue berries (Fig. 48). The only study on a Tasmanian rainforest species *D. tasmanica* yielded the anthocyanin, delphinidin 3-*O*-*β*-D-glucopyranoside-7,3′,5′-tri-*O*-(6-*O*-*p*-coumaroyl-*β*-glucopyranoside (**536**) as the major color-producing constituent of the fruits, with delphinidin 3,7-di-*O*-*β*-glucopyranoside-3′,5′-di-*O*-(6-*O*-*p*-coumaroyl-*β*-glucopyranoside (**537**) and delphinidin 3,7,3′,5′-tetra-*O*-(6-*O*-*p*-coumaroyl-*β*-glucopyranoside (**538**) being minor pigments (Fig. 49) [230].

The Cardwell lily *Proiphys amboinensis* (Amaryllidaceae) occurs in North Queensland and has an extended range into Malesia. It contains the widespread Amaryllidaceae alkaloids lycorine (**539**) and the crinine analogue, haemanthamine (**540**) [231]. Trimethoxyacetophenone (**541**) and **116** have also been reported [232].

The four *Cordyline* (Asparagaceae) species (Fig. 48) studied from Australian rainforests contain a rich variety of steroidal glycosides, with the majority containing a spirofuranopyran moiety (Figs. 50 and 51). These steroids comprise brisbagenin (**542**), cordylagenin (**543**), cannigenin (**544**), pompeygenin (**545**), chenogenin (**546**), 3-*epi*-tigogenin (**547**), tigogenin (**548**), 3-*epi*-neotigogenin (**549**), (25*S*)-5*α*-spirostan-1*β*,3*α*-diol 3-*O*-*β*-D-glucopyranoside (**550**), (25*S*)-5*α*-spirostan-1*β*,3*α*-25-triol 3-*O*-*β*-D-glucopyranoside (**551**), brisbenone (**552**), suttonigenins A–J (**553–562**), australigenin (**563**), crabbogenin (**564**), 1*β*-hydroxycrabbogenin (**565**), 5*α*-spirost-25(27)-ene-1*β*,3*α*-diol 3-*O*-*β*-D-glucopyranoside (**566**), 1*β*,3*α*-dihydroxyfurost-5-ene (**567**), 1*β*,3*α*,26-trihydroxy-5*α*-furostane (**568**), strictagenin (**569**), rubragenin (**570**), wallogenin (**571**), diosgenin (**572**), yamogenin (**573**), 5*α*-spirost-25(27)-ene-1*β*,3*α*-diol 1-*O*-(*O*-*α*-L-rhamnopyranosyl-(1 → 2)-*O*-[*β*-D-xylopyranosyl-(1 → 3)]-*β*-D-fucopyranoside

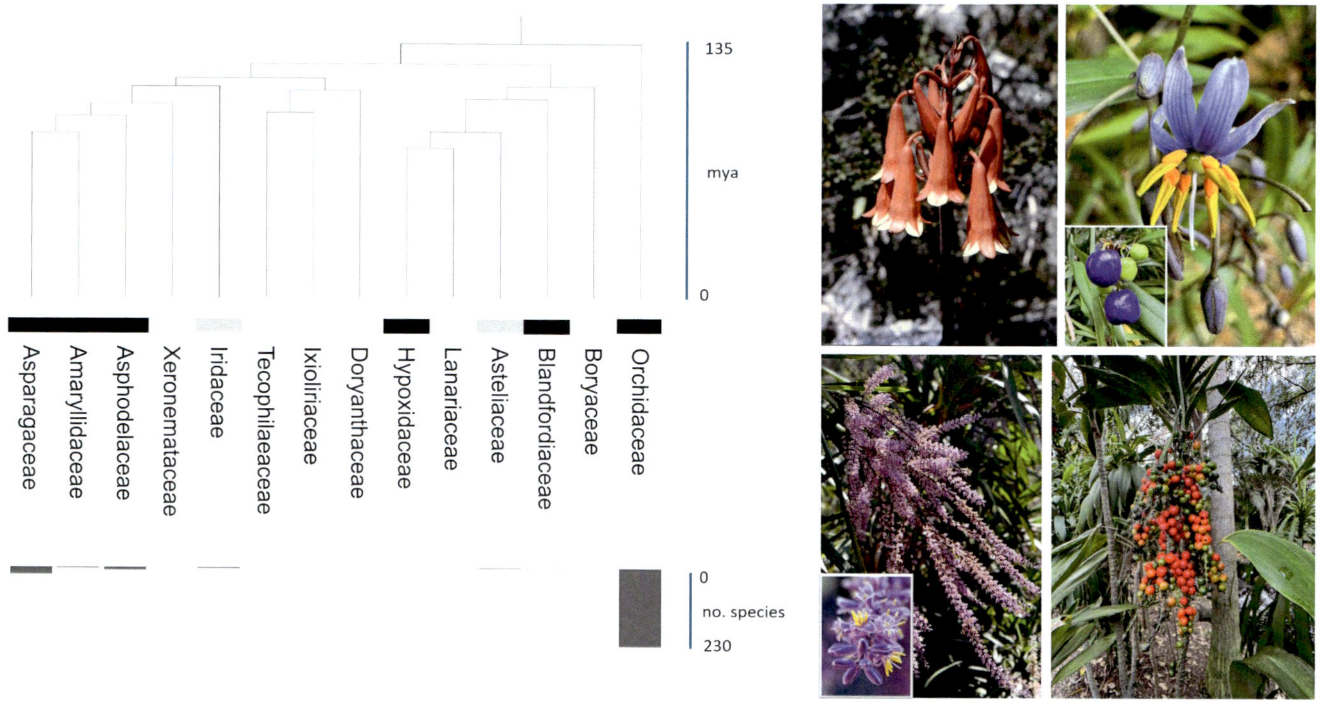

Fig. 48 Dated molecular phylogeny of the Asparagales, showing those families that have species present in Australian rainforests (grey) and studied (black) and the number of species occurring in Australian rainforest: flowers of *Blandfordia punicea* (top left), flowers and fruits of *Dianella caerulea* (top right) and flowers and fruit of *Cordyline manners-suttoniae* (bottom). (Photographs AR Carroll)

Table 13 Distribution of natural products isolated from the Asparagales

Species	Collection location[a,b]	Plant part[c]	Compound class	Compound(s)	Refs.
Orchidaceae					
Liparis nugentiae	NQ	ap	Pyrrolizidine	**528**	[39]
Blandfordiaceae					
Blandfordia punicea	Tas	s	Steroid	**529–535**	[229]
Asphodelaceae					
Dianella tasmanica	Tas	fr	Anthocyanin	**537–538**	[230]
Amaryllidaceae					
Proiphys amboinensis	CY (cult. (Thailand))	tub	*Amaryllis* alkaloid	**539**, **540**	[231]
	CY (cult. (Thailand))	lf	Ketone	**541**	[232]
	CY (cult. (Thailand))	lf	Phenyl propane	**116**	[232]
Asparagaceae					
Cordyline cannifolia	NQ	lf	Steroid	**542–544**, **552**	[233–235]
Cordyline manners-suttoniae	NQ	lf	Steroid	**542**, **553–563**	[236, 237]
Cordyline rubra	SEQ	lf	Steroid	**545–549**, **567–573**	[236, 238, 239]
Cordyline stricta	SEQ, NNSW	lf	Steroid	**543–545**, **547**, **549–551**, **564–566**, **569**, **574–583**	[236, 240–242]

[a] SEQ = Southeast Queensland, NQ = North Queensland, CQ = central Queensland, CY = Cape York, Queensland, NNSW = Northern New South Wales, Tas = Tasmania, cult. = cultivated, [b] natural location shown for those species studied from cultivated material, [c] lf = leaves, s = seeds, tub = tubers, ap = aerial parts, fr = fruits

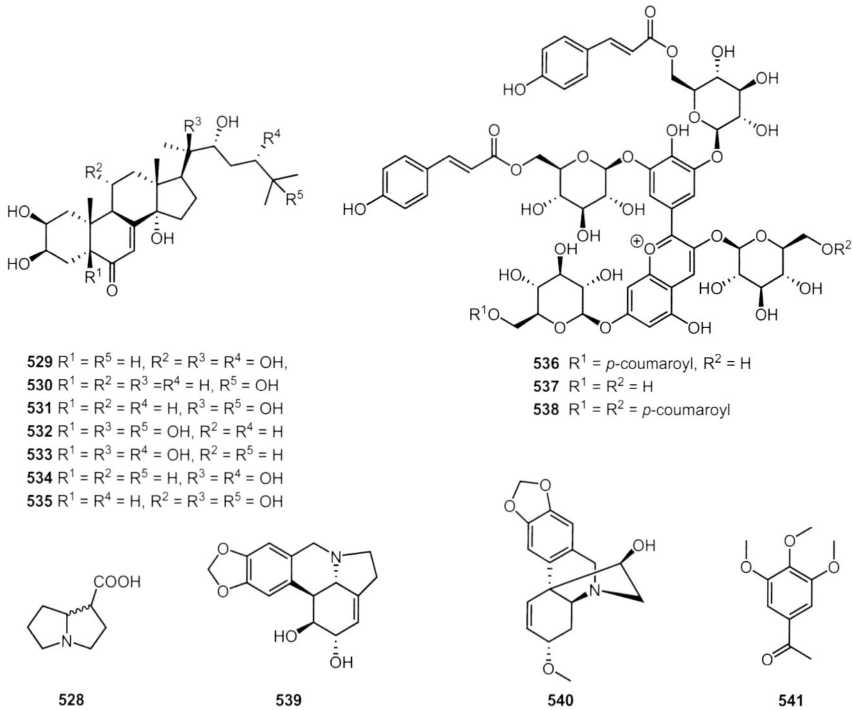

Fig. 49 Structures of compounds **528–541** reported from *Liparis nugentiae*, *Blandfordia punicea*, *Dianella tasmanica,* and *Proiphys amboinensis*

574), 5*α*-spirost-25(27)-ene-1*β*,3*α*-diol 1-*O*-(*O*-*α*-L-rhamnopyranosyl-(1 → 2)-*O*-[*β*-D-xylopyranosyl-(1 → 3)]-*β*-D-xylopyranoside (**575**), (25*S*)-5*α*-spirost-20-ene-1*β*,3*α*-diol 3-*O*-*β*-D-glucopyranoside (**576**), 1*β*,3*α*-dihydroxy-5*α*-pregn-16-en-20-one 3-*O*-*β*-D-glucopyranoside (**577**), 26-*O*-*β*-D-glucopyranosyl-22-*O*-methyl-(25*S*)-5*α*-furostan-3*α*,22*ξ*,26-triol 3-*O*-*β*-D-glucopyranoside (**578**), 26-*O*-*β*-D-glucopyranosyl-22-*O*-methyl-(25*S*)-5*α*-furostan-1*β*,3*α*,22*ξ*,26-tetrol 3-*O*-*β*-D-glucopyranoside (**579**), 26-*O*-*β*-D-glucopyranosyl-5*α*-furost-20(22)-ene-1*β*,3*α*,26-triol 3-*O*-*β*-D-glucopyranoside (**580**), 26-*O*-*β*-D-glucopyranosyl-22-*O*-methylfurost-5,25(27)-diene-1*β*,3*β*,22*ξ*,26-tetrol 1-*O*-(O-*α*-L-rhamnopyranosyl-(1 → 2)-*O*-[*β*-D-xylopyranosyl-(1 → 3)]-*β*-D-fucopyranoside (**581**), (25*S*)-5*α*-spirostan-1*β*,3*α*-diol 1-*O*-(*O*-*α*-L-rhamnopyranosyl-(1 → 2)-*O*-[*β*-D-xylopyranosyl-(1 → 3)]-*β*-D-xylopyranoside (**582**), and 26-*O*-*β*-D-glucopyranosyl-22-*O*-methyl-5*α*-furost-25(27)-ene-1*β*,3*β*,22*ξ*,26-tetrol 1-*O*-(*O*-*α*-L-rhamnopyranosyl-(1 → 2)-*O*-[*β*-D-xylopyranosyl-(1 → 3)]-*β*-D-fucopyranoside (**583**) [233–242]. Related compounds are found in a range of other monocots, particularly *Dioscorea* species. The suttonigenins **553–562** showed weak cytotoxic and antibacterial activities [236, 237].

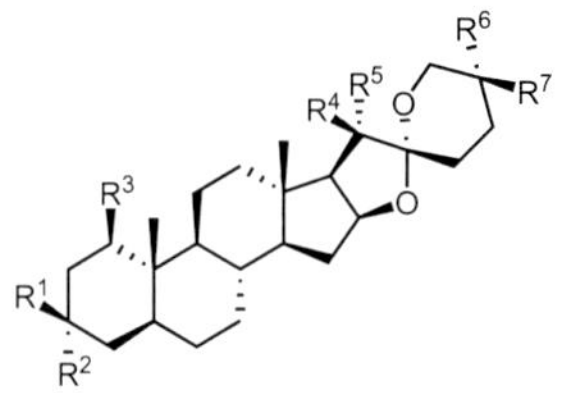

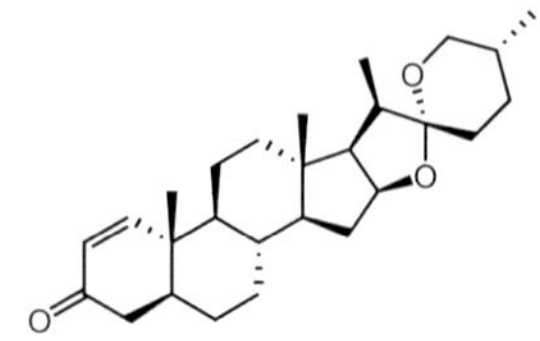

552

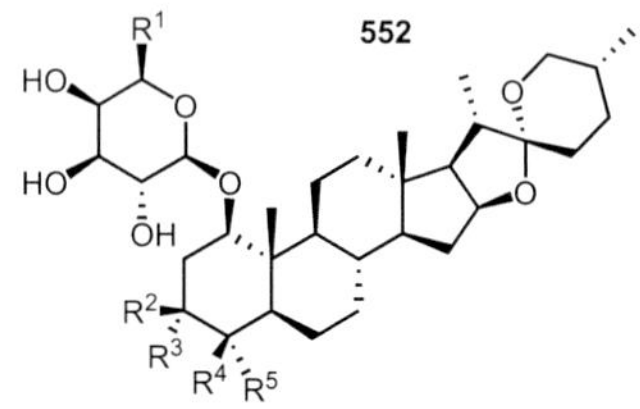

542 $R^1 = R^3 = OH, R^2 = R^5 = R^7 = H, R^4 = R^6 = Me$
543 $R^1 = R^5 = R^6 = H, R^2 = R^3 = OH, R^4 = R^7 = Me$
544 $R^1 = R^3 = OH, R^2 = R^5 = R^6 = H, R^4 = R^7 = Me$
545 $R^1 = R^4 = H, R^2 = R^3 = R^7 = OH, R^5 = R^6 = Me$
546 $R^1 = R^4 = H, R^2 = R^3 = R^6 = OH, R^5 = R^7 = Me$
547 $R^1 = R^3 = R^4 = R^7 = H, R^2 = OH, R^5 = R^6 = Me$
548 $R^1 = OH, R^2 = R^3 = R^4 = R^7 = H, R^5 = R^6 = Me$
549 $R^1 = R^3 = R^4 = R^6 = H, R^2 = OH, R^5 = R^7 = Me$
550 $R^1 = R^4 = R^6 = H$, R^2 = *O*-*β*-D-glc, $R^3 = OH, R^5 = R^7 = Me$
551 $R^1 = R^4 = H$, R_2 = *O*-*β*-D-glc, $R^3 = R^6 = OH, R^5 = R^7 = Me$

553 $R^1 = Me, R^2 = OH, R^3 = R^4 = R^5 = H$
554 $R^1 = R^3 = R^4 = R^5 = H, R^2 = OH$
555 $R^1 = Me, R^2 = R^4 = R^5 = H, R^3 = OH$
556 $R^1 = Me, R^2 = R^5 = OH, R^3 = R^4 = H$
557 $R^1 = Me, R^2 = R^4 = H, R^3 = R^5 = OH$

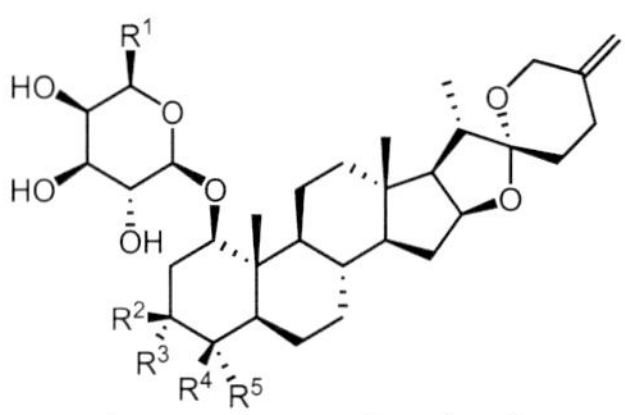

558 $R^1 = Me, R^2 = OH, R^3 = R^4 = R^5 = H$
559 $R^1 = R^3 = R^4 = R^5 = H, R^2 = OH$
560 $R^1 = Me, R^2 = R^4 = R^5 = H, R^3 = OH$
561 $R^1 = Me, R^2 = R^5 = OH, R^3 = R^4 = H$
562 $R^1 = Me, R^2 = R^4 = H, R^3 = R^5 = OH$

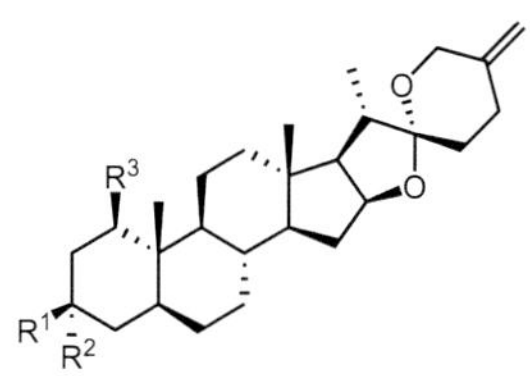

563 $R^1 = OH, R^2 = R^3 = H$
564 $R^1 = R^3 = H, R^2 = OH$
565 $R^1 = H, R^2 = R^3 = OH$
566 $R^1 = H$, R^2 = O-*β*-D-glc, $R^3 = OH$

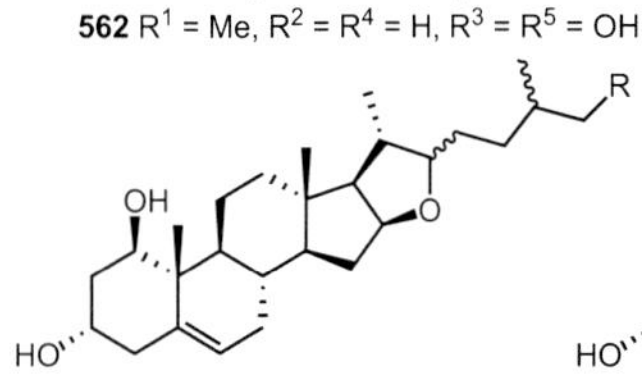

567 R = H
568 R = OH

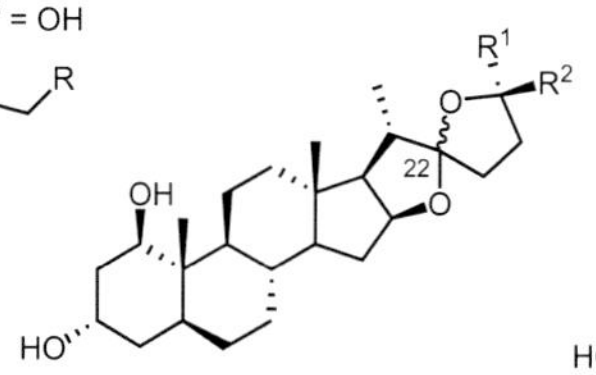

569 $R^1 = CH_2OH, R^2 = Me$, (22*S*)
570 $R^1 = Me, R^2 = CH_2OH$, (22*S*)
571 $R^1 = Me, R^2 = CH_2OH$, (22*R*)

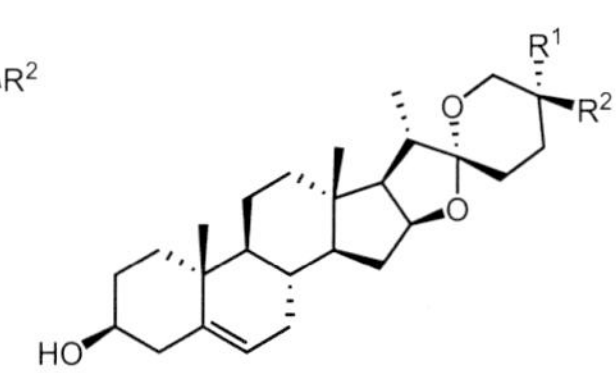

572 $R^1 = Me, R^2 = H$
573 $R^1 = H, R^2 = Me$

Fig. 50 Structures of steroids **542–573** reported from *Cordyline* spp.

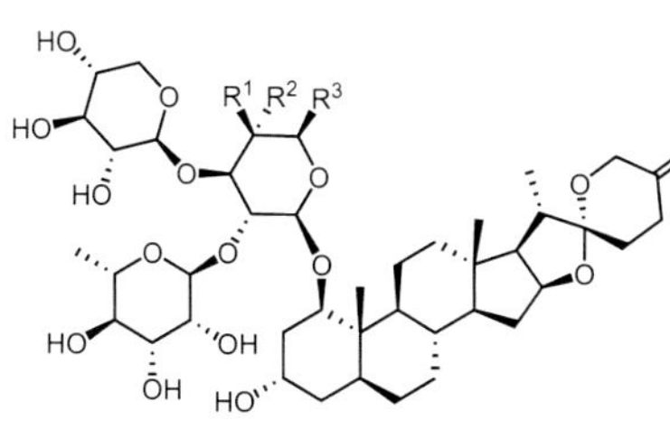

574 R¹ = OH, R² = H, R³ = Me
575 R¹ = H, R² = OH, R³ = H

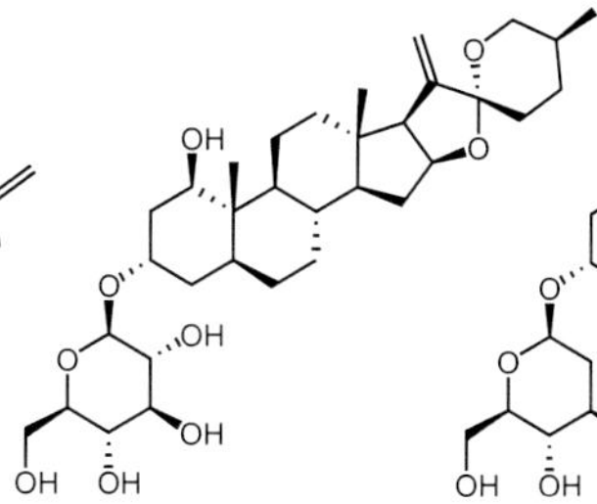

576

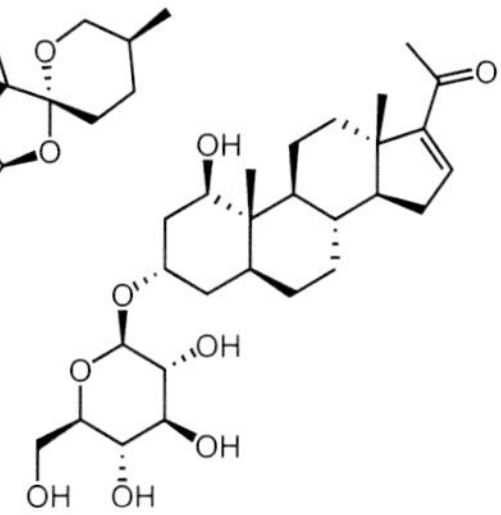

577

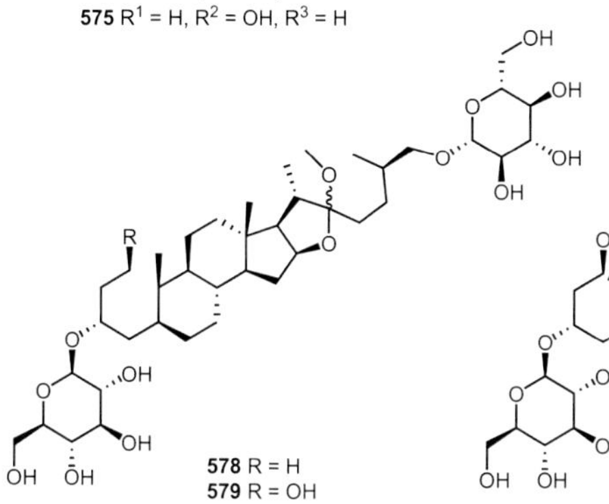

578 R = H
579 R = OH

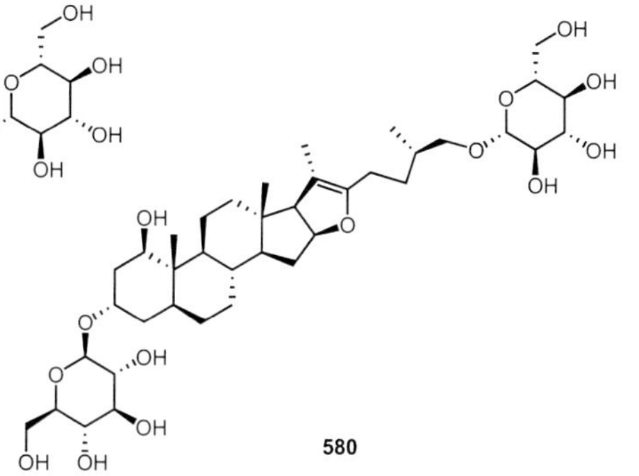

580

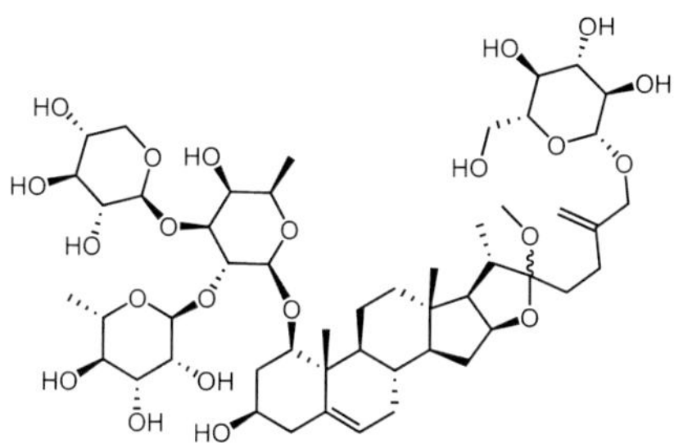

581

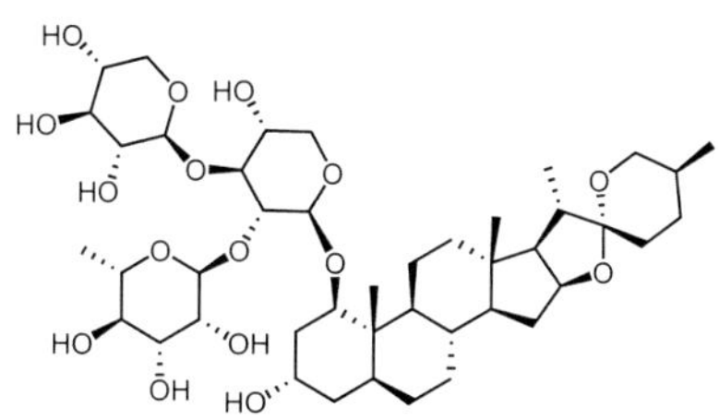

582

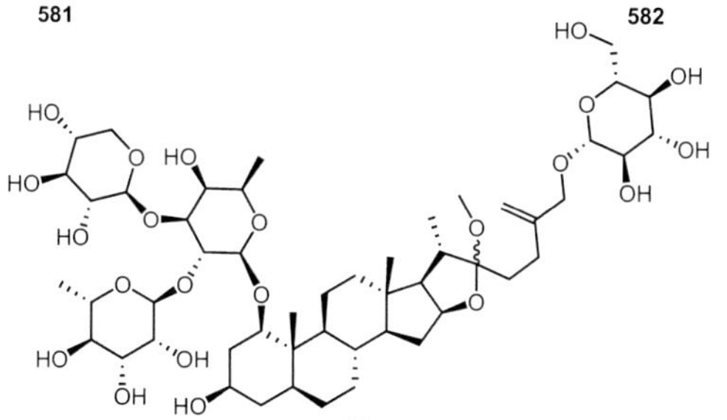

583

Fig. 51 Structures of steroids **574–583** reported from *Cordyline* spp.

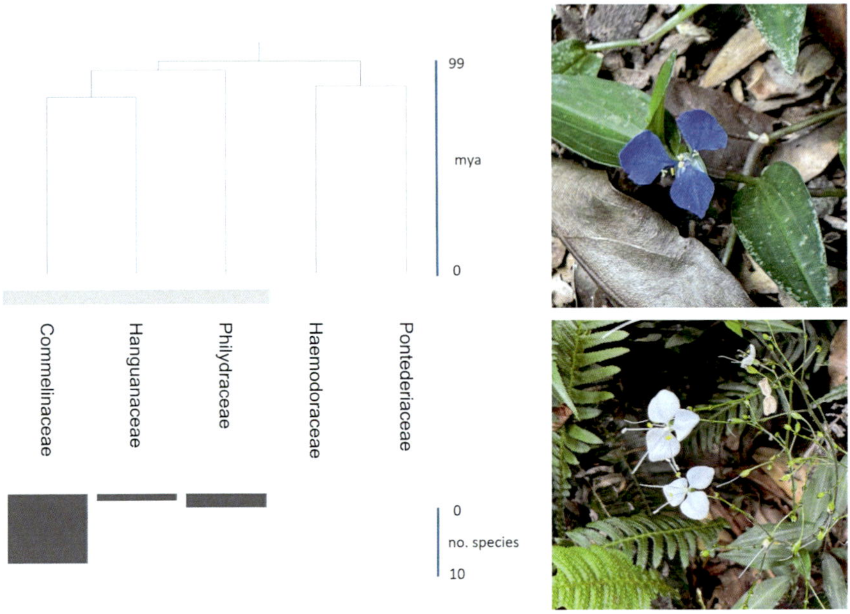

Fig. 52 Dated molecular phylogeny of the Commelinales [7], showing those families that have species present in Australian rainforests (grey), and the number of species occurring in Australian rainforest: flowers of *Commelina cyanea* (top) and *Aneilema acuminatum* (bottom). (Photographs AR Carroll)

4.3.4 Commelinids

The Commelinids are a clade of monocots and includes the Orders Commelinales, Poales, Arecales, and Zingiberales. Only a small number of chemical investigations have been undertaken on Commelinid species (Fig. 52, Table 14).

Commelinales

There are 13 species in three families within the Commelinales, but none of these has been investigated chemically.

Poales

Although there are numerous grasses and sedges (Poales) found in Australian rainforests (Fig. 53), only one study has appeared on *Cyperus haspan* subsp. *juncoides* (Cyperaceae), which identified the bis-furanoquinone, cyperaquinone (**584**) and its hydroxy derivative hydroxycyperaqunínone (**585**) (Fig. 55) [243].

Table 14 Distribution of natural products isolated from the Commelinids

Species	Collection location[a,b]	Plant part[c]	Compound class	Compound(s)	Refs.
Cyperaceae					
Cyperus haspan subsp. *juncoides*	NQ	lf	Furan	**584**, **585**	[243]
Arecaceae					
Livistona australis	SEQ (cult. (Egypt))	lf	Flavonoid	**586**	[244]
	SEQ (cult. (Egypt))	lf	γ-Pyrone	**593**	[245]
Wodyetia bifurcata	SEQ (cult. (Egypt))	lf	Benzoic acid	**591**, **592**	[246]
	SEQ (cult. (Egypt))	lf	Flavonoid	**587**, **588**	[246]
	SEQ (cult. (Egypt))	lf	Triterpenoid	**589**, **590**	[246]
Zingiberaceae					
Pleuranthodium racemigerum	NQ	wp	Curcumin	**594–597**	[247, 248]
Alpinia caerulea	NQ	fr	Diterpenoid	**598**, **599**	[249]

[a] SEQ = Southeast Queensland, NQ = North Queensland, cult. = cultivated, [b] natural location shown for those species studied from cultivated material, [c] lf = leaves, fr = fruits, wp = whole plant

Arecales

The Arecales contains only two families (Arecaceae and Dasypogonaceae) and the palms (Arecaceae) are an omnipresent feature of rainforests worldwide. Surprisingly, even though there are 44 Australian rainforests species, they have barely been investigated and the only reports come from three Egyptian studies on cultivated *Livistona australis* (cabbage tree palm) and *Wodyetia bifurcata* (foxtail palm) (Plate 14) that reported flavonoids (tricin 7-*O*-*β*-glucopyranoside-2″-sulfate sodium salt (**586**), apigenin (**587**) and kaempferol (**588**)), the triterpenoids lupeol (**589**), and β-amyrin (**590**), *p*-hydroxybenzoic (**542**) and gallic acids (**592**), and 3-hydroxy-2-(4-hydroxyphenyl)-6-methyl-4*H*-pyran-4-one (**593**) (Fig. 55) [244–246].

Zingiberales

Species from five Zingiberales families are found in Australian rainforests (Fig. 54). Three species of Australian ginger (Zingiberaceae) have yielded four curcumin derivatives (1-(4-methoxyphenyl)-7-(4-hydroxyphenyl)-(*E*)-hept-2-ene (**594**), 1-(4-methoxyphenyl)-7-(3,4-dihydroxyphenyl)-(*E*)-hept-2-ene (**595**),

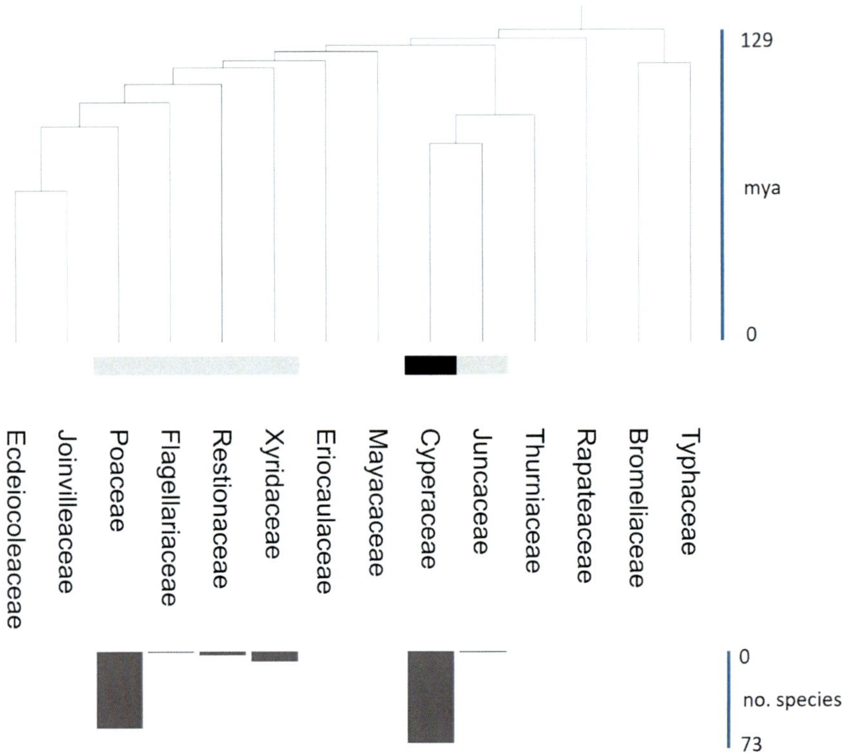

Fig. 53 Dated molecular phylogeny of the Poales [7], showing those families that have species present in Australian rainforests (grey) those that have species chemically investigated (black), and the number of species occurring in Australian rainforests

Plate 14 Foxtail palm (*Wodyetia bifurcata*) and cabbage tree palm (*Livistona australis*). (Photographs AR Carroll)

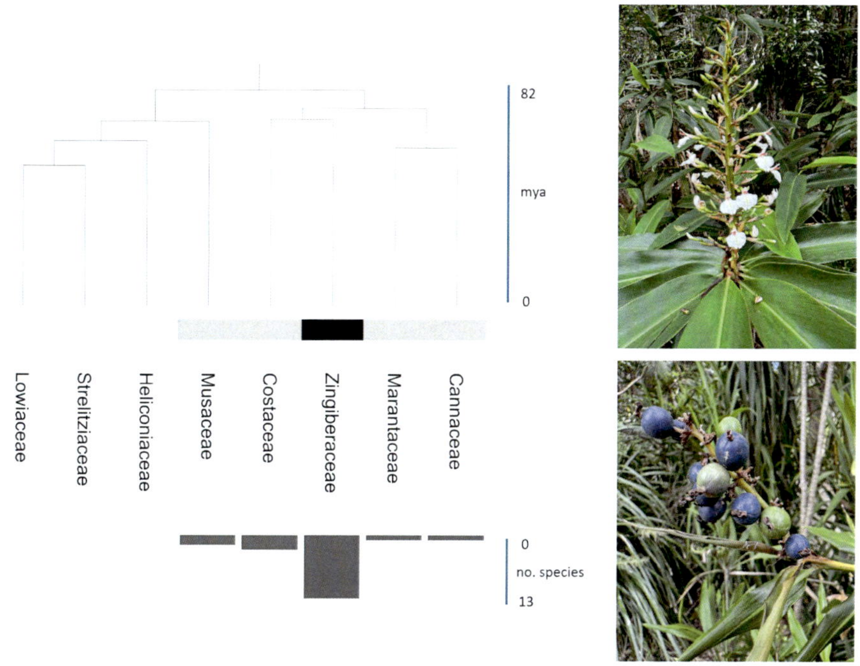

Fig. 54 Dated molecular phylogeny of the Zingiberales [7] showing those families that have species present in Australian rainforests (grey) those that have species chemically investigated (black) and the number of species occurring in Australian rainforests. Flowers and fruits of *Alpinia caerulea* (Photographs AR Carroll)

1-(4-methoxyphenyl)-7-(3,4-di-methoxyphenyl)-(*E*)-hept-2-ene (**596**), and 1-(4-methoxyphenyl)-7-(4-methoxyphenyl)-(*E*)-hept-2-ene (**597**)) from *Pleuranthodium racemigerum* rhizomes [247, 248] and the two labdane diterpenoids zerumin A (**598**) and (*E*)-8(17),12-labdadiene-15,16-dial (**599**) from the fruits of *Alpinia caerulea* from north Queensland [249]. The curcumin derivatives are weakly anti-inflammatory.

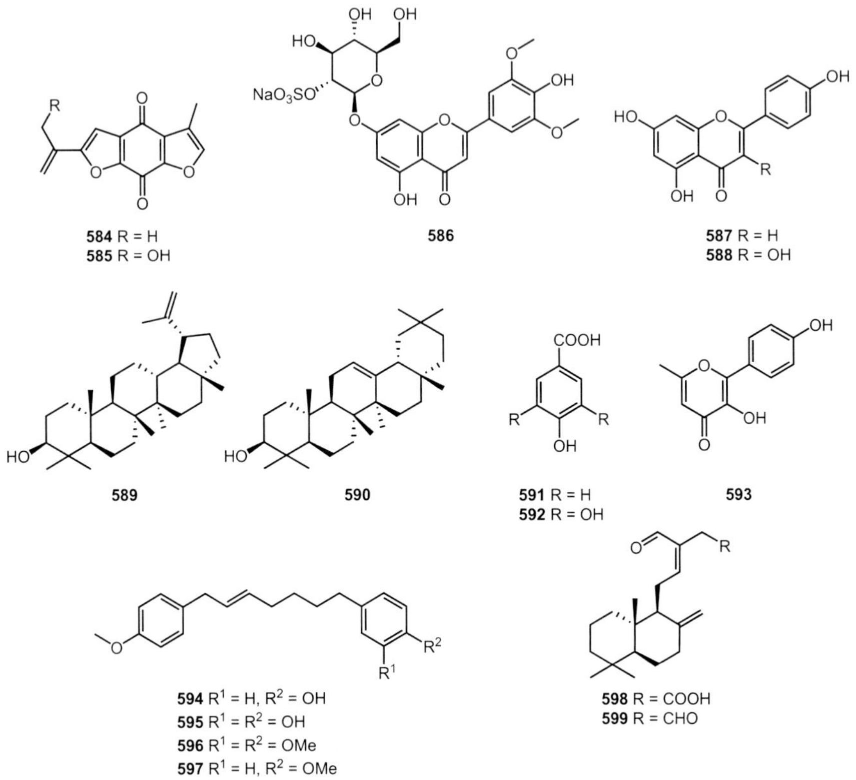

Fig. 55 Structures of compounds **584–599** from Poales, Arecales and Zingiberales species

4.4 Chemistry of Australian Rainforest Basal Eudicotyledon Angiosperms

The eudicots represent the largest clade of angiosperms and are characterized by the presence of two seed leaves (cotyledons). The basal eudicots are a group of 13 families in four orders. These species diverged from the Magnoliids 150 mya and are the ancestors of all eudicots [250]. They show ancestral characteristics such as numerous stamens and separate carpels. Species from two of the orders (Ranunculales (families: Menispermaceae and Ranunculaceae) and Proteales (family: Proteaceae)) occur Australian rainforests.

4.4.1 Ranunculales

Only two of seven families in the Ranunculales are found in Australian rainforests (Fig. 56) and most natural product chemistry investigations have been on species from the Menispermaceae, a family renowned for producing alkaloids. In addition, four *Clematis* species from the Ranunculaceae have also been studied (Table 15). The majority of Australian rainforest Menispermaceae species that have been investigated have restricted distributions in Australia, New Caledonia, and New Guinea, with subspecies of the vine *Stephania japonica* (Plate 15) having an extended distribution to northern Asia (Fig. 56).

Menispermaceae

Eleven of the 26 Australian rainforest Menispermaceae species have been chemically investigated. All but five of the reported natural products are benzyl isoquinoline alkaloids (Figs. 57 and 58). There are 16 macrocyclic bis-BIAs (**437**, **458**, (+)-2-norobamegine (**600**), (+)-2-norberbamine (**601**), (+)-bisnorobamegine (**602**), (+)-bisnorthalrugosine (**603**), (+)-2-northalrugosine (**604**), 6′,7-di-*O*-demethyltetrandrine (**605**), (+)-pycnazanthine (**606**), (+)-isotetrandrine (**607**), 2-*N*-norberbamine (**608**), 2-*N*-norobamegine (**609**), isotrilobine (**610**), isochondrodendrine (**611**), cycleanine (**612**), berberine (**613**)) and 12 of these congeners are mostly reported from two *Pycnarrhena* species (**437**, **458**, **600–609**) [39, 251–254]. This is not surprising, since the tribe Tiliacoreae to which these species belong also contains the South American species *Chondrodendron tomentosum* that produces the curare poison tubocurarine (**614**) [255]. The *Pycnarrhena* alkaloids all possess isoquinoline/isoquinoline and benzyl/benzyl ether linkages in common with the Atherospermataceae bis-BIAs. However, only two bis-BIAs, **437** and **458**, are found in Australian species in the two families. The (+)-enantiomer of isotetrandrine (**607**) occurs in the Menispermaceae while its (−)-enantiomer (**348**) occurs in the Atherospermataceae. Four additional bis-BIAs (**610–613**) have been reported from *Pleogyne australis* (a mono-specific genus endemic to Australia) and from *Stephania bancroftii* and *Pachygone ovata* [39, 251, 252].

Other benzyl isoquinoline alkaloids include husbananes (longanine (**615**), *N*-methylstephuline (**616**), 6-cinnamoylhernandine (**617**), stephalonine (**618**), *N*-methylstephisoferulin (**619**), prostephabyssine (**620**), delavayine (**621**), 7,8-didehydro-2,3,7,8-tetramethoxy-17-methylhasubanan-6-one (**622**), aknadinine (**623**), and aknadilactam (**624**)) [39, 256], two unusual modified husbananes, acutumine (**625**) and acutumidine (**626**) [251] and morphinanes (**393**, sinococuline (**627**), (+)-sebiferine (**628**), (−)-amurine (**629**), and tannagine (**630**)) mostly from *Stephania* species [39, 256–258], aporphines ((−)-stephanine (**631**), (−)-crebanine (**632**), (−)-ayuthianine (**633**), isocorydine methiodide (**634**), nuciferoline (**635**), magnoflorine (**636**), and proaporphine (+)-stepharine (**637**)) [39, 257–259] protoberberines (cyclanoline (**638**), (−)-tetrahydropalmatine (**639**), and (−)-corydalmine (**640**)) [39, 257], and benzyl isoquinoline alkaloids (**343**, **344**, **384**, (*N*-oxycodamine

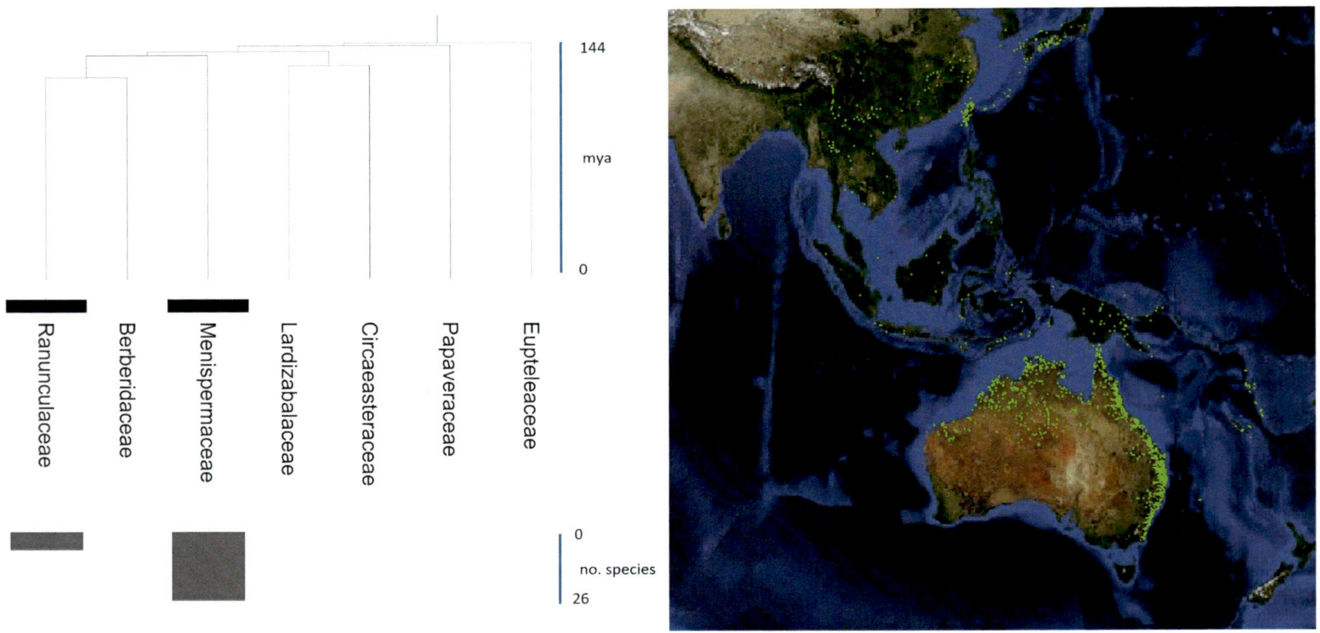

Fig. 56 Dated molecular phylogeny of the Ranunculales [7], showing those families that have species present in Australian rainforests (grey) and those that have species chemically investigated (black) and the number of species occurring in Australian rainforests. The map shows the collection locations of herbarium records for Menispermaceae species that have been investigated chemically [5, 43]

Table 15 Distribution of natural products isolated from the Ranunculales

Species	Collection location[a]	Plant part(s)[b]	Compound class	Compound(s)	Refs.
Menispermaceae					
Carronia pedicellata	NQ	wp	Aporphine	**636**	[39]
	NQ	wp	BIA	**384**	[39]
Echinostephia aculeata	SEQ	st	Aporphine	**635**	[39]
	SEQ	st, wp, r	BIA	**344**	[39, 257]
	SEQ	r	Morphinane	**629**	[257]
Legnephora moorei	SEQ	b, rb	Aporphine	**636**, **634**	[39, 259]
	SEQ	b	BIA	**384**	[39]
Pachygone ovata	NQ	r	Bis BIA	**610**	[251]
	NQ	r	Hasbanane	**625**, **626**	[251]
Parapachygone longifolia	NQ	lf	Morphinane	**393**	[39]
Pleogyne australis	SEQ	r	Bis BIA	**611**, **613**	[252]
Pycnarrhena novoguineensis	NQ	lf	Aporphine	**636**	[39]
	NQ	wp, lf	Bis BIA	**607**, **437**, **605**, **608**, **609**	[39, 253]
Pycnarrhena ozantha	VAN	st, b	Bis BIA	**458**, **600–604**, **606**	[254]
Stephania bancroftii	NQ	r, tub	Aporphine	**631–633**, **637**	[257, 258]
	NQ	wp	BIA	**343**	[39]
	NQ	wp	Bis BIA	**612**	[39]
	NQ	wp	Hasbanane	**621**	[39]
	NQ	tub, wp	Morphinane	**628**, **630**	[39, 258]
	NQ	wp, r, st	Protoberberine	**638–640**	[39, 257]
Stephania japonica var. *japonica*	NQ	ap	Hasbanane	**615–620**, **623**, **624**	[256]
	NQ	ap	Morphinane	**627**	[256]
Stephania renifolia	SEQ	b, wp	Hasbanane	**622**, **623**	[39]
Tinospora smilacina	SEQ	wp	Diterpenoid	**643–645**	[260]
	SEQ	wp	Lignan	**646**	[260]
	SEQ	wp	Phenyl propane	**647**	[260]
Ranunculaceae					
Clematis aristata	ACT	lf	Furan	**648**	[261]
Clematis fawcettii	SEQ	lf	Furan	**648**	[261]
Clematis glycinoides	SEQ	lf	Furan	**648**	[261]
Clematis leptophylla	ACT	lf	Furan	**648**	[261]

[a] SEQ = Southeast Queensland, NQ = North Queensland, VAN = Vanuatu, ACT = Australian Capital Territory, [b] lf = leaves, rb = root bark, ap = aerial parts, r = roots, tub = tubers, wp = whole plant, b = bark, st = stems

Plate 15 *Stephania japonica* growth form (left), male (top left) and female (top right) flowers and fruits (bottom right). (Photographs AR Carroll)

(**641**), and (+)-laudanidine (**642**)) [39, 257]. Several of the morphinane and husbanane alkaloids showed weak δ-opioid receptor binding affinities [256].

Tinospora smilacina from southeast Queensland contains three clerodane diterpenoids (columbin (**643**), 13*β*,14*β*-dihydroxy-15*β*,16*β*-dimethoxycolumbin (**644**), and 13*α*,14*β*-dihydroxy-15*β*,16*β*-dimethoxycolumbin (**645**)), the lignan isolariciresinol (**646**), and the phenyl propanol, dihydrosyringenin (**647**) [260].

Ranunculaceae

The furanone, protoanemonin (**648**) is the only compound reported from Australian Ranunculaceae (Figs. 58 and 59). It was found in the leaves of four species of *Clematis* [261].

4.4.2 Proteales

The Proteales comprises four families but only the Proteaceae is found in Australian rainforests. The Proteaceae is a Gondwanan family with almost exclusive Southern Hemispheric species distribution and having centers of diversity in South Africa and Australia (Fig. 60) [262]. Although most of the diversity of this species now resides in a small number of genera in temperate heathlands, the Australian rainforests contain

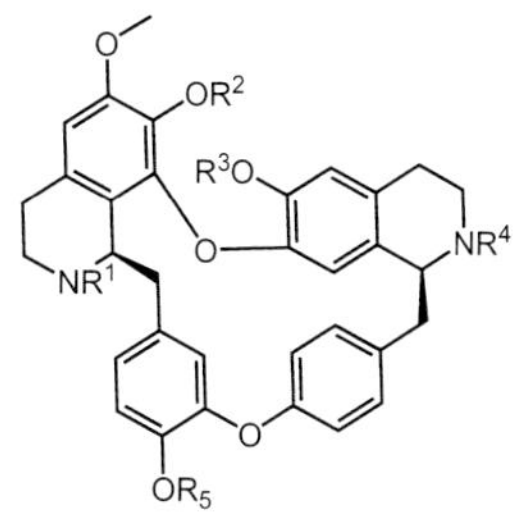

600 $R^1 = R^2 = R^5 = H$, $R^3 = R^4 = Me$
601 $R^1 = R^5 = H$, $R^2 = R^3 = R^4 = Me$
602 $R^1 = R^2 = R^4 = R^5 = H$, $R^3 = Me$
603 $R^1 = R^2 = H$, $R^3 = R^4 = R^5 = Me$
604 $R^1 = R^2 = H$, $R^3 = R^4 = R^5 = Me$
605 $R^1 = R^4 = R^5 = Me$, $R^2 = R^3 = H$

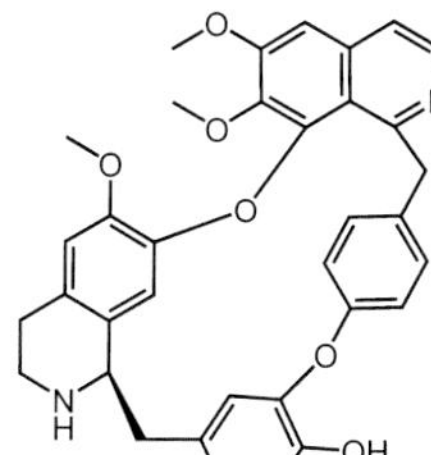

606

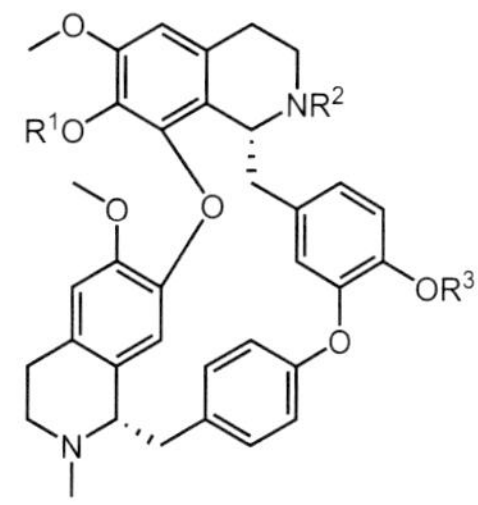

607 $R^1 = R^2 = R^3 = Me$
608 $R^1 = Me$, $R^2 = R^3 = H$
609 $R^1 = R^2 = H$, $R^3 = H$

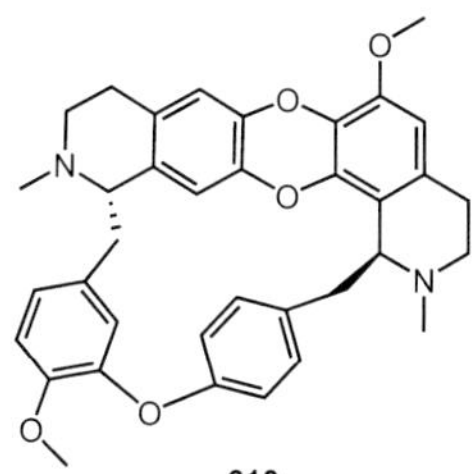

610

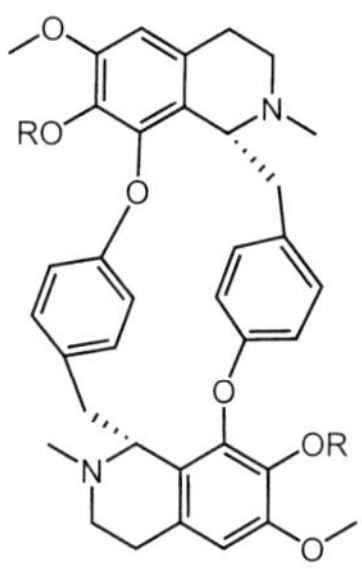

611 R = H
612 R = Me

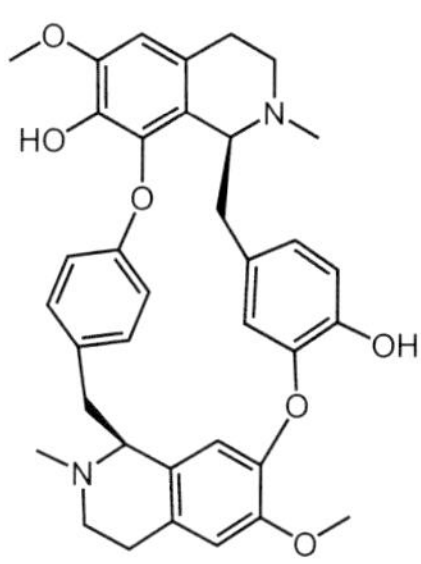

613

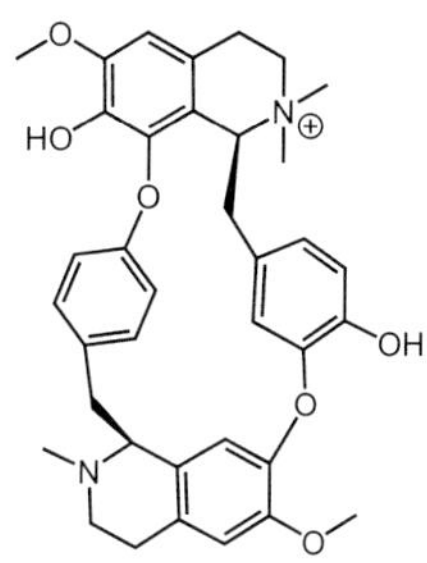

614

Fig. 57 Structures of bis-BIAs **600–614** from Menispermaceae species

99 species across 35 genera (42% of total genera) (Fig. 61). Many Proteaceae species have spectacular flowering events (Plates 16 and 17).

Only a few chemical studies on Australian rainforest Proteaceae have been undertaken (13 genera) but the results have provided some interesting trends (Table 16). The genera *Bellendena*, *Agastachys* (Tasmanian), *Darlingia* (NQ), *Triunia* (SEQ and NQ), and *Floydia* (SEQ) (Plate 16) have all yielded tropane alkaloids of unusual structure since most possess a side chain of two or more carbons attached to either C-2 or C-4 (Fig. 62). The simple oxygenated tropanes, 3α-cinnamoyloxytropane-6β-ol (**649**), (−)-6β-acetoxy-3α-tigloyloxytropane (**650**), 3α-acetoxy-6β-isobutoxytropane (**651**), 6β-acetoxy-3α-isobutoxytropane (**652**), 3α-acetoxy-6β-hydroxytropane (**653**), tropine (**654**), 3,5-dihydroxytropane (**655**), and 3α-(*p*-hydroxybenzoyloxy)trop-6-ene (**656**), are related to alkaloids found in various angiosperms mainly from the

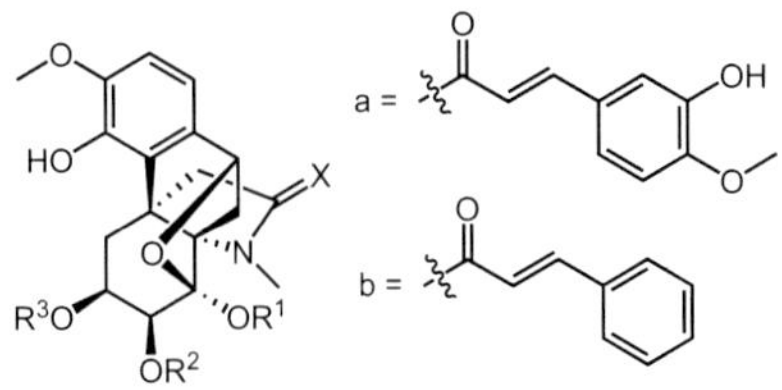

615 R[1] = R[3] = H, R[2] = Me, X = H,H
616 R[1] = R[2] = Me, X = H,H, R[3] = H
617 R[1] = Me, R[2] = H, X = H,H, R[3] = a
618 R[1] = R[2] = Me, X = O, R[3] = a
619 R[1] = R[2] = Me, X = H,H, R[3] = b

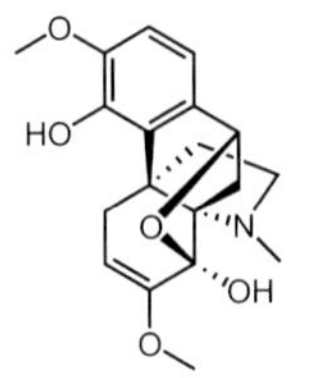

620

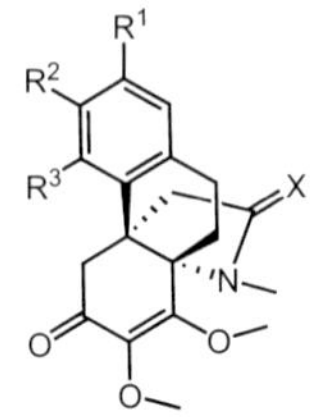

621 R[1]–R[2] = OCH_2O, R[3] = H, X = H,H
622 R[1] = R[2] = OMe, R[3] = H, X = H,H
623 R[1] = H, R[2] = OMe, R[3] = OH, X = H,H
624 R[1] = H, R[2] = OMe, R[3] = OH, X = O

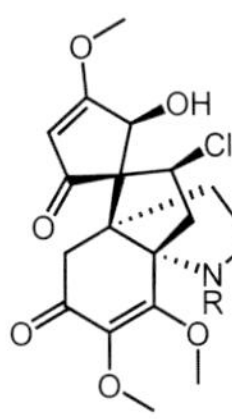

625 R = Me
626 R = H

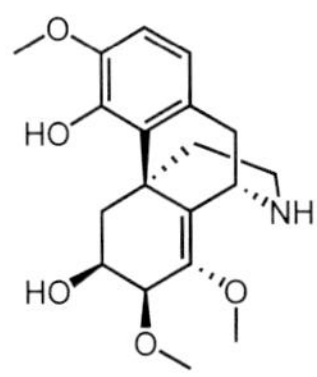

627

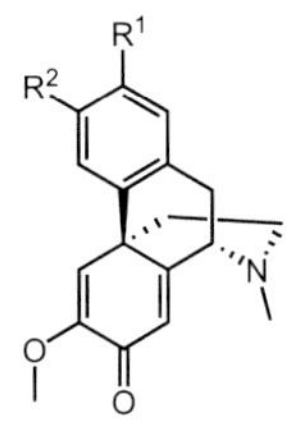

628 R[1] = R[2] = OMe
629 R[1]–R[2] = OCH_2O

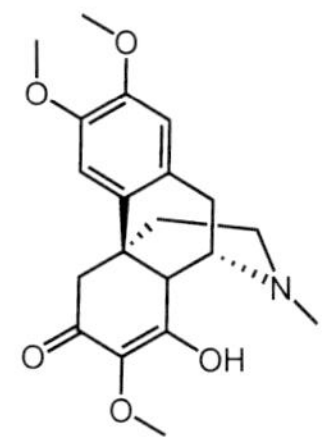

630

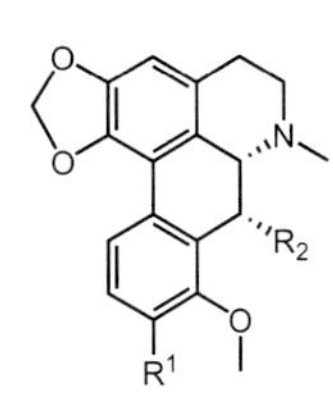

631 R[1] = H, R[2] = H
632 R[1] = OMe, R[2] = H
633 R[1] = H, R[2] = OH

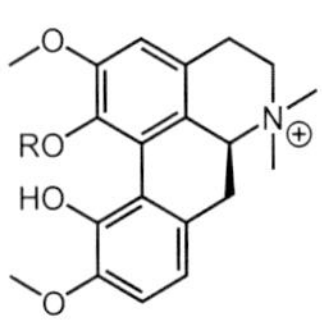

634 R = Me
636 R = H

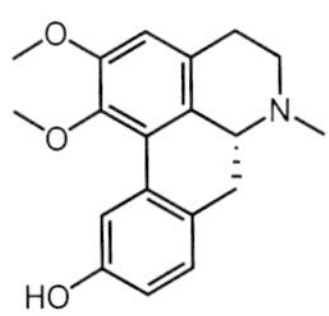

635

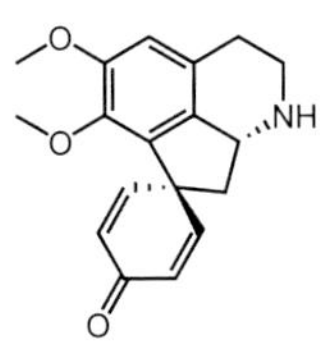

637

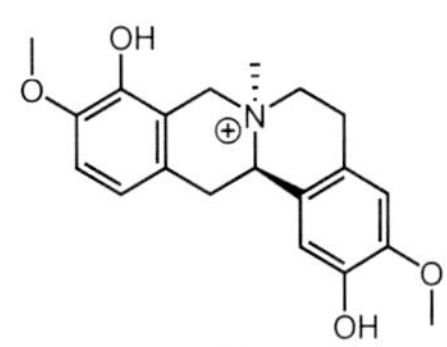

638

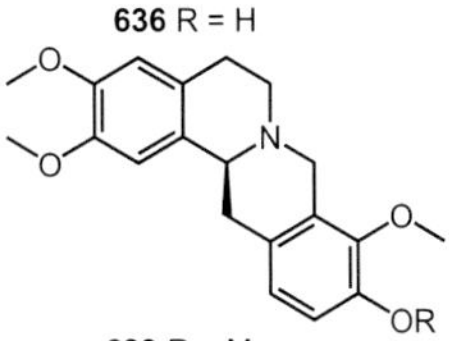

639 R = Me
640 R = H

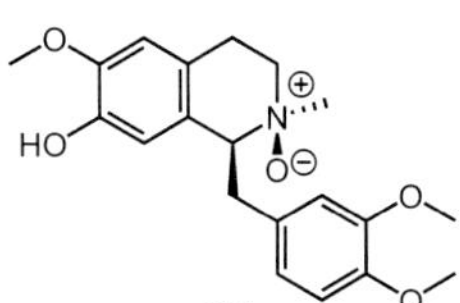

641

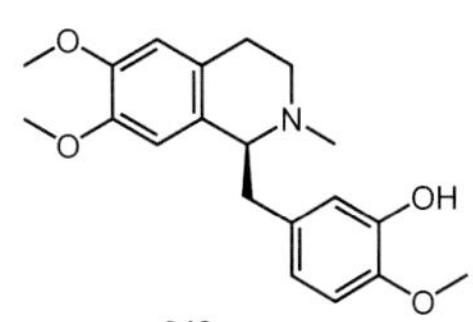

642

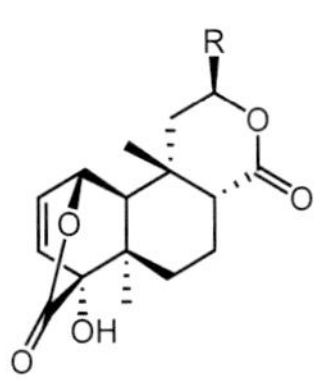

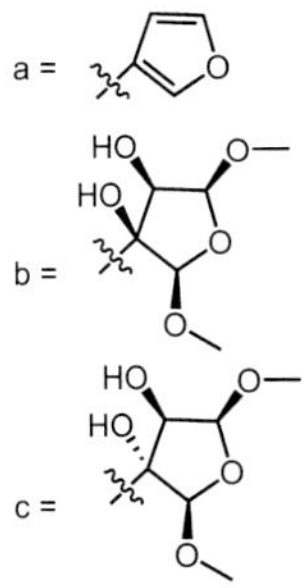

643 R = a
644 R = b
645 R = c

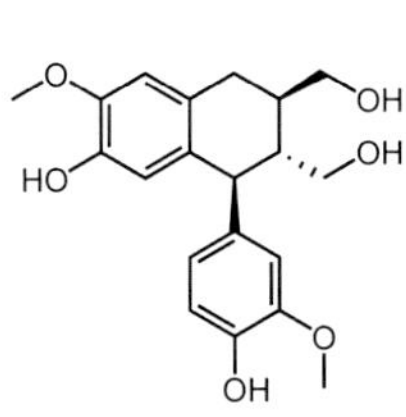

646

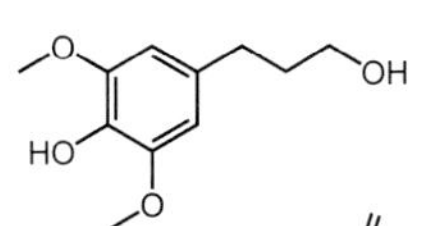

647

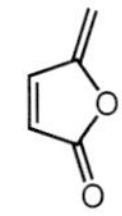

648

Fig. 58 Structures of compounds **615**–**648** from Menispermaceae and Ranunculaceae species

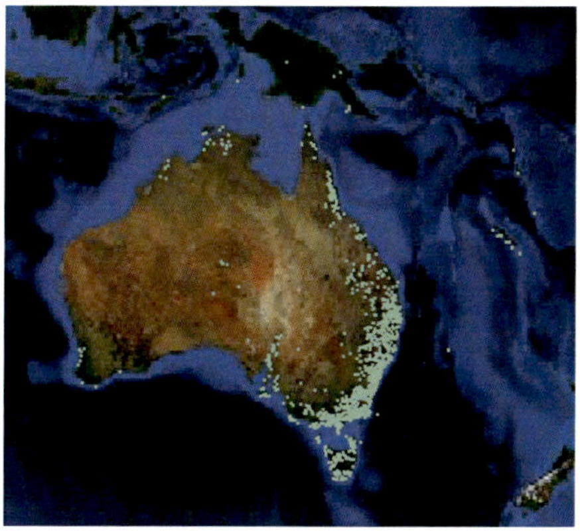

Fig. 59 A map of the locations of *Clematis* species that occur in Australian rainforests based on herbarium records [5, 43]. Female flowers of *Clematis aristata*. (Photograph AR Carroll)

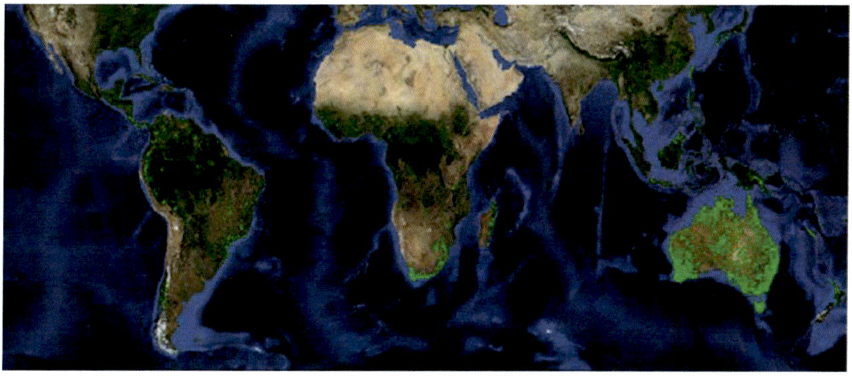

Fig. 60 A map of the locations of Proteaceae species based on herbarium records [5, 43]

families Solanaceae and Erythroxylaceae [39, 263, 264]. The remainder of the tropanes are characteristic of the Proteaceae. These compounds have either ethyl, benzyl isobutyl, isopentyl or butyl side chains attached to either C-2 or C-4 of the tropane ring. These side chains are mostly oxidized at the carbon directly attached to the tropane and in many examples the side chain cyclizes back into the tropane to generate a pyran ring. The pyranotropane congeners are darlingine (**657**), 10-hydroxydarlingine (**658**), bellendine (**659**), isobellendine (**660**), darlingine *N*-oxide (**661**), 5,11-dihydroisobellendine (**662**), 5,11-dihydrodarlingine (**663**), 2,3-dihydrobellendine (**664**), and 2,3-*epi*-dihydrobellendine (**665**) [39, 264–269]. Additional acylated tropanes include ferruginine (**666**), norecgonidine methyl ester (**667**), ferrugine (**668**), 3α-benzoyloxy-2α-hydroxybenzyltropane (**669**), and triuniamine A (**670**) [39, 267, 270]. Another series of alkaloids that are likely to be precursors to the

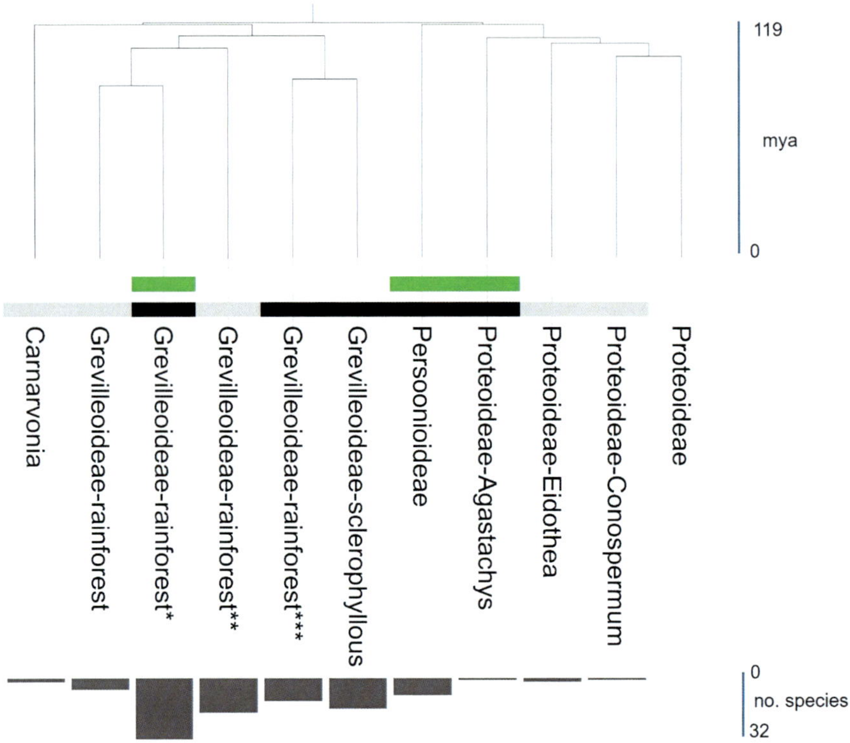

Fig. 61 Dated molecular phylogeny of the Proteaceae [262] showing those subfamilies that have species present in Australian rainforests (grey), alkaloid producing (green) and studied (black) and the number of species occurring in Australian rainforest. The rainforest genera found in each subfamily group are listed herein. Grevilleoideae (i) Rainforest—*Alloxylon, Telopea*; (ii) Rainforest*—*Sphalmium, Musgravea, Austromuellera, Triunia, Neorites, Orites, Megahertzia, Hollandaea, Helicia, Eucarpha, Floydia, Darlingia*; (iii) Rainforest**—*Banksia, Athertonia, Cataepidia, Cardwellia, Hicksbeachia, Bleasdelea, Macadamia, Nothorites*; (iv) Rainforest***—*Buckinghamia, Lomatia, Strangea, Stenocarpus*; (v) Sclerophyllous—*Grevillea, Hakea, Opisthiolepis*; Persoonioideae—*Bellendena, Placospermum, Persoonia,* Proteoideae (i) Agastachys—*Agastachys, Symphionema*; (ii) Eidothea—*Eidothea, Beauprea*; (iii) Conospermum—*Conospermum, Cenarrhenes*; (iv) African and southwest Australian species

alkyl/acyl tropanes are the pyrrolidines dehydrodarlinine (**671**), darlinine (**672**), *epi*-darlinine (**673**), dehydrodarlingianine (**674**), darlingianine (**675**), isodarlingianine (**676**), dihydrodarlingianine (**677**), tetrahydrodarlingianine (**678**), and 1-methyl-2-pyrrolidine acetic acid (**679**) [265]. Tri-(α-methylenylbutyrolactonyl)amine (**680**) was also reported in *B. montana* [271]. These tropane and pyrrolidine alkaloids are unlike any other tropanes reported from species of the families Solanaceae, Convolvulaceae, Erythroxylaceae, or Rhizophoraceae, suggesting that they are biosynthesized via a unique route.

A clue to their biogenesis is provided by the co-isolation of the pyrroline series from *D. darlingiana*. These compounds are reminiscent of the aryl-substituted

Plate 16 Proteaceae species that contain alkaloids. *Darlingia darlingiana* (tree and flowers left and top left), *D. ferrugenia* flowers (top middle), *Triunia robusta* fruits (top right) and flowers (bottom middle), *T. youngiana* flowers (bottom left), and *Bellendena montana* flowers (bottom right). (Photographs AR Carroll)

polyketides found in Lauraceae and Winteraceae species. Assuming the common tropane biosynthetic precursor pyrroline, derived from ornithine, is the starting point for Proteaceae tropane biosynthesis [272], a variety of carboxylic acid precursors (instead of malonate) would be required to generate the C-2 substituted tropanes (Scheme 3). These could be generated from condensation of simple carboxylic acids with malonate followed by decarboxylation. Condensation of these β-diketones with pyrroline followed by further cyclization would generate the 2-acyl tropanes. If malonate, butyrate, isobutyrate, 2-methylbutyrate, benzoate, or cinnamate are condensed with malonate, the full range of 2-acyl tropanes found in the Proteaceae could be generated. Condensation of two malonates with pyrroline generates the simple tropanes that are found in *Floydia*, both *Darlingia* species, and all four species of *Triunia, Bellendena*, and *Agastachys.* The common ester substituents found in derivatives of these tropanes also points to their availability as building blocks. Interestingly, the two *Eucarpha* species endemic to New Caledonia also contain C-2 alkyl substituted tropanes [273]. Among the Proteaceae tropanes, both the naturally occurring and unnatural enantiomers of ferruginine, *ent*-ferruginine (**681**), darlingine, and ferrugine, have been synthesized and tested for nicotinic acetylcholine receptor activity [274]. Interestingly, the unnatural ferruginine is a potent agonist of α4β2 and α7 subtypes, but the natural enantiomer is 700-fold less active. In 1879, *Triunia robusta* was documented as having toxic seeds from an account in a letter by F.M. Bailey to Baron von Mueller indicating that the consumption of half a seed was enough to render the consumer so ill that he had to "get off his horse and lie down for some time before he could go home" [275]. The tropanes found in the seeds are likely to be responsible for this toxicity. *Triunia robusta* was thought to be extinct until its rediscovery in southeastern Queensland in 1989. In another account, consumption

Table 16 Distribution of natural products isolated from the Proteaceae

Species	Natural location[a,b]	Plant part(s)[c]	Compound class	Compound(s)	Refs.
Bellendena montana	Tas	fl	Alkyl phenol	**756**	[295]
	Tas	fl	Amine	**680**	[271]
	Tas	lf	Flavonoid	**753**	[292]
	Tas	fl	Tropane	**651–653**, **657**, **659**, **660**, **662–665**	[264]
	Tas	lf	β-Arbutin	**726–730**	[292]
Cardwellia sublimis	NQ	w	5-Alkyl resorcinol	**682**, **684**, **685**, **690**, **691**	[278]
	NQ	w	Inositol	**766**	[278]
Cardwellia sublimis	NQ	w	5-Alkyl resorcinol	**682**, **684**, **685**, **690**, **691**	[278]
Persoonia linearis	Vic	w	5-Alkyl resorcinol	**687–689**	[278]
Opisthiolepis heterophylla	NQ	w	5-Alkyl resorcinol	**682–685**	[278]
	NQ	w	Bis 5-alkyl resorcinol	**698**	[278]
Agastachys odorata	Tas	lf, st	Alkyl phenol	**755**, **756**	[263]
	Tas	lf, st	Tropane	**650**, **656**	[263]
Bleasdalea bleasdalei	NQ	b	Triterpenoid	**767**	[296]
Cenarrhenes nitida	Tas	lf	β-Arbutin	**731**	[292]
Darlingia darlingiana	NQ	lf, st, b	Tropane	**657**, **658**, **661**, **663**, **666**, **667**	[39, 265, 266]
	NQ	lf, st	Pyrrolidine	**671–677**	[39, 265]
Darlingia ferruginea	NQ	lf, st, r, s	Tropane	**649**, **657**, **666**, **681**, **668**, **669**, **674**	[39, 267]

(continued)

Table 16 (continued)

Species	Natural location[a,b]	Plant part(s)[c]	Compound class	Compound(s)	Refs.
Floydia praealta	SEQ	fr, lf	Tropane	**654**, **655**, **666**, **668**	[39]
Triunia erythrocarpa	NQ	w, b, lf, st	Tropane	**657**, **658**, **668**	[39, 269]
Triunia montana	NQ	st	Tropane	**657**, **663**, **670**	[270]
Triunia robusta	SEQ	lf	Tropane	**657**, **658**, **663**	[39]
Triunia youngiana	SEQ	lf	Tropane	**657**, **658**, **663**	[39]
Lomatia arborescens	NNSW	s	Naphthoquinone	**751**	[294]
Lomatia polymorpha	Tas	lf	Alkyl phenol	**757**, **758**	[292]
	Tas	lf	β-Arbutin	**732–735**	[292]
Stenocarpus salignus	NNSW	w	Naphthoquinone	**749**, **750**	[293]
Grevillea hilliana	NQ	w	5-Alkyl resorcinol	**682–685**, **690–693**	[278]
Grevillea pteridifolia	NQ	w	5-Alkyl resorcinol	**683–686**, **691**	[278]
Grevillea robusta	SEQ	w	5-Alkyl resorcinol	**683–686**	[279–281]
	SEQ (cult. (Taiwan, India, Japan))	lf	5-Alkyl resorcinol	**715–724**	[282–284]
	SEQ (cult. (India, Taiwan))	w, lf	Bis 5-alkyl resorcinol	**694–700**	[285–287]
	SEQ	lf	Cinnamate	**759**, **760**	[281]
	SEQ (cult. (Taiwan, Japan))	lf	Cinnamate	**761–764**	[284, 290, 291]

(continued)

Table 16 (continued)

Species	Natural location[a,b]	Plant part(s)[c]	Compound class	Compound(s)	Refs.
	SEQ	lf	Flavonoid	**754**	[281]
	SEQ (cult. (Japan))	lf	Glycoside	**765**	[282]
	SEQ (cult. (India, Taiwan))	lf	Macrocyclic bis-5-alkyl resorcinol	**701**, **703**, **704**, **706–714**	[286, 287]
	SEQ	lf	Macrocyclic bis-5-alkyl resorcinol	**701–705**	[281]
	SEQ (cult. (Japan))	lf	Carotenoid	**725**	[283]
	SEQ (cult. (India, Japan))	lf	β-Arbutin	**727**, **736–748**	[286, 289–291]

[a] SEQ = Southeast Queensland, NQ = North Queensland, NNSW = Northern New South Wales, Tas = Tasmania, Vic = Victoria, cult. = cultivated, [b] natural location shown for those species studied from cultivated material, [c] lf = leaves = seeds, ap = aerial parts, r = roots, fl = flowers, fr = fruits, b = bark, w = wood, st = stems

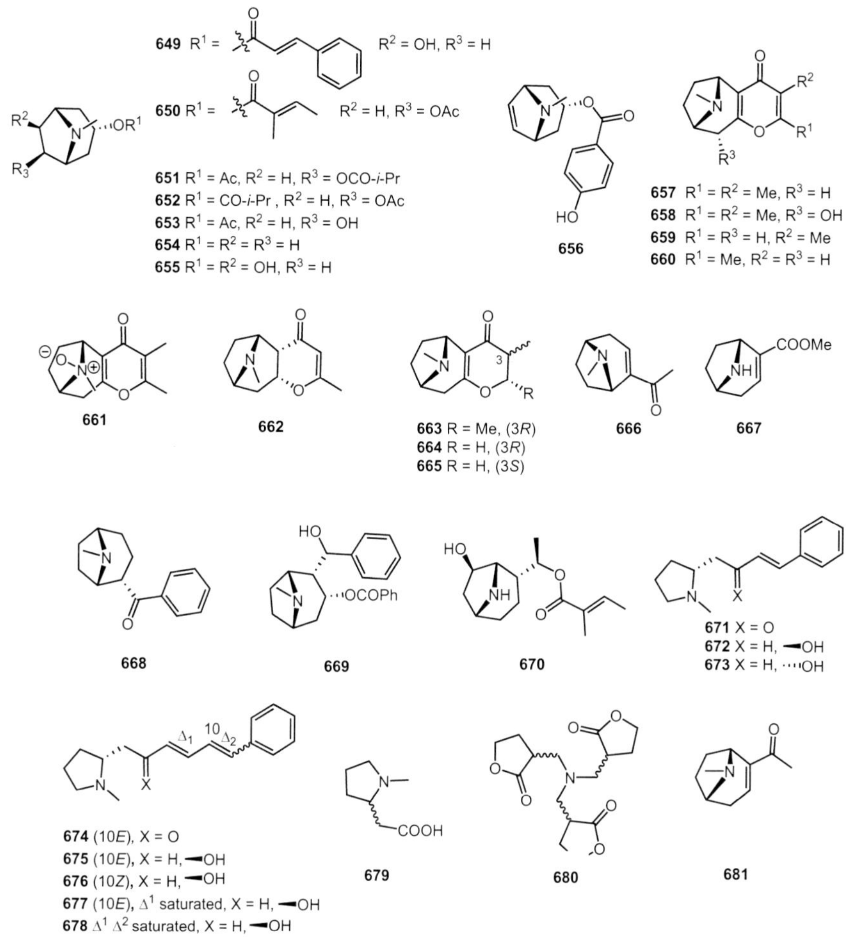

Fig. 62 Structures of alkaloids **649–681** reported from Proteaceae species

of a small amount of the seeds of *T. erythrocarpa* from north Queensland led to the consumer sweating so profusely that it was like they had taken a swim in their clothes and their heart rate dropped so low that they were admitted to hospital and given several doses of atropine [276]. These symptoms suggest that the alkaloids are likely to be M2/M3 muscarinic receptor agonists. Another interesting feature of Proteaceae tropanes is that some of the compounds have been reported to have a cocaine-type configuration while others have the opposite configuration. These absolute configurations were confirmed by total synthesis of **666** and **668** or application of the Mosher ester method on **658** [269, 277].

Another major group of compounds reported from the rainforest Proteaceae are long-chain 5-alkyl resorcinols and long-chain alkyl compounds substituted at each end by a resorcinol (bis-5-alkylresorcinols) unit (Fig. 63). Of the 43 compounds

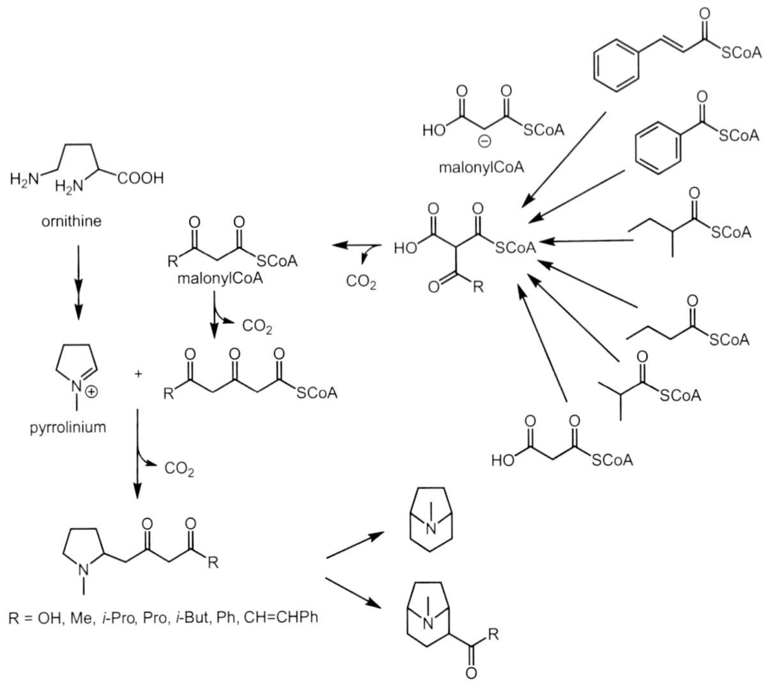

Scheme 3 Biogenesis of the Proteaceae tropane alkaloids

that have been isolated from rainforest species most are from *Grevillea robusta* (Plate 17). Other congeners have been reported from *G. hilliana*, *G. pteridifolia, Opisthiolepis heterophylla, Persoonia linearis*, and *Cardwellia sublimis*. The 5-alkylresorcinols are heptadec-8-enylresorcinol (**682**), grevillol (**683**), pentadecylresorcinol (**684**), pentadecenylresorcinol (**685**), (5*E*)-pentadec-10-en-1-ylresorcinol (**686**), 5-undecylresorcinol (**687**), 5-nonylresorcinol (**688**), 5-undec-3-enylresorcinol (**689**), nonadecenylresorcinol (**690**), heptadecylresorcinol (**691**), heptadec-10-enylresorcinol (**692**), and nonadec-10-enylresorcinol (**693**) [278–284]. In turn, the linear bis-5-alkylresorcinols are mono-norstriatol (**694**), bis-norstriatol (**695**), grevirobstol A (**696**), 5,5′-(tetradecane-1′,14′-diyl)bisresorcinol (**697**), 5,5′-(tetradec-6-ene-1′,14′-diyl)bisresorcinol (**698**), bisgravillol (**699**), and dehydrobisgravillol (**700**) [278, 285–287]. Many of the *G. robusta* compounds are macrocyclic, containing ether or carbon–carbon linkages between the two resorcinols. These include robustol (**701**), 2-oxa-1(1,3),3(1,4)-dibenzenacyclononadecaphane-1^5,3^2,3^6-triol (**702**), (*Z*)-2-oxa-1(1,3),3(1,4)-dibenzenacycloheptadecaphan-11-ene-1^5,3^2,3^6-triol (**703**), (*Z*)-2-*oxa*-1(1,3),3(1,4)-dibenzenacycloheptadecaphan-13-ene-1^5,3^2,3^6-triol (**704**), (*Z*)-2-oxa-1(1,3),3(1,4)-dibenzenacycloheptadecaphan-13-ene-1^5,3^2,3^6-triol (**705**), grevirobstol C (**706**), 2-oxa-1(1,3),3(1,4)-dibenzenacyclopentadecaphane-1^5,3^2,3^6-triol (**707**), gravicycle (**708**), dehydrogravicycle (**709**), grevirobstol B (**710**), dehydrograviphane (**711**), methyldehydrograviphane (**712**), graviphane (**713**), and

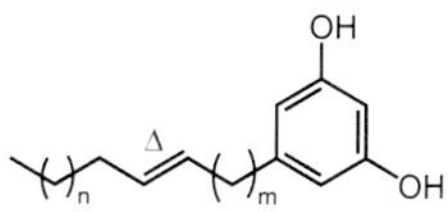

682 m = 7, n = 6, Δ (*Z*)
683 m = 1, n = 8, Δ saturated
684 m = 1, n = 10, Δ saturated
685 m = 7, n = 4, Δ (*E*)
686 m = 9, n = 2, Δ (*E*)
687 m = 1, n = 6, Δ saturated
688 m = 1, n = 4, Δ saturated
689 m = 2, n = 5, Δ saturated
690 m = 1, n = 14, Δ saturated
691 m = 1, n = 12, Δ saturated
692 m = 9, n = 4, Δ (*Z*)
693 m = 9, n = 6, Δ (*Z*)

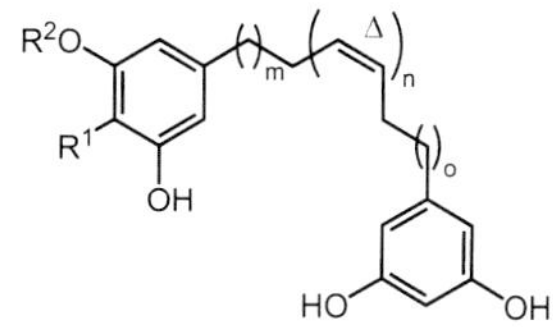

694 R^1 = Me, R^2 = H, m = 4, n = 1, o = 6, Δ saturated
695 R^1 = R^2 = H, m = 4, n = 1, o = 6, Δ saturated
696 R^1 = R^2 = H, m = 4, n = 2, o = 5
697 R^1 = R^2 = H, m = 6, n = 1, o = 4, Δ saturated
698 R^1 = R^2 = H, m = 4, n = 1, o = 6
699 R^1 = R^2 = Me, m = 6, n = 1, o = 3, Δ saturated
700 R^1 = R^2 = Me, m = 6, n = 1, o = 3

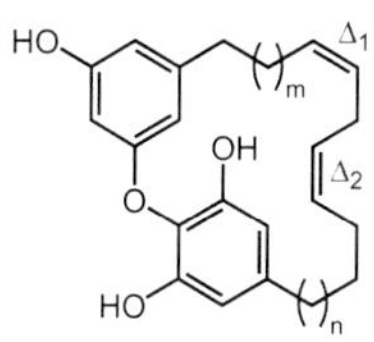

701 m = 3, n = 3, Δ_1, Δ_2 saturated
702 m = 5, n = 3, Δ_1, Δ_2 saturated
703 m = 5, n = 1, Δ_2 saturated
704 m = 2, n = 4, Δ_2 saturated
705 m = 6, n = 2, Δ_2 saturated
706 m = 3, n = 3
707 m = 3, n = 1, Δ_1, Δ_2 saturated

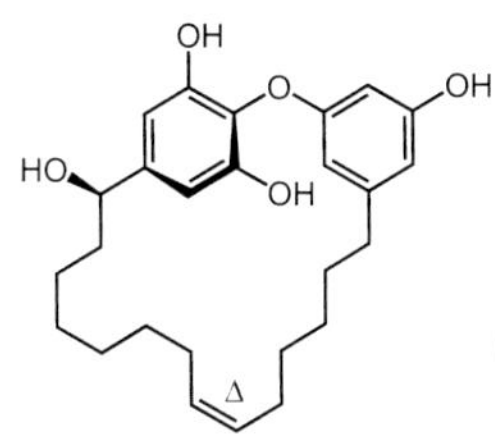

708 Δ saturated
709

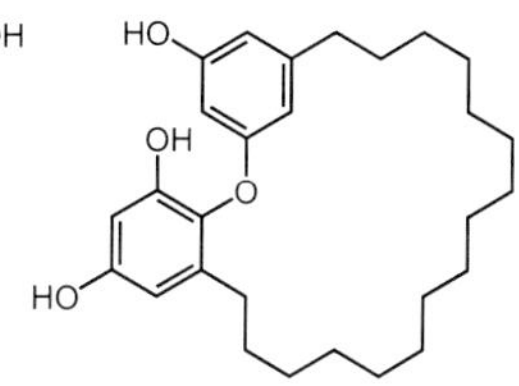

710

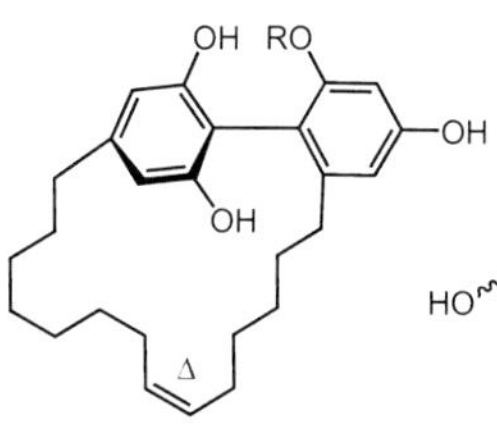

711 R = H
712 R = Me
713 R = H, Δ saturated
714 R = Me, Δ saturated

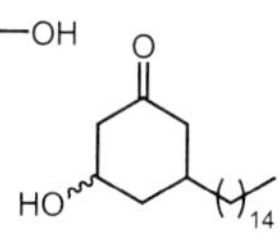

715

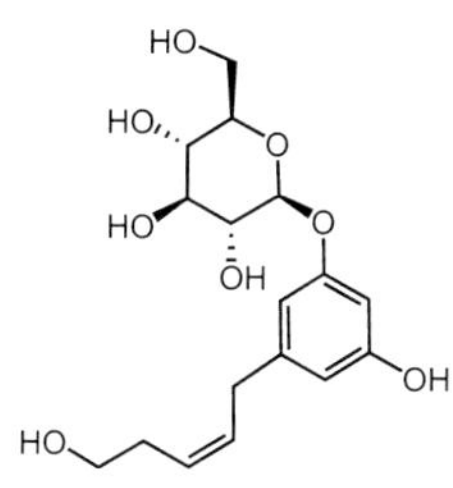

716

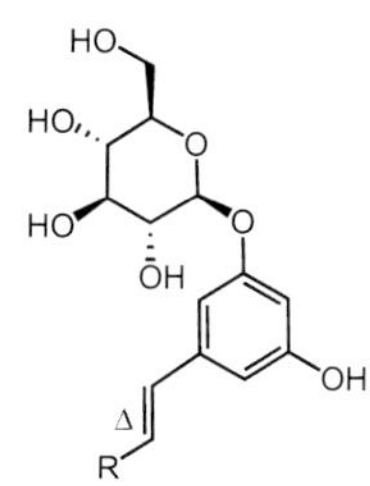

717 R = $(CH_2)_9OH$, Δ saturated
718 R = $(CH_2)_2COOH$, Δ saturated
719 R = $(CH_2)_2OH$, Δ saturated
720 R = CH_2COOH, Δ saturated
721 R = COOH
722 R = OH, Δ saturated
723 R = $(CH_2)_4OH$, Δ saturated
724 R = CH_2COOMe, Δ saturated

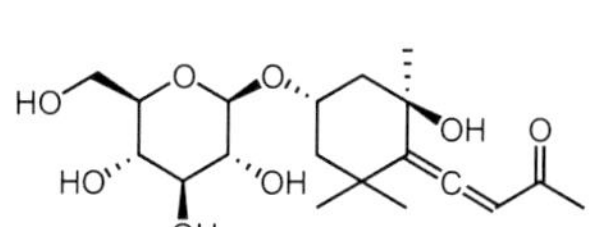

725

Fig. 63 Structures of alkyl resorcinols **682–724** and an oxidized carotenoid **725** reported from Proteaceae species

Plate 17 *Grevillea robusta* (left and bottom middle), *Stenocarpus salignus* (top middle), *Lomantia arborescens* (top right), and *Persoonia adenantha* (bottom right) flowers. (Photographs AR Carroll)

methylgraviphane (**714**) [281, 286, 287]. This class of compound has also been reported from non-rainforest *Grevillea*, *Hakea*, and *Petrophile* species. The alkyl resorcinols have been shown to be responsible for allergic contact dermatitis produced by many *Grevillea* species [288]. Some of the compounds are cytotoxic through DNA cleavage [287]. Bis-5-alkylresorcinol derivatives are almost exclusively found in Proteaceae species from the subfamily Grevilleoideae. *Maesa* species in the Rubiaceae are the only other source. Six of the cyclophanes are chiral atropisomers. Finally, an oxidized alkyl resorcinol derivative, (3*Z*)-hydroxy-5-pentadecylcyclohexanone (**715**), was isolated from *G. robusta* [284]. Grevillosides A–H (**716–721**, **722**, **723**) and grevilloside E methyl ether (**724**) are shorter chain alkyl resorcinols with either alcohol, carboxylic acid or ester groups substituting the terminus of the alkyl chain. They also contain a glucose moiety attached to one of the aromatic oxygens. All these compounds were isolated from the leaves of *G. robusta* cultivated in Japan [282]. A short-chain glucosidic oxidation product of β-carotenoid staphylionoside D (**725**) was also co-isolated in these studies.

Rainforest Proteaceae are also a rich source of β-arbutin derivatives (Fig. 64) with 23 analogues reported: 6-*O*-(3-(4-hydroxyphenyl)-3-oxopropianoyl)arbutin

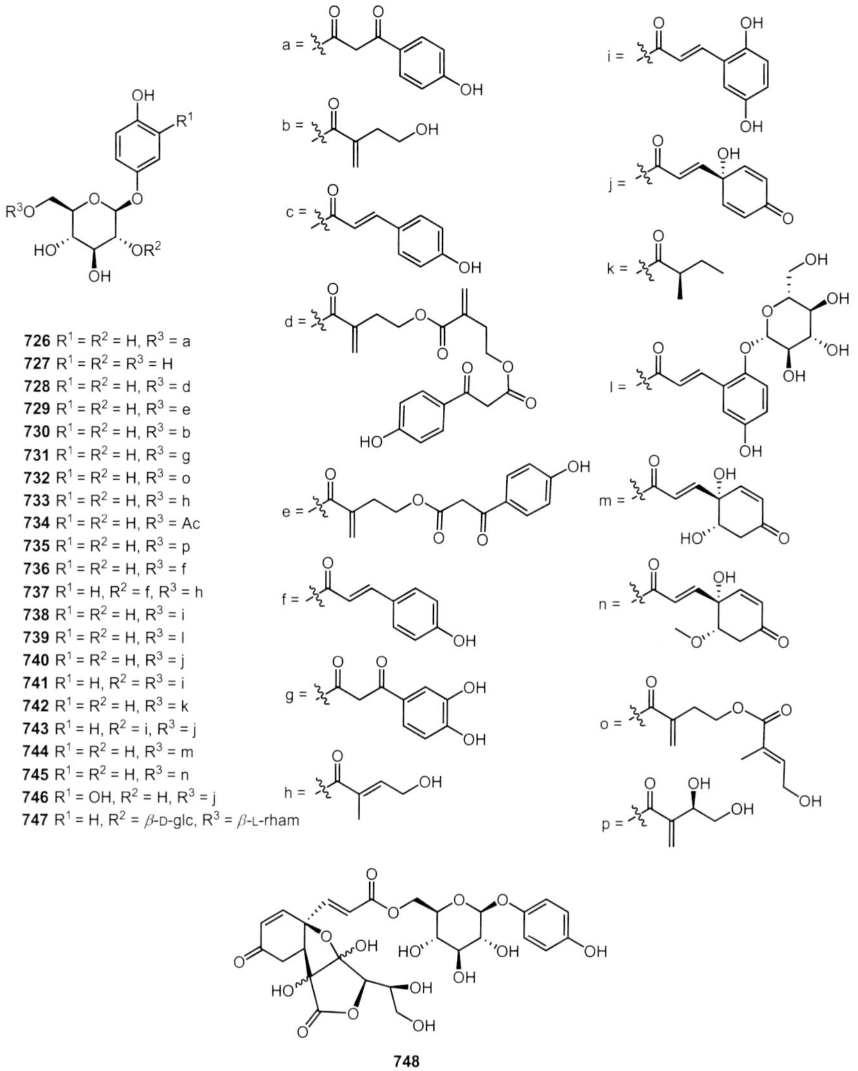

Fig. 64 Structures of arbutin derivatives **726**–**748** reported from Proteaceae species

(**726**), arbutin (**727**), 6-*O*-(4-((4-((3-(4-hydroxyphenyl)-3-oxopropianoyl)oxy)-2-methylenebutanoyl)oxy)-2-methylenebutanoyl)arbutin (**728**), 6-*O*-(4-((3-(4-hydroxyphenyl)-3-oxopropianoyl)oxy)-2-methylenebutanoyl)arbutin (**729**), 6-*O*-(4-hydroxy-2-methylenebutanoyl)arbutin (**730**), 6-*O*-(3-(3,4-dihydroxyphenyl)-3-oxopropianoyl)arbutin (**731**), 6-*O*-((*E*)-4-((4-hydroxy-2-methylbut-2-enoyl)oxy)-2-methylenebutanoyl)arbutin (**732**), 6-*O*-((*E*)-4-hydroxy-2-methylbut-2-enoyl)arbutin (**733**), 6-*O*-acetoxyarbutin (**734**), 6-*O*-((*R*)-3,4-dihydroxy-2-methylenebutanoyl)arbutin (**735**), 6-*O*-*p*-coumarylarbutin (**736**), 2-*O*-*p*-coumarylarbutin (**737**), robustaside B (**738**), robustaside C (**739**), robustaside D (**740**), grevillosides I–M (**741–745**), grevilloside O (**746**), grevilloside P (**747**), and grevilloside N (**748**) [289–292]. The compounds can possess a variety of ester groups attached to the glucose moiety. These include 1′-hydroxycyclohexadien-4′-one acrylate, 2,5-dihydroxyphenyl acrylate, coumaroyl, 3-(4-hydroxyphenyl)-3-oxopropionate and 4-hydroxy-2-methylenebutyrate units, either alone or conjugated together. Such phytochemicals have been isolated from *G. robusta* cultivated in Japan, Taiwan, and India, as well as from *Lomatia polymorpha* and *Bellendena montana* collected in Tasmania. Those that have been tested show marginal cytotoxicity and/or antioxidant activity.

Naphthoquinones have been reported from the northern New South Wales *Stenocarpus salignus* (stenocarpoquinone B (**749**) and A (**750**)), and *Lomatia arborescens* (lomatiol (**751**) and juglone (**752**)) (Plate 16, Fig. 65) [293, 294]. A flavonoid glycoside, polifolisode (**753**), and the widespread rutin (**754**) have been reported from *Bellendena montana* and *G. robusta*, respectively [281, 292]. Several simple alkyl phenols (*p*-hydroxyacetophenone (**755**), methyl-(*p*-hydroxybenzoyl) acetate (**756**), 1-(4-hydroxyphenyl)propane-1,3-diol (**757**), and ethyl 3-hydroxy-3-(4-hydroxyphenyl)propanoate (**758**) and cinnamates (2,5-dihydroxycinnamic acid, (**759**), methyl-*p*-coumarate (**760**), graviquinone (**761**), methyl 5-ethoxy-2-hydroxycinnamate (**762**), 6-*O*-((*E*)-3-(2,5-dihydroxyphenyl)acrylyl)-glucose (**763**), and grevilloside Q (**764**)) and a glycoside of benzyl alcohol (icariside F2 (**765**)) have also been reported from either *B. montana*, *A. odorata*, *L. polymorpha*, or *G. robusta* [263, 282, 284, 290–292, 295]. *Cardwellia sublimis* also was found to contain high concentration levels of the inositol, (−)-borneitol (**766**) [278]. The triterpenoid carboxylic acid, (24*E*)-3*β*-hydroxy-7,24-euphadien-26-oic acid (**767**), from *Bleasdalea* bleasdalei, was found to have DNA polymerase inhibitory activity [296].

4.5 *Chemistry of Australian Rainforest Core Eudicotyledon Angiosperms*

All the remaining angiosperms in Australian rainforest are classified as core eudicots. Most species fall into two major clades, the Superasterids or the Superrosids, with

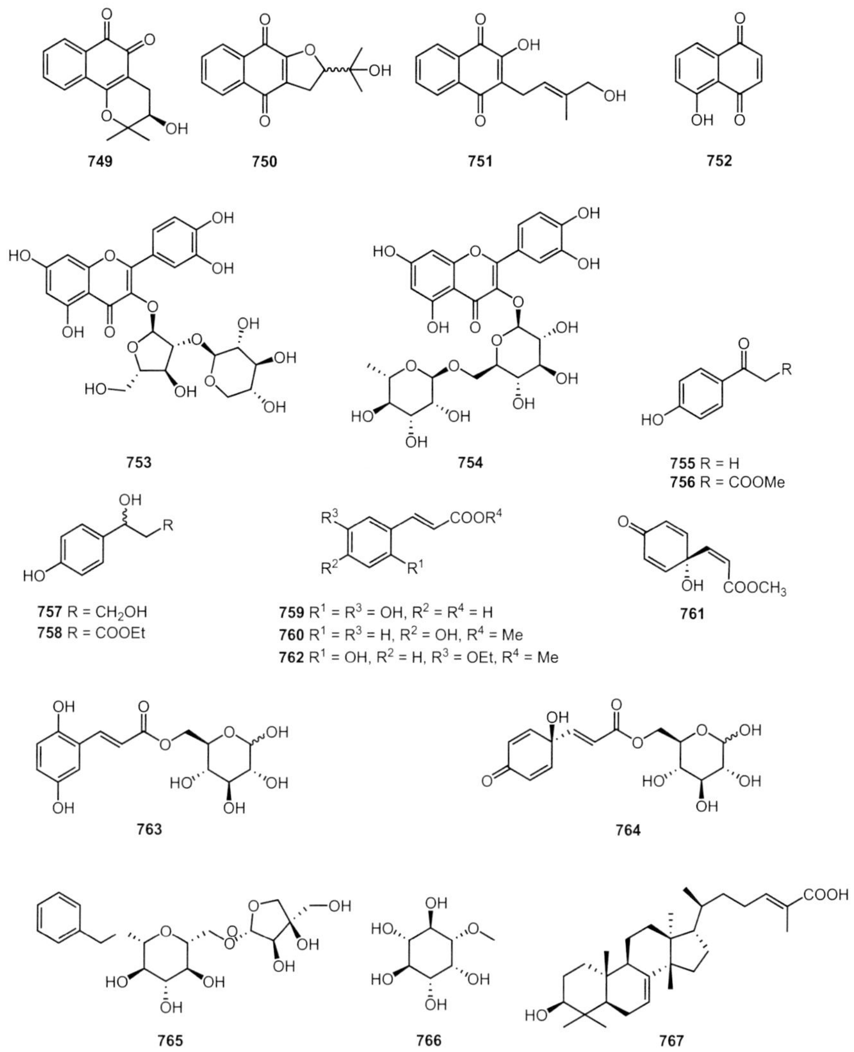

Fig. 65 Structures of compounds **749–767** reported from Proteaceae species

a very small number of species in the orders Caryophyllales and Santalales making up the rest.

4.5.1 Superasterids

The Superasterids are a monophyletic clade that encompasses 21 orders that are well represented in Australian rainforests.

Table 17 Distribution of natural products isolated from the Berberidopsodales and Cornales

Species	Collection location[a]	Plant part[b]	Compound class	Compound(s)	Refs.
Berberidopsidaceae					
Berberidopsis beckleri	NNSW	lf	Cyanogenic glycoside	**768**	[297]
Cornaceae					
Alangium villosum subsp. *polyosmoides*	FIJI	b	Triterpenoid	**778**, **779**	[299]
Alangium villosum subsp. *tomentosum*	SEQ	wp	Tubulosine alkaloid	**769–777**	[298]

[a] SEQ = Southeast Queensland, NNSW = Northern New South Wales, [b] lf = leaves, wp = whole plant, b = bark

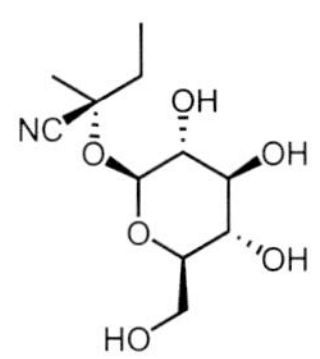

768

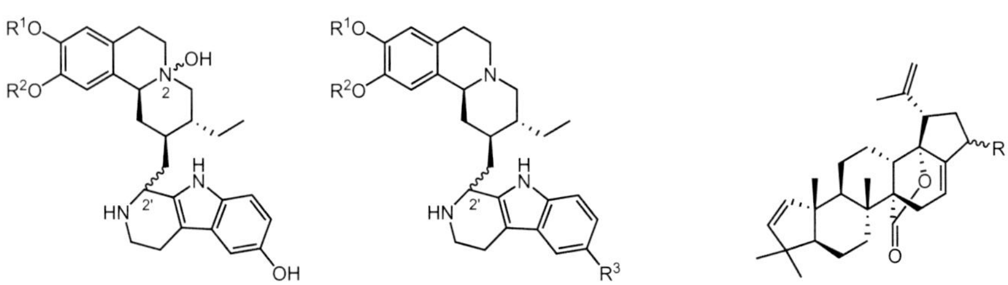

769 R^1 = Me, R^2 = H, (2*S*,2′*R*)
770 R^1 = Me, R^2 = H, (2*R*,2′*S*)
771 R^1 = R^2 = H, (2*S*,2′*R*)
772 R^1 = R^2 = H, (2*R*,2′*S*)

773 R^1 = Me, R^2 = H, R^3 = OH, (2′*R*)
774 R^1 = Me, R^2 = H, R^3 = OH, (2′*S*)
775 R^1 = R^2 = H, R^3 = OH, (2′*R*)
776 R^1 = R^2 = H, R^3 = OH, (2′*S*)
777 R^1 = R^2 = R^3 = H, (2′*R*)

778 R = H
779 R = OH

Fig. 66 Structures of compounds **768–779** reported from *Berberidopsis* and *Alangium* species

Berberidopsodales

This very small order is Southern Hemispheric containing only a handful of species in two families and three genera native to Chile and Australia. Leaves of the Australian *Berberidopsis beckleri* contain the cyanogenic glycoside lotaustralin (**768**) (Table 17, Fig. 66) [297].

Cornales

The Cornales are an early diverging order within the Superasterids and the family Cornaceae is considered to have a Northern Hemispheric origin. Two subspecies of *Alangium villosum* have been studied, *A. villosum* subsp. *polyosmoides* and *A. villosum* subsp. *tomentosum* (Table 17, Fig. 66). From Southeast Queensland collections of *A. villosum* subsp. *tomentosum*, nine tubulosine-type alkaloids (tubulosine *N*-5-oxide (**769**), isotubulosine *N*-5-oxide (**770**), 9-demethyltubulosine *N*-5-oxide (**771**), 9-demethylisotubulosine *N*-5-oxide (**772**), tubulosine (**773**), isotubulosine (**774**), 9-demethyltubulosine (**775**), 9-demethylisotubulosine (**776**), and deoxytubulosine (**777**)), were obtained [298]. These compounds are hybrids of monoterpenoid indole alkaloids and isoquinoline alkaloids. They are found in disparate plants from the Rubiaceae, Icacinaceae, and Cornaceae. Two A ring-contracted lupane triterpenoids, namely, 22-deoxyemmolactone (**778**) and emmolactone (**779**) were reported from *Alangium villosum* subsp. *polyosmoides* collected in Fiji [299].

Ericales

The Ericales contains 22 families including eight represented in Australian rainforests (Fig. 67). Only species from the families, Lecythidaceae, Ebenaceae, Primulaceae, Pentaphylaceae, and Ericaceae have been investigated chemically (Table 18). A chemotaxonomic meta-analysis of the Ericales has highlighted a high incidence of triterpenoid glycosides and their aglycones, and studies on Australian rainforest species follow this trend [300].

Polyoxygenated triterpenoid glycosides and their aglycones are characteristic molecules from the Lecythidaceae species *Planchonia careya* (collected in central Queensland) and *Barringtonia acutangula* (collected in the Kimberley, Western Australia) (Fig. 68). These saponins include 16-deoxybarringtogenol C (**780**), 16α,21β,22α,28-tetrahydroxyolean-12-en-3-one (**781**), barringtogenol C (**782**), hyptatic acid-A (**783**), 3β-*O*-*cis*-*p*-coumaroyltormentic acid (**784**), and 3β-*O*-*trans*-*p*-coumaroyltormentic acid (**785**) from *P. careya* and acutanguloside A (**786**), acutanguloside B (**787**), acutanguloside C (**788**), acutanguloside D (**789**), acutanguloside E (**790**), acutanguloside F (**791**), acutanguloside D methyl ester (**792**), acutanguloside E methyl ester (**793**), and acutanguloside F methyl ester (**794**) from *B. acutangula* [301–303]. *Barringtonia acutangula* is used traditionally to treat pain and the saponins that it contains are regarded as the active principles [304]. Ongoing research is being undertaken to understand their mode of action suggesting dual actions against inflammation and nerve injury. *Barringtonia acutangula* has also been extensively studied outside of Australia. In addition, the flavonols kaempferol 3-*O*-[α-rhamnopyranosyl(1 $\rightarrow$ 3)-(2-*O*-*p*-coumaroyl)]-β-glucopyranoside, 7-*O*-[α-rhamnopyranosyl-(1 $\rightarrow$ 3)-(4-*O*-*p*-coumaroyl)]-α-rhamnopyranoside (**795**), (+)-gallocatechin (**796**), and gallocatechin-(4α $\rightarrow$ 8)-gallocatechin (**797**) were reported from *P. careya* [302, 305].

Fig. 67 Dated molecular phylogeny of the Ericales [7] showing those families that have species present in Australian rainforests (grey) and studied (black) and the number of species occurring in Australian rainforest. Flowers of Australian rainforest Ericales species *Trochocarpa laurina* (Ericaceae, top left), *Diospyros australis* (Ebenaceae, bottom left), and *Barringtonia calyptrata* (Lecythidaceae, right)

Table 18 Distribution of natural products isolated from the Ericales

Species	Collection location[a]	Plant part(s)[b]	Compound class	Compound(s)	Refs.
Lecythidaceae					
Planchonia careya	NQ	lf	Flavonoid	**795–797**	[302, 305]
	NQ	b, lf	Triterpenoid	**780–785**	[301, 302]
Barringtonia acutangula	NWA	b	Triterpenoid	**786–794**	[303]
Primulaceae					
Tapeinosperma pseudojambosa	NQ	lf, b	Flavonoid	**804**	[307]
	NQ	lf, b	Triterpenoid	**803**, **799–801**	[307]
Pentaphylacaceae					
Ternstroemia cherryi	SEQ	b, lf, fr	Triterpenoid	**802**, **805–813**	[308–310]
Ericaceae					
Trochocarpa laurina	SEQ	lf	Iridoid	**814**	[311]
	SEQ	lf	Shikimate	**815**	[311]
Ebenaceae					
Diospyros hebecarpa	NQ	lf, fr	Naphthoquinone	**798**	[306]

[a] SEQ = Southeast Queensland, NQ = North Queensland, NWA = North Western Australia, [b] lf = leaves, fr = fruits, b = bark

The simple naphthoquinone, plumbagin (**798**) is the only constituent reported from the Ebenaceae species *Diospyros hebecarpa* [306]. The Primulaceae species *Tapeinosperma pseudojambosa* contains six triterpenoids including **590**, primulagenin A (**799**), echinocystic acid, (**800**) bauerenol (**801**), betulinic acid (**802**) and the widespread compound, α-amyrin (**803**) [307]. The common flavonoid quercitrin (**804**) was also reported [307]. The leaves and fruits of the Pentaphylacaceae species *Ternstroemia cherryi* from north Queensland contains nine polyoxygenated triterpenoid saponins (ternstroenol A (**805**), ternstroenol B (**806**), ternstroenol C (**807**), ternstroenol D (**808**), ternstroenol E (**809**), 3-*O*-[α-L-rhamnopyranosyl-(1 → 2)]-[*β*-D-galactopyranosyl-(1 → 3)]-[*β*-D-glucopyranosyl-(1 → 2)]-[*β*-D-glucuronopyranosyl]-22-*O*-[2*E*,4*E*]-octa-2,4-dienoyl-barrigenol-R1 (**810**), ternstroenol F (**811**), sanchakasaponin G (**812**), and yuchasaponin A (**813**)), and showed homology to the *Barringtonia* saponins [308–310] (Fig. 69). They differ most notably by the presence of polyunsaturated fatty acid esters attached to the E ring. These compounds show potent anti-inflammatory, moderate cytotoxic and TNFα down-regulation activities. The Ericaceae species *Trochocarpa laurina* contains the iridoid glycoside schismoside (**814**) and the furanone glycoside trochocarposide (**815**) [311].

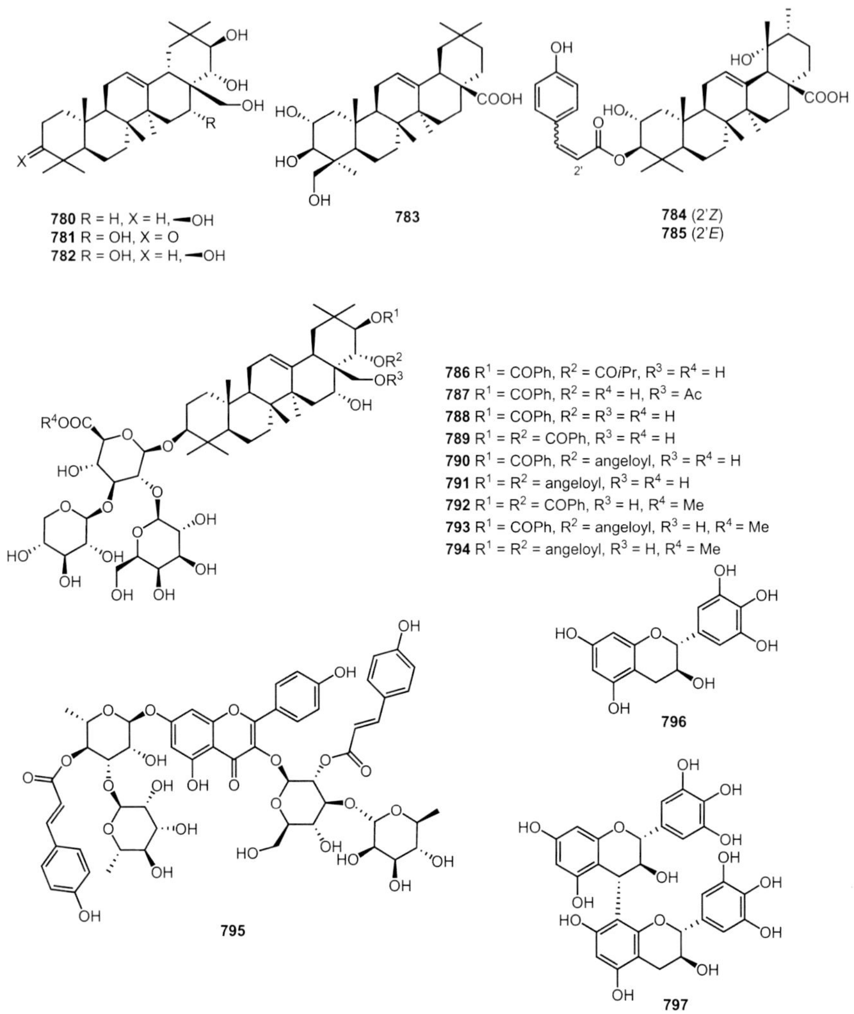

Fig. 68 Structures of compounds **780–797** reported from Lecythidaceae species

Campanulids

The Campanulids represent a major sub-clade within the Superasterids and includes 12 of its orders.

Escallionales

The order Escallionales, containing only one family, Escallionaceae, is almost exclusively Southern Hemispheric, indicative of its Gondwanan origins [312]. In

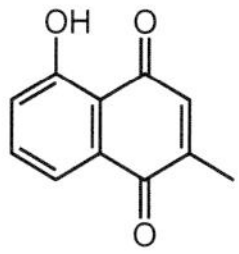

798

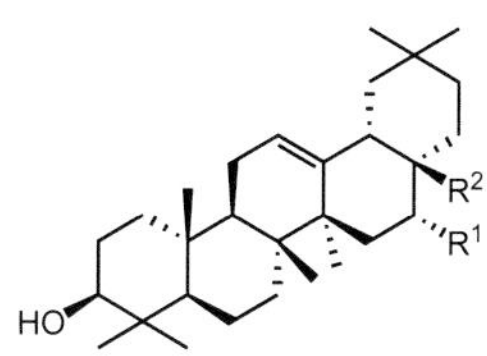

799 R^1 = OH, R^2 = CH_2OH
800 R^1 = OH, R^2 = COOH

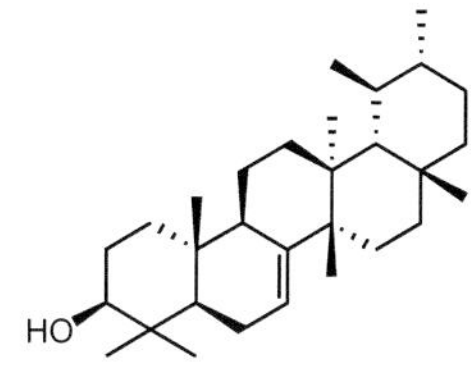

801

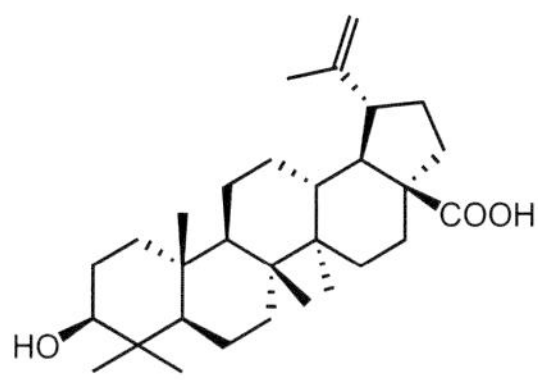

802

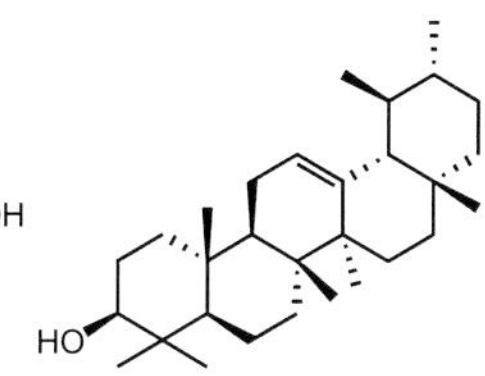

803

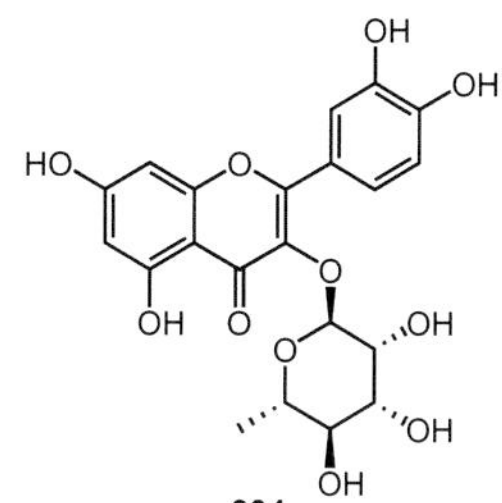

804

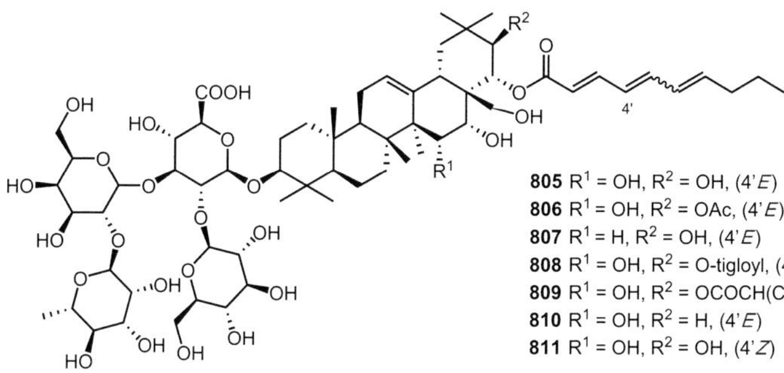

805 R^1 = OH, R^2 = OH, (4'*E*)
806 R^1 = OH, R^2 = OAc, (4'*E*)
807 R^1 = H, R^2 = OH, (4'*E*)
808 R^1 = OH, R^2 = O-tigloyl, (4'*E*)
809 R^1 = OH, R^2 = $OCOCH(CH_3)CH_2CH_3$, (4'*E*)
810 R^1 = OH, R^2 = H, (4'*E*)
811 R^1 = OH, R^2 = OH, (4'*Z*)

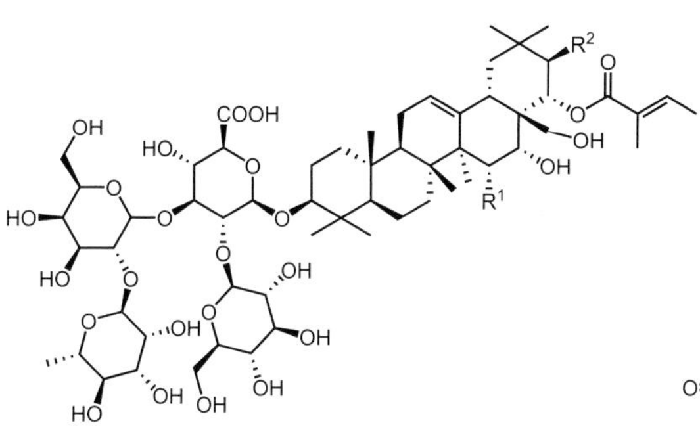

812 R^1 = H, R^2 = H
813 R^1 = OH, R^2 = O-tigloyl

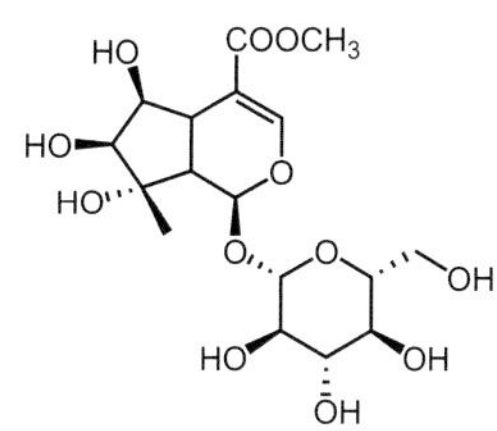

814

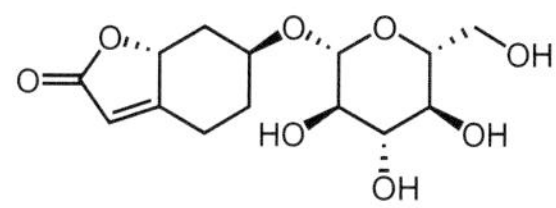

815

Fig. 69 Structures of compounds **798–815** reported from Ebenaceae, Primulaceae, Pentaphylacaceae, and Ericaceae species

Australian rainforests, two species of *Anopterus* have been studied (Plate 18), one from Tasmania and the other from northern New South Wales/southeastern Queensland (Table 19).

Thirteen highly cytotoxic diterpenoid alkaloids have been reported (Fig. 70) (6*α*-acetoxyanopterine (**816**), 4*α*-hydroxy-6*α*-acetoxyanopterine (**817**), 4*α*-hydroxyanopterine (**818**), anopterine (**819**), 7*β*-hydroxyanopterine (**820**), 7*α*,4*β*-dihydroxyanopterine (**821**), 11*α*-benzoylanopterine (**822**), 7*β*-hydroxy-11*α*-benzoylanopterine (**823**), anopteryl 11*α*-4′-hydroxybenzoate 12*α*-tiglate (**824**), 7*β*-hydroxyanopteryl 11*α*-4′-hydroxybenzoate 12*α*-tiglate (**825**), anopteryl *12α*-tiglate (**826**), anopterimine (**827**), and anopterimine *N*-oxide (**828**)) [313–319]. These compounds possess a unique skeleton derived from *ent*-kaurane diterpenoids. Related *ent*-kaurane diterpenoid alkaloids have been isolated from unrelated Garryaceae and Ranunculaceae species, but none of these compounds has a carbon–carbon bond between C-14 and C-20. The *Anopterus* alkaloids also contain more hydroxy and esters substituents compared to the related compounds. The cytotoxicity of the *Anopterus* alkaloids is attributed to their ability to disrupt mitosis through the generation of mitotic spindle defects that results in asymmetric cell division and apoptosis in PC-3 and LNCaP prostate cancer, and HeLa cervical cancer cells [314].

Plate 18 *Anopterus macleaylanus* flowers and fruit. (Photographs AR Carroll)

Table 19 Distribution of natural products isolated from the Escallionaceae

Species	Collection location[a]	Plant parts[b]	Compound class	Compounds	Refs.
Anopterus glandulosus	Tas	lf, tw	Anopterine alkaloid	**819–821**, **823**, **826**	[313]
Anopterus macleayanus	NNSW	lf, tw, b	Anopterine alkaloid	**816–828**	[314–319]

[a] = northern New South Wales, Tas = Tasmania, [b] lf = leaves, tw = twigs, b = bark

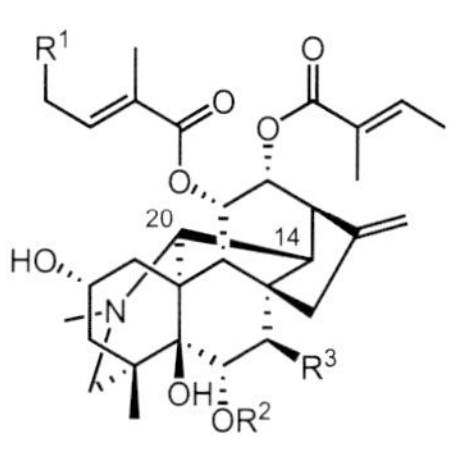

816 $R^1 = R^3 = H$, $R^2 = Ac$
817 $R^1 = OH$, $R^2 = Ac$, $R^3 = H$
818 $R^1 = OH$, $R^2 = R^3 = H$
819 $R^1 = R^2 = R^3 = H$
820 $R^1 = R^2 = H$, $R^3 = OH$
821 $R^1 = R^3 = OH$, $R^2 = H$

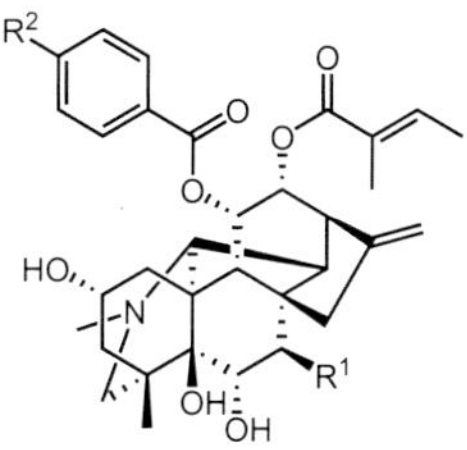

822 $R^1 = R^2 = H$
823 $R^1 = OH$, $R^2 = H$
824 $R^1 = H$, $R^2 = OH$
825 $R^1 = R^2 = OH$

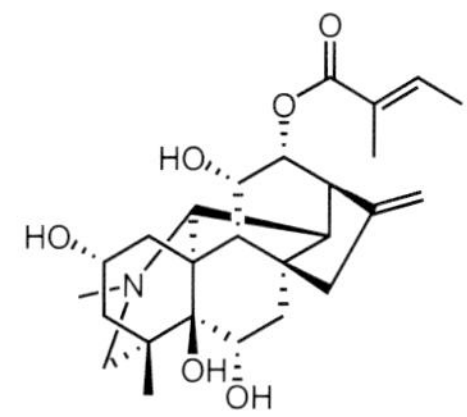

826

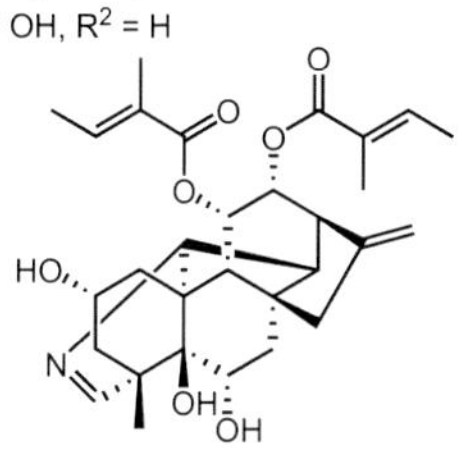

827

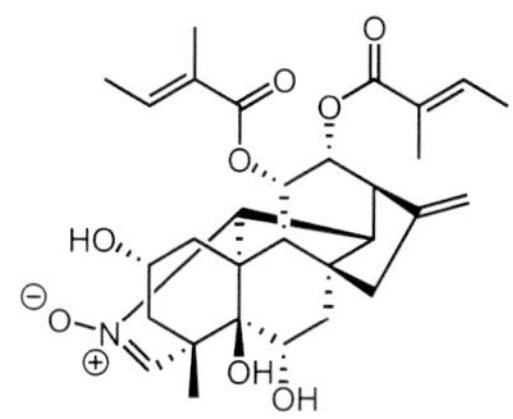

828

Fig. 70 Structures of alkaloids **816–828** reported from *Anopterus* species

Apiales

The Apiales contains seven families (Fig. 71) and five have species in Australian rainforests. Species from the Pittosporaceae, Araliaceae, and Apiaceae families have been investigated chemically (Table 20).

Pittosporaceae

Pittosporum angustifolium has yielded a series of triterpenoid saponins (Fig. 72) (pittangretosides N–P (**829–831**), 21*β*-angeloyloxy-22*α*-acetoxy-3*β*-[*β*-D-glucopyranosyl-1 → 2)]-[*α*-L-arabinopyranosyl-(1 → 3)]-[*α*-L-arabinofuranosyl-(1 → 4)]-*β*-D-glucuronopyranosyloxyolean-12-ene-15*α*,16*α*,28-triol (**832**), 22*α*-(2-methylbutyroyloxy)-3*β*-[*β*-D-glucopyranosyl-(1 → 2)]-[*α*-L-arabinopyranosyl-(1 → 3)]-[*α*-L-arabinofuranosyl-(1 → 4)]-*β*-D-glucuronopyranosyloxyolean-12-ene-15*α*,16*α*,28-triol (**833**), pittangretoside A (**834**), 21*β*-angeloyloxy-22*α*-angeloyloxy-3*β*-[*β*-D-glucopyranosyl-(1 → 2)]-[*α*-L-arabinopyranosyl-(1 → 3)]-[*α*-L-arabinofuranosyl-(1 → 4)]-*β*-D-glucuronopyranosyloxyolean-12-ene-15*α*,16*α*,28-triol (**835**), pittangretoside J (**836**), pittangretoside K (**837**), pittangretoside M (**838**), pittangretoside Q (**839**), pittangretoside B (**840**), pittangretosides R–Z (**841–849**),

Fig. 71 Dated molecular phylogeny of the Apiales [7] showing those families that have species present in Australian rainforests (grey) and studied (black) and the number of species occurring in Australian rainforests. The umbrella tree *Schefflera actinophylla* (top) and flowers of Australian rainforest *Pittosporum* species *P. undulatum* (bottom left) and *P. revolutum* (bottom right). (Photographs AR Carroll)

pittangretoside A1 (**850**), pittangretoside B1 (**851**), pittangretoside E (**852**), pittangretoside F (**853**), pittangretoside L (**854**), pittangretoside C1 (**855**), pittangretoside C (**856**), pittangretoside D (**857**), and pittangretosides G–I (**858–860**)) related to the Lecythraceae and Pentaphylacaceae compounds [320–322]. Some of these saponins are weakly and non-selectively cytotoxic to cancerous and non-cancerous cell lines.

Glycosylated flavonoids (Fig. 73) **754**, **861**, and 6″-(3-hydroxy-3-methylglutaroyl)-isoquercitrin (**862**), and dihydrocaffeic acid esters of shikimic acid (3,4-dicaffeoylquinic acid (**863**), and 4,5-dicaffeoyquinic acid (**864**)) were also reported from *P. augustifolium* [323]. *Pittosporum venulosum* yielded venuloside B (**865**), venuloside A (**866**), venuloside C (**867**), and venuloside D (**868**) that are monoterpenoids conjugated to fucose esters. These compounds were discovered as part of a function-based high-throughput (HTS) assay to find inhibitors of L-configurated amino acid transporters implicated in anti-prostate cancer therapy [324, 325].

Table 20 Distribution of natural products isolated from the Apiales

Species	Collection location[a,b]	Plant part(s)[c]	Compound class	Compound(s)	Refs.
Pittosporaceae					
Pittosporum angustifolium	CQ	lf	Flavonoid	**527**, **754**, **862**	[323]
	CQ	lf	Shikimate	**863**, **864**	[323]
	CQ	lf, s	Triterpenoid	**829–860**	[320–322]
Pittosporum venulosum	NQ	lf	Monoterpenoid glycoside	**865–868**	[324, 325]
Araliaceae					
Polyscias australiana	NQ	lf	Cyanohydrin	**869**, **870**	[326]
Polyscias murrayi	SEQ	fl	Cinnamate	**871**, **873**, **874**	[327]
	SEQ	fl	Shikimate	**51**, **872**	[327]
Schefflera actinophylla	NQ (cult. (Egypt))	lf	Triterpenoid	**875–890**	[328]
Apiaceae					
Mackinlaya macrosciadea	NQ	lf	Pyridine	**895**	[332]
	NQ	lf	Quinolone	**891–894**	[330–332]

[a] SEQ = Southeast Queensland, NQ = North Queensland, CQ = central Queensland, cult. = cultivated, [b] natural location shown for those species studied from cultivated material, [c] lf = leaves, s = seeds

Araliaceae

Two cyanogenic glycosides, proteacin (**869**) and dhurrin (**870**) from *Polyscias australiana*, five dihydrocinnamates (**51**, *N*,*N*,*N*-trimethylethyl-1-aminium *p*-hydroxycinnamate (**871**), 3-(3,4-dihydroxyphenyl)propanoic acid (**872**), 3,4-di-*O*-3-(4-hydroxyphenyl) propionyl-1,5-dihydroxycyclohexanecarboxylic acid (**873**), and 3,5-di-*O*-3-(4-hydroxyphenyl)propionyl-1,4-dihydroxycyclohexanecarboxylic acid (**874**)) from *P. murrayi*, and 16 triterpenoid saponins and aglycones (3*β*-hydroxylup-20(29)-en-28-oic acid 3-*O*-*β*-D-glucuronopyranoside (**875**), 3*α*-hydroxylup-20(29)-en-30-ol-23,28-dioic acid (**876**), 3*α*-hydroxylup-20(29)-en-23,28-dioic acid 3-*O*-*β*-D-glucopyranoside (**877**), 3*α*-hydroxylup-20(29)-ene-23,28-dioic acid 23-*O*-*β*-D-glucopyranosyl ester (**878**), 3*α*-hydroxylup-20(29)-ene-23,28-dioic acid (**879**), 3-*epi*-betulinic acid 3-*O*-sulfate (**880**), 3-*epi*-betulinic acid 3-*O*-*β*-D-glucopyranoside (**881**), 3*α*-hydroxylup-20(29)-ene-23,28-dioic acid 28-*O*-[4″-*O*-*α*-L-rhamnopyranosyl-6′-*O*-*β*-D-glucopyranosyl]-*β*-D-glucopyranosyl ester (**882**), 3*α*,23-dihydroxylup-20(29)-en-28-oic acid 28-*O*-[4″-*O*-*α*-L-rhamnopyranosyl-6′-*O*-*β*-D-glucopyranosyl]-*β*-D-glucopyranosyl ester (**883**), 3-*epi*-betulinic acid 3-*O*-sulphate, 28-*O*-[4″-*O*-*α*-L-rhamnopyranosyl-6′-*O*-*β*-D-glucopyranosyl]-*β*-D-glucopyranosyl ester (**884**), 3-*epi*-betulinic acid 28-*O*-[4″-*O*-*α*-L-rhamnopyranosyl-6′-*O*-*β*-D-glucopyranosyl]-*β*-D-glucopyranosyl ester (**885**),

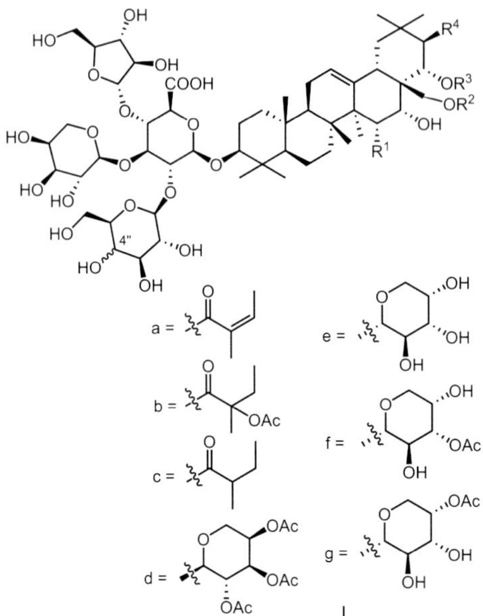

829 R^1 = OH, R^2 = H, R^3 = a, R^4 = OAc, (4''*S*)
830 R^1 = OH, R^2 = H, R^3 = b, R^4 = OAc, (4''*S*)
831 R^1 = OH, R^2 = H, R^3 = Ac, R^4 = Oc, (4''*S*)
832 R^1 = OH, R^2 = H, R^3 = Ac, R^4 = Oa, (4''*S*)
833 R^1 = OH, R^2 = R^4 = H, R^3 = c, (4''*S*)
834 R^1 = OH, R^2 = R^4 = H, R^3 = a, (4''*S*)
835 R^1 = OH, R^2 = H, R^3 = a, R^4 = Oa, (4''*S*)
836 R^1 = OH, R^2 = R^3 = R^4 = H, (4''*R*)
837 R^1 = OH, R^2 = R^3 = R^4 = H, (4''*S*)
838 R^1 = R^4 = OH, R^2 = H, R^3 = d, (4''*S*)
839 R^1 = R^2 = R^4 = H, R^3 = a, (4''*R*)
840 R^1 = OH, R^2 = R^4 = H, R^3 = a, (4''*R*)
841 R^1 = R^4 = OH, R^2 = H, R^3 = e, (4''*S*)
842 R^1 = OH, R^2 = H, R^3 = a, R^4 = Oa, (4''*R*)
843 R^1 = R^2 = H, R^3 = a, R^4 = Oa, (4''*R*)
844 R^1 = R^4 = OH, R^2 = H, R^3 = d, (4''*R*)
845 R^1 = R^4 = OH, R^2 = H, R^3 = f, (4''*S*)
846 R^1 = R^4 = OH, R^2 = H, R^3 = g, (4''*S*)
847 R^1 = OH, R^2 = R^4 = H, R^3 = d, (4''*R*)
848 R^1 = OH, R^2 = R^4 = H, R^3 = d, (4''*S*)
849 R^1 = OH, R^2 = R^3 = H, R^4 = Oa, (4''*S*)
850 R^1 = R^2 = H, R^3 = Ac, R^4 = Oa, (4''*R*)
851 R^1 = R^2 = H, R^3 = Ac, R^4 = Oa, (4''*S*)
852 R^1 = OH, R^2 = a, R^3 = R^4 = H, (4''*R*)
853 R^1 = OH, R^2 = a, R^3 = R^4 = H,(4''*S*)

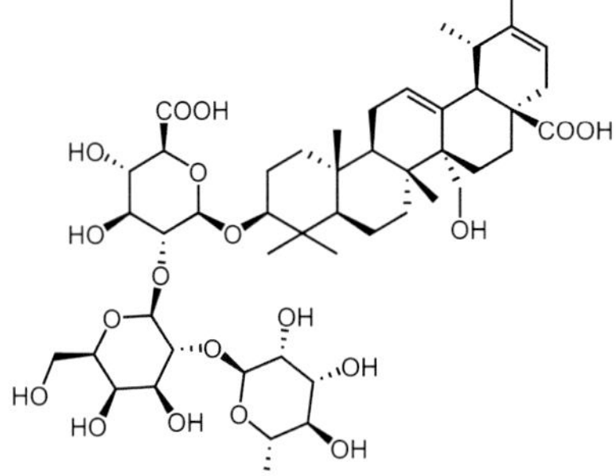

854

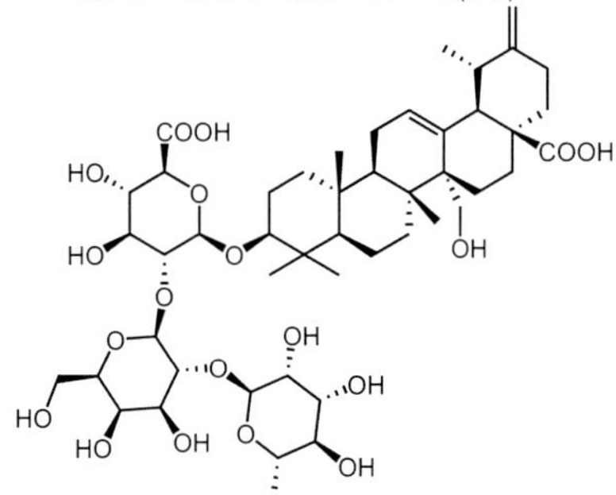

855

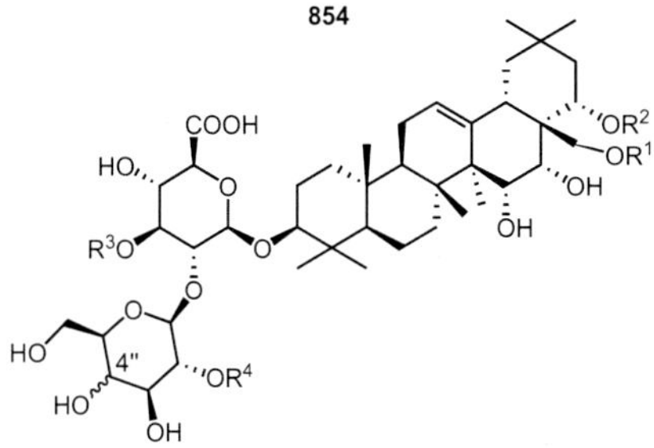

856 R^1 = R^4 = H, R^2 = a, R^3 = e, (4''*R*)
857 R^1 = R^3 = H, R^2 = a, R^4 = *β*-L-rham, (4''*R*)
858 R^1 = a, R^2 = R^4 = H, R^3 = e, (4''*S*)

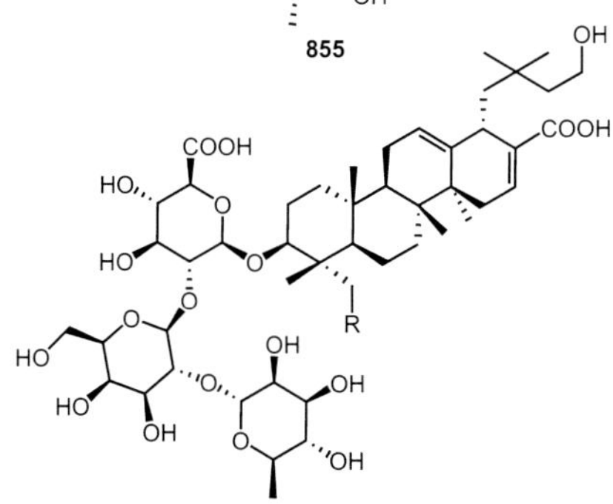

859 R = H
860 R = OH

Fig. 72 Structures of triterpenoid saponins **829**–**860** reported from *Pittosporum angustifolium*

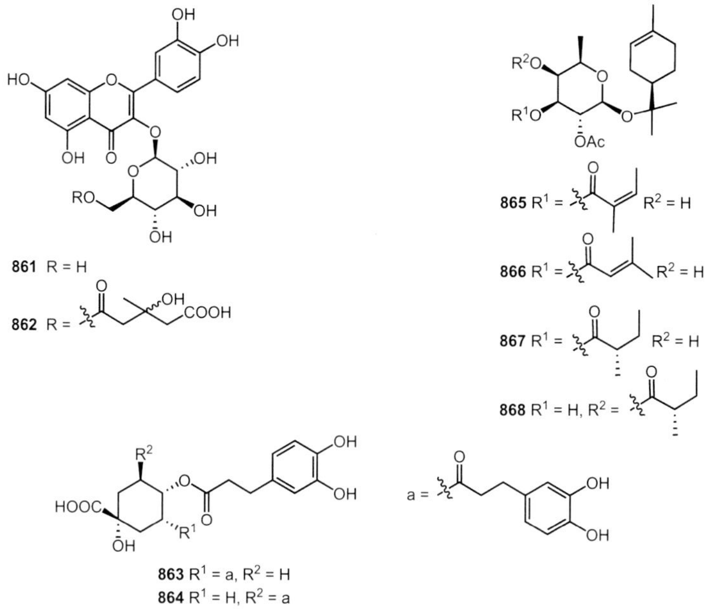

Fig. 73 Structures of compounds **861**–**868** reported from *Pittosporum* spp.

3-*epi*-betulinic acid 3-*O*-β-D-glucopyranoside, 28-*O*-[4″-*O*-α-L-rhamnopyranosyl-6′-*O*-β-D-glucopyranosyl]-β-D-glucopyranosyl ester (**886**), 3-*epi*-betulinic acid 3-*O*-β-D-6′-acetylglucopyranoside, 28-*O*-[4″-*O*-α-L-rhamnopyranosyl-6′-*O*-β-D-glucopyranosyl]-β-D-glucopyranosyl ester (**887**), 3β-hydroxyurs-12-en-28-oic acid 3-*O*-[β-D-galactopyranosyl-(1 → 2′)]-β-D-glucuronopyranoside (**888**), 3β-hydroxyurs-12-en-28-oic acid 3-*O*-β-D-glucuronopyranoside-6′-*O*-methyl ester (**889**), and 3β-hydroxyurs-12-en-28-oic acid 3-*O*-β-D-glucuronopyranoside 6′-*O*-butyl ester (**890**)) from *Schefflera actinophylla* cultivated in Japan are the only compounds reported from Australian rainforest Araliaceae species (Fig. 74) [326–328]. This family has an Asian origin with the likely recent migration of species into Australia.

Apiaceae

The carrot family, Apiaceae, is widely distributed in the Northern Hemisphere, but molecular studies indicate that it evolved in Australasia about 65 mya [329]. Much of the extant Australian Apiaceae is found outside of rainforests in grassland, woodland and alpine areas. The Apiaceae is represented in Australian rainforests by four species in the basal subfamily Mackinlayoideae, but only *Mackinlaya macrosciadea* has been investigated chemically (Fig. 75). It contains a series of quinoline

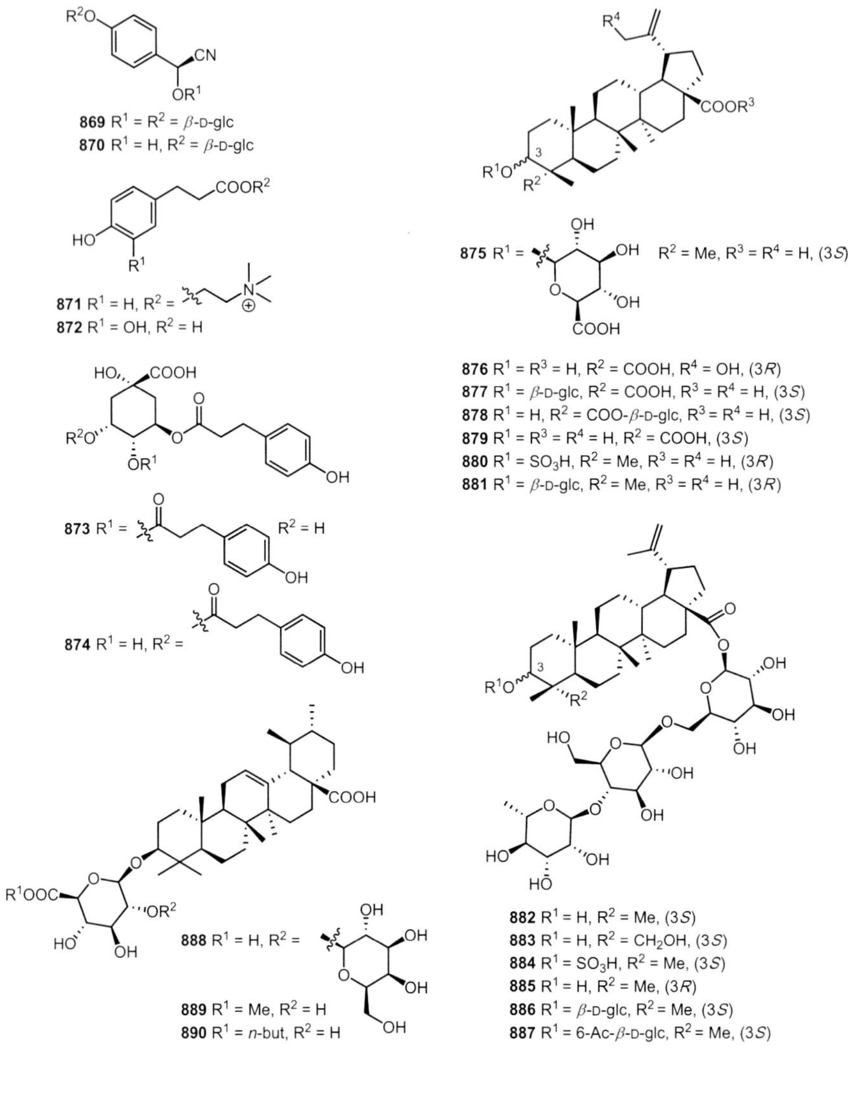

Fig. 74 Structures of compounds **869–890** reported from *Polyscias* species and *Schefflera actinophylla*

alkaloids, 3,4,5,6,7,8-hexahydro-1,7-benzodiazecin-2(1*H*)-one (**891**), deoxyvasicinone (**892**), 6,7,8,9-tetrahydropyrido[2,1-*b*]-quinazolin-11-one (**893**), and 6,7,8,9-tetrahydro-11*H*-pyrido[2,1-*b*]quinazoline (**894**) as well as the nicotine analogue anabasine (**895**) [330–332].

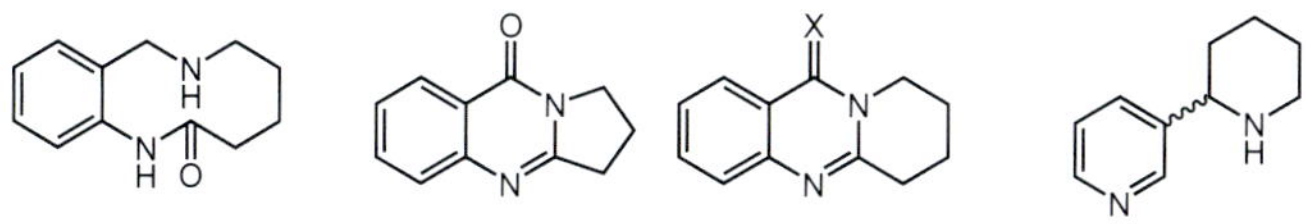

Fig. 75 Structures of alkaloids **891–895** reported from *Mackinlaya macrosciadea*

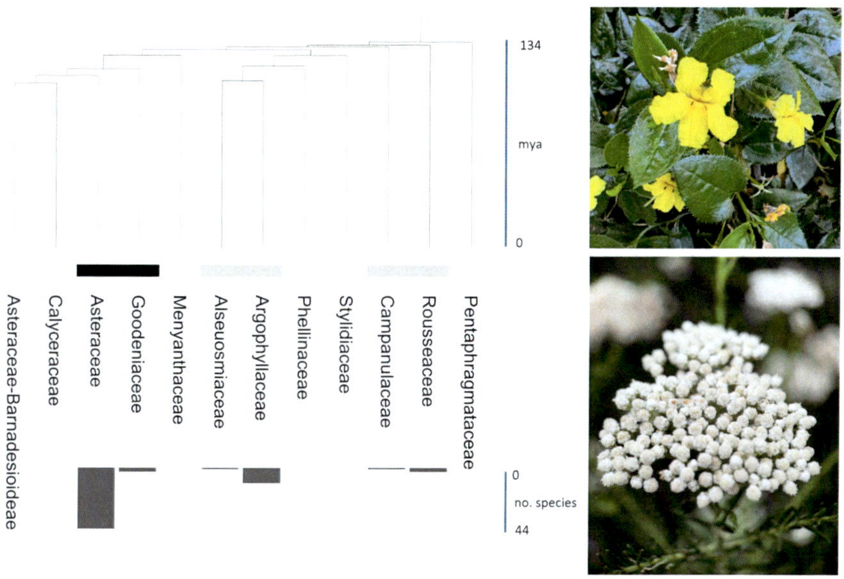

Fig. 76 Dated molecular phylogeny of the Asterales [7] showing those families that have species present in Australian rainforests (grey) and studied (black) and the number of species occurring in Australian rainforests. Flowers of *Goodenia ovata* (top) and *Ozothamnus diosmifolius* (bottom) (Photographs AR Carroll)

Asterales

The Asterales contain a huge diversity of genera and species worldwide (~14% of all angiosperms). The Asterales is Gondwanan in origin [333]. Although genus and species diversity within the order is high in Australia, only one species from the Campunulaceae, two Rousseaceae, 11 Argrophylaceae, one Alseuosmiaceae, three Goodeniaceae, and 98 species in 50 genera of Asteraceae are Australian rainforest natives (Fig. 76).

Only species from the Goodeniaceae and Asteraceae have been studied phytochemically (Table 21, Fig. 77). *Goodenia ovata* (Goodeniaceae) contains the widespread triterpenoid ursolic acid (**896**) [334]. Of the seven Asteraceae species studied, the leaves of *Bedfordia arborescens* contain seven sesquiterpenoid furanones, (guaianolide (**897**), 8-*epi* inuviscolide

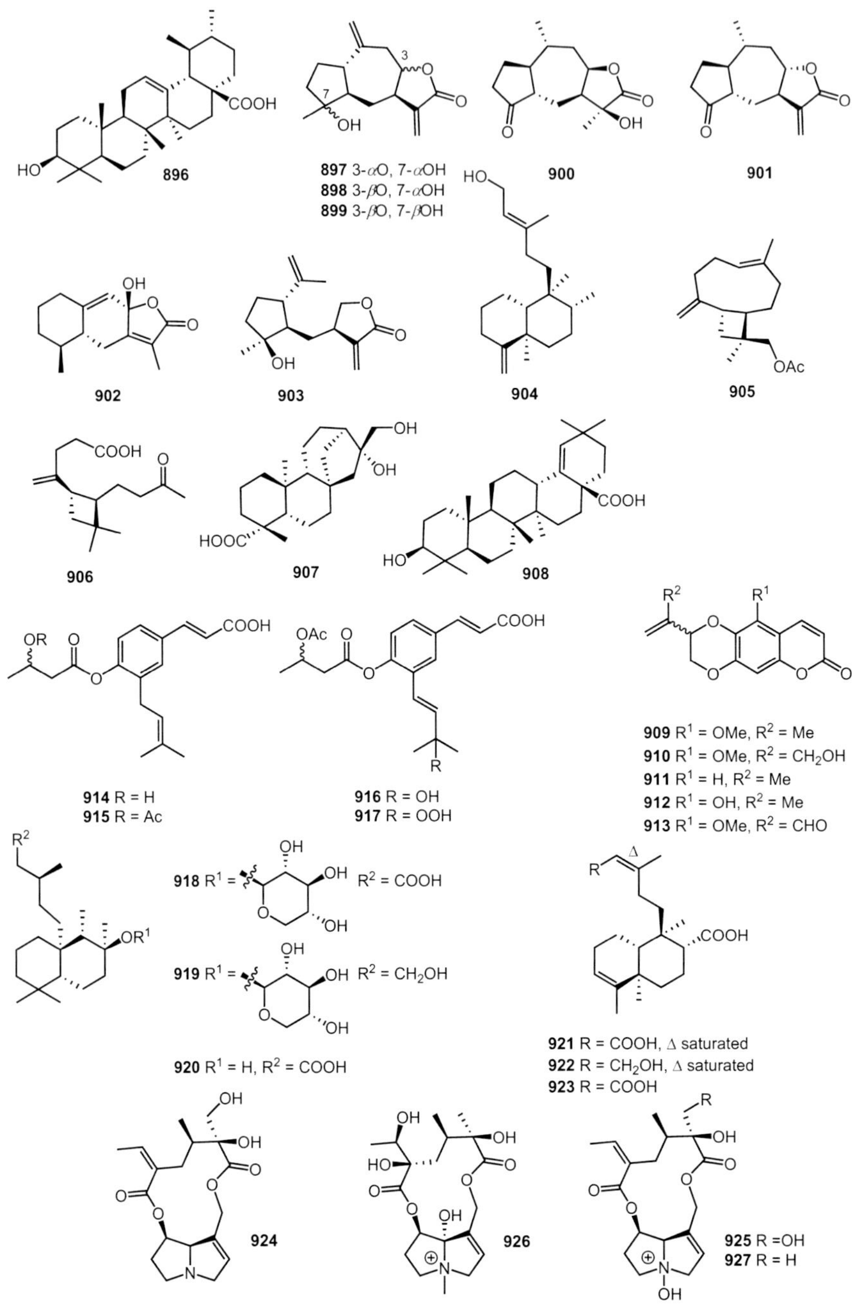

Fig. 77 Structures of compounds **896–927** reported from Asterales species

Table 21 Distribution of natural products isolated from the Asterales

Species	Collection location[a,b]	Plant part[c]	Compound class	Compound(s)	Refs.
Goodeniaceae					
Goodenia ovata	WA	lf	Triterpenoid	**896**	[334]
Asteraceae					
Bedfordia arborescens	ACT	lf	Diterpenoid	**904**	[335]
	ACT	lf	Sesquiterpenoid	**897–903**	[335]
Erechtites valerianifolius	NQ	r	Pyrrolizidine	**926, 927**	[39]
Gynura drymophila var. *drymophila*	CQ	r	BIA	**382**	[39]
Olearia fulgens	NNSW	wp	Diterpenoid	**918–923**	[338]
Ozothamnus diosmifolius	SEQ	lf	BIA	**382**	[39]
	ACT	wp	Cinnamate	**914–917**	[336]
	ACT	wp	Coumarin	**909–913**	[336]
	NSW	wp	Diterpenoid	**907**	[337]
	ACT	wp	Sesquiterpenoid	**905, 906**	[336]
	NSW	wp	Triterpenoid	**896, 908**	[337]
Senecio bipinnatisectus	NQ (cult. (New Zealand))	wp	Pyrrolizidine	**924**	[339]
	NQ	fl	Pyrrolizidine	**925**	[39]
Tithonia diversifolia	NQ	s	BIA	**382**	[39]

[a] SEQ = Southeast Queensland, NQ = North Queensland, CQ = central Queensland, NNSW = Northern New South Wales, NSW = New South Wales, ACT = Australian Capital Territory, WA = Western Australia, cult. = cultivated, [b] natural location shown for those species studied from cultivated material, [c] lf = leaves, s = seeds, r = roots, fl = flowers, wp = whole plant

(**898**), 4,8-bis-*epi*-inuviscolide (**899**), 11-β-hydroxy-11,13-dihydro-8-*epi*-confertin (**900**), 2,3-dihydroaromaticin (**901**), eremophilanolide (**902**), and bedfordiolide (**903**)) and a clerodane diterpenoid (*E*)-3-methyl-5-((1*S*,2*R*,4a*R*,8a*R*)-1,2,4a-trimethyl-5-methylenedecahydronaphthalen-1-yl)pent-2-en-1-ol (**904**) [335]. *Ozothamnus diosmifolius* has yielded cyclobutyl-containing sesquiterpenoids (13-acetoxycaryophellene (**905**), and methyl-4-oxo-seco-caryophyllene-5-oate (**906**)), the kaurene diterpenoid 16,17-dihydroxy-16β-(−)-kauran-19-oic acid (**907**), the triterpenoid carboxylic acids **896** and morolic acid (**908**), coumarins (5-methoxyobliquin (**909**), 5-methoxy-13-hydroyxobliquin (**910**), obliquin (**911**), 5-hydroxyobliquin (**912**), and 5-methoxy-13-*oxo*-obliquin (**913**)), isoprenylcinnamic acid derivatives (methyl-3-prenyl-4-*O*-[β-hydroxybutryl]-coumarate (**914**), methyl-3-prenyl-4-*O*-[*β*-acetoxybutryl]-coumarate (**915**), methyl-3-[3-hydroxy-3,3-dimethylprop-1-enyl]-4-*O*-[*β*-acetoxybutryl]-coumarate (**916**), and methyl-3-[3-peroxy-3,3-dimethylprop-1-enyl]-4-*O*-[*β*-acetoxybutryl]-coumarate (**917**)), and the benzyl isoquinoline alkaloid **382** [258, 336, 337]. *Olearia fulgens* contains *ent*-labdane and clerodane diterpenoids (8-*O*-*β*-D-xylopyranosyl-*ent*-8*β*-hydroxylabdan-15-oic acid (**918**), 8-*O*-*β*-D-xylopyranosyl-*ent*-labdan-8*β*,15-diol (**919**), *ent*-8*β*-hydroxylabdan-15-oic acid (**920**), *ent*-clerod-3-en-15,17-dioic acid (**921**), *ent*-15-hydroxyclerod-3-en-17-oic acid (**922**), and *ent*-cleroda-3,13-dien-15,17-dioic acid, (**923**)) [338]. Pyrrolizidine alkaloids have been reported from *Senecio bipinnatisectus* (retrorsine (**924**) and retrorsine-*N*-oxide (**925**)) [39, 339], and *Erechtites valerianifolius* (onetine (**926**) and deoxyretrorsine-*N*-oxide (**927**)) [39]. *Tithonia diversifolia* and *Gynura drymophila* var. *drymophila* both contain the benzyl isoquinoline alkaloid oblongine (**382**) [39].

Lamiids

The second major sub-clade of the Superasterids is the Lamiids and it includes five orders. This sub-clade includes the Orders Gentianales and Lamiales and includes many species that produce alkaloids or terpenoids, respectively.

Solanales

The order Solanales has a Gondwanan origin [340]. The Solanaceae is mainly represented by *Solanum* species in Australian rainforest and this group are proposed to have dispersed to Australia from Africa via Asia [341]. By contrast, the earliest fossil records of the Convolvulaceae are from India suggesting an east Gondwanan origin [342] (Fig. 78).

Only two of the 39 Convolvulaceae species from Australian rainforests, *Ipomoea muelleri* and *I. velutina*, have been studied (Table 22) and both contain calystegine-type tropane alkaloids (Fig. 79) (ipvelutine (**928**), calystegine A3 (**929**), calystegine B1 (**930**), calystegine B2 (**931**), calystegine C1 (**932**), calystegine B3 (**933**), and calystegine B5 (**934**)), from whole plant material [343, 344]. *Ipomoea muelleri* also produces diterpenoid indole alkaloids (11-hydroxy-12,13-epoxyterpendole K (**935**), terpendole E (**936**), paspaline (**937**), and terpendole C (**938**)), in the leaf and seed material, and ergoline-type indole alkaloids (9,10-didehydro-*N*-[(1*S*)-2-hydroxy-1-methylethyl]-6-methyl-(8)-ergoline-8-carboxamide (**939**), 9,10-didehydro-8-hydroxy-6-methyl-(8)-ergoline-8-methanol (**940**), (8)-9,10-didehydro-6-methyl-ergoline-8-methanol (**941**), (8)-9,10-didehydro-6-methyl-ergoline-8-carboxamide

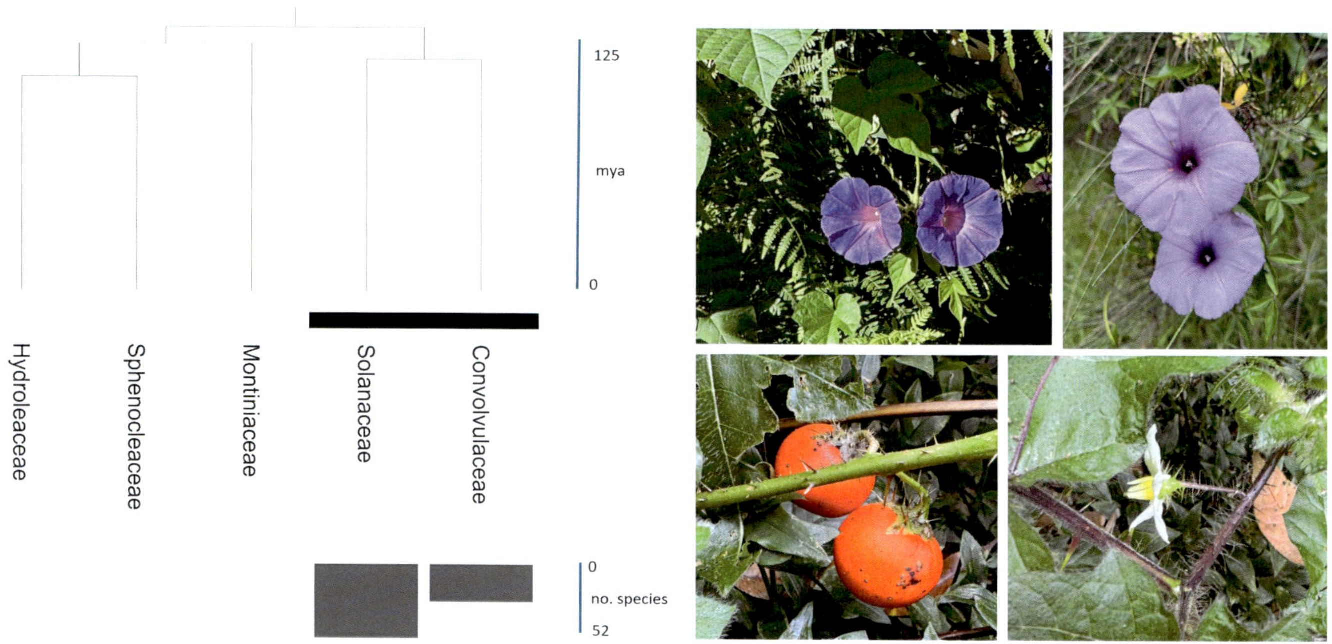

Fig. 78 Dated molecular phylogeny of the Solanales [7] showing those families that have species present in Australian rainforests (grey) and studied (black), and the number of species occurring in Australian rainforests. Flowers of the naturalized species *Ipomoea indica* (top left), *I. cairica* (top right) and flowers and fruits of the naturalized species *Solanum capsicoides* (bottom). (Photographs AR Carroll)

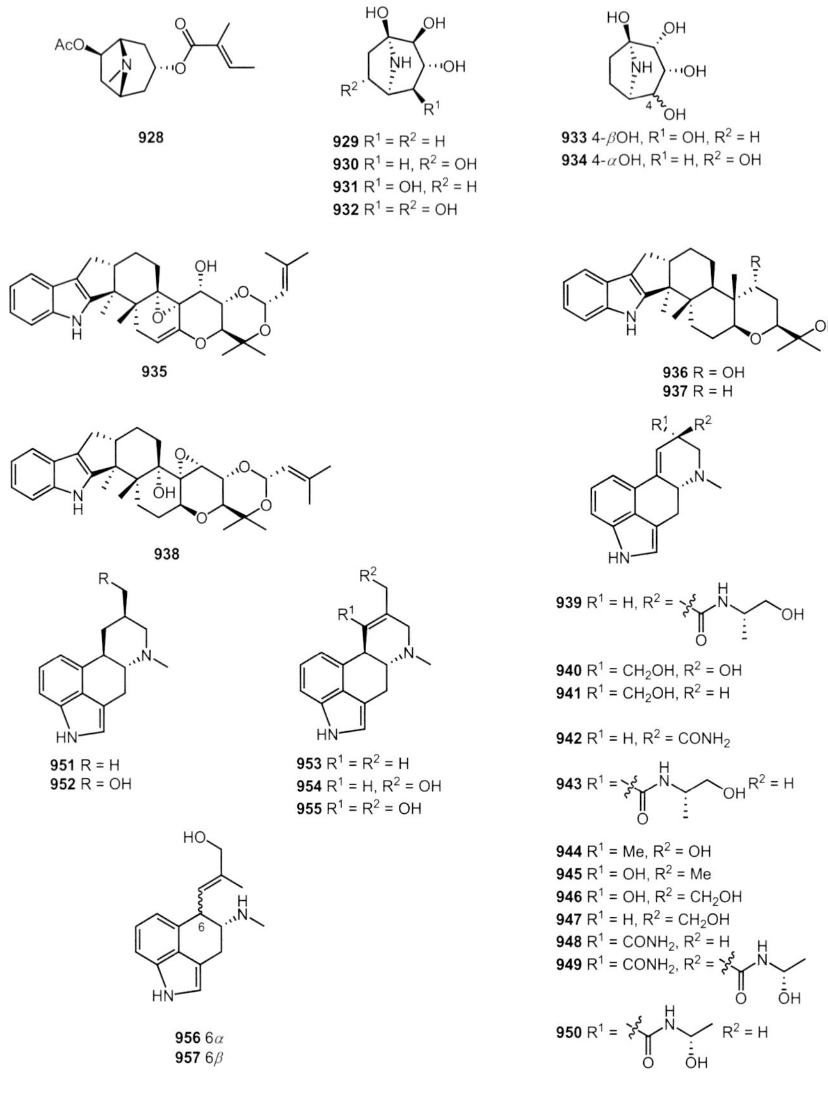

Fig. 79 Structures of alkaloids **928–967** reported from *Ipomoea* species

(**942**), (8)-9,10-didehydro-*N*-[(1*S*)-2-hydroxy-1-methylethyl]-6-methyl-ergoline-8-carboxamide (**943**), (8)-9,10-didehydro-6,8-dimethyl-ergolin-8-ol (**944**), (8)-9,10-didehydro-6,8-dimethyl-ergolin-8-ol (**945**), (8)-9,10-didehydro-8-hydroxy-6-methyl-ergoline-8-methanol (**946**), (8)-9,10-didehydro-6-methyl-ergoline-8-methanol (**947**), (8)-9,10-didehydro-6-methyl-ergoline-8-carboxamide (**948**), (8)-9,10-didehydro-*N*-[(1*S*)-1-hydroxyethyl]-6-methyl-ergoline-8-carboxamide (**949**), (8)-9,10-didehydro-*N*-[(1*S*)-1-hydroxyethyl]-6-methyl-ergoline-8-carboxamide

Table 22 Distribution of natural products isolated from the Solanales

Species	Collection location[a,b]	Plant part(s)[c]	Compound class	Compound(s)	Refs.
Convolvulaceae					
Ipomoea muelleri	NWA	wp	Calystegine	**929–934**	[343]
	NWA	lf, s	Diterpenoid indole alkaloid	**935–938**	[345]
	NWA	s	Ergot alkaloid	**939–957**	[346]
Ipomoea velutina	NT	wp	Calystegine	**928**	[344]
Solanaceae					
Capsicum annuum var. *glabriusculum*	SEQ	r	Tropane	**968**	[39]
Duboisia leichhardtii	SEQ	b	Tropane	**654**	[39]
Duboisia myoporoides	SEQ	wp	Piperidine	**973**	[352]
	SEQ	lf, wp	Pyridine	**895**, **971**	[39, 349]
	SEQ	lf	Pyrrolidine	**972**	[349]
	SEQ (cult. (Japan))	wp	Triterpenoid	**896**, **974**	[353]
	SEQ (cult. (Japan, India))	lf	Tropane	**654**, **896**, **974**, **958–962**, **965–967**	[348, 350]
	SEQ	lf, b	Tropane	**654**, **959–964**, **970**	[39, 349]
Lycianthes shanesii	NQ	lf	Steroidal alkaloid	**975**	[356]
Nicotiana forsteri	SEQ	lf	Sesquiterpenoid	**989**, **990**	[357]
Solanum aviculare	NQ	lf, wp	Steroidal alkaloid	**978–983**	[39]
	NQ	s	Tropane	**960–962**, **969**	[39]
Solanum callium	SEQ	wp	Steroidal alkaloid	**985**, **986**	[354]

(continued)

Table 22 (continued)

Species	Collection location[a,b]	Plant part(s)[c]	Compound class	Compound(s)	Refs.
Solanum dunalianum	NQ	lf	Steroidal alkaloid	**976**, **977**	[355]
Solanum seaforthianum	CQ	st	Spermidine	**987**, **988**	[39]
Solanum torvum	NQ	r	Steroidal alkaloid	**984**	[39]
Boraginaceae					
Cynoglossum australe	SEQ	lf	Pyrrolizidine	**991**, **992**	[358]
Ehretia saligna var. *membranifolia*	CQ	r	Piperidine	**993**, **994**	[39]
	CQ	r	Spermidine	**995**	[39]

[a] SEQ = Southeast Queensland, NQ = North Queensland, CQ = central Queensland, NT = Northern Territory, NWA = North Western Australia, cult. = cultivated, [b] natural location shown for those species studied from cultivated material, [c] lf = leaves, s = seeds, r = roots, wp = whole plant, b = bark, st = stems

(**950**), (8)-6,8-dimethyl-ergoline (**951**), (8)-6-methyl-ergoline-8-methanol (**952**), 8,9-didehydro-6,8-dimethyl-ergoline (**953**), 8,9-didehydro-6-methyl-ergoline-8-methanol (**954**), 8,9-didehydro-9-hydroxy-6-methyl-ergoline-8-methanol (**955**), (2*E*)-*rel*-2-methyl-3-[(4*R*,5*S*)-1,3,4,5-tetrahydro-4-(methylamino)benz[*cd*]indol-5-yl]-2-propen-1-ol (**956**), and (2*E*)-2-methyl-3-[(4*R*,5*R*)-1,3,4,5-tetrahydro-4-(methylamino)benz[*cd*]indol-5-yl]-2-propen-1-ol, (**957**)), from the seeds [345, 346]. The seed-containg compounds are reminiscent of natural products reported from fungi, suggesting that a fungal endophyte may be responsible for their production [347].

The two *Duboisia* species, *Solanum aviculare* and *Capsicum annum* var. *glabriusculum* from the Solanaceae have yielded a diversity of tropane alkaloids (Fig. 80). These include 7*β*-hydroxyhyoscyamine (**958**), 6*β*-hydroxyhyoscyamine (**959**), hyoscyamine (**960**), norhyoscyamine (**961**), scopolamine (**962**), aposcopolamine (**963**), butropine (**964**), valtropine (**965**), apoatropine (**966**), tropine nonanoate (**967**), phyllalbine (**968**), teloidine (**969**), the 3-isovalerate ester of 3*α*,6*β*-tropanediol, (**970**) [39, 348–350]. *Duboisia myoporoides* is a high-yielding producer of hyoscyamine and scopolamine and other tropanes and it is now cultivated widely, particularly in India for tropane alkaloid production for motion sickness and ophthalmic and anti-spasmodic indications [351]. *Duboisia myoporoides* also contains nicotine (**971**), **895**, hygrine (**972**) isopelletierine (**973**) and two triterpenoids, **896** and betulonic acid (**974**) [349, 352, 353]. Five Australian rainforest *Solanum* species have been studied, with *S. aviculare, S. callium*, *S. dunalianum*, and *S. torvum* containing steroidal alkaloids (soladunalinidine (**976**), tomatidine (**977**), tomatidenol *β*-D-glucopyranoside (**978**), solasodine (**979**), β1-solamargine (**980**), α-solasonine (**981**), (12*β*,25*S*)-12-hydroxyspirosol-4-en-3-one (**982**), solanaviol (**983**), spirostan-3-amine (**984**), 25-isosolafloridine (**985**), and solacallinidine (**986**)). *Solanum seaforthianum* produces feruloylputrescine (**987**) and caffeoylputrescine (**988**) [39, 354, 355], and *Lycianthes shanesii* the steroid alkaloid soladulcidine (**975**) [356]. *Nicotinia forsteri* biosynthesizes two related sesquiterpenoids debneyol (**989**) and cyclodebneyol (**990**), with cyclodebneyol being a potent antifungal agent [357].

Boraginales

The Boraginales contains only the single family Boraginaceae, and the only two prior studies (Fig. 81) on Australian rainforest Boraginaceae species have yielded the pyrrolizidine alkaloids cynaustraline (**991**) and cynaustine (**992**) from *Cynoglossum australe*, and the pipecolic acid derivatives *trans*-4-hydroxy-L-pipecolic acid (**993**) and methyl-4-hydroxypipecolate (**994**) as well as *N*1,*N*10-dicoumaroylspermidine (**995**) from *Ehretia saligna* var. *membranifolia* [39, 358].

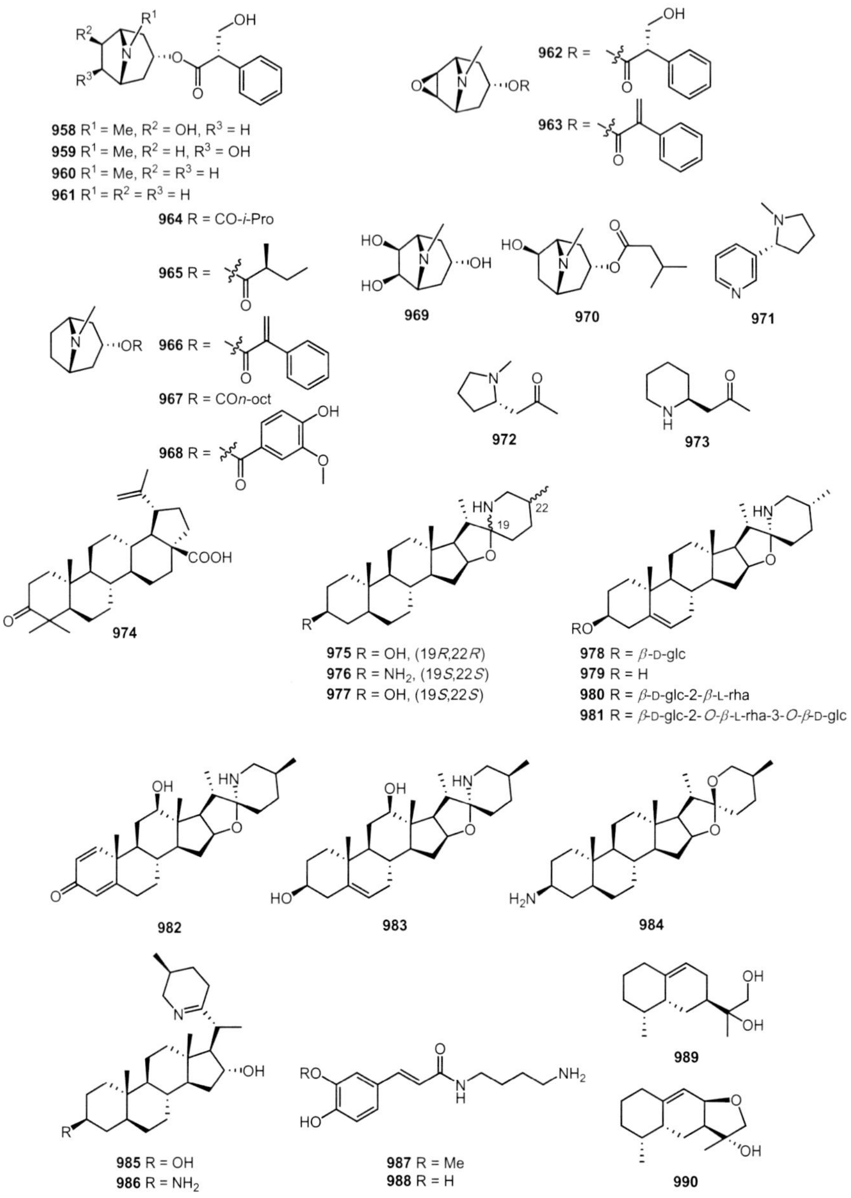

Fig. 80 Structures of compounds **958–990** reported from Solanaceae species

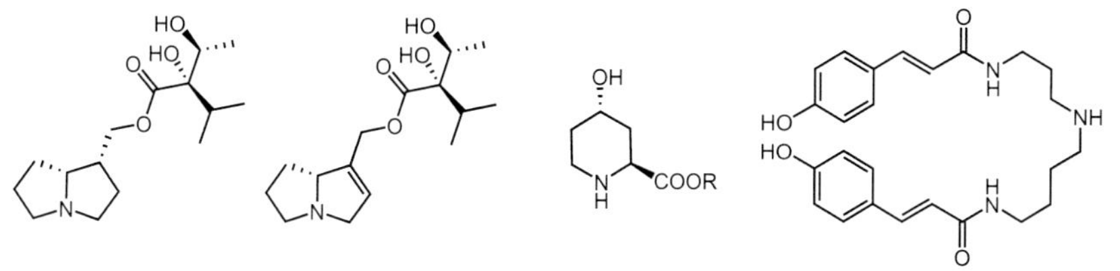

Fig. 81 Structures of alkaloids **991**–**995** reported from *Cynoglossum australe* or *Ehretia saligna* var. *membranifolia*

Gentianales

The Gentianales families Rubiaceae (43 genera, 156 species), and Apocynaceae (33 genera, 143 species) are well represented in Australian rainforests. Two other families, Loganiaceae (three genera and nine species) and Gentianaceae (two genera and five species) are also present in the rainforests, but both these families have Australian species more commonly occurring outside of rainforests. Within the Gentianales, the Rubiaceae diverged from a common ancestor first, followed by the Loganiaceae and Apocynaceae (sister families) and then the Gentianaceae and Gelsemiaceae (sister families) (Fig. 82) [359]. Species within the Gentianales are almost exclusive producers of monoterpenoid indole alkaloids derived from the conjugation of tryptamine with *seco*-loganin.

The Rubiaceae family is divided into two subfamilies, Rubioideae and Cinchonoideae, with two-thirds of Australia's rainforest species being in the Cinchonoideae. The Rubioideae has an African origin and is basal, with Cinchonoideae diverging later. Rubioideae species contain a more diverse variety of alkaloids compared to Cinchonoideae species since they produce monoterpenoid indole, emetine, and pyrrolidinoindoline alkaloids, while the Cinchonoideae are the main source of monoterpenoid indole alkaloids and their derived quinoline alkaloids. There is strong evidence that the Australian Rubiaceae are likely to be migrants from Africa via south America and into Asia [360]. This points to their relatively recent arrival into Australia's rainforests with fewer species found outside of the tropical rainforests. Although there is a rich diversity of Australian Rubiaceae rainforest species, it is a little surprising that only five species have been studied for their natural product constituents (Table 23).

Within the Australian rainforest Cinchonoideae, *Antirhea putaminosa* contains the monoterpenoid indole alkaloid antirhine (**996**) [361] and *Uncaria ferrea* from the Daintree region of north Queensland contains six diastereomeric monoterpenoid indole alkaloids uncarines A–F (**997**–**1002**) (Fig. 83) [362]. These compounds have five stereogenic centers including a spiroindolizidine-indolone, and diastereomers result from epimerization about this spiro center, the bridgehead of the indolizidine or the oxygenated methine of a dihydropyran moiety. Uncarine D is cytotoxic and causes DNA damage. *Psydrax montigena* contains the iridoid arboriside E (**1003**) that

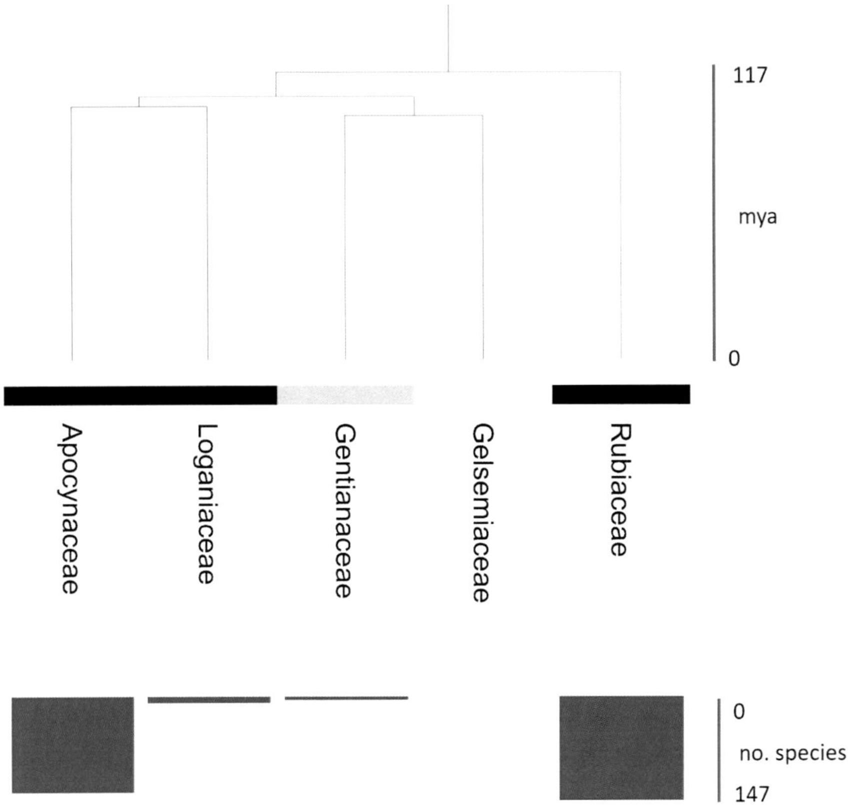

Fig. 82 Dated molecular phylogeny of the Gentianales [7] showing those families that have species present in Australian rainforests (grey) and studied (black) and the number of species occurring in Australian rainforests

was shown, by affinity MS, to bind to the *Plasmodium falciparum* protein PfRab11a involved in endosome recycling [363]. For malaria treatment, PfRab11a is a potential novel drug target.

The two Australian rainforest Rubioideae species studied contained a series of tri- or tetrameric pyrrolidinoindoline alkaloids, namely, hodgkinsine (**1004**), quadrigemines A **1005**) and B (**1006**) from *Eumachia frutescens* [364] and an iridoid, asperuloside (**1007**) from the Tasmanian *Coprosma quadrifida* [365]. The pyrrolidinoindoline alkaloids are strikingly like idiospermuline (**337**) found in the ancient Calycanthaceae species *Idiospermum australiense* collected in North Queensland, but these differ in the level of *N*-methylation and the configurations at the attachment point between the pyrrolidinoindoline monomers a′ and a″. The absolute configuration of hodgkinsine (**1004**) was established by X-ray crystallography [366].

Table 23 Distribution of natural products isolated from the Rubiaceae and Loganiaceae

Species	Collection location[a]	Plant part(s)[b]	Compound class	Compound(s)	Refs.
Rubiaceae					
Antirhea putaminosa	CQ	lf, tw	MIA	**996**	[361]
Coprosma quadrifida	Tas	lf	Iridoid	**1007**	[365]
Eumachia frutescens (*Hodgkinsonia frutescens*)	NQ	lf	Indole	**1004–1006**	[364]
Psydrax montigena	NQ	wp	Iridoid	**1003**	[363]
Uncaria ferrea	NQ	b	MIA	**997–1002**	[362]
Loganiaceae					
Strychnos lucida	NQ	fr	Iridoid	**1012**	[367]
	NQ	fr	MIA	**1010, 1011**	[367]
Strychnos psilosperma	SEQ	lf	MIA	**1008, 1009**	[39]

[a] SEQ = Southeast Queensland, NQ = North Queensland, CQ = central Queensland, Tas = Tasmania, [b] lf = leaves tw = twigs, fr = fruits, wp = whole plant, b = bark

The Loganiaceae species *Strychnos psilosperma* and *S. lucida* have yielded four strychnine alkaloids (*N*-desacetylspermostrychnine (**1008**), 1-deacetyl-1-methyl-diaboline (**1009**), strychnine (**1010**) and brucine (**1011**)) and an iridoid, loganin (**1012**) [39, 367]. None of the Australian rainforest Gentianaceae species have been investigated.

Apocynaceae

The Australian rainforest Apocynaceae species (Plate 19) have been a particularly prolific source of natural products, yielding mostly monoterpenoid indole alkaloids (Table 24). The molecular phylogeny of the Apocynaceae splits the family into five subfamilies: the basal Rauvolfioideae, followed by Apocynoideae, Periplocoideae, Secamonoideae, and then the Asclepiadoideae (Fig. 84) [368]. The Apocynaceae is likely to have originated in a tropical equatorial land belt that encompasses areas in Africa-South America (West Gondwana), Southeast Asia (Laurasia), and possibly Oceania (East Gondwana) [369]. The Rauvolfioideae contain the basal species from which widespread radiation has occurred. This subfamily is the only one to contain monoterpenoid indole alkaloids indicating that this alkaloid structural class has an ancient origin shared with ancestors of the Rubiaceae and Loganiaceae and that this biosynthetic capacity may have been lost in successive subfamily branch points in later diversification of the Apocynaceae. Speciation and pan-tropical dispersal of the Rauvolfioideae after the late Eocene (~35 mya) has resulted in a diverse but small group of species inhabiting Australian rainforests. The 19 Australian Rauvolfioideae species are likely to be relatively recent arrivals into Australia, but ancestral monoterpenoid indole alkaloids (MIA) biosynthetic diversification is reflected in the rich

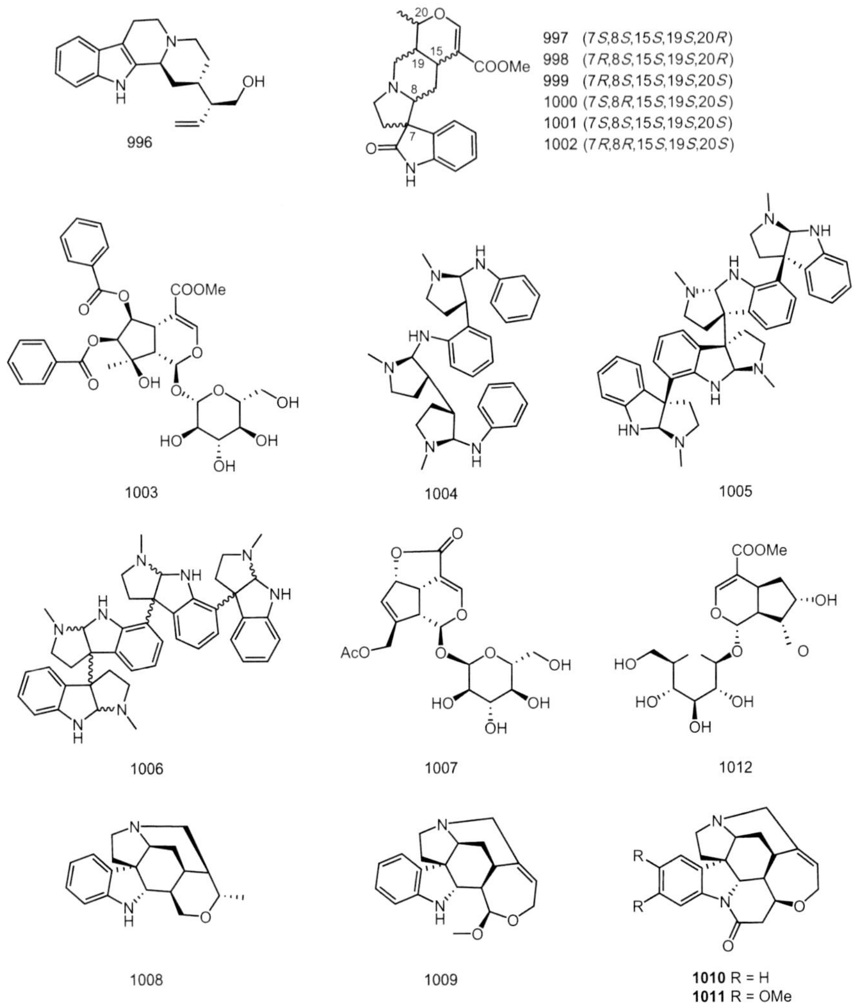

Fig. 83 Structures of compounds **996–1012** reported from Rubiaceae and Loganiaceae species

variety of MIAs that the Australian species contain. The 19 species are spread across the tribes Alstonieae, Vinceae, Tabernaemontaneae, and Melodineae and 15 of these species have been investigated chemically. Although these species are likely to be derived from southern migrations into Australia their extant distribution is mostly either in Australia or with extended distributions only into Malesia or New Caledonia (Fig. 84). The Rauvolfioideae tribes contain approximately 430 species worldwide and are all major producers of monoterpenoid indole alkaloids.

Twenty-seven monoterpenoid indole alkaloid subclasses are represented in Australian rainforest species and include 124 different monoterpenoid indole alkaloids (Table 24, Figs. 85, 86, 87, 88, 89, 90, and 91). Since over 3,000 monoterpenoid

Plate 19 Australian rainforest Apocynaceae species, flowers of *Melodinus australis* (top and bottom left), *Tabernaemontana pandacaqui* (top two middle panels), *Hoya australis* (bottom middle), and flowers and fruit of *Parsonsia straminea* (top and bottom right). (Photographs AR Carroll)

indole alkaloids have been reported worldwide and there are 1,000 species in the Rauvolfiodeae, this richness both in number and diversity of monoterpenoid indole alkaloids in Australian rainforest species is remarkable. The alkaloid actinophyllic acid (**1013**) from the Cape York *Alstonia actinophylla* even contains an unprecedented monoterpenoid indole alkaloid skeleton. This compound is potently active in a coupled enzyme assay used to discover carboxypeptidase U inhibitors that could be useful for stroke treatments [370]. Four strictosidine alkaloids (14-ketoalstonidine (**1014**), alstonidine (**1015**), desocycordifoline (**1016**), and strictosidinic acid (**1017**)) [39, 371–373], and nine yohimbine alkaloids (alstonilidine (**1018**), alstoniline (**1019**), yohimbine (**1020**), powerine (**1021**), 11-methoxyyohimbine (**1022**), β-yohimbine (**1023**), poweridine (**1024**), 10-methoxydihydrocorynantheol (**1025**), and reserpine (**1026**)) have been reported [39, 371, 372, 374, 375] from *Alstonia*, *Neisosperma*, and *Ochrosia* species.

Fifteen ajmalicine alkaloids (serpentine (**1027**), *N*-methylarcinine (**1028**), holeinine (**1029**), tetrahydroalstonine (**1030**), aricine (**1031**), reserpinine (**1032**), isoreserpiline (**1033**), (−)-reserpiline (**1034**), ochropposine (**1035**), 10,11-dimethoxy-picraphylline (**1036**), methoxy-11-picraphylline (**1037**), 10,11-dimethoxy-ajmalicinine (**1038**), *N*-oxy-de-(3*R*,4*R*)-reserpiline (**1039**), *N*-oxy-de-(3*R*,4*S*)-reserpiline (**1040**), and angustine (**1041**)) [376–378], three β-carbolines (1-carbomethoxy-β-carboline (**1042**), 1-carbamoyl-carboline (**1043**), and harmic amide (**1044**)) [371, 378], and a monoterpenoid indole alkaloid/tyramine hybrid,

Table 24 Distribution of natural products isolated from the subfamily Rauvolfioideae within the Apocynaceae

Species	Collection location[a,b]	Plant part(s)[c]	Compound class	Compound(s)	Refs.
Alstonieae					
Alstonia actinophylla	CY	lf	MIA	**1013**	[370]
Alstonia constricta	NQ	rb	Benzoate	**1143**	[371]
	NQ, SEQ	rb, st, lf, b	MIA	**1014**, **1015**, **1018–1020**, **1023**, **1026**, **1027**, **1042**, **1077–1081**, **1089**, **1133**	[39, 371, 372, 374]
Alstonia muelleriana	NQ	b	MIA	**1075**, **1078**, **1082–1085**, **1119**, **1120**, **1123**, **1125**, **1134**	[379, 383–385, 394, 395]
Alstonia spectabilis subsp. *ophioxyloides*	PNG	b	MIA	**1078**, **1080**, **1084**, **1094**, **1123**, **1124**	[386]
Vinceae					
Kopsia arborea	NQ	b, lf	MIA	**1063–1066**	[39, 381]
Neisosperma kilneri	CQ	b, lf	MIA	**1017**, **1025**, **1033**, **1034**, **1041**, **1043–1047**, **1051**, **1089**, **1096**, **1142**	[39, 378]
Neisosperma poweri	NQ	st, b, lf, r	MIA	**1020–1022**, **1024**, **1026**, **1102–1104**	[39, 375, 391]
Ochrosia elliptica	NQ (cult. (China))	st, lf	Isoflavonoid	**1144–1151**	[401]
	NQ (cult. (China, Egypt))	st, lf	MIA	**1028**, **1029**, **1086**, **1088**, **1090**, **1092**, **1093**, **1112**	[376, 388, 392, 398]
	NQ, VAN	wp, b, r, lf	MIA	**1016**, **1025**, **1030–1040**, **1045**, **1048–1058**, **1096**, **1138–1141**	[39, 377, 399]
Ochrosia moorei	SEQ, VAN	st, lf, b	MIA	**1016**, **1037**, **1039**, **1051–1053**, **1058**, **1070**, **1071**, **1113**, **1138**, **1139**	[373, 377]
Tabernaemontaneae					
Tabernaemontana orientalis	NQ	lf, b, wp, st	MIA	**1097**, **1098**, **1100**, **1101**, **1104–1109**, **1113–1116**	[39, 389]

(continued)

Table 24 (continued)

Species	Collection location[a,b]	Plant part(s)[c]	Compound class	Compound(s)	Refs.
Tabernaemontana pandacaqui	NQ	wp	MIA	**1110**, **1111**	[39]
Voacanga grandifolia	INDO	wp	MIA	**1099**, **1121**, **1122**, **1126**, **1127**, **1136**	[390, 396]
Melodineae					
Alyxia buxifolia	WA	b	Triterpenoid	**802**	[402]
Melodinus acutiflorus	SEQ, NQ	lf, st	MIA	**1086**, **1087**, **1129–1132**	[39, 387, 397]
Melodinus australis	NNSW, SEQ, NQ	b, st, wp	MIA	**1059–1063**, **1067–1069**, **1072–1074**, **1076**, **1093**, **1095**, **1117**, **1118**	[39, 380, 393]
Melodinus baccellianus	NQ	lf, st	MIA	**1091**, **1128**, **1135**	[39]
Alxieae					
Alyxia buxifolia	WA	b	Triterpenoid	**802**	[402]
Carisseae					
Carissa lanceolata	WA	r	Acetophenone	**1156**	[403]
	WA	r	Lignan	**1155**	[403]
	NQ, WA	w, r	Sesquiterpenoid	**1152–1154**	[403, 404]
Plumerieae					
Cerbera floribunda	NQ	s	Steroid	**1157**	[405]

[a] SEQ = Southeast Queensland, NQ = North Queensland, CQ = central Queensland, CY = Cape York, NNSW = Northern New South Wales, WA = Western Australia, INDO = Indonesia, VAN = Vanuatu, PNG = Papua New Guinea, cult. = cultivated, [b] natural location shown for those species studied from cultivated material, [c] lf = leaves, s = seeds, r = roots, rb = root bark, wp = whole plant, b = bark, w = wood, st = stems

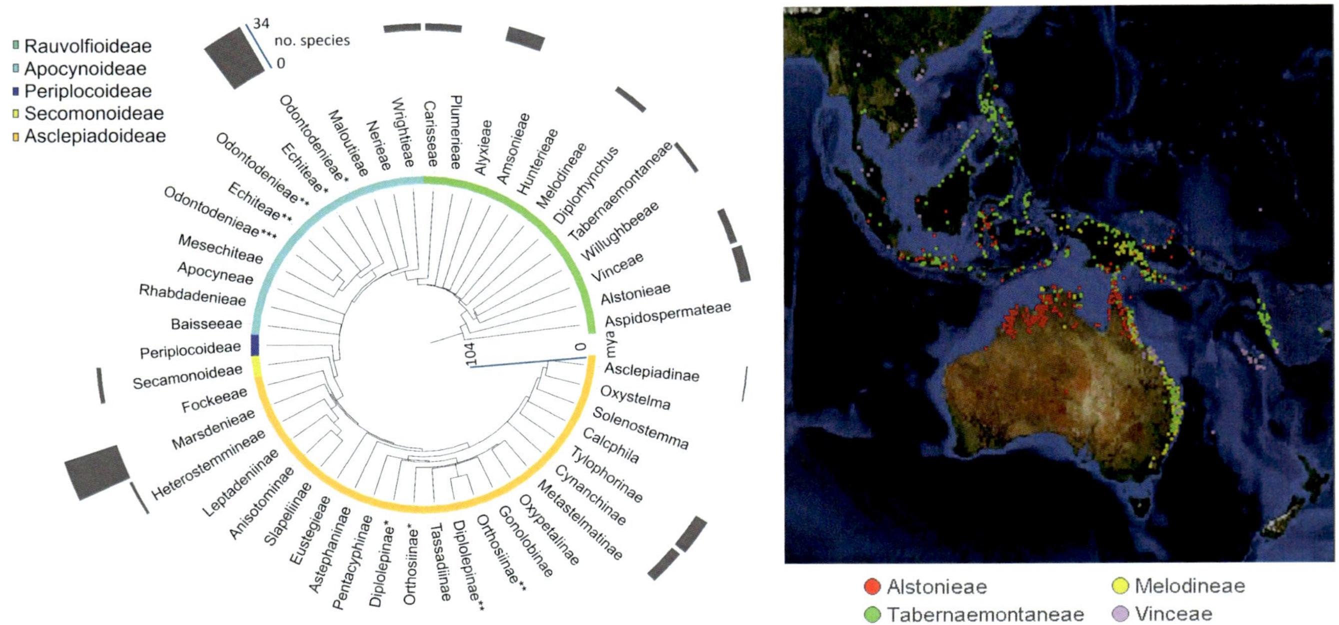

Fig. 84 Dated molecular phylogeny of the Apocynaceae [368] showing subfamilies and tribes and the number of species occurring in Australian rainforests. The map shows the herbarium record locations of Rauvolfioideae species that have been chemically investigated [5, 43]

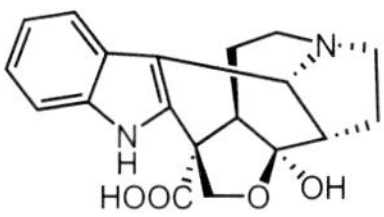

1013

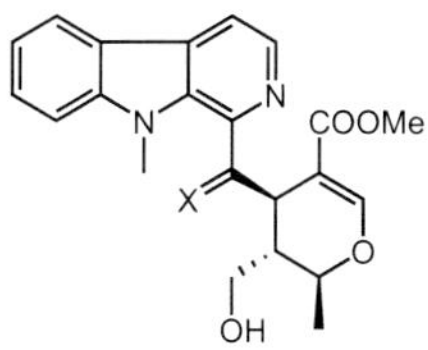

1014 X = O
1015 X = H, H

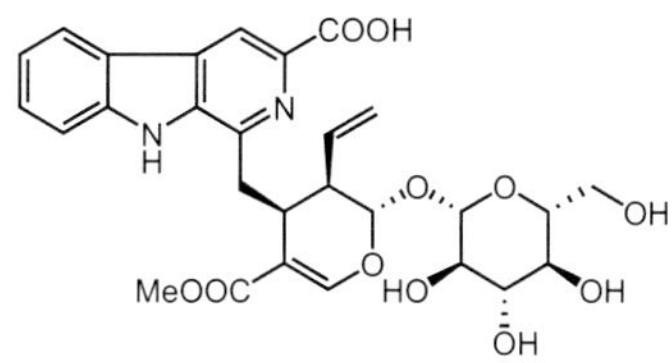

1016

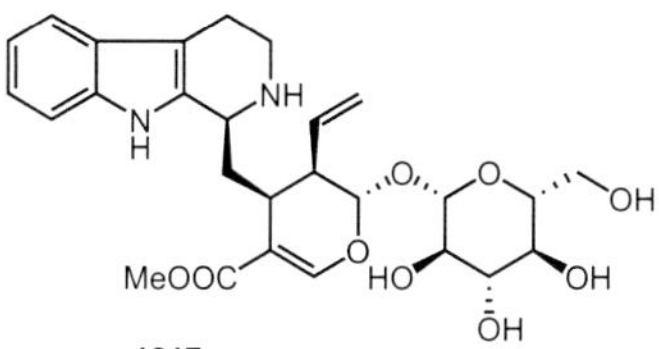

1017

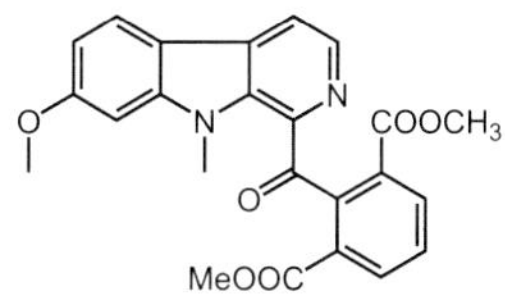

1018

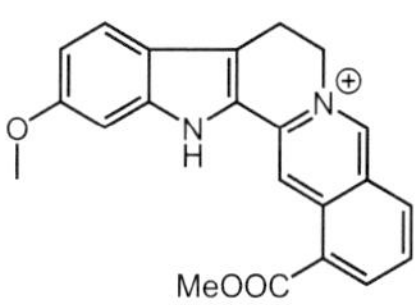

1019

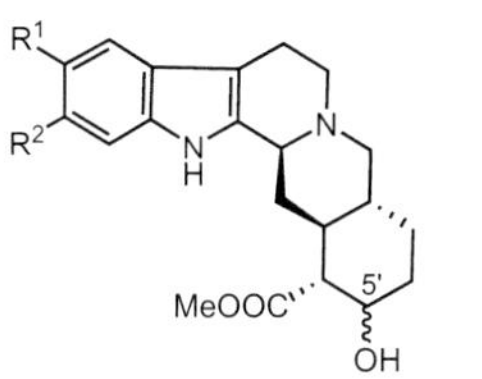

1020 $R^1 = R^2$ = H, 5' αOH
1021 R^1 = OH, R^2 = H, 5' αOH
1022 R^1 = H, R^2 = OMe, 5' αOH
1023 $R^1 = R^2$ = H, 5' βOH

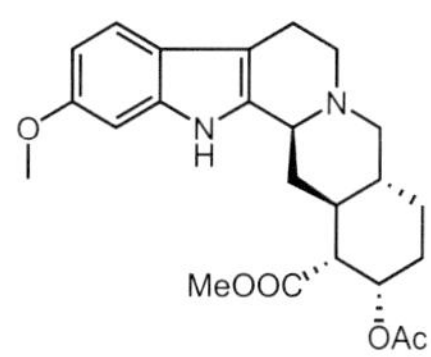

1024

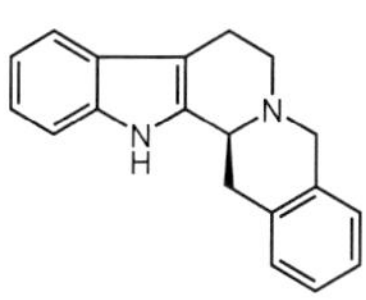

1025

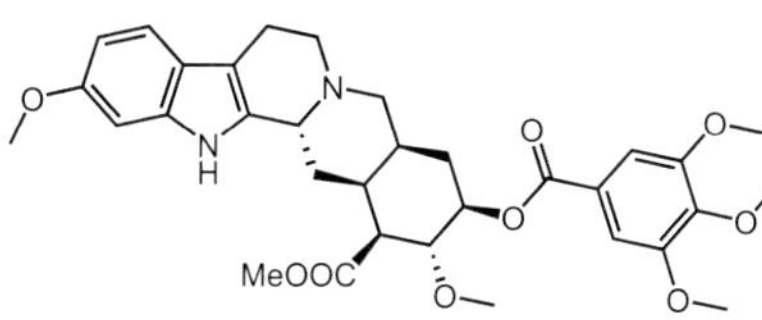

1026

Fig. 85 Structures of alkaloids **1013–1026** reported from Apocynaceae species

ochrolifuanine A (**1045**) [378] have been reported from *Ochrosia* and *Neisosperma* species.

Nine mitragynine alkaloids (geissoschizol (**1046**), 10-methoxygeissoschizol (**1047**), ochropposinine (**1048**), 10-methoxy-18,19-dihydro-corynanthanol (**1049**), 18,19-dihydro-corynanthanol (**1050**), corynan-10,17-diol (**1051**), (16*S*, 20*R*)-10,11-dimethoxy18,19-dihydro-sitsirikine (**1052**), *N*-oxy-de-(3*S*,4*R*)-dimethoxy-10-dihydrocorynanthanol (**1053**), and *N*-oxy-de-(3*S*,4*S*)-dimethoxy-10-dihydrocorynanthanol (**1054**)) [377, 378], four mitraphylline alkaloids (rauvoxine (**1055**), pseudoindoxyl-3-isoreserpiline (**1056**), isocarapanaubine (**1057**), and oxindole (7*S*,3*S*)-devochropposinine (**1058**)) [377] were all obtained from *Ochrosia* or *Neisosperma* species. Eleven kopsinine alkaloids (6-hydroxykopsinine (**1059**), 14-hydroxykopsinine (**1060**), pyrifoline (**1061**), refractidine (**1062**), kopsinine (**1063**),

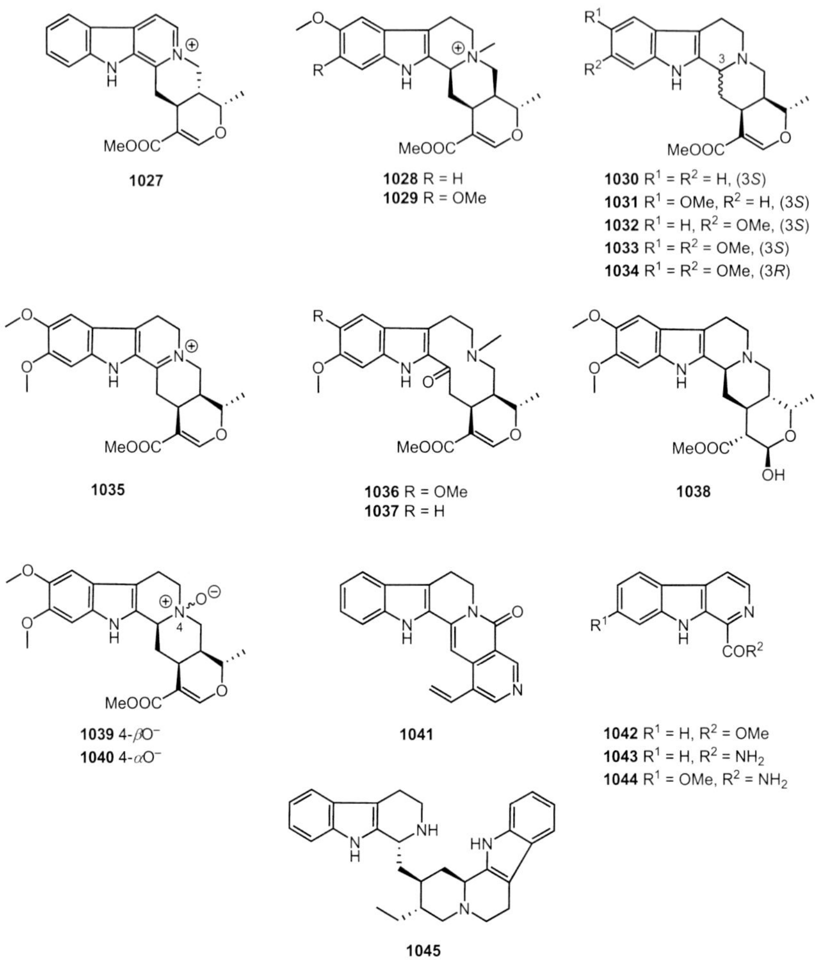

Fig. 86 Structures of alkaloids **1027–1045** reported from Apocynaceae species

kopsilongine (**1064**), kopsamine (**1065**), kopsiflorine (**1066**), venalstonidine (**1067**), 6,7-epoxykopsinine (**1068**), venalstonine (**1069**)), seven akuammine (ochrosamine B (**1070**), ochrosamine A (**1071**), akuammicine (**1072**), stemmadenine (**1073**), condylocarpine (**1074**), 11-methoxyakuammicine (**1075**), and (19*R*)-methoxytubotaiwine (**1076**)) [39, 372, 373, 379–381] have been reported from *Melodinus*, *Ochrosia*, and *Kopsia* species. Akuammicine is a submicromolar partial to full agonist of κ-opioid receptors in various rodent tissues [382].

Six ajmaline alkaloids were isolated from *Alstonia* species (*O*-3,4,5-trimethoxybenzoylquebrachidine (**1077**), quebrachidine (**1078**), vincamedine (**1079**), vincamajine (**1080**), *O*-3,4,5-trimethoxycinnamoylvincamajine (**1081**), and alstonisidine, which was originally assigned as **1082**, but later revised as **1083**)

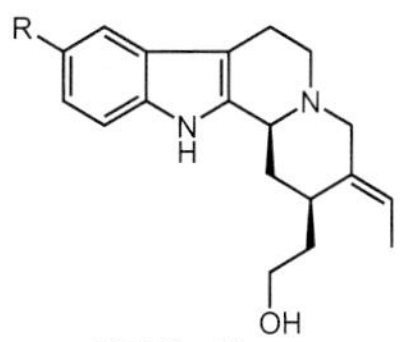

1046 R = H
1047 R = OMe

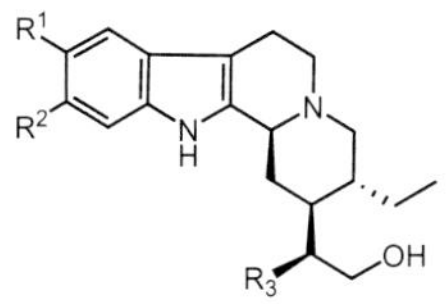

1048 $R^1 = R^2$ = OMe, R^3 = H
1049 R^1 = OMe, $R^2 = R^3$ = H
1050 $R^1 = R^2 = R^3$ = H
1051 R^1 = OH, $R^2 = R^3$ = H
1052 $R^1 = R^2$ = OMe, R^3 = COOMe

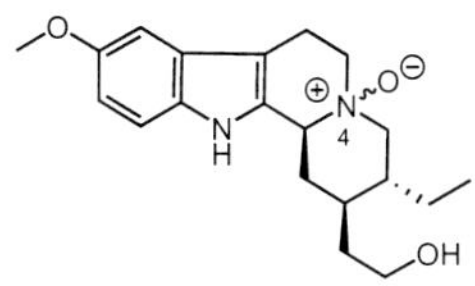

1053 4-αO^-
1054 4-βO^-

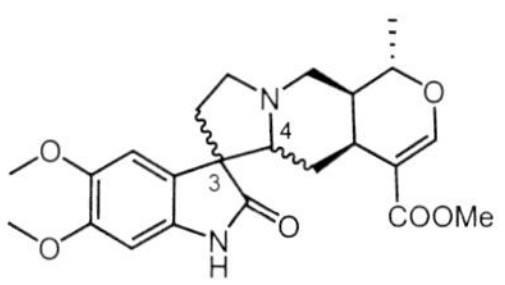

1055 (3*R*,4*R*)
1056 (3*R*,4*S*)
1057 (3*S*,4*S*)

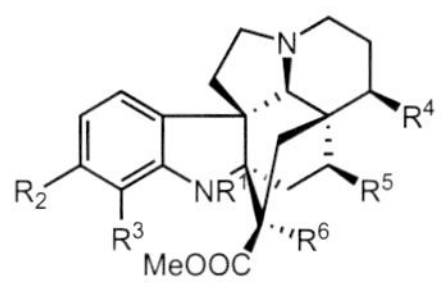

1059 $R^1 = R^2 = R^3 = R^5 = R^6$ = H, R^4 = OH
1060 $R^1 = R^2 = R^3 = R^4 = R^6$ = H, R^5 = OH
1061 R^1 = Ac, $R^2 = R^5 = R^6$ = H, $R^3 = R^4$ = OMe
1062 R^1 = CHO, $R^2 = R^3 = R^5 = R^6$ = H, R^4 = OMe
1063 $R^1 = R^2 = R^3 = R^4 = R^5 = R^6$ = H
1064 R^1 = COOMe, $R^2 = R^4 = R^5$ = H, R^3 = OMe, R^6 = OH
1065 R^1 = COOMe, R^2–R^3 = OCH_2O, $R^4 = R^5$ = H, R^6 = OH
1066 R^1 = COOMe, $R^2 = R^3 = R^4 = R^5$ = H, R^6 = OH

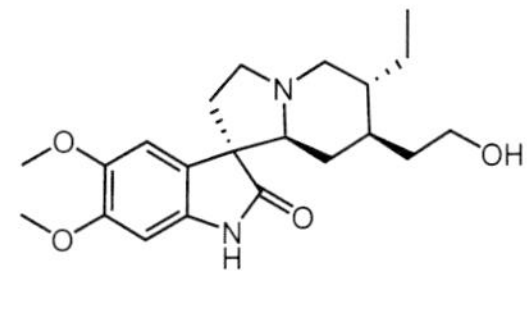

1058

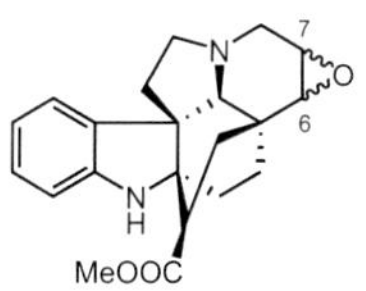

1067 6,7-αO
1068 6,7-βO

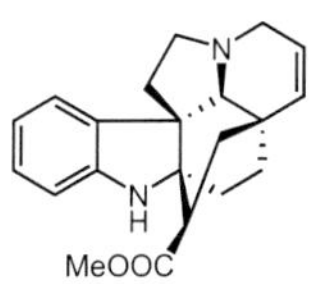

1069

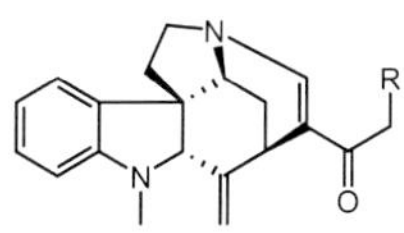

1070 R = H
1071 R = OH

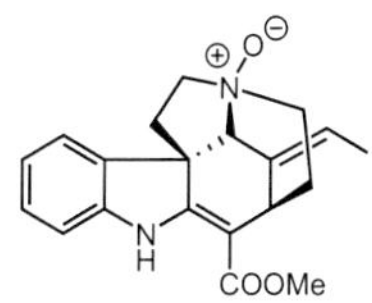

1072

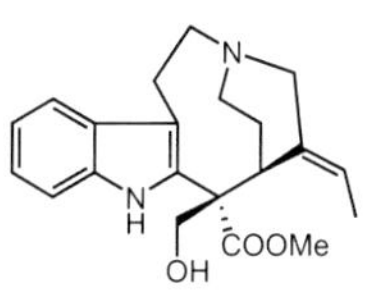

1073

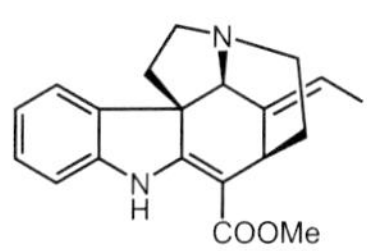

1074

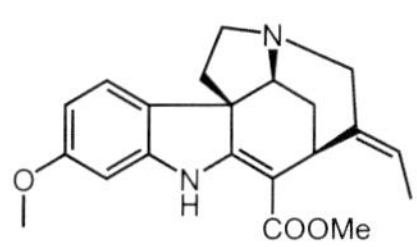

1075

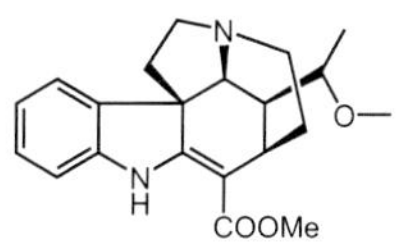

1076

Fig. 87 Structures of alkaloids **1046–1076** reported from Apocynaceae species

[371, 372, 383, 384], as were two pleiocarpamine alkaloids (pleiocarpamine (**1084**) and 2,7-dihydropleiocarpamine (**1085**)) [385, 386]. Six akuammiline alkaloids (deacetylakuammiline (**1086**), 16-*epi*-rhazinaline (**1087**), strictamine (**1088**), cathafoline (**1089**), picrinine (**1090**), picraline (**1091**)) [39, 387, 388] and five akuammidine alkaloids (10-methoxyakuammidine (**1092**), akuammidine (**1093**), *N*(α)-methylsarpagine (**1094**), voachalotine (**1095**), and akuammidine 17-*O*-*β*-D-glucopyranoside (**1096**)) [39, 386, 388] have been reported from *Melodinus*, *Ochrosia*, and *Alstonia* species.

Ten voacamine alkaloids (dregamine (**1097**), tabernaemontanine (**1098**), voacalgine F (**1099**), voacamine (**1100**), 16-demethoxycarbonylvoacamine

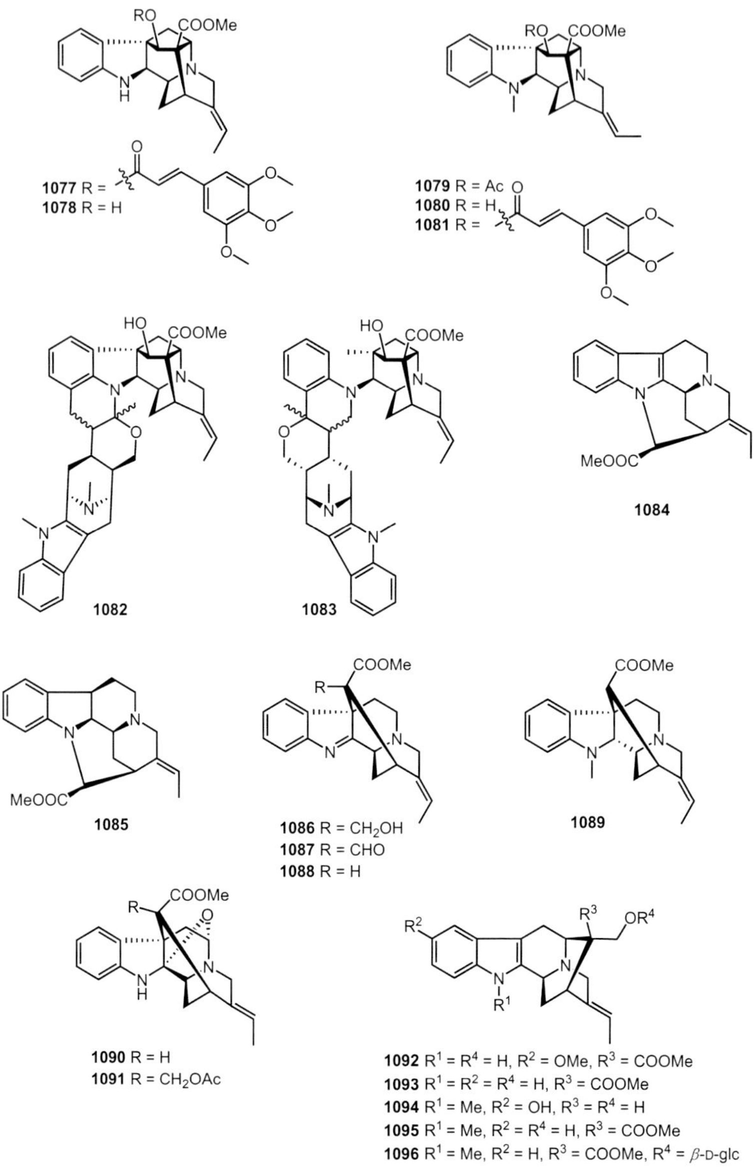

Fig. 88 Structures of alkaloids **1077**–**1096** reported from Apocynaceae species

(**1101**), ochropamine (**1102**), ochropine (**1103**), vobasine (**1104**), 16-demethoxycarbonyldihydrovoacamine (**1105**), and 16-demethoxycarbonyl-20′-*epi*-dihydrovoacamine (**1106**)), five of which are bis-monoterpenoid indole alkaloids [389–391], have been found in *Tabernaemontana*, *Voacanga*, and *Neisosperma* species. Five ibogaine alkaloids (ibogaine (**1107**), iboxygaine (**1108**),

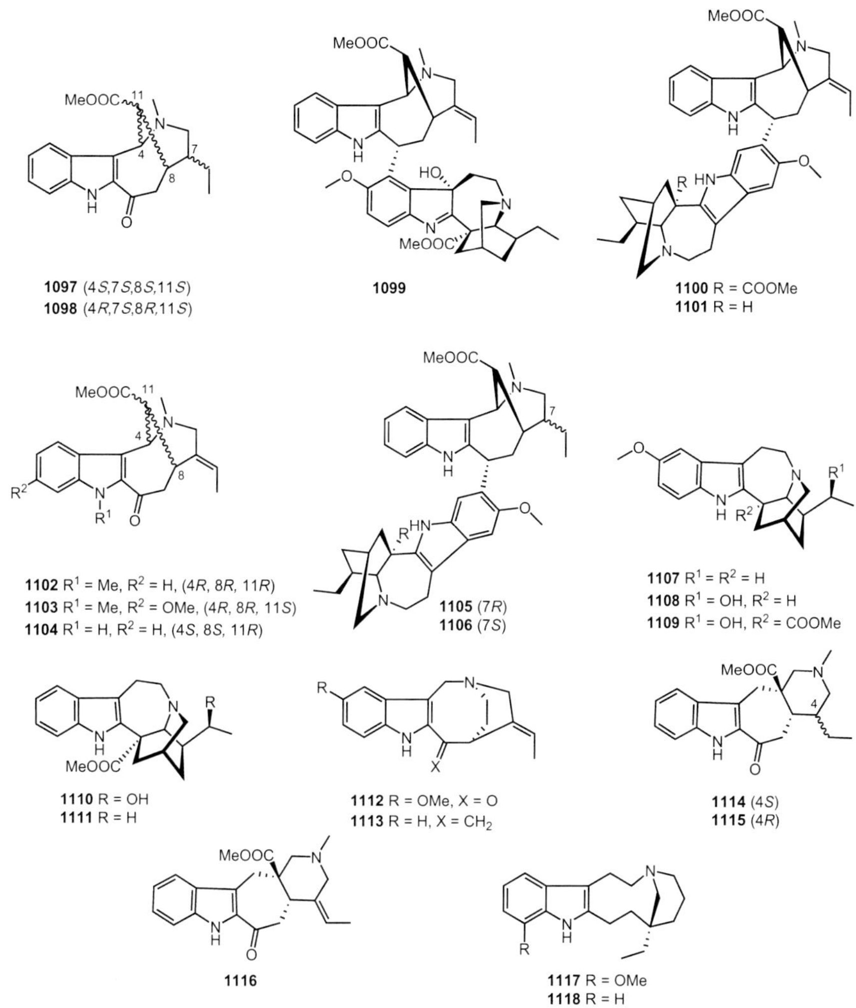

Fig. 89 Structures of alkaloids **1097–1118** reported from Apocynaceae species

voacristine (**1109**), heyneanine (**1110**), and coronaridine (**1111**)) [39, 389] from *Tabernaemontana orientalis*, two percine alkaloids (10-methoxyconolidine (**1112**) and apparicine (**1113**) [373, 392] from *Ochrosia* species, three ervatamine alkaloids (ervatamine (**1114**), 20-*epi*-ervatamine (**1115**), and 19-dehydroervatamine (**1116**)) [389] (*Tabernaemontana orientalis*) and two quebrachamine alkaloids (17-methoxyquebrachamine (**1117**) and quebrachamine (**1118**)) from *Melodinus australis* [393] also have been reported.

Nine macralstonine alkaloids (des-*N'*-methylanhydromacralstonine (**1119**), macroalstonine (**1120**), voacalgine D (**1121**), voacalgine E (**1122**), villastonine

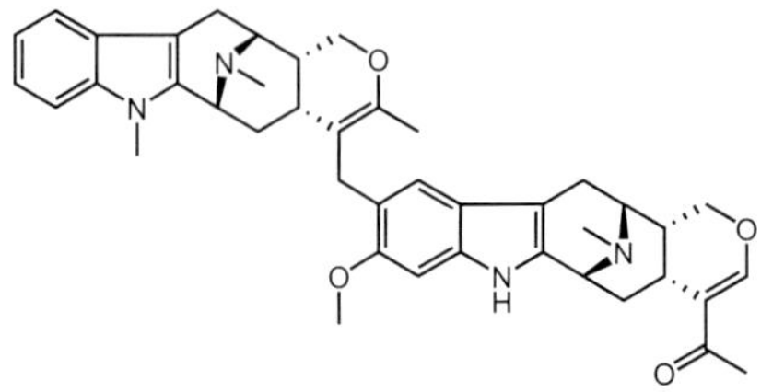

1119

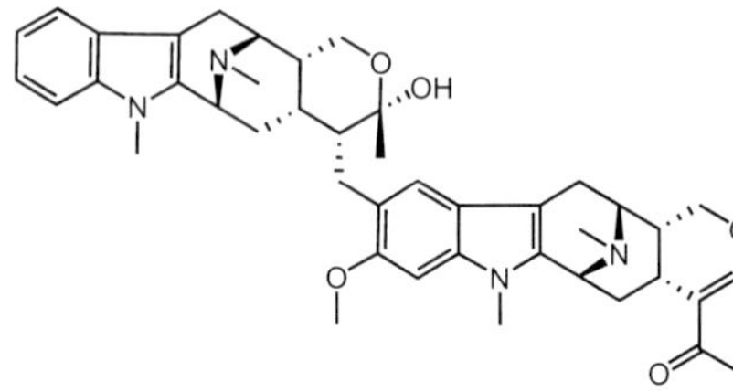

1120

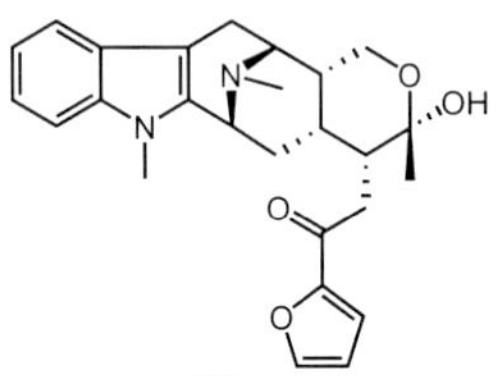

1121

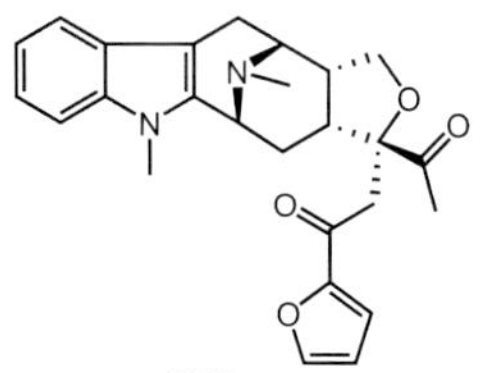

1122

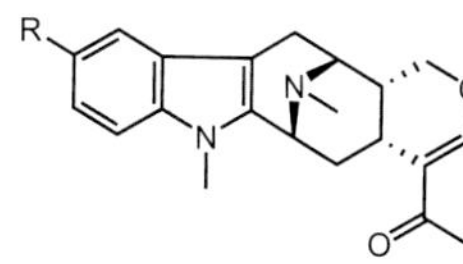

1125 R = H
1126 R = OH

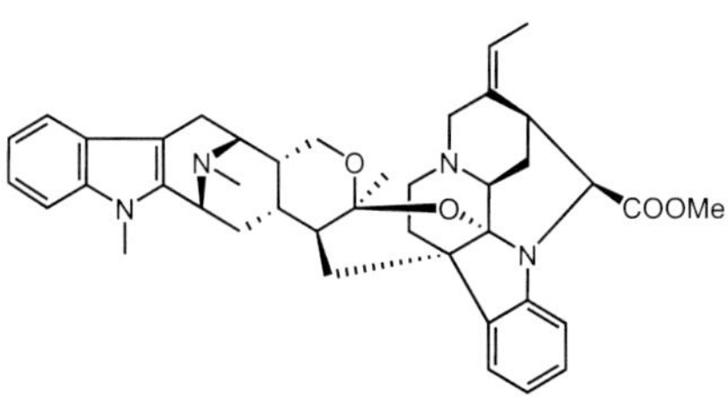

1123

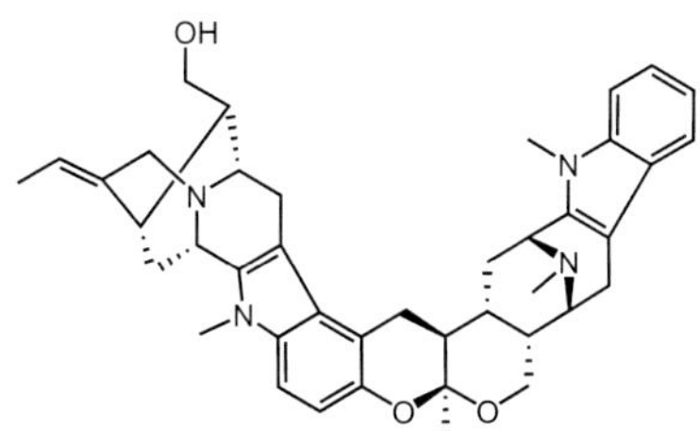

1124

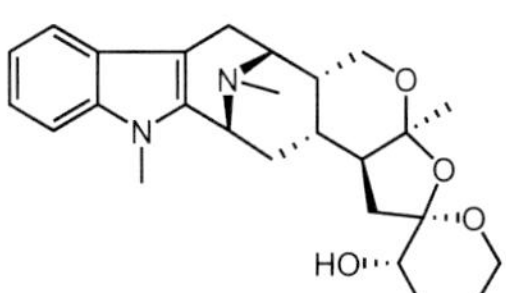

1127

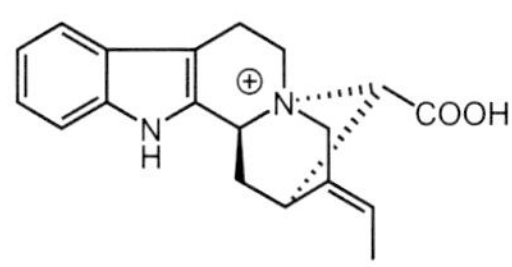

1128

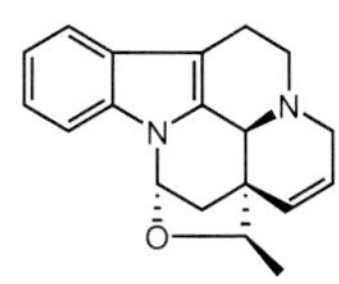

1129

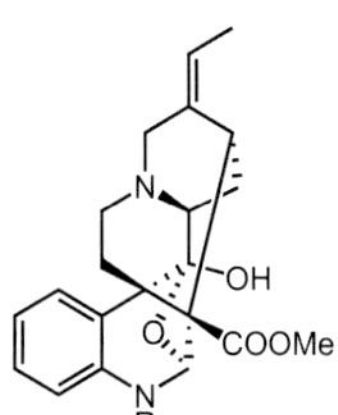

1130 R = H
1133 R = Me

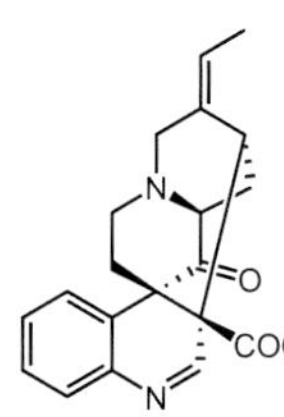

1131

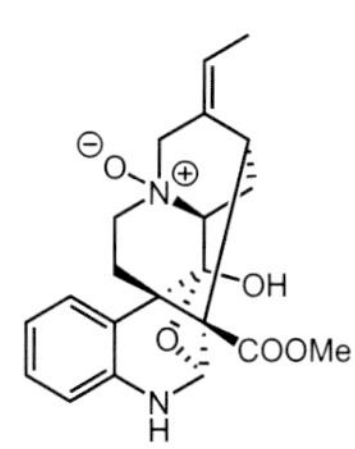

1132

Fig. 90 Structures of alkaloids **1119–1132** reported from Apocynaceae species

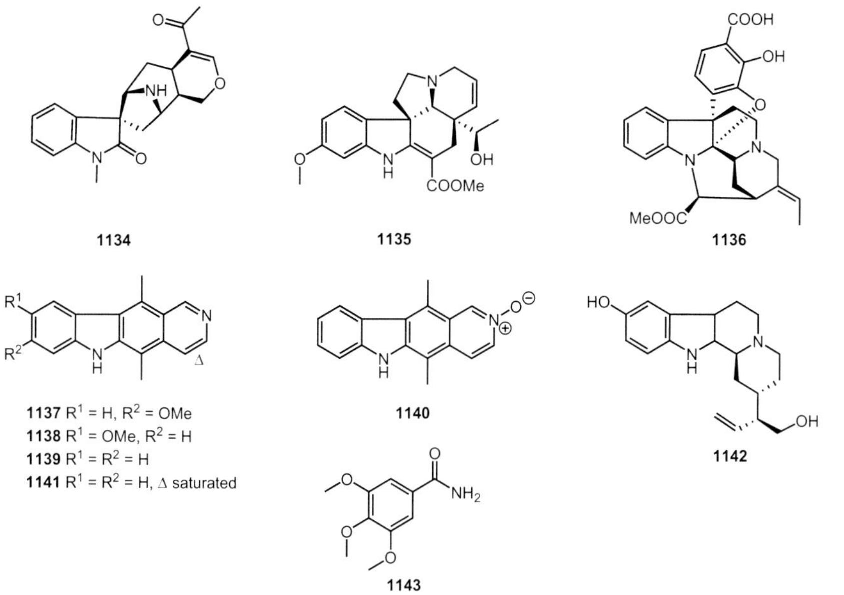

Fig. 91 Structures of alkaloids **1133–1142** and the benzamide **1143** reported from Apocynaceae species

(**1123**), macralstonidine (**1124**), alstonerine (**1125**), voacalgine B (**1126**), and voacalgine C (**1127**)), as well as singaporentinidine (**1128**), and 16,19-epoxy-Δ^{14}-vincanol (**1129**)) [39, 385, 386, 394–396] were obtained from *Alstonia*, *Voacanga*, and *Melodinus* species. Four rhazicine alkaloids (rhazicine (**1130**), rhazimine (**1131**), rhazicine *N*-oxide (**1132**), and lanceomigine (**1133**)) have been reported from *Melodinus acutiflorus* and *Alstonia constricta* [39, 387, 397].

Alstonisine (**1134**) was isolated from *Alstonia muelleriana* [394], vandrikidine (**1135**) from *Melodinus baccellianus* [39], voacalgine A (**1136**) from *Voacanga grandifolia* [396], and five ellipticine alkaloids (8-methoxyellipticine (**1137**), 9-methoxyellipticine (**1138**), ellipticine (**1139**), *N*4-oxyellipticine (**1140**), and dihydro-3,14-ellipticine (**1141**)) were all isolated from *Ochrosia* species [39, 373, 377, 398, 399]. The final monoterpenoid indole alkaloid, 10-hydroxyantirhine (**1142**), found in *Neisosperma kilneri*, is an analogue of **996** isolated from the Rubiaceae species *Antirhea putaminosa* [39]. Trimethoxybenzamide (**1143**) occurs in *Alstonia constricta* [371]. The ellipticines, when first discovered in Australian *Ochrosia* species, were actively pursued as potential novel anticancer agents. However, the toxicity and insolubility of the lead compound, ellpticine (**1139**), resulted in it later being abandoned. It was one of the first DNA intercalators to be discovered, with its mode of action being as a topoisomerase II inhibitor [400].

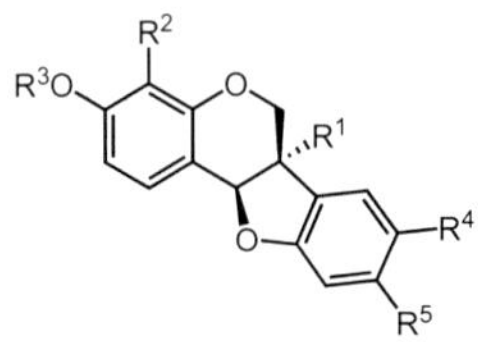

1144 $R^1 = R^2 = R^3 = R^4 = H$, $R^5 = OMe$
1145 $R^1 = R^4 = H$, $R^2 = OH$, $R^3 = Me$, $R^5 = OMe$
1146 $R^1 = R^3 = R^4 = H$, $R^2 = R^5 = OMe$
1147 $R^1 = OH$, $R^2 = R^3 = R^4 = H$, $R^5 = OMe$
1148 $R^1 = R^2 = R^3 = H$, $R^4 = R^5 = OMe$
1149 $R^1 = R^2 = R^3 = H$, $R^4–R^5 = OCH_2O$
1150 $R^1 = H$, $R^2 = OH$, $R^3 = Me$, $R^4–R^5 = OCH_2O$
1151 $R^1 = R^3 = H$, $R^2 = OMe$, $R^4–R^5 = OCH_2O$

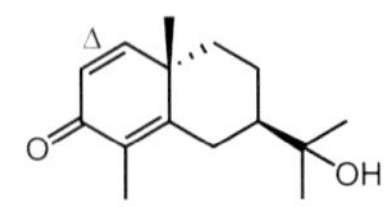

1152 Δ saturated
1153

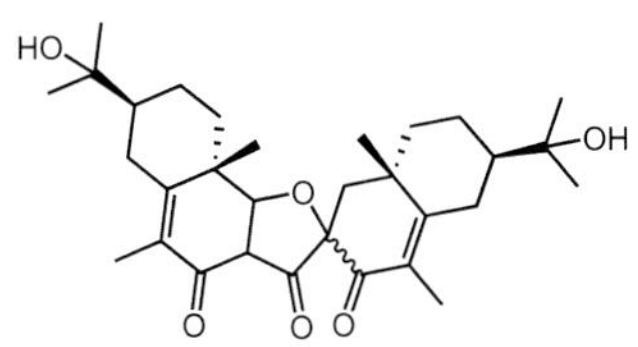

1154

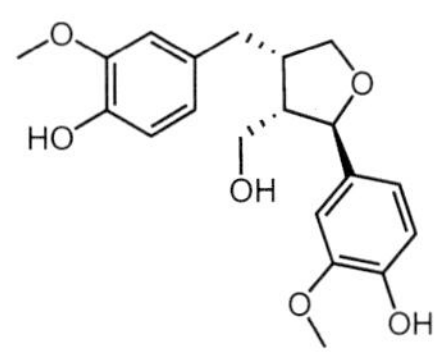

1155

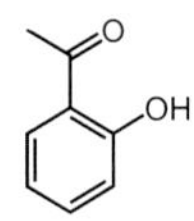

1156

Fig. 92 Structures of compounds **1144–1156** reported from Apocynaceae species

A paper reporting isoflavonoids (medicarpin (**1144**), melilotocarpan A (**1145**), 4-methoxymedicarpin (**1146**), 3-hydroxymedicarpin (**1147**), 3-hydroxy-8,9-dimethoxypterocarpan (**1148**), maackiain (**1149**), 4-hydroxy-3-methoxy-8,9-methylenedioxy pterocarpan (**1150**), and 4-methoxymaackiain (**1151**)) from the Australian *Ochrosia eliptica*, cultivated in southern China, could be a case of plant misidentification since the pterocarpans described are compounds most closely associated with Fabaceae species (Fig. 92) [401]. The other non-alkaloid constituents of Rauvolfioideae species are a triterpenoid (**802**) from western Australian *Alyxia buxifolia* [402], the sesquiterpenoids, carissone (**1152**), dehydrocarissone (**1153**) and carindone (**1154**), a lignan, carinol (**1155**), and α-hydroxyacetophenone (**1156**) from *Carissa lanceolata* [403, 404], and a steroid glycoside, cerbertin (**1157**) obtained from *Cerbera floribunda* [405].

There is a greater representation of species from the Apocynoideae and Asclepiadoideae subfamilies in Australian rainforests, but the number of these species inveistigated chemically to date is small (Table 25). Five *Parsonsia* species, vines in the Apocynoideae, contain pyrrolizidine alkaloids (lycopsamine (**1158**), intermedine (**1159**), indicine (**1160**), indicine *N*-oxide (**1161**) and platynecine trimethoxybenzoate ester (**1162**)) and the carboxylic acid, trachelanthic acid (**1163**) [39, 406, 407], while species of *Hoya* and *Marsdenia* from the Asclepiadoideae contain A-ring oxidized α- and β-amyrins (australinal A (**1164**), australinol A (**1165**), methyl dihydronyctanthate (**1166**), australinal B (**1167**), australinol B (**1168**), methyl dihydroaustralinoate (**1169**), australinal D (**1170**), australinol D (**1171**), australinal E (**1172**), australinol E (**1173**), australinal F (**1174**), and australinol F **1175**)), and

Table 25 Distribution of natural products isolated from the subfamilies Apocynoideae and Asclepiadoideae within the Apocynaceae

Species	Collection location[a,b]	Plant part[c]	Compound class	Compound(s)	Refs.
Apocynoideae					
Parsonsia densivestita	SEQ	lf	Pyrrolizidine	**1162**	[39]
Parsonsia eucalyptophylla	SEQ	lf	Carboxylic acid	**1163**	[39]
	NQ, SEQ	lf	Pyrrolizidine	**1158–1161**	[39, 406]
Parsonsia fulva	SEQ	st	Pyrrolizidine	**1161**	[39]
Parsonsia straminea	NQ	lf	Pyrrolizidine	**1158**, **1159**	[407]
Asclepiadoideae					
Hoya australis subsp. *australis*	SEQ (cult. (Netherlands))	lf	Triterpenoid	**1164–1169**, **1170–1177**	[408]
Marsdenia flavescens	NSW	wp	Steroid	**1178**	[409]
Marsdenia rostrata	NNSW	wp	Pyridine	**895**	[410]
Vincetoxicum polyanthum (*Tylophora benthamii*)	NQ	wp	Tylophorine	**1183–1190**	[411, 412]

[a] SEQ = Southeast Queensland, NQ = North Queensland, NNSW = Northern New South Wales, NSW = New South Wales, cult = cultivated, [b] natural location shown for those species studied from cultivated material, [c] lf = leaves, wp = whole plant, st = stems

lupeol-type triterpenoids (australinal C (**1176**) and australinol C (**1177**)), and steroids (flavescin (**1178**), rostratine (**1179**), metaplexigenin (**1180**), rostratamine (**1181**), and dihydrorostratine (**1182**)), respectively (Fig. 93) [408–410]. One species of *Vincetoxicum* (previously *Tylophora*, Asclepiadoideae) contains eight vesicant indolizidine-phenanthrene alkaloids (tylocrebrine (**1183**), tylophorine (**1184**), and alkaloids A–F (**1185–1190**)) that are related to natural products isolated from *Cryptocarya* (Lauraceae), *Ficus* (Moraceae), and *Boehmeria* (Urticaceae) species (Fig. 94) [411, 412]. The triterpenoids, steroids, and sesquiterpenoids are widely distributed classes of natural products, while the pyrrolidine and indolizidine-phenanthrene alkaloids are of restricted occurrence in unrelated families. This suggests a convergent evolution, with pyrroline derived from ornithine being the precursor to the indolizidine alkaloid, and arginine the precursor to the pyrrolizidine alkaloids.

Lamiales

Although the Lamiales are represented by ten families and 157 species in Australian rainforests, only 12 species in the Lamiaceae (five species), Scrophulariaceae (three species), and Oleaceae (four species) have been studied (Fig. 95, Table 26).

The three *Myoporum* species (Scrophulariaceae) all produce the furanosesquiterpenoids ngaione (**1191**), myodesmone (**1192**), 10,11-dehydromyodesmone (**1193**),

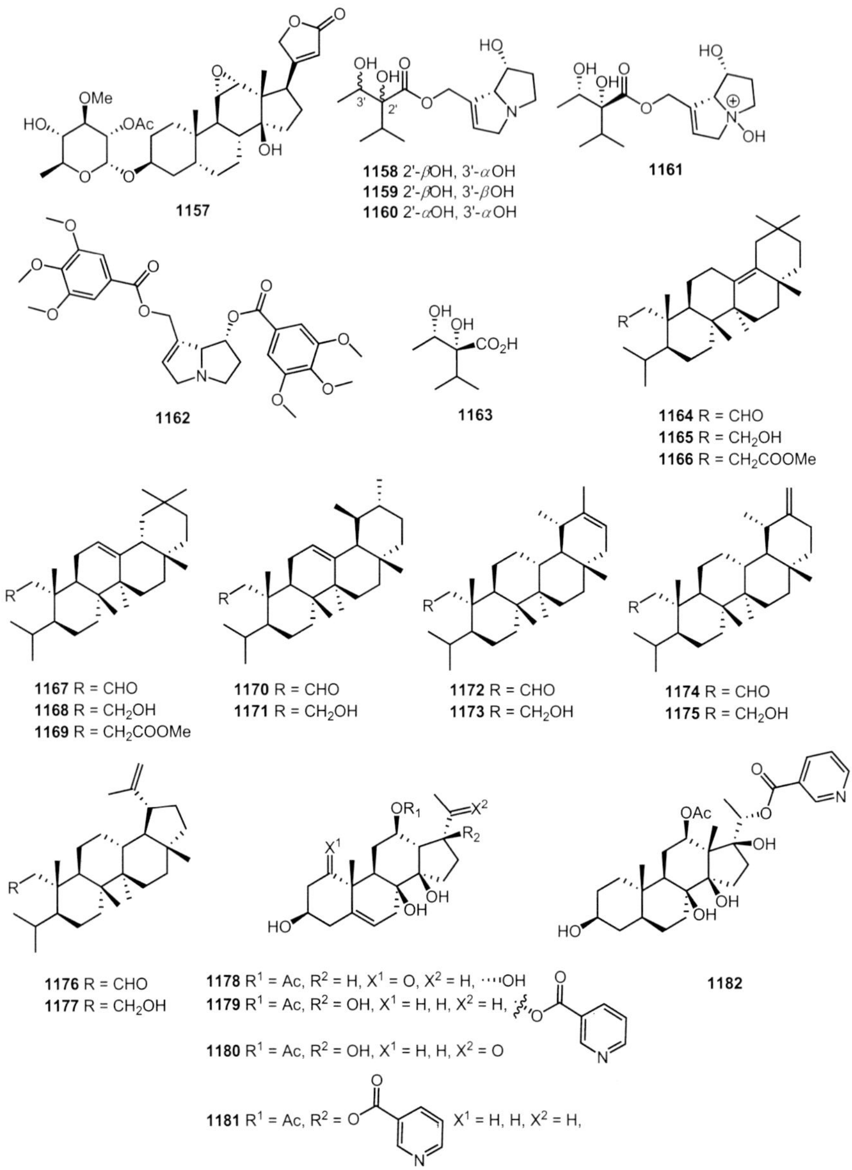

Fig. 93 Structures of compounds **1156–1182** reported from Apocynaceae species

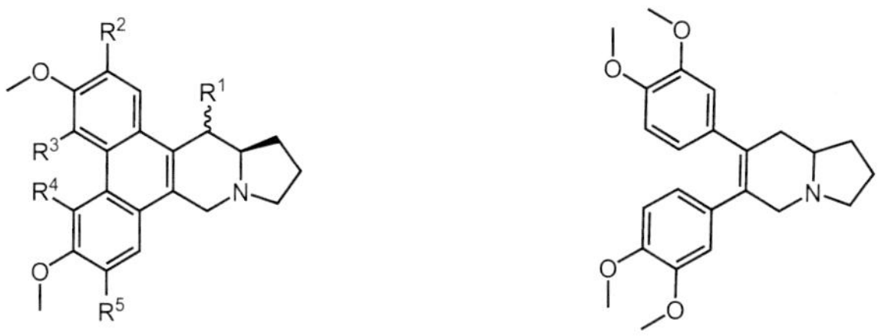

1183 $R^1 = R^3 = R^5 = H, R^2 = R^4 = OMe$
1184 $R^1 = R^3 = R^4 = H, R^2 = R^5 = OMe$
1185 $R^1 = OH, R^2 = R^4 = H, R^3 = R^5 = OMe$
1186 $R^1 = R^2 = R^4 = H, R^3 = OH, R^5 = OMe$
1187 $R^1 = R^3 = OH, R^2 = R^4 = H, R^5 = OMe$
1188 $R^1 = OH, R^2 = R^3 = R^5 = OMe, R^4 = H$
1189 $R^1 = R^4 = H, R^2 = R^3 = R^5 = OMe$

1190

Fig. 94 Structures of alkaloids **1183–1190** reported from *Vincetoxicum polyanthum*

isomyodesmone (**1194**), 10,11-dehydroisomyodesmone (**1195**), and myoporone (**1196**) [413–416], and the two *Jasminium* species (Oleaceae) contain the commonly occurring lignan pinoresinol (**1197**) [417]. *Jasminium lineare* and *J. gracile* accumulate the terpenoid alkaloid jasminine (**1198**) (Fig. 96) [418].

Although *Bascilicum polystachyon* (Lamiaceae) is a rainforest species, only collections from arid Queensland have been studied, and these contain the labdane, stachyonic acid A (**1199**), and three primarane diterpenoids (basimarol A (**1200**), basimarol B (**1201**), and basimarol C (**1202**)). Stachyonic acid A displays a spectrum of antiviral activities [419, 420]. Two cyanogenic glycosides lucumin (**1203**), and prunasin (**1204**) have been isolated from *Clerodendron grayi* [421]. *Glossocarya calcicola* from north Queensland contains the clerodane diterpenoids calcicolins A–C (**1205–1207**) that are cytotoxic to insects [422]. *Oxera splendida* produces two oleanane triterpenoids (3-*O*-acetyloleanolic aldehyde (**1208**) and 3-*O*-acetylerythrodiol (**1209**)) [423]. The leaves of *Plectranthus parviflorus* cultivated in Switzerland have afforded parviflorons A–F (**1210–1215**), which are bright red colored abietane diterpenoids [424].

Caryophyllales

The Caryophyllales are poorly represented in Australian rainforests, with 41 species in 13 genera (Fig. 97). Only one of these, *Rhagodia candolleana* subsp. *candolleana* (Amaranthaceae), has been reported to contain the steroid ecdysone (**1216**), in cultivated seeds studied in the U.K. (Table 27, Fig. 96) [425].

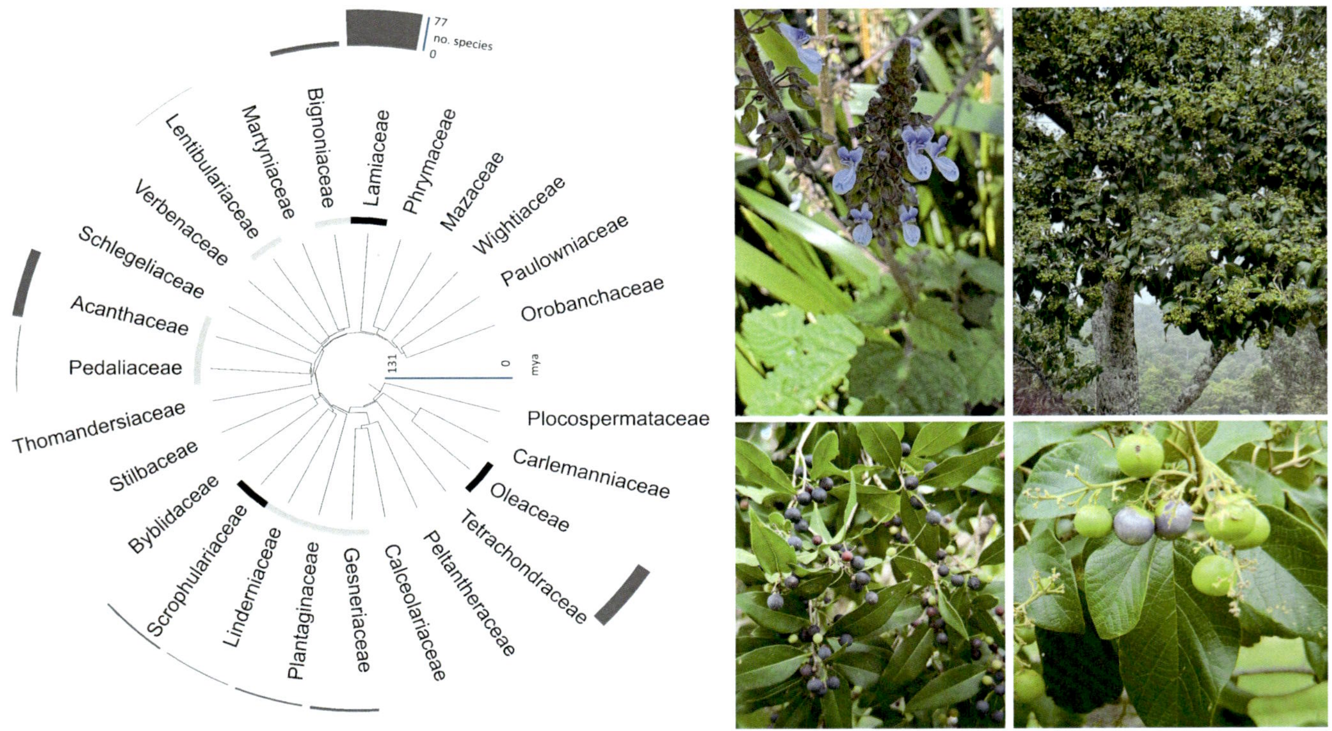

Fig. 95 Dated molecular phylogeny of the Lamiales [7], showing those families that have species present in Australian rainforests (grey) and studied (black), and the number of species occurring in Australian rainforests. The flowers of *Plectranthus parviflorus* (Lamiaceae, top left), *Notelaea longifolia* fruits (Oleaceae, bottom left) and *Gmelina leichardtii* tree and fruits (Lamiaceae, top and bottom right) (Photographs AR Carroll)

Table 26 Distribution of natural products isolated from the Lamiales

Species	Natural location[a,b]	Plant part(s)[c]	Compound class	Compound(s)	Refs.
Scrophulariaceae					
Myoporum acuminatum	SEQ	lf	Sesquiterpenoid	**1191–1195**	[413, 414, 416]
Myoporum betcheanum	SEQ	lf, st	Sesquiterpenoid	**1196**	[415]
Myoporum montanum	SEQ	lf	Sesquiterpenoid	**1196**	[416]
Oleaceae					
Jasminum dallachii	NQ	lf, st	Lignan	**1197**	[417]
Jasminum simplicifolium subsp. *australiense*	NQ	lf, st	Lignan	**1197**	[417]
Jasminum lineare	CQ	wp	Alkaloid	**1198**	[418]
Jasminum gracile	NQ	wp	Alkaloid	**1198**	[418]
Lamiaceae					
Basilicum polystachyon	WQ	wp	Diterpenoid	**1199–1202**	[419, 420]
Clerodendrum grayi	NQ	lf	Cyanohydrin	**1203**, **1204**	[421]
Glossocarya calcicola	NQ	lf	Diterpenoid	**1205–1207**	[422]
Oxera splendida	NQ	r	Triterpenoid	**1208**, **1209**	[423]
Plectranthus parviflorus	NNSW (cult. (United Kingdom))	lf	Diterpenoid	**1210–1215**	[424]

[a] SEQ = Southeast Queensland, NQ = North Queensland, CQ = central Queensland, NNSW = Northern New South Wales, WQ = Western Queensland, cult. = cultivated, [b] natural location shown for those species studied from cultivated material, [c] lf = leaves, wp = whole plant, st = stems, r = roots

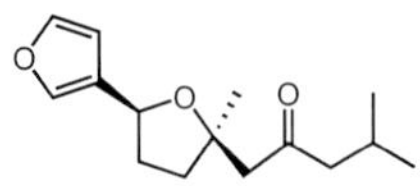

1191

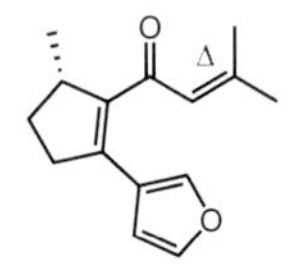

1192 Δ saturated
1193

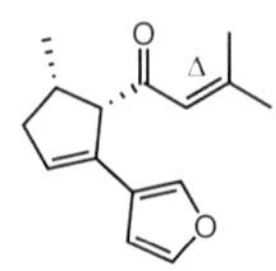

1194 Δ saturated
1195

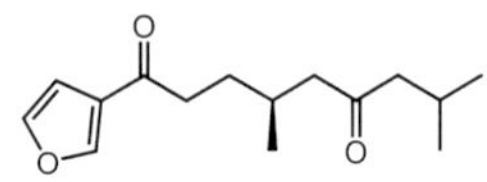

1196

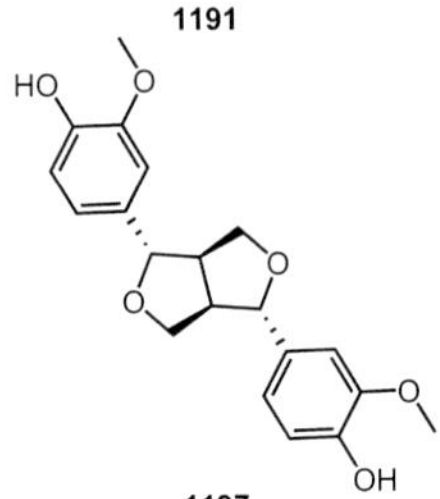

1197

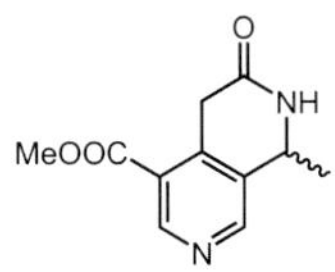

1198

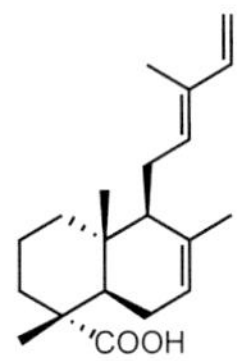

1199

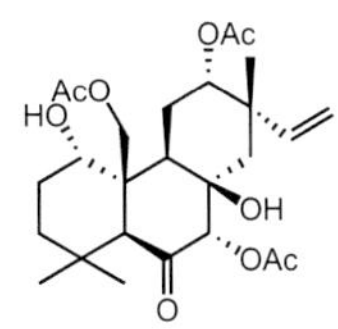

1200

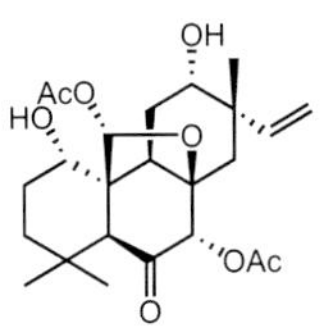

1201

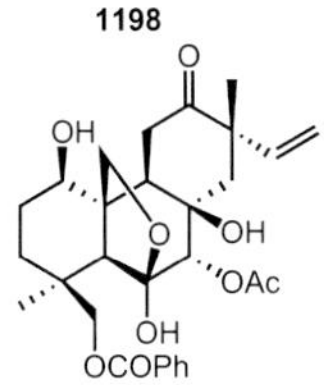

1202

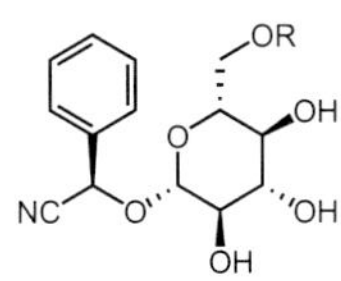

1203 R =
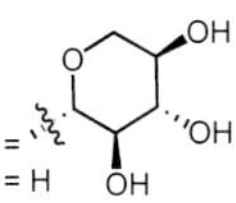
1204 R = H

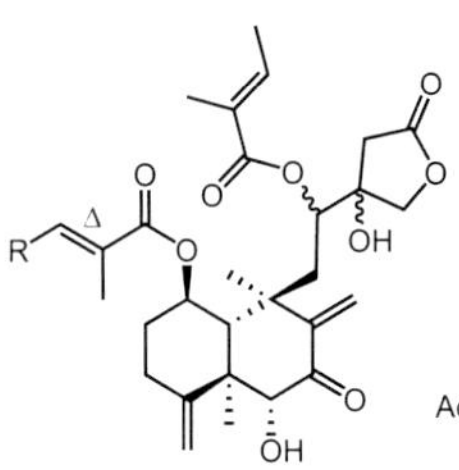

1205 R = H, Δ saturated
1206 R = Me
1207 R = Me, Δ saturated

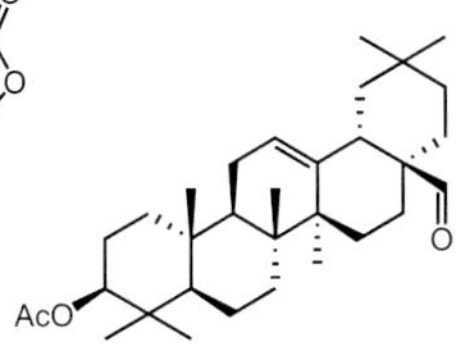

1208 R = CHO
1209 R = CH_2OH

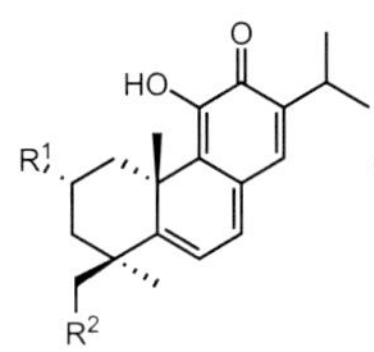

1210 R^1 = H, R^2 = a
1211 R^1 = H, R^2 = d
1212 R^1 = H, R^2 = b
1213 R^1 = H, R^2 = c
1214 R^1 = b, R^2 = H
1215 R^1 = c, R^2 = H

a =
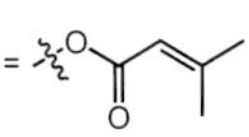

b =
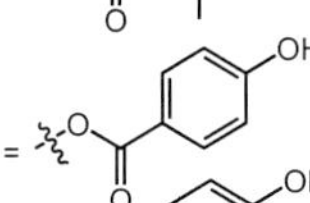

c =
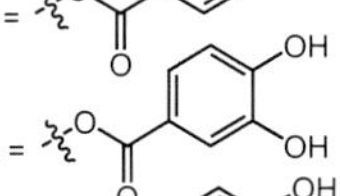

d =
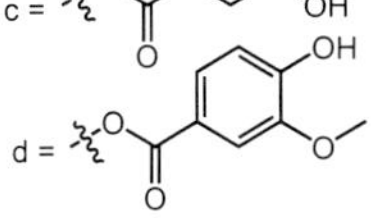

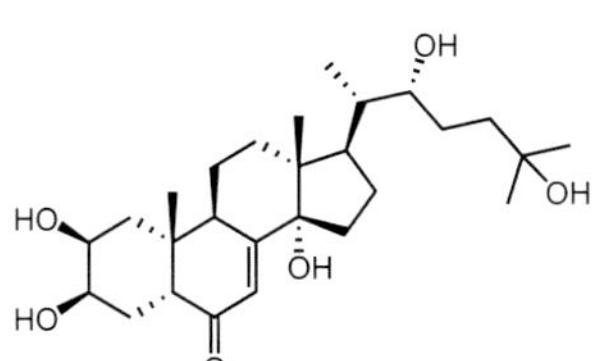

1216

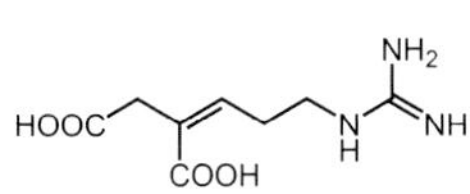

1217

Fig. 96 Structures of compounds **1191–1217** reported from Lamiales, Caryophyllales and Santalales species

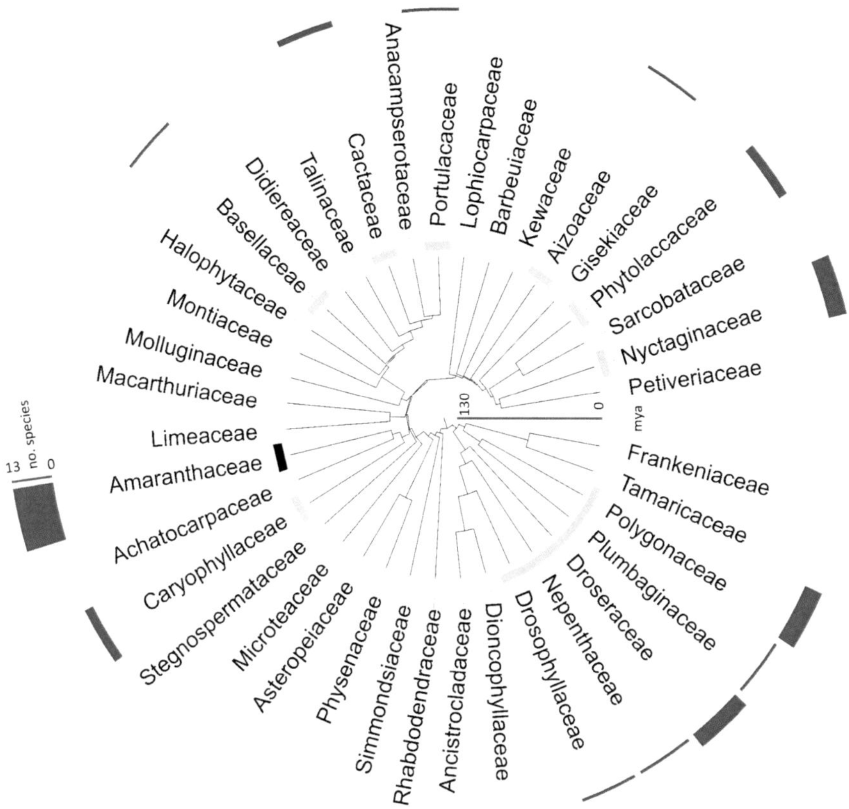

Fig. 97 Dated molecular phylogeny of the Caryophyllales [7] showing those families that have species present in Australian rainforests (grey) and studied (black), and the number of species occurring in Australian rainforests

Santalales

The order Santalales contains many species that are root and aerial parasites of other plants. Australia's rainforests have species in five of its seven families (Fig. 98). The Loranthaceae, the most speciose family within the Santalales are also called mistletoes, with their seed dispersal being facilitated by birds. The family has an Australasian Gondwanan origin around 59 mya and diversified across Gondwana into South America with a second wave of diversification occurring into Southeast Asia through bird migration from Australia. Although the Loranthaceae has 39 species in Australian rainforests, only one has been studied. *Lysiana subfalcata* from south-eastern Queensland contains the arginine mimic lysianadioic acid (**1217**). This simple molecule is a potent inhibitor of the coagulation cascade enzyme carboxypeptidase B [426]. Outside of Australia, the family is also poorly studied.

Table 27 Distribution of natural products isolated from the Amaranthaceae and Loranthaceae

Species	Collection location[a,b]	Plant part[c]	Compound class	Compound	Refs.
Amaranthaceae					
Rhagodia candolleana subsp. *candolleana*	NNSW (cult. (Spain))	s	Steroid	**1216**	[425]
Loranthaceae					
Lysiana subfalcata	SEQ	wp	Alkaloid	**1217**	[426]

[a] SEQ = Southeast Queensland, NNSW = Northern New South Wales, cult. = cultivated, [b] natural location shown for those species studied from cultivated material, [c] s = seeds, wp = whole plant

4.5.2 Superrosids

The second major subgroup of core eudicots are the Superrosids. A 2024 paper updated phylogenetic relationships within the angiosperms with molecular evidence from almost 8,000 genera and 353 nuclear genes, and this has led to a restructuring of relationships within the Superrosids [7]. The Vitales are sisters to all Rosids, and the Saxifragales are then a sister to the remaining Rosids, and these are subtended into the Crossosomatales/Geraniales sister pairing, and then the sister pairing of Myrtales/ Zyophyllales, and finally the division of the remaining orders into two major clades the Fabids (four orders) and Malvids (eight orders).

Vitales and Saxifragales

Although the Vitales and Saxifragales are found in Australia rainforests, none of the Vitaceae (35 species) and Saxifragales (five species) have been investigated chemically.

Myrtales

The Myrtales (containing nine families) is believed to have a west Gondwanan (south America/African) origin about 127 mya [7, 427]. The Combretaceae is sister to the other families and has 21 species in Australian rainforests. Within the other eight families, Onagraceae (three species), Lythraceae (two species), Melastomataceae (11 species) are also found in Australian rainforests, but the Myrtaceae is dominant and includes one of the highest number of species of any family with 230 species (Fig. 99).

The Myrtaceae is believed to have also originated in Africa (or maybe Eastern Gondwana) with subsequent long-distance dispersal into Australia in the Paleocene (65 mya) [428]. The Myrtaceae is split into two subfamilies—the Psiloxyloideae and the Myrtoideae. Psiloxyloideae contains only two genera, and four species restricted to southern Africa and the Mascarene Islands off the east coast of Africa, and these

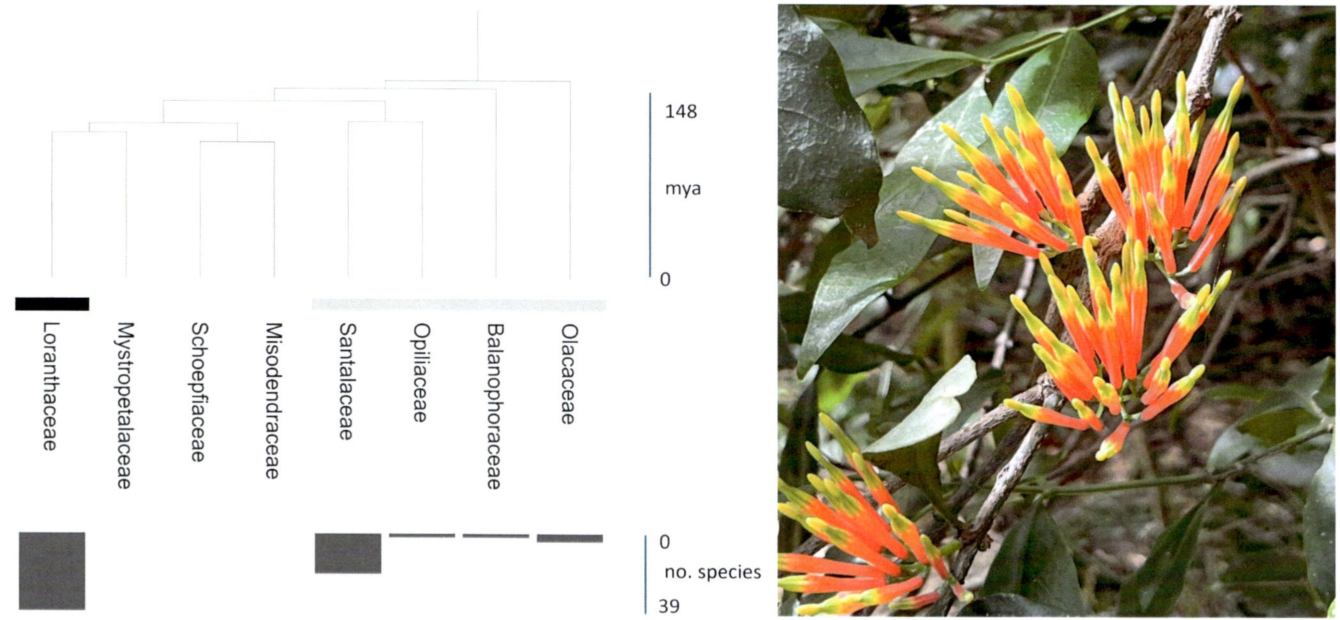

Fig. 98 Dated molecular phylogeny of the Santalales showing those families that have species present in Australian rainforests (grey) and studied (black) and the number of species occurring in Australian rainforests. Photograph of a rainforest mistletoe *Amylotheca dictopheleba.* (Photograph AR Carroll)

are basal members of the Myrtaceae. The rest of the family comprises 140 genera and about 5,500 species [428]. The Australian rainforest Myrtaceae species are the ancestors of the species that now dominate the Australian dryland landscapes. Australia's rainforests are home to many species that are representatives of the early divergence within the family in the late Cretaceous including those in the genera *Xanthostemon*, *Lophostemon*, and *Melaleuca*/*Callistemon*. A branch point in the early Paleocene has led to a lineage that includes the mostly sclerophyllous Eucalypteae (seven Eucalypt genera), the Syncarpieae (the turpentines) and the Leptospermeae (including the *Leptospermum* and *Kunzea*), and another group that diverged to produce two clades of fleshy fruited species. One, the Myrteae, includes Australian rainforest species in the genera *Rhodomyrtus* and *Pilidiostigma* as well as *Myrtus communis* (from the Mediterraneum) and *Psidium* spp. (guavas from America), while the other mostly includes the rainforest *Syzygium* species. There is ongoing debate regarding the taxonomy of *Callistemon* [429]. Several Australian herbaria have followed the paper by Craven et al. in 2014 and subsumed *Callistemon* into the larger genus *Melaleuca*,

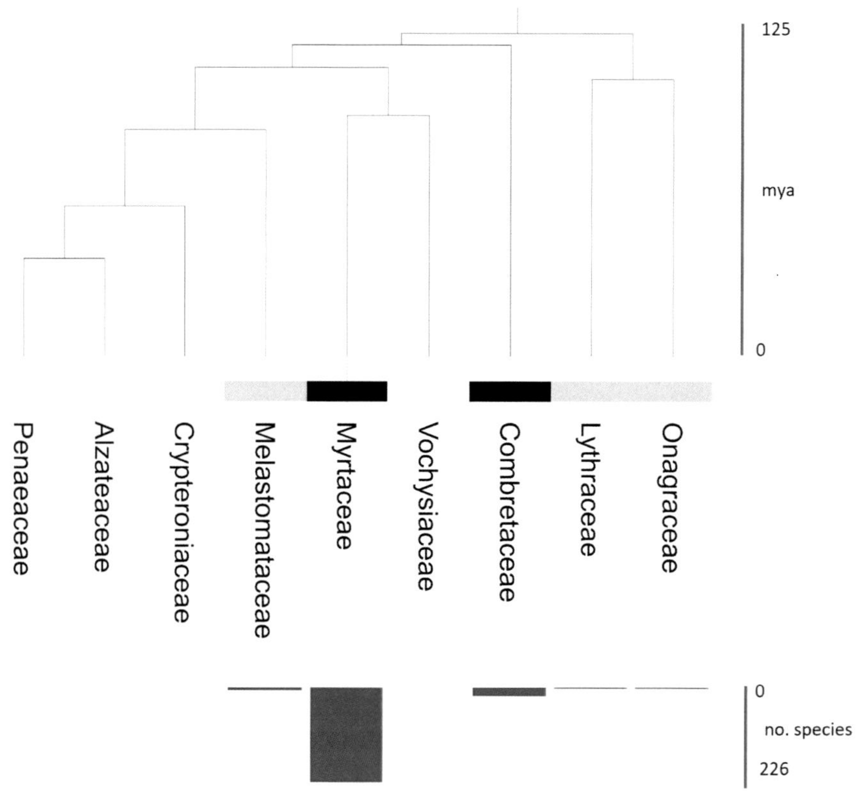

Fig. 99 Dated molecular phylogeny of the Myrtales [7] showing those families that have species present in Australian rainforests (grey) and studied (black) and the number of species occurring in Australian rainforests

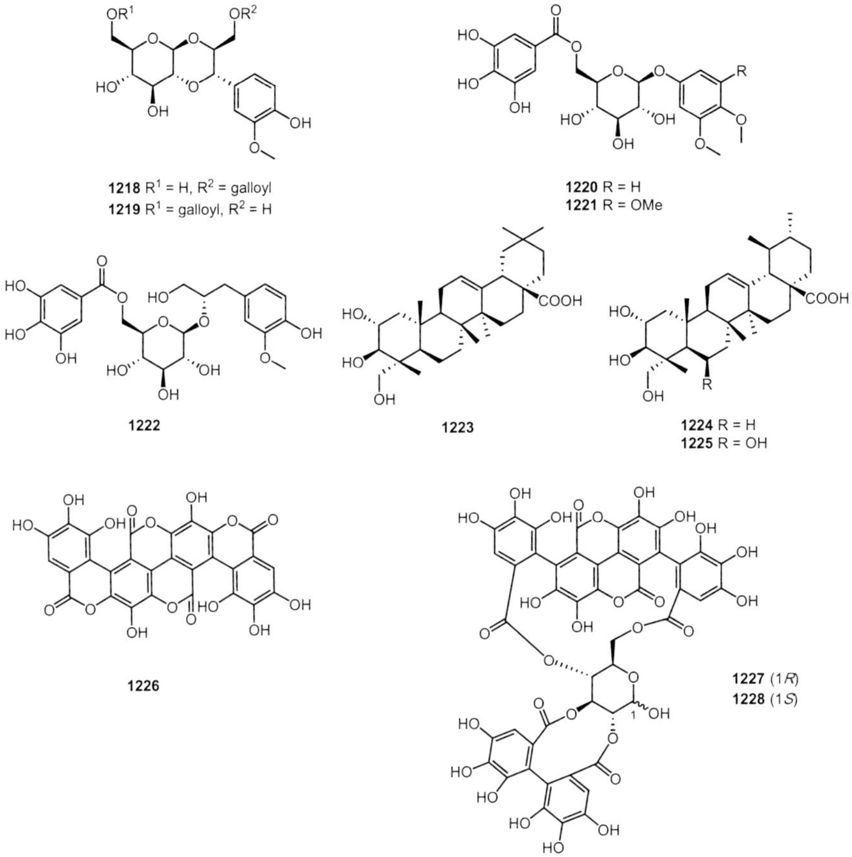

Fig. 100 Structures of compounds **1218–1228** reported from Combretaceae species

while others still recognize *Callistemon* as a valid genus [430]. Molecular evidence suggests that the genus *Melaleuca* (including *Callistemon* species) is polyphyletic [429], but the World Plant Index now recognizes all previously named *Callistemon* species as *Melaleuca* species. The present author has adopted this assignment and so those papers reporting on the chemistry of the Australian rainforest endemics *Callistemon viminalis* and *C. salignus* are discussed herein under the names *Melaleuca viminalis* and *M. salicinai*, respectively (Fig. 100).

The study of the chemistry of Australian rainforest Myrtales outside of the Myrtaceace is restricted to three investigations of Combretaceae species (Table 28, Fig. 101). *Macropyteranthes leichhardtii* contains five gallic acid esters of glucose (macropteranthol (**1218**), mallophenol A (**1219**), 3,4-dimethoxyphenol-1-D-(6-*O*-galloyl)glucopyranoside (**1220**), 3,4,5-trimethoxyphenol-1-D-(6-*O*-galloyl)glucopyranoside (**1221**), and 3-(4-hydroxy-3-methoxyphenyl)propane-1,2-diol 2-D-(6-*O*-galloyl)glucopyranoside (**1222**)), which in the three congeners is also

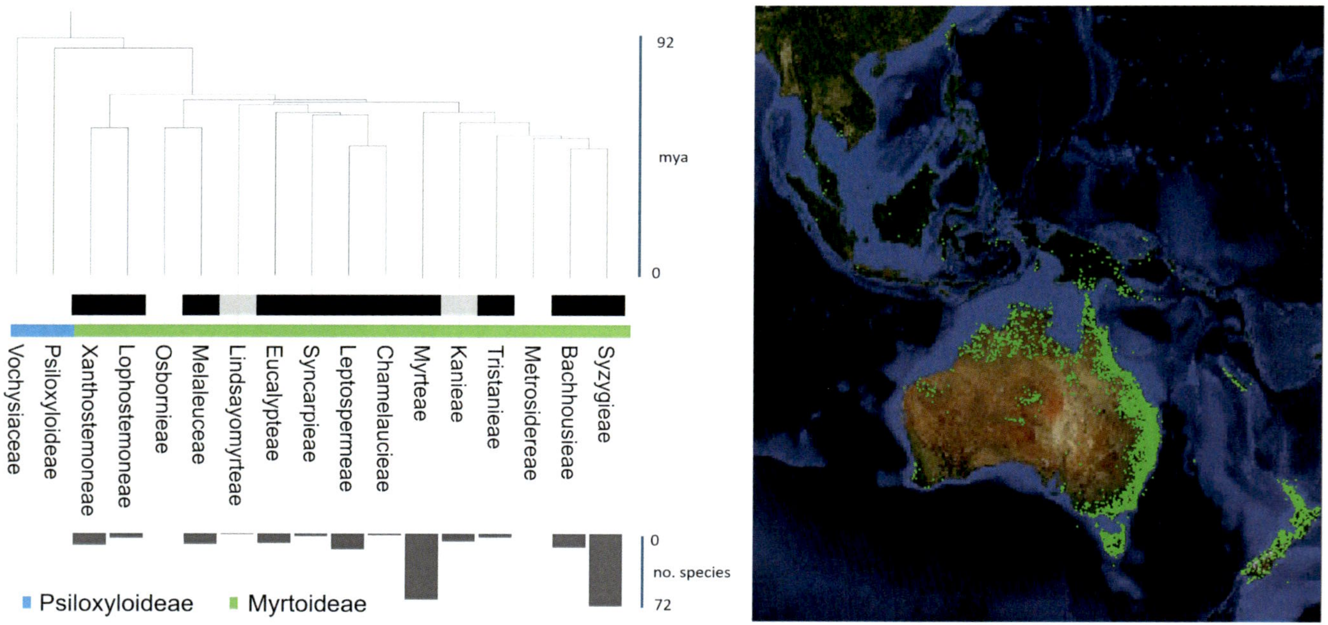

Fig. 101 Dated molecular phylogeny of the Myrtaceae showing subfamilies and tribes with species occurring (grey) and studied (black) and the number of species in Australian rainforests. The map shows the collection locations for herbarium records of Australian rainforest Myrtaceae species that have been chemically investigated

Table 28 Distribution of natural products isolated from the Combretaceae

Species	Collection location[a]	Plant part[b]	Compound class	Compounds	Refs.
Macropteranthes leichhardtii	SEQ	wp	Gallic acid ester	**1218–1222**	[431]
Terminalia complanata	PNG	lf	Triterpenoid	**1223–1225**	[432]
Terminalia oblongata	CQ	lf	Tannin	**1226–1228**	[433, 434]

[a] SEQ = Southeast Queensland, CQ = central Queensland, PNG = Papua New Guinea, cult = cultivated, [b] lf = leaves, wp = whole plant

conjugated to phenylpropanes (**1218**, **1219**, **1222**) [431]. *Terminalia complanata* contains three triterpenoid carboxylic acids (arjunolic acid (**1223**), asiatic acid (**1224**), and terminolic acid (**1225**)) and *T. oblongata* produces three tannins (terminalin (**1226**), β-punicalagin (**1227**), and α-punicalagin (**1228**)) [432–434]. Two of the gallic acid esters (**1220** and **1222**) inhibit tyrosyl-DNA phosphodiesterase 1 and two of the tannins (**1227** and **1228**) cause liver necrosis.

Myrtaceae

The chemistry of Australian rainforest Myrtaceae has been studied extensively (Table 29) and the structures identified can be classified into five major groups: triterpenoids; phloroglucinols, and β-triketones and their meroterpenoid adducts; flavonoids and chalcones; neolignans, and tannins. There has been considerable interest in Australian rainforest Myrtaceae species in China since the early 2000's and many studies on plant material from endemic Australian species cultivated in China have appeared in the literature. Some of these species are reported to be "traditional" Chinese medicines, but these claims seem to stretch the definition of traditional because the plants are Australian endemics that are likely to have been cultivated in China only relatively recently [435].

The 30 triterpenoids (Fig. 102) reported from Australian rainforest Myrtaceae are derivatives of either betulinic acid, ursolic acid or oleanolic acid. Their occurrence in the Australian rainforest Myrtaceae is widespread taxonomically. Hydroxylation of the A ring is common, as is lactonization of the carboxylic acid at C-28 and esterification of the C-3 hydroxy group with either a cinnamate or acetate substituent. These triterpenoids include **1223**, 20-taraxastene-3*α*,28-diol (**1229**), 3*α*,27-dihydroxy-28,20*β*-taraxastanolide (**1230**), 3*α*-hydroxy-13(18)-oleanene-27,28-dioic acid (**1231**), ursolaldehyde (**1232**), ursolic acid (**1233**), 1*α*-hydroxyursolic acid (**1234**), ursolic acid-*p*-coumarate (**1235**), ursolic acid 3-acetate

Table 29 Distribution of natural products isolated from the Myrtaceae

Species	Collection location[a,b]	Plant part(s)[c]	Compound class	Compound(s)	Refs.
Backhousia angustifolia	NQ	lf	Acyl BTK	**1427**, **1428**, **1434**	[493]
	SEQ	lf, tw	Triterpenoid	**1245**	[444]
Backhousia bancroftii	NQ	lf	Acylphloroglucinol	**1273–1275**	[448]
Backhousia citriodora	NQ	lf	Monoterpenoid	**1270**	[448]
Backhousia myrtifolia	NSW	lf	Flavonoid	**1310–1315**, **1317**	[450]
	NSW	lf	Lignan	**1261**, **1262**	[450]
	NQ	lf	Phenyl propane	**1260**	[448]
Baeckea gunniana	Tas	lf, st	Triterpenoid	**1242**	[439]
Corymbia intermedia	SEQ	fl	Monoterpenoid BTK	**1451–1453**, **1468**	[500]
Corymbia torelliana	SEQ	fl	Acyl BTK	**1441**	[497]
	SEQ	fl	Monoterpenoid BTK	**1454**, **1455**	[497]
	SEQ	fl	Phloroglucinol BTK	**1491–1496**, **1502**, **1503**	[497]
Eucalyptus grandis	SEQ	lf	Acyl BTK	**1437–1439**	[495]
	SEQ (cult. (Japan))	lf	Bisformylphloroglucinol	**1337**	[467]
	SEQ (cult. (Japan))	lf	Monoterpenoid formyl phloroglucinol	**1332–1336**, **1338–1344**	[466, 468, 469]
Eucalyptus resinifera	SEQ (cult. (Japan))	lf	Flavonoid	**1325**, **1326**	[461]

(continued)

Table 29 (continued)

Species	Collection location[a,b]	Plant part(s)[c]	Compound class	Compound(s)	Refs.
Eucalyptus terreticornis	SEQ (cult. (China))	lf	Chromone	**1291**	[443]
	SEQ (cult. (China))	lf	Flavonoid	**1316**, **1317**	[443]
	SEQ (cult. (China))	lf, tw	Monoterpenoid formyl phloroglucinol	**1345–1357**	[470, 471]
	SEQ (cult. (China))	lf, tw	Sesquiterpenoid formyl phloroglucinol	**1358–1366**	[471]
	SEQ (cult. (China))	lf	Triterpenoid	**896**, **974**, **1243**, **1244**, **1246**	[443]
Kunzea ericoides	NZ	lf, st	Acyl BTK	**1423**	[489]
	Tas (cult. (United Kingdom))	ap	Monoterpenoid BTK	**1455**	[501]
	NZ	lf, st	Phloroglucinol BTK	**1506**, **1507**	[489]
Leptospermum polygalifolium	NZ	lf, st	Acyl BTK	**1419**, **1420**, **1423**	[445]
	NZ	lf, st	Flavonoid	**1319–1324**	[445]
	NZ	lf, st	Triterpenoid	**1249**	[445]
Leptospermum scoparium	NZ, Tas	lf, st	Acyl BTK	**1415–1425**	[490, 491]
	Tas (cult. (China))	lf	Flavonoid	**1328**, **1329**	[462]
	Tas (cult. (China))	lf	Sesquiterpenoid chalcone	**1392–1396**	[462, 480]
	Tas (cult. (China))	fl	Triterpenoid	**1250–1252**	[446]

(continued)

Table 29 (continued)

Species	Collection location[a,b]	Plant part(s)[c]	Compound class	Compound(s)	Refs.
Lophostemon confertus	SEQ	w	Triterpenoid	**1223**	[436]
Lophostemon suaveolens	NNSW	lf	Flavonoid	**1318**	[463]
	NNSW	lf	Triterpenoid	**802**	[463]
Melaleuca bracteata	NQ (cult. (China))	lf	Phenyl propane	**1257–1259**	[449]
Melaleuca leucadendra	NQ (cult. (Taiwan))	lf	Phloroglucinol BTK	**1497–1500**	[506]
	NQ (cult. (China))	lf, st	Terpenoid benzyl linked phloroglucinol	**1390**, **1391**	[479]
	NQ (cult. (China))	hw	Triterpenoid	**1229–1234**, **1237–1241**, **1247**	[437, 438]
Melaleuca quinquenervia	SEQ	lf	Acylphloroglu-cinol	**1283**, **1284**	[455]
	SEQ	lf	Methylphloroglu-cinol	**1290**	[455]
	SEQ	lf	Monoter-penoid	**1268**, **1269**	[455]
Melaleuca salicina (syn. *Callistemon saligna*)	SEQ (cult. (China))	lf, tw	Acyl BTK	**1437–1439**	[453]
	SEQ (cult. (China))	lf, tw	Acylphloroglu-cinol	**1277–1280**	[453]
	SEQ (cult. (China))	lf, tw	Monoter-penoid acylphloroglu-cinol	**1374**, **1378**, **1387–1389**, **1448**, **1450**, **1458–1464b** **1466**, **1471**, **1472**	[453, 476]
	SEQ (cult. (China))	lf, tw	Phloroglucinol BTK	**1481–1485**, **1487**, **1510**	[453, 476]

(continued)

Table 29 (continued)

Species	Collection location[a,b]	Plant part(s)[c]	Compound class	Compound(s)	Refs.
Melaleuca styphelioides	SEQ (cult. (Egypt))	lf	Flavonoid	**1327**	[447]
	SEQ (cult. (Egypt))	lf	Tannin	**1255**, **1256**	[447]
Melaleuca viminalis (syn. *Callistemon viminalis*)	NQ (cult. (China))	lf	Acyl BTK	**1430**, **1431**	[454, 483]
	NQ (cult. (China))	lf	Acylphloroglu-cinol	**1277–1282**	[454]
	NQ (cult. (China))	lf	Bisphloroglu-cinol	**1401**, **1402**	[482, 483]
	NQ (cult. (China, United Kingdom))	lf, fr	Monoter-penoid acylphloroglu-cinol	**1371–1373**, **1375**, **1376**, **1378–1389**, **1448**, **1449**, **1456–1459**, **1463–1467**, **1469**, **1470**, **1473–1480**	[477, 478, 499, 502–504]
	NQ (cult. (China))	fr, lf	Phloroglucinol BTK	**1481**, **1482**, **1484–1486**, **1488**, **1490**, **1504**, **1505**, **1508**	[454, 505]
	SEQ	fl	Phloroglucinol BTK	**1509**	[39]
	NQ (cult. (Egypt))	b, fr	Pyrrolidine	**1272**	[452]
	NQ (cult. (China))	fr, lf	Sesquiter-penoid BTK	**1442–1446**	[499]
Pilidiostigma glabrum	NNSW	lf	Bisphloroglu-cinol	**1399**, **1400**, **1403–1405**, **1408–1410**	[484–486]
Pilidiostigma tropicum	NQ	b	Bisphloroglu-cinol	**1410**	[440]
	NQ	b	Triterpenoid	**1235**	[440]

(continued)

Table 29 (continued)

Species	Collection location[a,b]	Plant part(s)[c]	Compound class	Compound(s)	Refs.
Rhodomyrtus macrocarpa	NQ	unripe fr	Bisphloroglucinol	**1406**, **1407**, **1411**, **1410**	[487]
Rhodomyrtus psidioides	NNSW	lf	Sesquiterpenoid BTK	**1447**	[472]
	NNSW	lf	Sesquiterpenoid formyl phloroglucinol	**1367–1370**	[472]
Stockwellia quadrifida	NQ	lf	Methylphloroglucinol	**1285–1289**	[456]
Syncarpia glomulifera	NSW	lf	Acyl BTK	**1435**, **1436**	[464]
	NQ (cult. (USA))	b	Acylphloroglucinol	**1276**	[442]
	NSW	lf	Flavonoid	**1316**	[464]
	NQ (cult. (USA))	b	Monoterpenoid BTK	**1479**	[442]
	NQ (cult. (USA))	b	Phloroglucinol BTK	**1489**	[442]
	NSW	lf	Phloroglucinol BTK	**1501**	[464]
	NQ	lf	Triterpenoid	**802**, **1236**, **1248**	[441]
	NQ (cult. (USA))	b	Triterpenoid	**802**, **1236**	[442]
Syncarpia hillii	SEQ	lf	Acyl BTK	**1440**	[496]
Tristaniopsis laurina	NSW	lf	Acyl BTK	**1424**	[492]
Uromyrtus metrosideros	NQ	lf	Carotenoid	**1271**	[451]
	NQ	lf	Gallic acid ester	**1263**, **1264–1267**	[451]
Waterhousia mulgraveana	NQ	lf	Benzyl linked bis-phloroglucinol	**1412–1414**	[488]

(continued)

Table 29 (continued)

Species	Collection location[a,b]	Plant part(s)[c]	Compound class	Compound(s)	Refs.
Xanthostemon chrysanthus	NQ (cult. (China))	lf	Chalcone	**1300–1309**, **1426**, **1432**, **1433**	[457, 459, 460]
	NQ (cult. (China))	lf	Chalcone alkaloid	**1292–1299**	[457, 458]
	NQ (cult. (China))	lf	Flavonoid	**1330**, **1331**	[465, 466]
	NQ (cult. (China))	lf	Triterpenoid	**1253**, **1254**	[446]
Xanthostemon oppositifolius	SEQ	lf	Acyl BTK	**1429**	[494]

[a] SEQ = Southeast Queensland, NQ = North Queensland, NNSW = Northern New South Wales, NSW = New South Wales, Tas = Tasmania, NZ = New Zealand, cult. = cultivated, [b] natural location shown for those species studied from cultivated material, [c] lf = leaves, tw = twig, fl = flowers, fr = fruits, b = bark, hw = heartwood, w = wood, st = stems

(**1236**), 3*β*-*cis*-*p*-coumaroyloxy-2*α*-hydroxyurs-12-en-28-oic acid (**1237**), 3*β*-*trans*-*p*-coumaroyloxy-2*α*-hydroxyurs-12-en-28-oic acid (**1238**), 3*β*-*cis*-caffeoyloxy-2*α*-hydroxyurs-12-en-28-oic acid (**1239**), 3*β*-*trans*-caffeoyloxy-2*α*-hydroxyurs-12-en-28-oic acid (**1240**), 3*β*-*cis*-*p*-*O*-coumaroyloxy-2*α*-hydroxyursa-12,20(30)dienoic acid (**1241**), 3*β*-hydroxyurs-18,20(30)-dien-28-oic acid (**1242**), tereticornate A (**1243**), tereticornate B (**1244**), 2*α*-hydroxyoleanolic acid (**1245**), ursolic acid lactone (**1246**), eupha-7,24-diene-3*β*,22*β*-diol (**1247**), oleanolic acid-3-acetate (**1248**), and oleanolic acid (**1249**) [436–445]. Three acyl phloroglucinol adducts of betulinic acid, myrtphlotritin C (**1250**), myrtphlotritin A (**1251**), and myrtphlotritin B (**1252**), have been reported from the flowers of *Leptospermum scoparium* cultivated in China and two acyl β-triketone adducts of the triterpenoid dammarane myrtphlotritin D (**1253**), myrtphlotritin E (**1254**) were obtained from the leaves of the Chinese-cultivated *Xanthostemon chrysanthus* [446].

The only tannins from rainforest Myrtaceae are pedunculagin (**1255**) and casuarinin (**1256**), as reported from *Melaleuca styphelioides* cultivated in Egypt (Fig. 103). Other tannins are known to be widespread in non-rainforest Myrtaceae [447]. The phenylpropanoids, melaleucins A–C (**1257–1259**) and isoelemicin (**1260**), and the neolignans, 4-*O*-methylcedrusin (**1261**) and 7-*O*-methylcedrusin (**1262**), have been reported from *Melaleuca bracteata* and *Backhousia myrtifolia*, respectively [448–450] and several gallic acid derivatives (norbergenin (**1263**), bergenin (**1264**), galloyl-lawsoniaside A (**1265**), α-terpineol 8-*O*-*β*-D-(6-*O*-galloyl)glucopyranoside (**1266**), and uromyrtoside (**1267**)) have been reported from *Uromyrtus metrosideros* [451]. Two trioxygenated monoterpenoids (melacoside A (**1268**) and 4-(1-hydroxy-1-methylethyl)-1-methyl-1,3-cyclohexanediol (**1269**)), citronellal (**1270**), and an oxidized carotenoid reoseoside (**1271**) have been found in *Backhousia*, *Melaleuca*,

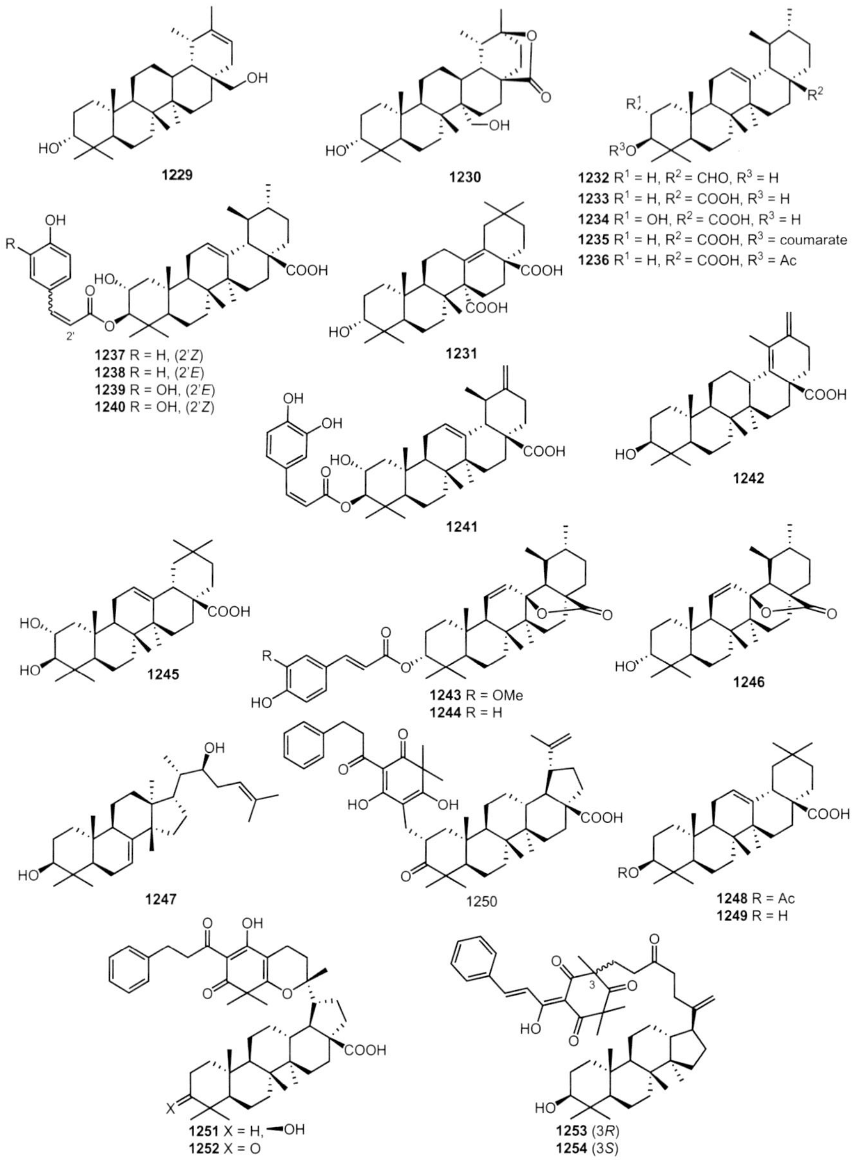

Fig. 102 Structures of triterpenoids **1228–1254** reported from Myrtaceae species

and *Uromytus* species [448, 451, 455]. The simple pyrrolidine, 2-(hydroxymethyl)-4-methyl-3,4-dihydro-2*H*-pyrrol-2-ol (**1272**), was reported from *Melaleuca viminalis* [452].

The Myrtaceae rainforest phloroglucinol adducts can be categorized into several groups. Simple phloroglucinols generally contain an acyl side chain and are

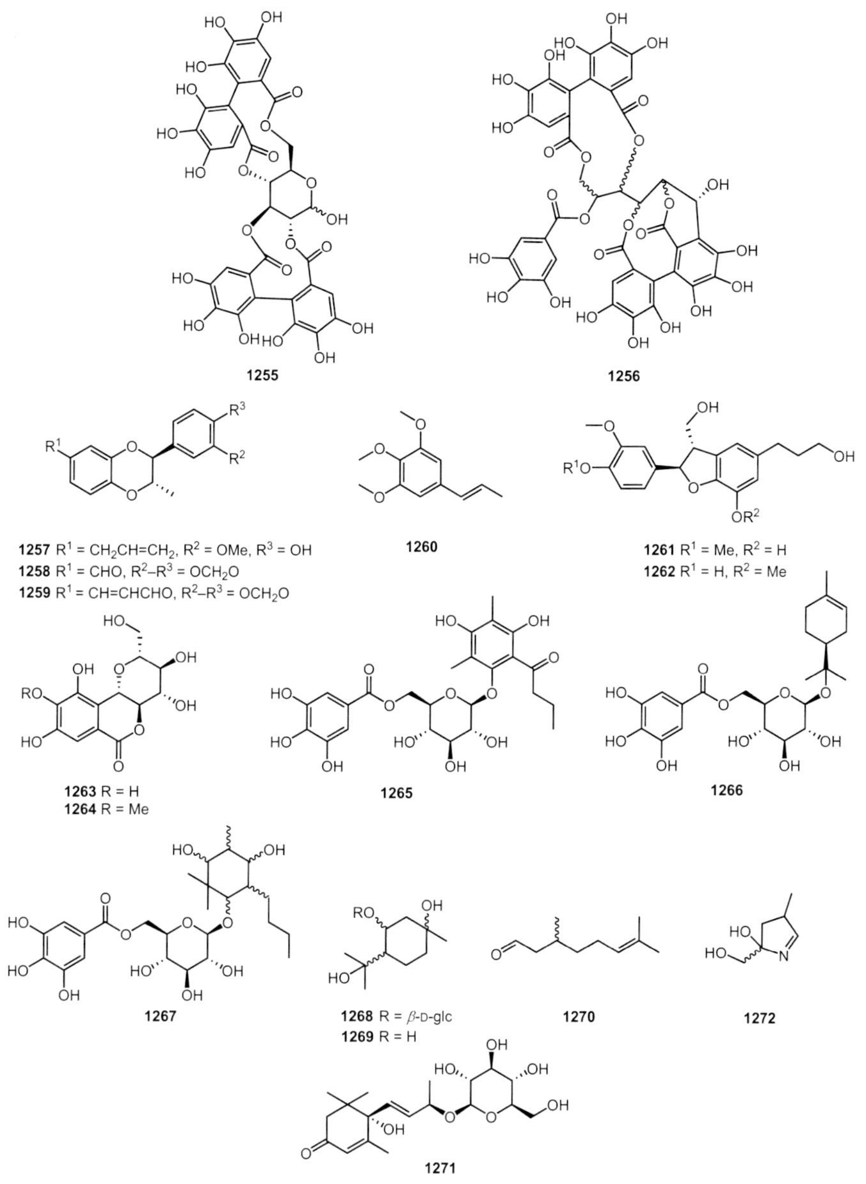

Fig. 103 Structures of compounds **1255–1272** reported from Myrtaceae species

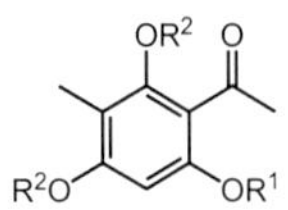

1273 $R^1 = R^2 =$ Me
1274 $R^1 = R^2 =$ H
1275 $R^1 =$ H, $R^2 =$ Me

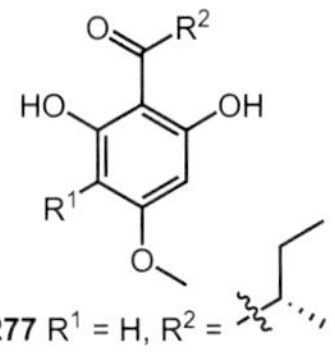

1277 $R^1 =$ H, $R^2 =$
1278 $R^1 =$ H, $R^2 =$ *i*-Pro
1279 $R^1 =$ Me, $R^2 =$ *i*-Pro
1280 $R^1 =$ H, $R^2 =$ *i*-But
1281 $R^1 =$ CHO, $R^2 =$ *i*-But
1282 $R^1 =$ Me, $R^2 =$ *i*-But

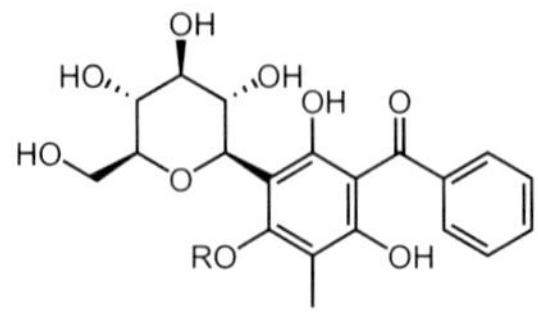

1283 R = *β*-D-glc
1284 R = H

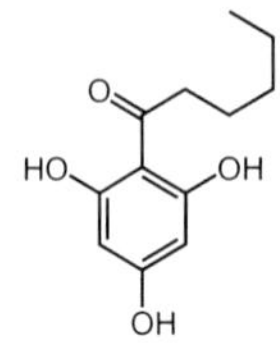

1276

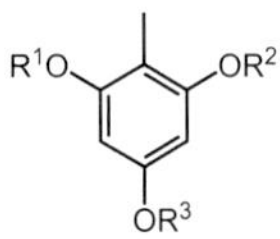

1285 $R^1 = R^2 = R^3 =$ H
1286 $R^1 =$ Me, $R^2 = R^3 =$ H
1287 $R^1 = R^2 =$ H, $R^3 =$ Me
1288 $R^1 = R^3 =$ Me, $R^2 =$ H
1289 $R^1 = R^2 = R^3 =$ Me

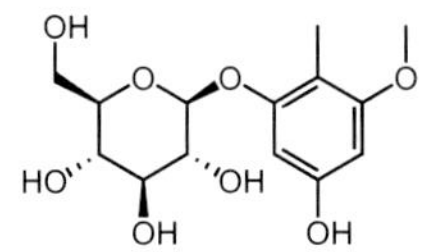

1290

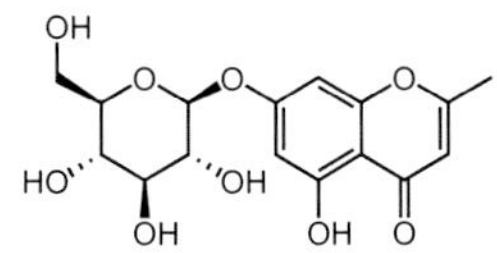

1291

Fig. 104 Structures of compounds **1273–1291** reported from Myrtaceae species

often C-methylated on the phloroglucinol ring (Fig. 104). Eighteen congeners (3-methyl-2,4,6-trimethoxy-acetophenone, (**1273**), 3-methyl-2,4,6-trihydroxy-acetophenone (**1274**), 3-methyl-2-hydroxy-4,6-dimethoxy-acetophenone (**1275**), 1-(2,4,6-trihydroxyphenyl)-1-hexanone (**1276**), callisalignone A (**1277**), 2,6-dihydroxy-4-methoxyisobutyrophenone (**1278**), 2,6-dihydroxy-4-methoxy-3-methylisopropiophenone (**1279**), 2,6-dihydroxy-4-methoxyisovalerophenone (**1280**), callistenone J (**1281**), 3-methyl-2,6-dihydroxy-4-methoxyisovalerophenone (**1282**), [3-*β*-D-glucopyranosyl-4-(*β*-D-glucopyranosyloxy)-2,6-dihydroxy-5-methylphenyl]phenyl-methanone (**1283**), [3-*β*-D-glucopyranosyl-2,4,6-trihydroxy-5-methylphenyl]phenyl-methanone (**1284**), 2,4,6-trihydroxytoluene (**1285**), 2,4-dihydroxy-6-methoxytoluene (**1286**), 2,6-dihydroxy-4-methoxytoluene (**1287**), 2-hydroxy-4,6-dimethoxytoluene (**1288**), 2,4,6-trimethoxytoluene (**1289**), and 3-hydroxy-5-methoxy-4-methylphenyl *β*-D-glucopyranoside (**1290**)) have been reported from *Backhousia*, *Syncarpia*, *Melaleuca*, and *Stockwellia* species [442, 448, 453–456]. A chromone, undulatoside (**1291**) isolated from *Eucalyptus tereticornis* is also a simple acyl phloroglucinol [443].

Dihydrochalcones are common across the taxonomic range (Fig. 105) and the only alkaloids reported from Myrtaceae rainforest species are a series of eight oxazole derivatives (xanchryone A (**1292**), xanchryone B (**1293**), and xanchryones I–N (**1294–1299**)) derived from 3-methyl-dihydrochalcone that were reported from the North Queensland endemic *Xanthostemon chrysanthus* (Plate 20), and cultivated in China [457, 458]. A series of dihydrochalcones, xanchryone C (**1300**), xanchryone D (**1301**), [457] myrigalone G (**1302**) [459] and others conjugated to an oxidized

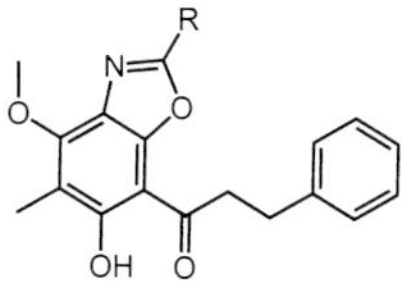

1292 R = H
1293 R = Me
1294 R = CH_2Ph
1295 R = CH_2CH_2Ph
1296 R = *i*-Pro
1297 R = *i*-But
1298 R = 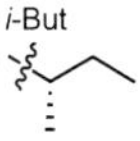
1299 R = *n*-Pent

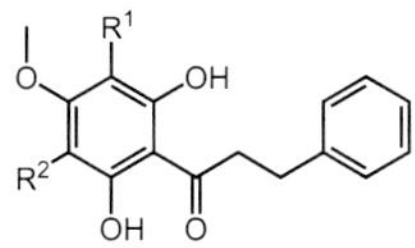

1300 R^1 = OMe, R^2 = Me
1301 R^1 = OH, R^2 = CHO
1302 R^1 = H, R^2 = Me

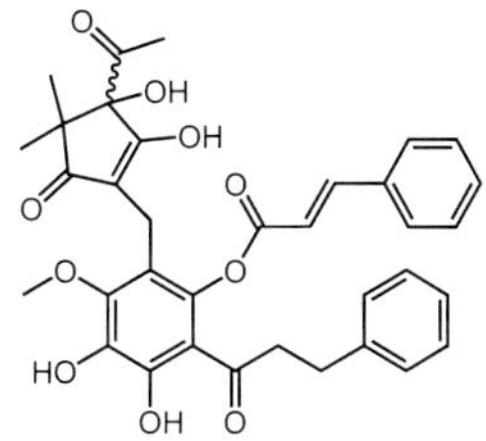

1303

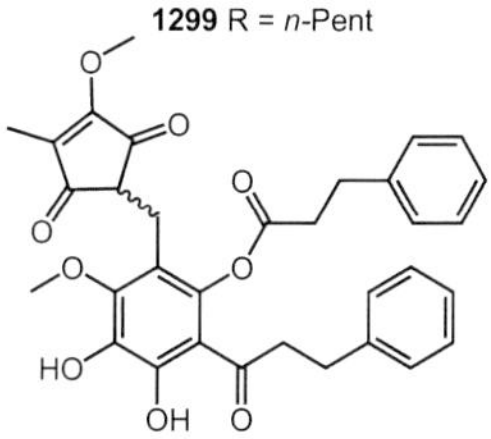

1304

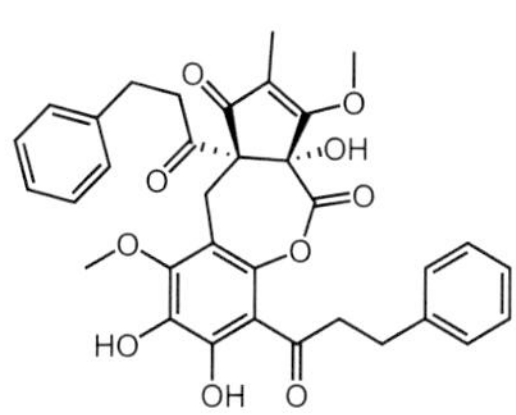

1305

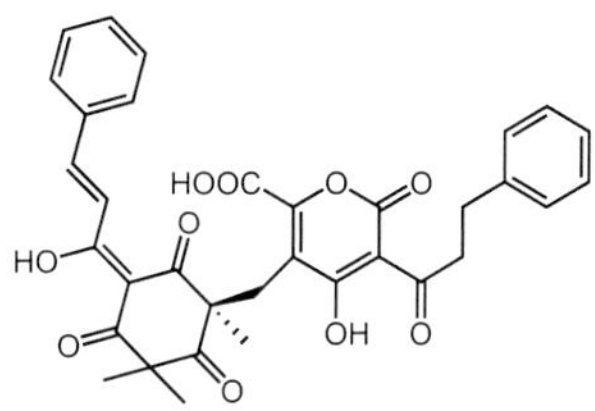

1306

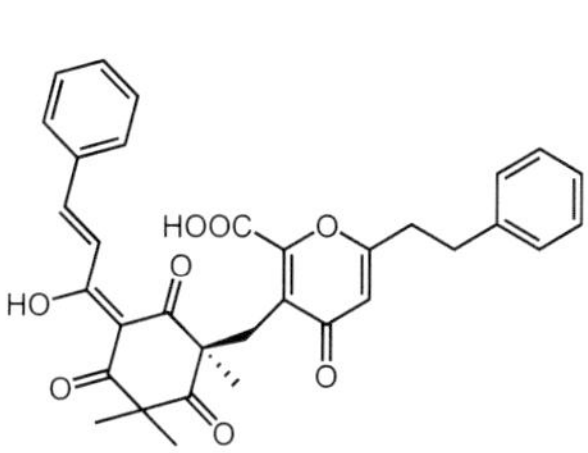

1307

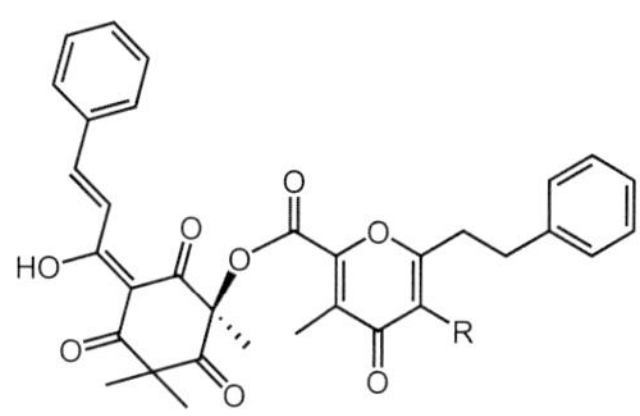

1308 R = H
1309 R = COOH

Fig. 105 Structures of compounds **1292–1309** reported from Myrtaceae species

phloroglucinol (xanthchrysone C (**1303**), and xanthchrysone B (**1304**) [459] or to a second chalcone (xanthchrysone A (**1305**), xanthostone B (**1306**), xanthostone A (**1307**), xanthostone C (**1308**), and xanthostone D (**1309**) [459, 460], have also been reported from *X. chrysanthus.*

Flavonoids (Fig. 106) (myrtinols A–F (**1310–1315**), sideroxylin (**1316**), 8-demethylsideroxylin (**1317**), eucalyptin (**1318**), 6-methyl-7-methoxyflavone (**1319**), 8-methyl-7-methoxyflavone (**1320**), *cis*-2,3,5-trihydroxy-6-methyl-7-methoxyflavanone (**1321**), *trans*-2,3,5-trihydroxy-6-methyl-7-methoxyflavanone (**1322**), *cis*-2,3,5-trihydroxy-8-methyl-7-methoxyflavanone (**1323**), *trans*-2,3,5-trihydroxy-8-methyl-7-methoxyflavanone (**1324**), resinoside A (**1325**), resinoside B (**1326**), afzelin (**1327**), leptosperol G (**1328**), leptosperol F (**1329**), xanchryone F (**1330**), and xanchryone G (**1331**)) have been reported from *Lophostemon* (Plate

Plate 20 Flowers of *Xanthostemon chrysanthus* (left and bottom middle) *Lophostemon confertus* (right), and *Syncarpia glumulifera* (top middle). (Photographs AR Carroll)

19), *Melaleuca, Eucalyptus, Syncarpia* (Plate 19), *Backhousia, Leptospermum*, and *Xanthostemon* (Plate 19) species [445, 450, 461–466].

Conjugates of acyl phloroglucinol to either a monoterpenoid or a sesquiterpenoid are mostly found in *Melaleuca* and *Eucalyptus* species and those found in the eucalypts almost exclusively have the aryl methyl groups oxidized to formyl moieties (Figs. 107 and 108). The rainforest eucalypts, *E. tereticornis, E. resinifera*, and *E. grandis* (Plate 21) contain 35 formyl phloroglucinol meroterpenoids, with 26 conjugated to a monoterpenoid (euglobals G1–G5 (**1332–1366**), grandinal (**1337**), euglobals G6–G12 (**1338–1344**), eucateretin A (**1345**), eucateretin B (**1346**), eucateretin E (**1347**), eucateretin C (**1348**), eucateretin D (**1349**), robustadial A (**1350**), robustadial B (**1351**), robusone X (**1352**), eucalrobusone X (**1353**), euglobal B1-1 (**1354**), euglobal Ic (**1355**), euglobal IIa (**1356**), and euglobal Ib (**1357**)) [466–471] and nine from *E. tereticornis* being conjugated to a sesquiterpenoid (eucalteretials A–D (**1358–1361**), eucalrobusone J (**1362**), euglobal IX (**1363**), eucalteretial E (**1364**), euglobal VII (**1365**), and euglobal IVa (**1366**)) [471]. The critically endangered *Rhodomyrtus psidioides* that is highly susceptible to the introduced myrtle rust fungus biosynthesizes an additional four congeners, rhodomyrtal B (**1367**), rhodomyrtal A (**1368**), rhodomyrtal C (**1369**), and eucalyptin A (**1370**) [472].

Since Myrtaceae species are renowned for their diverse mono- and sesquiterpenoid-contining essential oils, these building blocks reflect a huge potential for diverse meroterpenoid adducts to occur in many Myrtaceae species. For example, there are over 500 *Eucalyptus* species the grow outside of rainforest habitats in Australia. Using covariance 2D NMR analysis of HMBC data in combination with principal component analysis (PCA) of published ^{1}H and ^{13}C NMR data reported for formylated meroterpenoids, the present author and his associates have been able to identify and classify structures classes of formylated meroterpenoids in eucalypt extracts and this has allowed the prediction of the presence of a variety of formylated

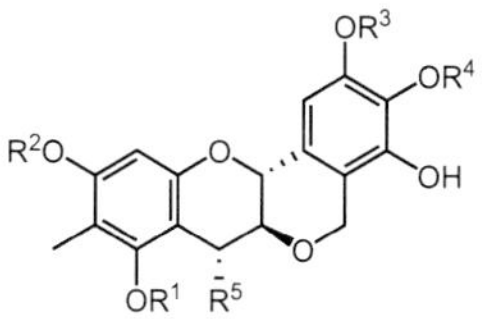

1310 R^1 = Me, R^2 = R^5 = H, R^3–R^4 = OCH_2O
1311 R^1 = Me, R^2 = R^5 = H, R^3 = R^4 = OMe
1312 R^1 = H, R^2 = Me, R^3 = R^4 = OMe, R^5 = OH
1313 R^1 = R^5 = H, R^2 = Me, R^3 = R^4 = OMe
1314 R^1 = Me, R^2 = R^5 = H, R^3–R^4 = OCH_2O
1315 R^1 = R^2 = R^3 = R^4 = OMe, R^5 = H

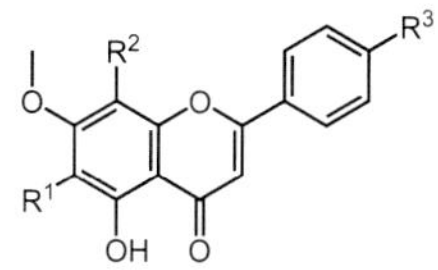

1316 R^1 = R^2 = Me, R^3 = OH
1317 R^1 = Me, R^2 = R^3 = OH
1318 R^1 = Me, R^2 = R^3 = OMe
1319 R^1 = Me, R^2 = R^3 = H
1320 R^1 = R^3 = H, R^2 = Me

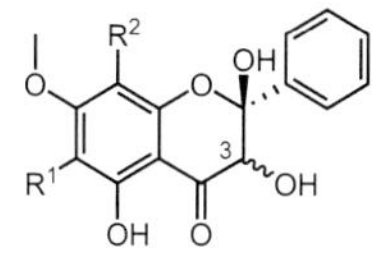

1321 R^1 = Me, R^2 = H, (3*S*)
1322 R^1 = Me, R^2 = H, (3*R*)
1323 R^1 = H, R^2 = Me, (3*S*)
1324 R^1 = H, R^2 = Me, (3*R*)

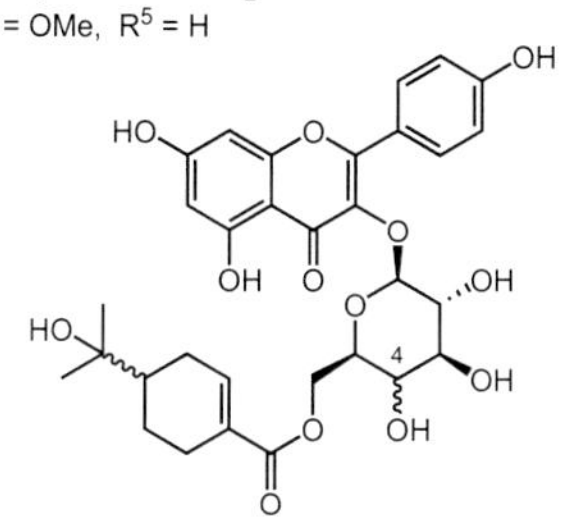

1325 (4*S*)
1326 (4*R*)

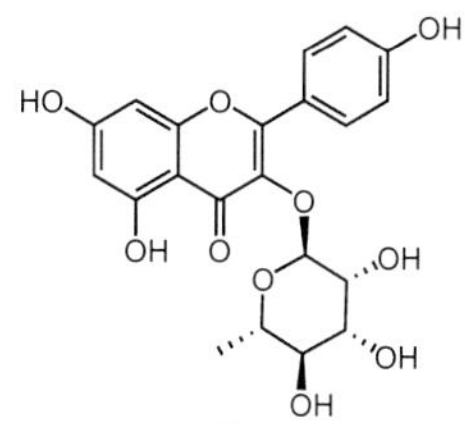

1327

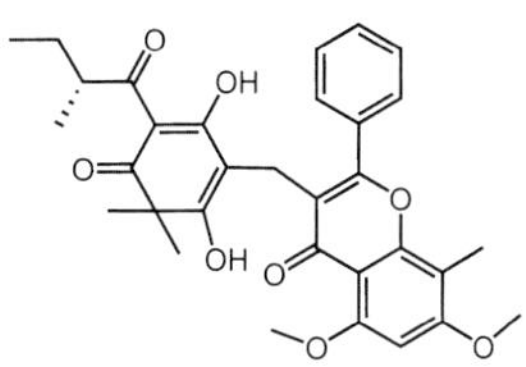

1328

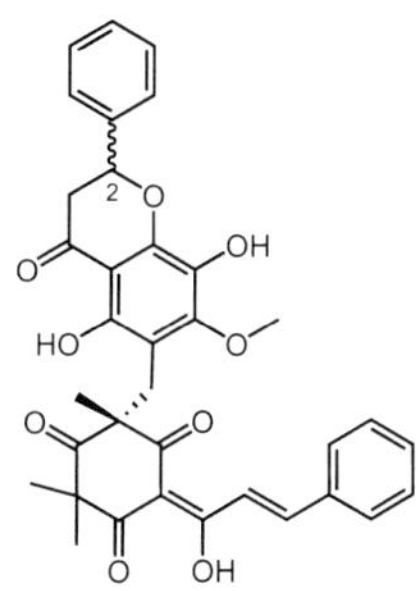

1330 2β
1331 2α

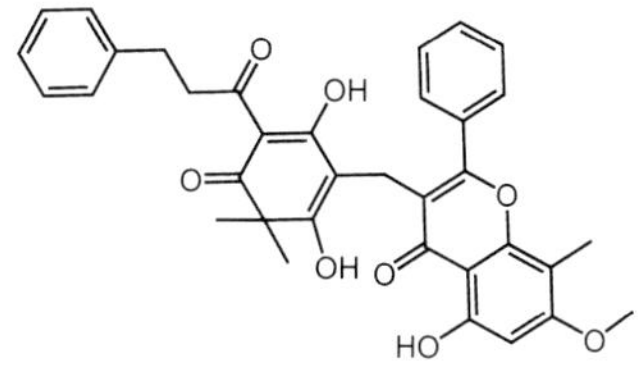

1329

Fig. 106 Structures of compounds **1310–1331** reported from Myrtaceae species

meroterpenoid sub-classes in complex mixtures (Fig. 109) [473]. This analysis technique was used to study extracts from over 300 taxonomically diverse *Eucalyptus* species and indicated that formylated meroterpenoids are widely distributed across *Eucalyptus* species that grow outside of Australian rainforest habitats [474]. Many of the formyl phloroglucinols isolated from rainforest Myrtaceae have antiviral activity against Epstein Bar virus, while others are inhibitors of ATP citrate lyase [475].

Melaleuca viminalis (syn. *Callistemon viminalis*) and *Melaleuca salicina* (syn. *C. salignus*) contain 19 meroterpenoid congeners (Fig. 110) (callviminol A/viminalin A (**1371**), callviminol B/viminalin B (**1372**), viminalin C (**1373**), viminalin B

Plate 21 *Eucalyptus grandis* (left) and *E. tereticornis* (middle) are tall emergent tree species found in Australian rainforests. The flowers of *E. grandis* (bottom right) and *E. tereticornis* (top right). (Photographs AR Carroll)

(reported with the wrong structure) (**1374**), viminalin D (**1375**), viminalin L (**1376**), callisalignene A (**1377**), viminalin N (**1378**), viminalin M (**1379**), viminalin O (**1380**), callviminol C (**1381**), callviminol D (**1382**), viminalins E–G (**1383–1385**), callviminol E/viminalin H (**1386**), callisalignene C/viminalin K (**1387**), viminalin I (**1388**), and callisalignene B/viminalin J (**1389**)) that are mostly dihydrobenzofurans [453, 476–478]. Another seven meroterpenoids (melaleucadine A (**1390**), melaleucadine B (**1391**), leptosperols C–E (**1392–1394**), leptosperol A (**1395**), and leptosperol B (**1396**)), isolated from *M. leucadendra* or *Leptospermum scoparium*, contain mono- or sesquiterpenoids conjugated to either a benzophenone or a chalcone unit [462, 479, 480]. Several of the monoterpenoid phloroglucinol adducts have different trivial names for the same compounds. Furthermore, some of these compounds have been assigned opposite absolute configurations even through their ECD and optical rotations are identical. Viminalin N (**1378**) has identical NMR data to callisalignene A (**1377**), but has the configuration at C-4″ misassigned. In the same paper, two related structures, viminalins M (**1379**) and O (**1380**), which are proposed to possess the same C-4″ configuration as **1378** and likewise have large vicinal couplings indicative of a diaxial configuration for H-3″ and H-4″, should therefore have their configurations at C-4″ inverted to that depicted in **1397** and **1398**. The total synthesis of viminalin M confirmed that its structure is **1397** [481].

Dibenzofuran phloroglucinol dimers (Fig. 111) (pilidiostigmin (**1399**), originally proposed to have structure **1400**, callistemenonone A (**1401**), callistemonol A (**1402**), 1,3,7,9-tetrahydroxy-2,8-dimethyl-4-(2-methylbutanoyl)-6-(2-methylpropionyl)dibenzofuran (**1403**), 1,3,7,9-tetrahydroxy-2,8-dimethyl-4,6-di(2-methylpropionyl)dibenzofuran (**1404**), 1,3,7,9-tetrahydroxy-2,8-dimethyl-4,6-di(2-methylbutanoyl)dibenzofuran (**1405**), δ-rhodomyrtoxin

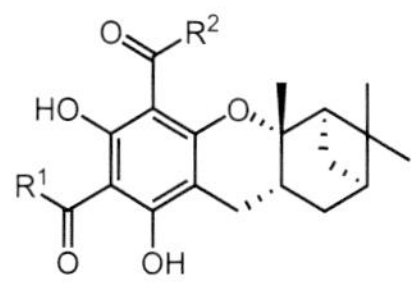

1332 R^1 = H, R^2 = *i*-But
1333 R^1 = *i*-But, R^2 = H

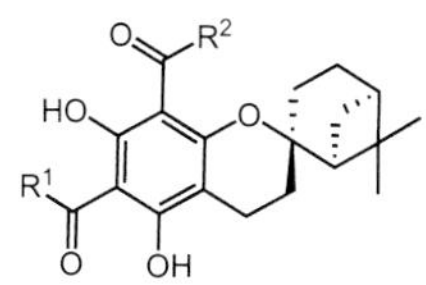

1334 R^1 = *i*-But, R^2 = H
1335 R^1 = H, R^2 = *i*-But

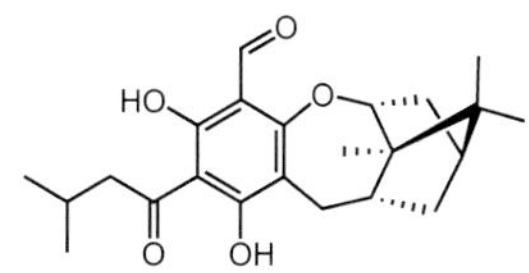

1336

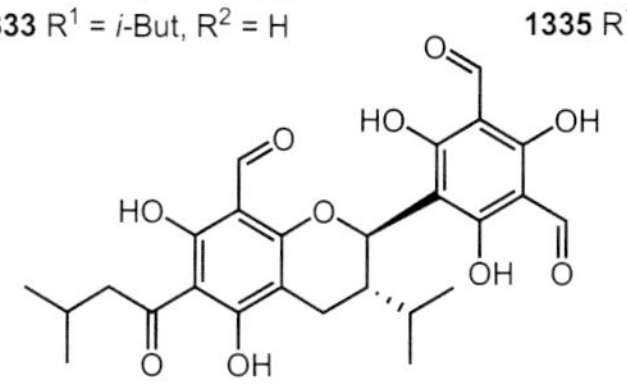

1337

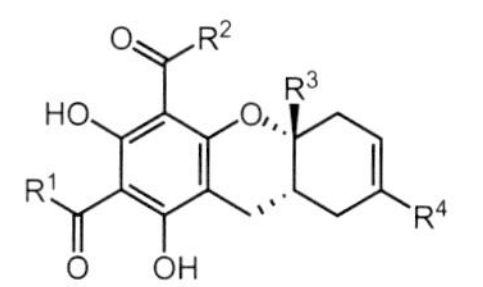

1338 R^1 = *i*-But, R^2 = H, R^3 = Me, R^4 = *i*-Pro
1339 R^1 = H, R^2 = *i*-But, R^3 = Me, R^4 = *i*-Pro
1340 R^1 = *i*-But, R^2 = H, R^3 = *i*-Pro, R^4 = Me

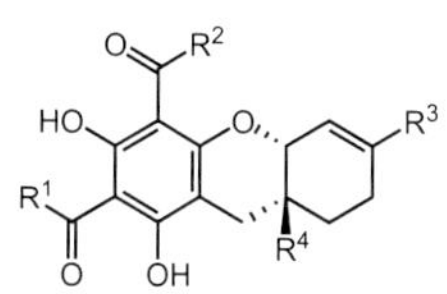

1341 R^1 = *i*-But, R^2 = H, R^3 = *i*-Pro, R^4 = Me
1342 R^1 = H, R^2 = *i*-But, R^3 = *i*-Pro, R^4 = Me
1343 R^1 = *i*-But, R^2 = H, R^3 = Me, R^4 = *i*-Pro

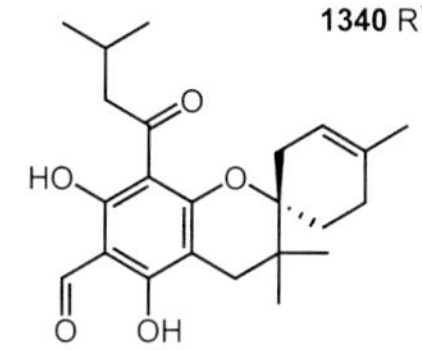

1344

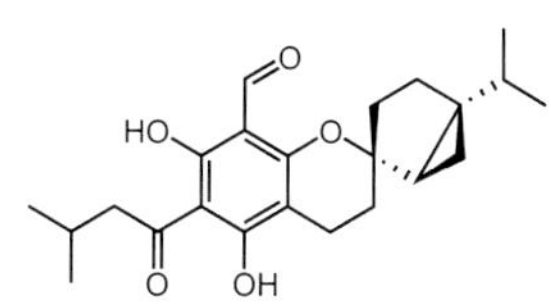

1345

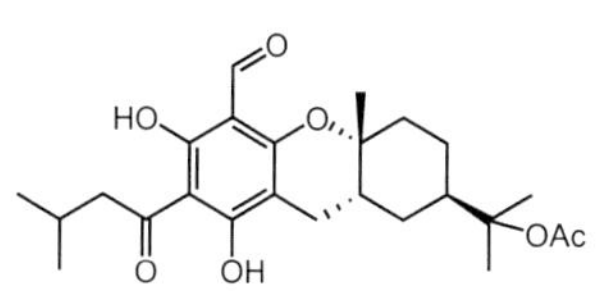

1346

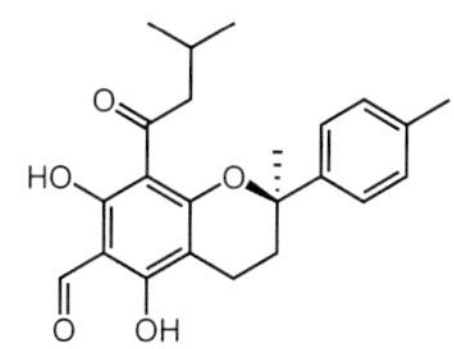

1347

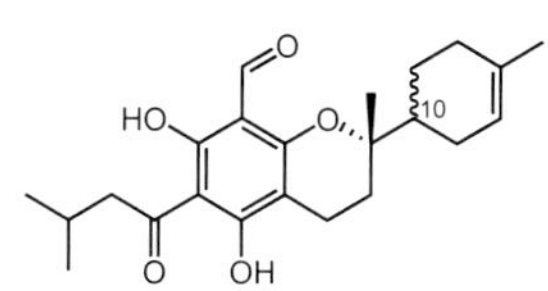

1348 (10*S*)
1349 (10*R*)

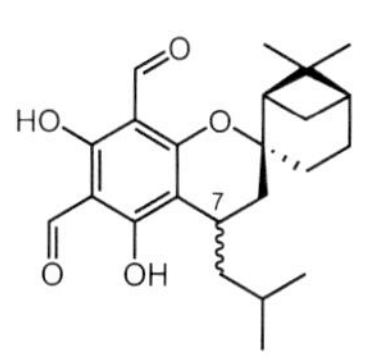

1350 7*β*
1351 7*α*

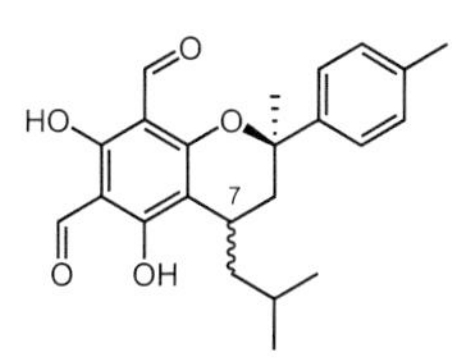

1352 7*β*
1353 7*α*

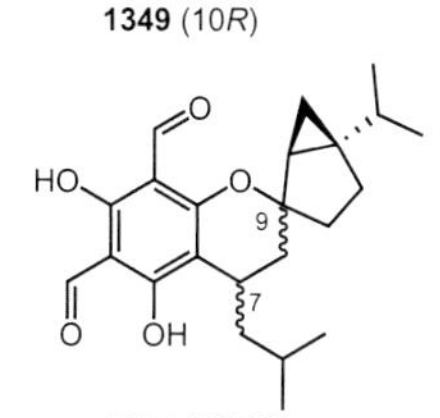

1354 (7*S*,9*S*)
1356 (7*S*,9*R*)
1357 (7*R*,9*S*)

Fig. 107 Structures of formyl phloroglucinols **1332–1357** reported from Myrtaceae species

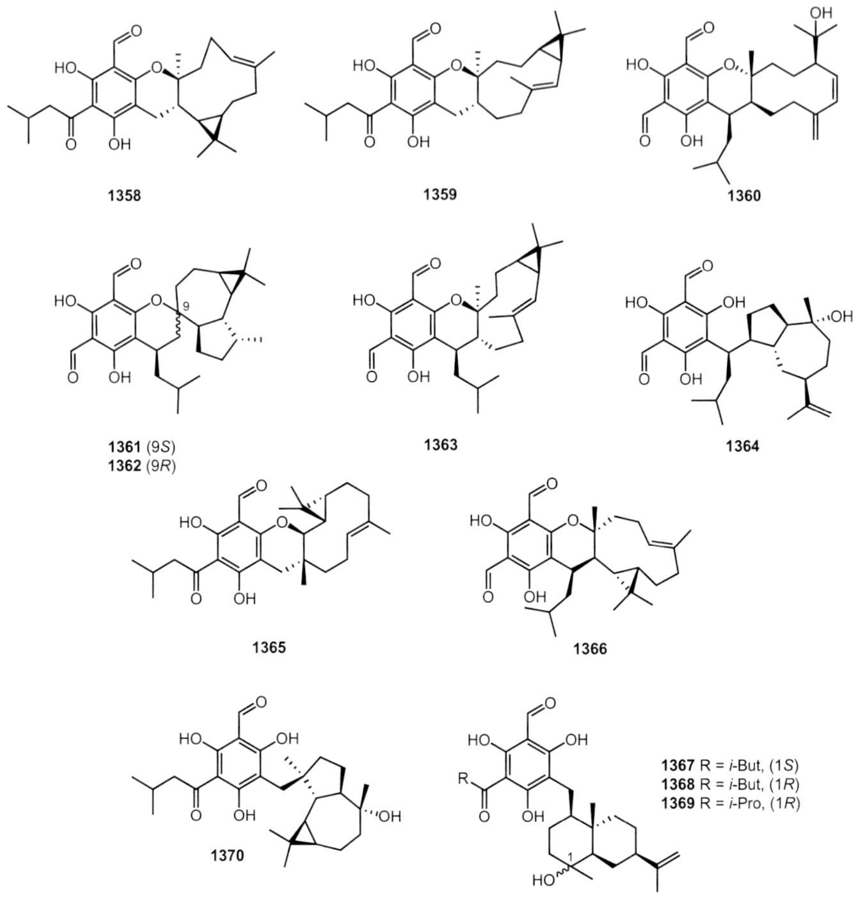

Fig. 108 Structures of formyl phloroglucinols **1358–1370** reported from Myrtaceae species

(**1406**), rhodomyrtoxin (**1407**), 1,3,7,9-tetrahydroxy-4,6-dimethyl-2,8-di(2-methylpropionyl)dibenzofuran (**1408**), 1,3,7,9-tetrahydroxy-4,6-dimethyl-2-(2-methylbutanoyl)-8-(2-methylpropionyl)dibenzofuran (**1409**), rhodomyrtoxin C (**1410**), and rhodomyrtoxin B (**1411**)) are restricted to *Pilidiostigma* (Plate 21), *Rhodomyrtus* and *Melaleuca* species [482–487]. In turn, phloroglucinol dimers conjugated through a bond between the benzylic carbon of one acylphloroglucinol to an aryl carbon of the second acylphloroglucinol (mulgravanols B (**1412**), C (**1413**), and A (**1414**)) have been reported from *Syzygium floribundum* (Plate 22) [488].

Multiple C-methylations on carbons *ortho* to the oxygenated aromatic carbons in the phloroglucinol ring generate β-triketones, and these compounds are common across more ancient lineages such as *Xanthostemon*, *Melaleuca* (Plate 23), *Corymbia*, *Leptospermum*, and *Kunzea*, but absent in the Syzygieae (Figs. 112 and 113). These β-triketones are either simple acylated derivatives or can be conjugated to either

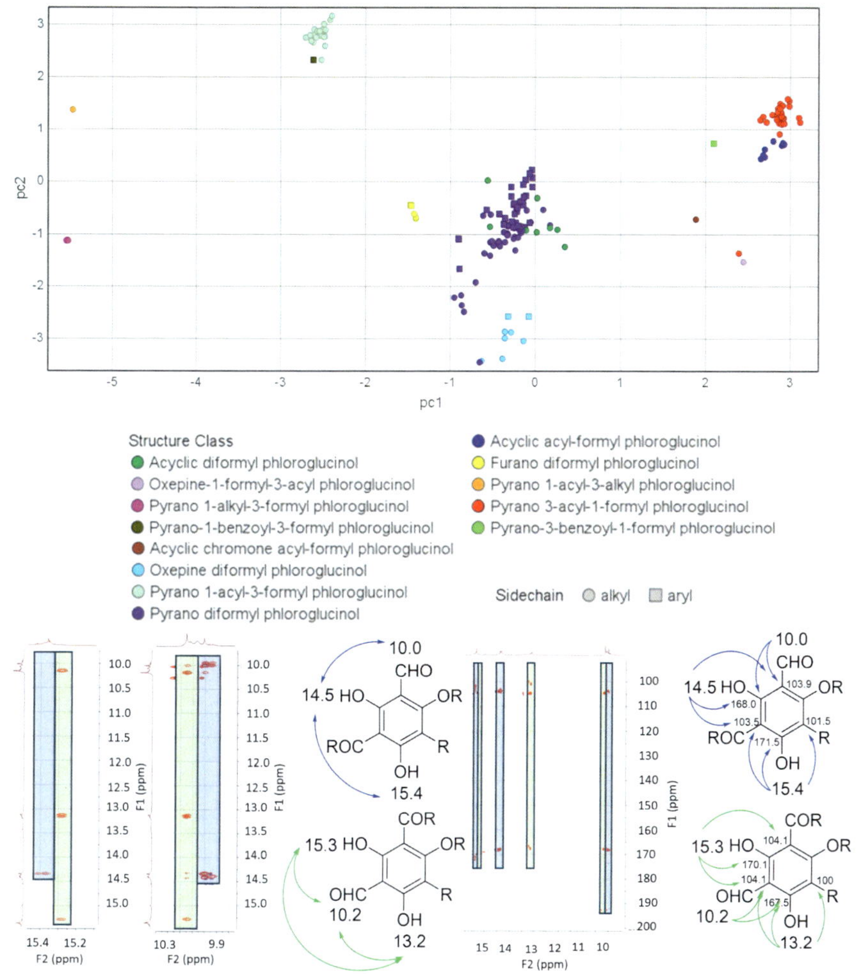

Fig. 109 Principal component analysis plot of substructure classes of formyl phloroglucinols based on ^{13}C and ^{1}H NMR data reported for all formyl phloroglucinol natural products. This data can be used in combination with classic and covariance analysis of HMBC data to classify individual natural products in these substructure classes in complex mixtures obtained from eucalypt extracts

a monoterpenoid, sesquiterpenoid or a phloroglucinol. Simple acylated β-triketones include norflavesone (**1415**), norisoleptospermone (**1416**), norleptospermone (**1417**), myrigalone A (**1418**), isoleptospermone (**1419**), flavesone (**1420**), grandiflorone (**1421**), nortriketone 9 (**1422**), leptospermone (**1423**), tristaenone A (**1424**), nortriketone 10 (**1425**), xanchryone H (**1426**), angustione (**1427**), dehydroangustione (**1428**), xanthostemone (**1429**), callistenone K (**1430**), callistemonol B (**1431**), xanchryone E (**1432**), champanone B (**1433**), angustifolenone (**1434**), tetragocarbone C (**1435**), tetragocarbone B (**1436**), G1 (**1437**), G2 (**1438**), G3 (**1439**), hillyl acetate (**1440**),

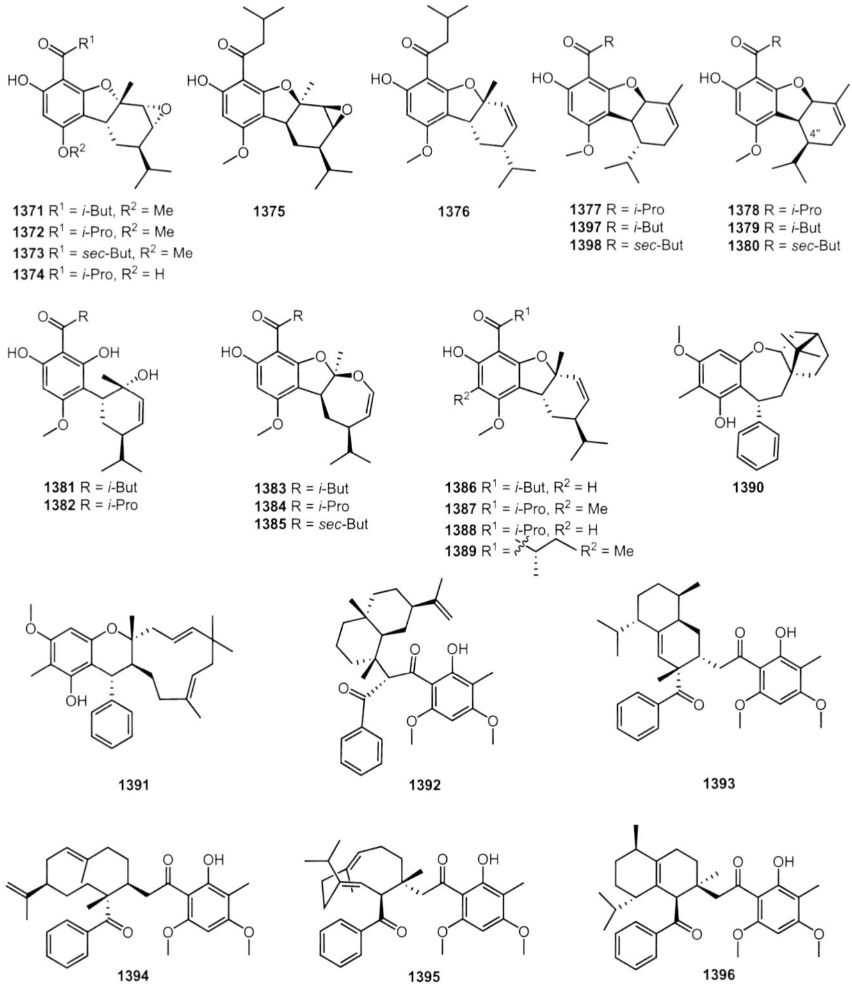

Fig. 110 Structures of acyl phloroglucinols **1371–1398** reported from Myrtaceae species

and torellianol A (**1441**) [445, 453, 454, 459, 460, 464, 483, 489–497]. The simple acyl β-triketones include the root growth inhibitors (G1–G3) that contribute to the invasiveness of *Eucalyptus* species in vast regions of the world where they have been naturalized from Australia [498]. There are 39 β-triketone-terpenoid adducts reported from Australian rainforest species, namely, callistiviminene E (**1442**), callistiviminene C (**1443**), callistiviminene D (**1444**), callistiviminene A (**1445**), callistiviminene B (**1446**), rhodomyrtal D (**1447**), callistiviminene I (**1448**), callistiviminene J (**1449**), callisalignene I (**1450**), intermediadiones A–C (**1451–1453**), (4*S*)-ficifolidione (**1454**), (4*R*)-ficifolidione (**1455**), viminadione A (**1456**), viminadione B (**1457**), callistiviminene M (**1458**), callistiviminene N (**1459**), callisalignene

Plate 22 Flowers of *Syzygium floribundum* (left), *Pilidiostigma glabrum* (middle), and *Leptospermum scoparium* (right). (Photographs AR Carroll)

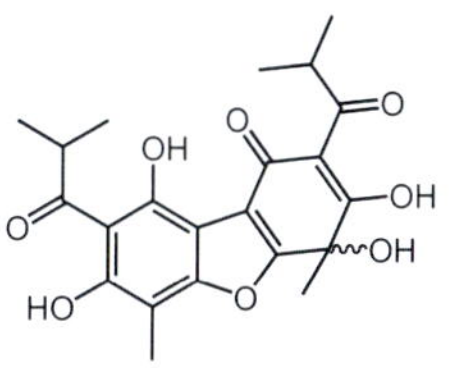

1399

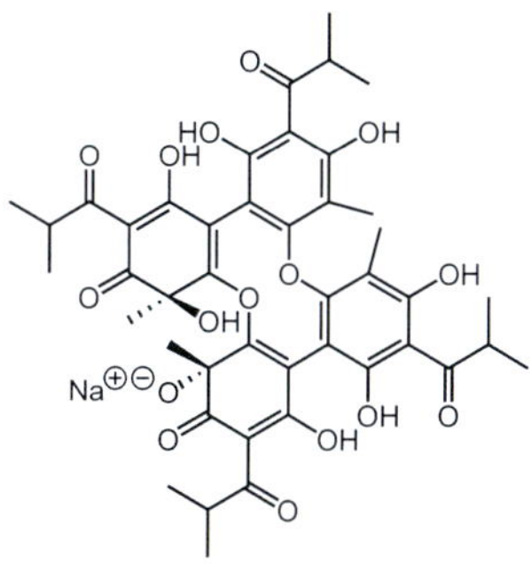

1400

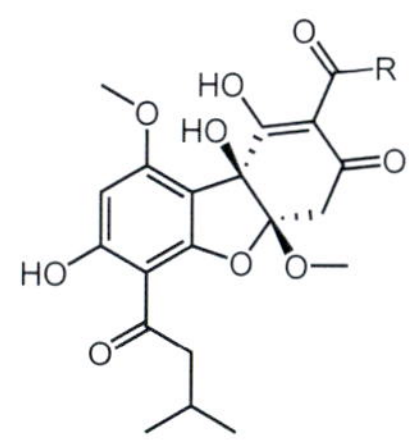

1401 R = *i*-But
1402 R = *i*-Pro

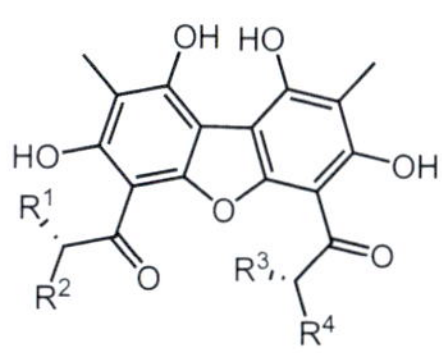

1403 $R^1 = R^2 = R^3 = Me, R^4 = Et$
1404 $R^1 = R^2 = R^3 = R^4 = Me$
1405 $R^1 = R^3 = Me, R^2 = R^4 = Et$
1406 $R^1 = Me, R^3 = H, R^2 = Et, R^4 =$ *i*-Pro
1407 $R^1 = R^3 = H, R^2 = R^4 =$ *i*-Pro

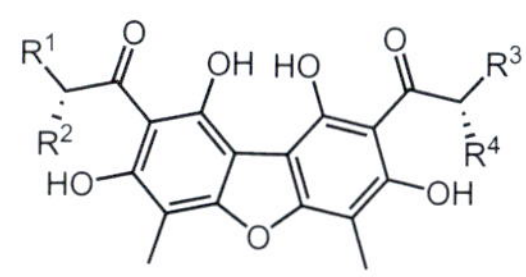

1408 $R^1 = R^2 = R^3 = R^4 = Me$
1409 $R^1 = R^3 = R^4 = Me, R^2 = Et$
1410 $R^1 =$ *i*-Pro, $R^2 = H, R^3 = Et, R^4 = Me$
1411 $R^1 = R^3 = Et, R^2 = R^4 = Me$

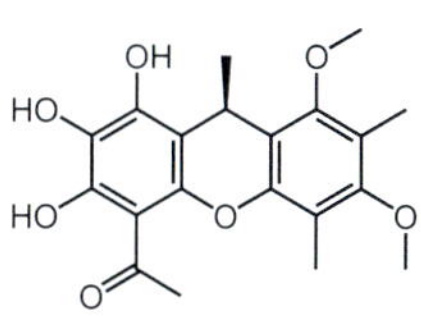

1412

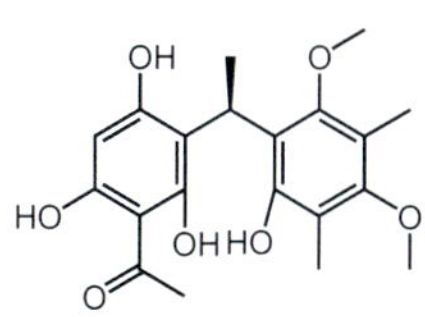

1413

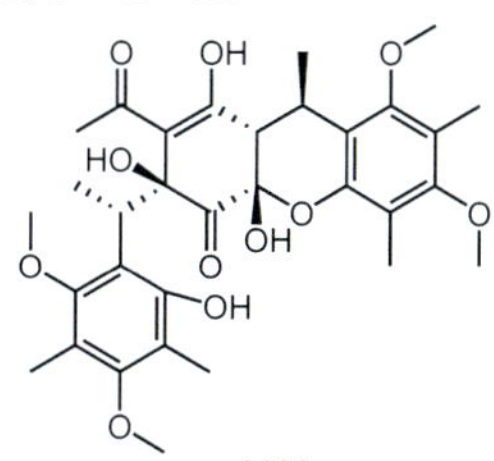

1414

Fig. 111 Structures of phloroglucinols **1399–1414** reported from Myrtaceae species

Plate 23 Flowers of *Melaleuca salicina* (top left), *M. viminalis* (bottom left and right) and *M. leucadendra* (middle). *Melaleuca viminalis* has many growth forms, from large trees (right) to dwarf shrubs (right inset). (Photographs AR Carroll)

G (**1460**), callisalignene D (**1461**), callisalignene H (**1462**), callistiviminene O (**1463**), callistiviminene F (**1464**), callistiviminene G (**1465**), callistiviminene H (**1466**), myrtucommulone L (**1467**), intermediadione D (**1468**), callistiviminene K (**1469**), callistiviminene L (**1470**), callisalignene E (**1471**), callisalignene F (**1472**), calliviminone A (**1473**), calliviminone E (**1474**), calliviminone G (**1475**), calliviminone H (**1476**), calliviminone D (**1477**), calliviminone B (**1478**), calliviminone F (**1479**), and calliviminone C (**1480**) [453, 472, 476, 497, 499–504].

Either one or two acyl β-triketones can be conjugated to an acylphloroglucinol and several congeners (Fig. 114) have been found in Australian rainforest *Xanthostemon*, *Melaleuca*, *Corymbia* (Plate 24), *Syncarpia*, *Kunzea*, and *Leptospermum* species, including callisalignone C/callistenone N (**1481**), callistenone F (**1482**), callistenone B (**1483**), callistenone H (**1484**), callisalignone B/callistenone O (**1485**), callistenone P (**1486**), isomyrtucommulone B (**1487**), callistenone G (**1488**), rhodomyrtosone F (**1489**), callistenone I (**1490**), torellianones A (**1491**) and B (**1492**), kunzeanones A (**1493**) and B (**1494**), torellianones C (**1495**), and D (**1496**), leucadenones A–D (**1497–1500**), lumaflavanone (**1501**), torellianones E (**1502**) and F (**1503**), callistenones L (**1504**) and M (**1505**), 4-[1-[2,6-dihydroxy-4-methoxy-3-(2-methyl-1-oxopropyl)phenyl]-3-methylbutyl]-5-hydroxy-2,2,6,6-tetramethyl-4-cyclohexene-1,3-dione (**1506**), 4-[1-[2,6-dihydroxy-4-methoxy-3-(3-methyl-1-oxobutyl)phenyl]-3-methylbutyl]-5-hydroxy-2,2,6,6-tetramethyl-4-cyclohexene-1,3-dione (**1507**), watsonianone A (**1508**), myrtucommulone A (**1509**), and myrtucommulone D (**1510**) [39, 442, 453, 454, 464, 476, 489, 497, 505, 506].

The β-triketones have some unusual NMR spectroscopic properties. Highly deshielded enol protons such as those in watsonianone B (**1511**) have been reported to resonate between 12 and 20 ppm, because they share hydrogen bonds with up to three oxygens [497, 507, 508]. Phloroglucinol-β-triketone adducts can adopt multiple alternative intramolecular hydrogen bonding networks as well as keto-enol tautomers in solution and this results in extremely complex NMR spectra. Those with one acylphloroglucinol conjugated to a β-triketone linked by a chiral benzylic carbon, for

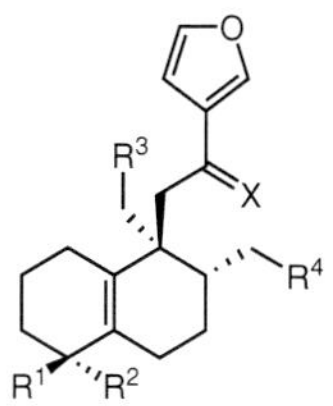

1771 R^1 = COOH, R^2 = Me, R^3 = OH, R^4 = H, X = H,H
1772 R^1 = COOH, R^2 = Me, R^3 = H, R^4 = OH, X = H,H
1773 R^1 = Me, R^2 = COOH, R^3 = R^4 = H, X = O
1774 R^1 = Me, R^2 = COOH, R^3 = R^4 = H, X = H,H
1775 R^1 = COOH, R^2 = Me, R^3 = R_4 = H, X = H,H

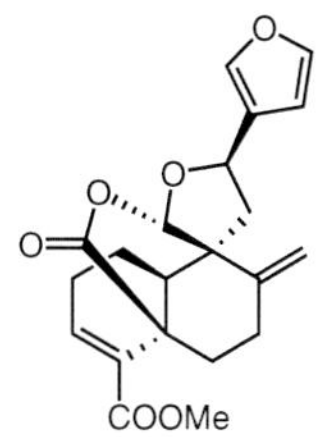

1776

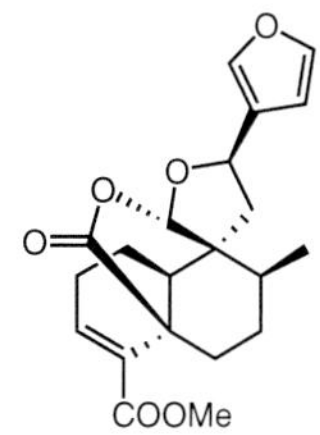

1777

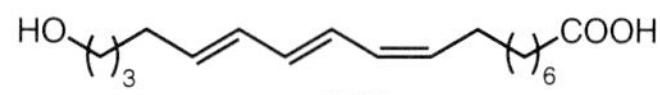

1780

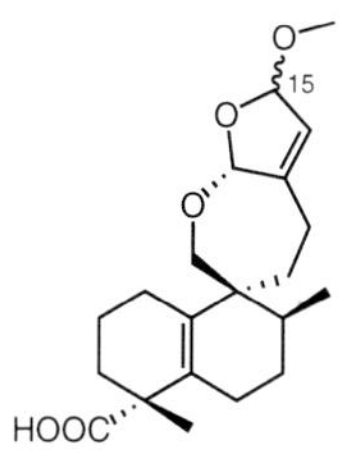

1778 (15*S*)
1779 (15*R*)

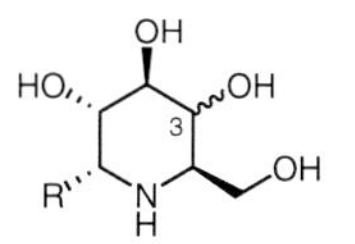

1781 R = CH_2OH, 3α
1782 R = H, 3α
1783 R = H, 3β

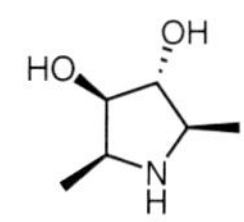

1784

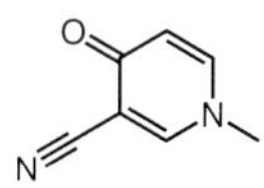

1785

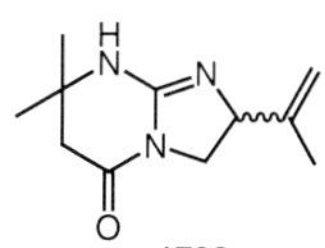

1786

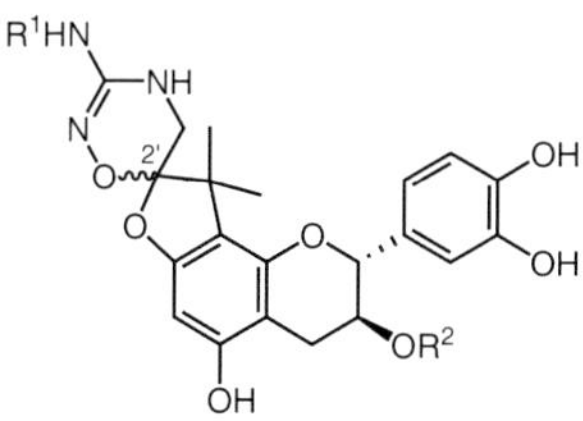

1787 R^1 = $CH_2CH{=}C(CH_3)_2$, R^2 = β-L-rham, (2'*R*)
1788 R^1 = $CH_2CH{=}C(CH_3)_2$, R^2 = β-L-rham, (2'*S*)
1789 R^1 = H, R^2 = β-L-rham, (2'*R*)
1790 R^1 = H, R^2 = β-L-rham, (2'*S*)
1791 R^1 = R^2 = H, (2'*R*)
1792 R^1 = R^2 = H, (2'*S*)

Fig. 112 Structures of compounds **1415–1441** reported from Myrtaceae species

example, can adopt two interconverting conformations/tautomers, and compounds containing a second alkyl β-triketone like that found in the myrtucommulones have the potential to generate four interconverting conformations/tautomers. The resulting NMR spectra have resonances for each of these conformers making their structural analysis extremely challenging. It is not surprising then that some structures have been assigned incorrectly, only to be rectified later through more detailed spectroscopic analysis or by total synthesis. ROESY NMR has been shown to be a very useful tool to identify interconverting structures since correlations antiphase with ROE correlations are diagnostic for chemical exchange (Fig. 115) [509]. Pilidiostigmin from *P. glabrum* was originally assigned as a dimer (**1400**), but the presence of exchange correlations in the ROESY data clearly refuted this structural proposal

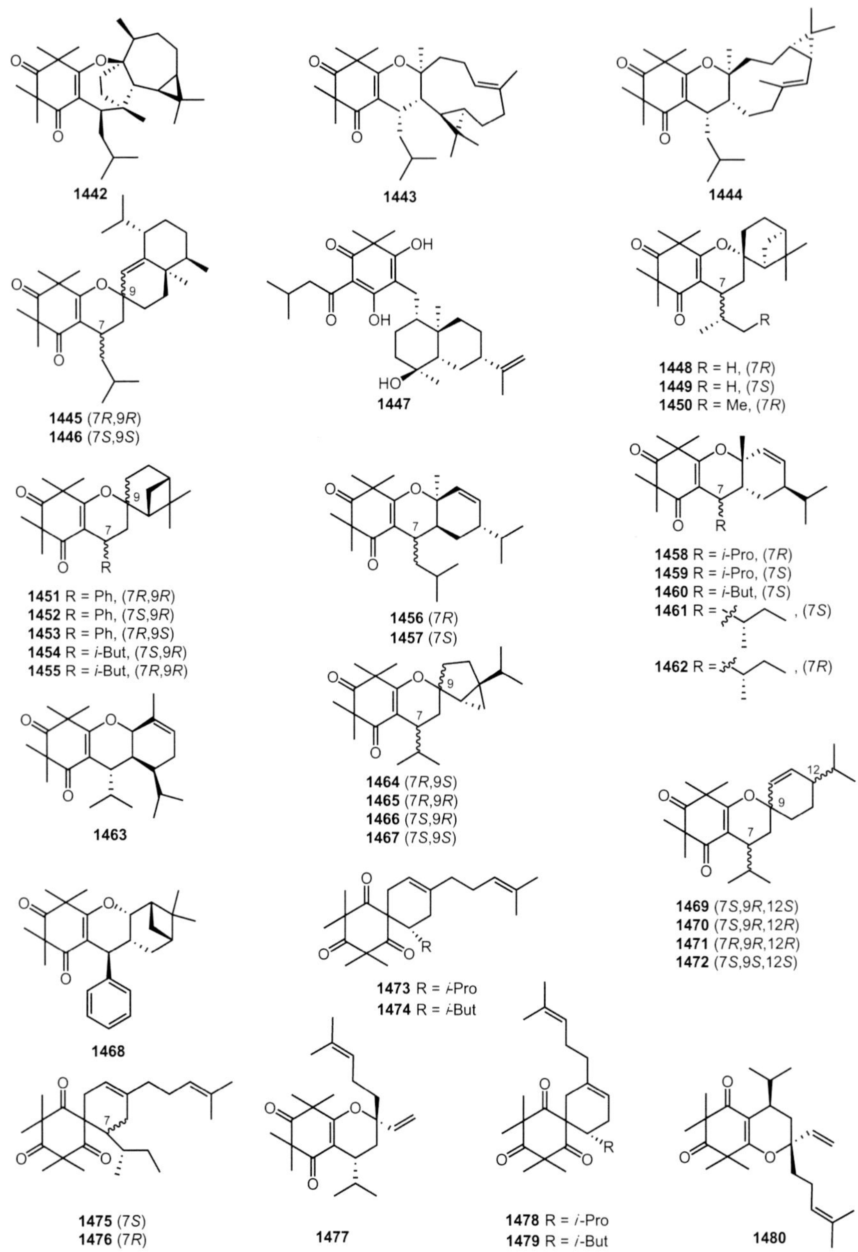

Fig. 113 Structures of compounds **1442–1480** reported from Myrtaceae species

Plate 24 Flowers of *Corymbia torelliana* (left), *C. intermedia* (right and bottom middle), and *Tristaniopsis laurina* (bottom left). (Photographs AR Carroll)

and its structure subsequently was revised to a monomeric dibenzofuran (**1399**) that adopts two interconverting keto-enol tautomers in solution [484, 485].

The β-triketones isolated from *Corymbia* species have moderate antiplasmodial activity [497, 500, 507, 508], while related compounds from *Kunzea* are antiviral [489]. The myrtucommulones show potent antibacterial activity [510]. Interestingly, the first congener of this series, myrtucommulone A (**1509**) was described from the Mediterranean species *Myrtus communis* and has since been found in *Melaleuca* and *Corymbia* species [39, 507, 510]. This highlights the antiquity of this structural class, since *M. communis* is an emigrant into Europe via long-distance dispersal from East Gondwana in the Eocene, and so it has inherited its biosynthetic capacity from ancestors that still survive in Australian rainforests. The *Melaleuca lophantha* β-triketones and phloroglucinols show Gram-positive antibacterial activity. The four monoterpenoid β-triketone adducts viminadiones A (**1456**) and B (**1457**) and (4*R*)- and (4*S*)-ficifolidiones ((**1454**) and (**1455**)) exhibit insecticidal activities [501, 502]. The dibenzofurans (**1399–1410**) reported from *Pilidiostigma glabrum* display antibacterial activity [486]. Callistemonols A and B (**1402**, **1431**) are bactericidal [483]. Mulgravanols A–C (**1414–1435**) from *Syzygium* (syn. *Waterhousia*) *mulgraveana*, myrtinols A–F (**1310–1315**), 4-*O*-methylcedrusin, 9-*O*-methylcedrusin (**1261**, **1262**), and 8-demethylsideroxylin (**1317**) from *Backhousia myrtifolia*, both collected in Sydney, and tetragocarbones B and C (**1435**, **1436**), sideroxylin (**1316**), and lumaflavanone (**1501**) from *Syncarpia glomulifera* subsp. *glomulifera*, collected in North Queensland, are all weakly anti-inflammatory [450, 464, 488].

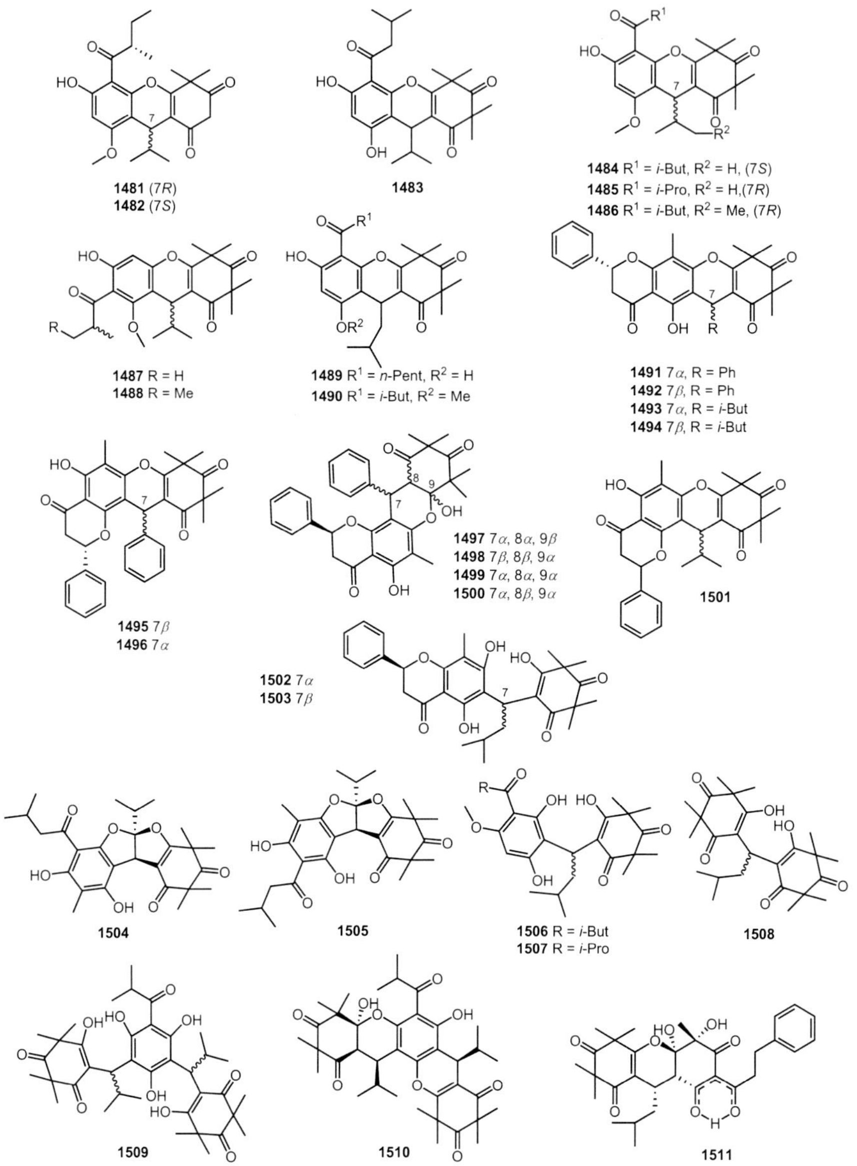

Fig. 114 Structures of compounds **1481–1511** reported from Myrtaceae species

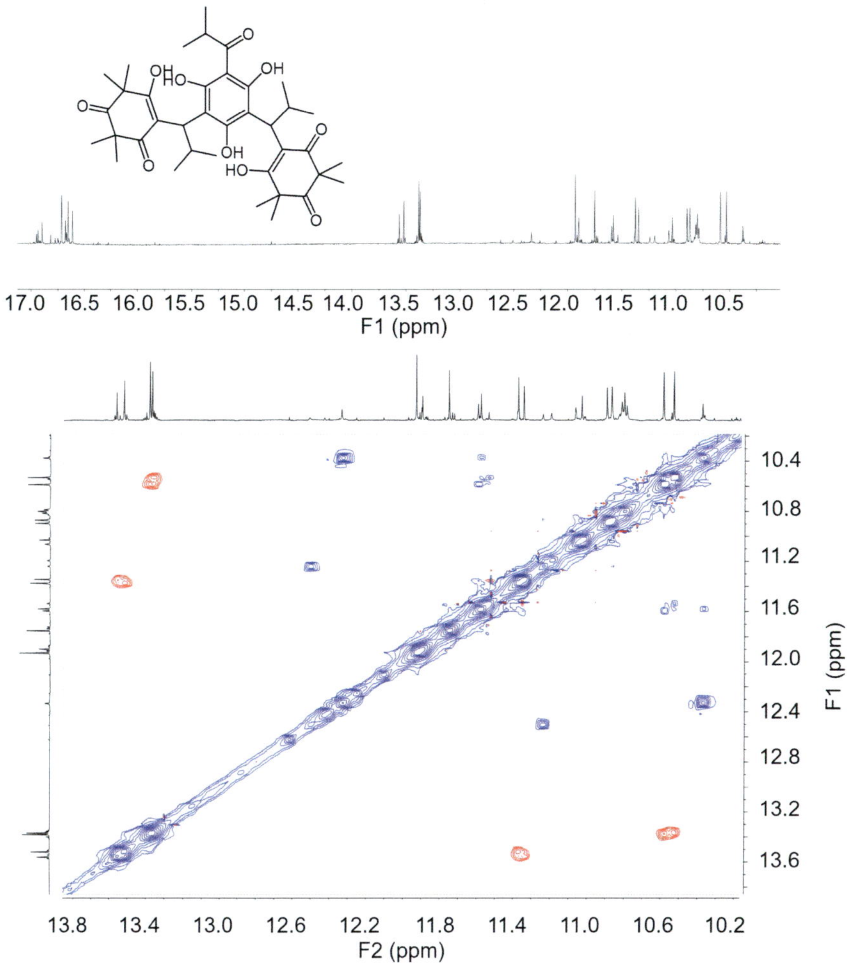

Fig. 115 ^{1}H NMR spectrum (top) of the deshielded phenol and enol resonances in myrtucommulone A (**1509**) in $CDCl_3$. There are 40 resonances associated with phenol and enol proton resonances of the four conformational isomers of the two diastereomers of myrtucommulone A. The ROESY spectrum (bottom) of myrtucommulone A showing blue exchange correlations and red ROE correlations

Fabids

Species from all four orders (Fabales, Cucurbitales, Fagales, and Rosales) that comprise the Fabids are found in Australian rainforests (13 of 27 families), but only five families (Fabaceae, Casuarinaceae, Rhamnaceae, Urticaceae, and Moraceae) have species that have been chemically investigated.

Fabales

Three of the four families in the Fabales have species in Australian rainforests, but the vast majority are from the family Fabaceae (Fig. 116).

Fabaceae

The Fabaceae (Plate 25) is the second largest family of angiosperms and its species have a worldwide distribution. The family has its origins in the regions surrounding the Tethys Sea that separated Laurasia from Gondwana [511]. The Fabaceae is well represented in Australian rainforests, with taxonomically dispersed genera being represented. This reflects their recent arrival into Australia mainly via Southeast

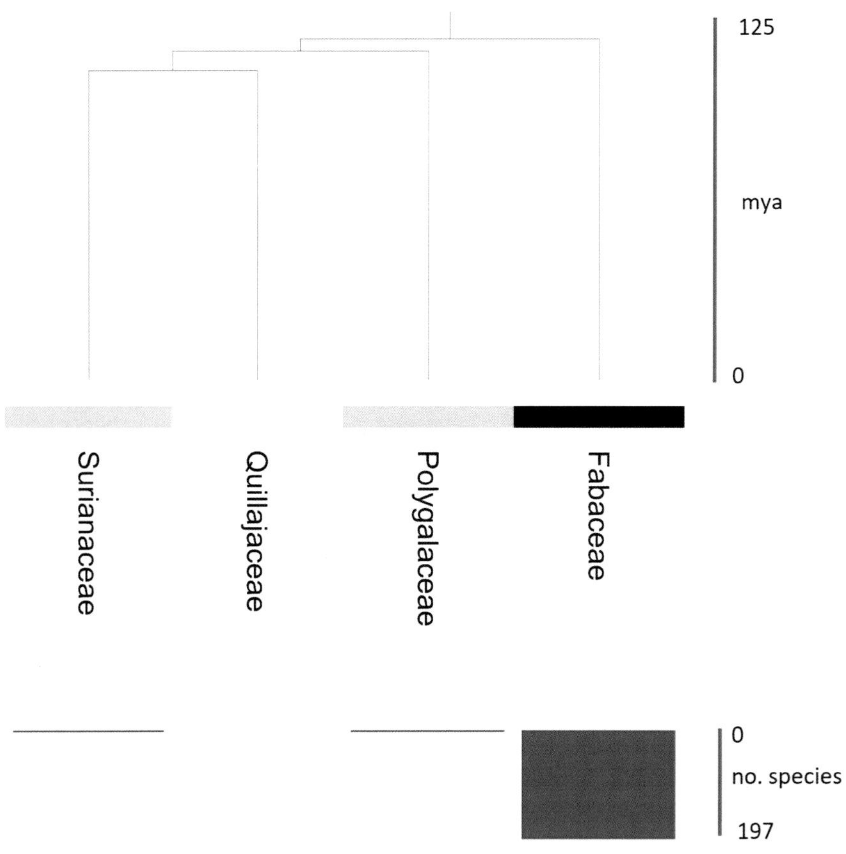

Fig. 116 Dated molecular phylogeny of the Fabales [7] showing those families that have species present in Australian rainforests (grey) and studied (black) and the number of species occurring in Australian rainforests

Plate 25 Flowers and fruits of *Castanospermum australe* (top left and middle), flowers of *Erythrina variegata* (top right), flowers of *Acacia melanoxylon* (bottom left), and *A. aulacocarpa* (bottom middle), flowers of *Senna* sp. (bottom middle right), and leaves of *Barklya syringifolia* (bottom right). (Photographs AR Carroll)

Asia [511]. Of the 230 endemic rainforest species within 64 genera, only 20 have been investigated chemically (Table 30).

Structurally diverse alkaloids, flavonoids, isoflavonoids and triterpenoids are the main constituents reported (Figs. 117, 118, 119 and 120). Oleanolic acid derivatives (castanogenin (**1512**), bayogenin (**1513**), medicagenic glucoside (**1514**), 3-[4-deoxy-5-carboxy-β-L-threo-hex-4-eno-pyranosyl]-bayogenin (**1515**), casteraleside F (**1516**), 3,21-dioxo-olean-18-en-oic acid (**1517**), casteraleside G (**1518**), and casteraleside H (**1519**)) have been reported from the leaves of *Castanospermum australe* and *Acacia aulacocarpa* [512–515]. The triterpenoid **1517** is the first non-protein Tie2 tyrosine kinase inhibitor. Inhibition of this kinase has been implicated with inhibition of angiogenesis, a key mechanism involved in the genesis of cancer. Isoflavonoids and related benzofurans and pterocarpans have been reported from *C. australe, Erythrina vespertilio*, and from *Sophora* species (alpiniumisoflavone (**1520**), formononetin (**1521**), afromosin (**1522**), koparin (**1523**), calycosin (**1524**), sophoracarpan B (**1525**), sophoraisoflavanone A (**1526**), fraserinone A (**1527**), phaseollidin (**1528**), phaseollin (**1529**), sophoronol (**1530**), and dihydroxyphenyl-5,6-methylenedioxybenzofuran (**1531**) [516–519]. Flavonoids (teracacidin (**1532**), 4-*epi*-teracacidin (**1533**), hydroxy teracacidin (**1534**), peltogynin (**1535**), 5 hydroxypeltogynin (**1536**), *epi*-catechin gallate

Table 30 Distribution of natural products isolated from the Fabaceae

Species	Collection location[a,b]	Plant part(s)[c]	Compound class	Compound(s)	Refs.
Acacia aulacocarpa	CY	lf	Triterpenoid	**1517**	[515]
Acacia dealbata	NSW (cult. (Italy))	fl	Chalcone	**1548**	[525]
Acacia fasciculifera	NSW	hw	Flavonoid	**1535**, **1536**	[520]
Acacia maidenii	SEQ	hw	Flavonoid	**1534**, **1540**	[521]
	SEQ	hw	Piperidine	**993**	[521]
	SEQ	b	Tryptamine	**1597**, **1598**	[546]
Acacia melanoxylon	SEQ	hw	Bisflavonoid	**1541–1544**	[522, 523]
	SEQ	hw	Flavonoid	**1532**, **1533**	[522]
Barklya syringifolia	SEQ	s	Gallic acid ester	**1545**	[39]
Bauhinia binata	NQ	lf	Flavonoid	**1538**	[39]
Bauhinia hookeri	NQ (cult. (Egypt))	lf	Flavonoid	**1537**	[524]
Castanospermum australe	SEQ	w	Flavonoid	**1539**	[517]
	SEQ	s	Indolizidine	**1549–1552**	[526–529]
	SEQ, NNSW	w	Isoflavonoid	**1521–1523**	[517, 518]
	SEQ (cult. (USA))	lf	Piperidine	**1560**	[530]
	SEQ (cult. (USA))	s	Pyrrolidine	**1554, 1555**	[530, 531]
	SEQ	s, lf	Pyrrolizidine	**1553**, **1556–1559**	[527, 529, 532, 533]
	SEQ	w	Triterpenoid	**1512**, **1513**	[512]
	SEQ (cult. (Pakistan))	lf	Triterpenoid	**1514–1516**, **1518**, **1519**	[513, 514]
Erythrina numerosa	SEQ	st	Erythrinan	**1572**	[39]
Erythrina variegata	SEQ	fl	Erythrinan	**1573–1575**	[39]
Erythrina vespertilio	SEQ	fr	Anthrinilic acid amide	**1600**	[516]

(continued)

Table 30 (continued)

Species	Collection location[a,b]	Plant part(s)[c]	Compound class	Compound(s)	Refs.
	SEQ	fr, w	Erythrinan	**1570**, **1571**	[39, 516]
	SEQ	fr	Isoflavonoid	**1520, 1528, 1529**	[516]
	SEQ	fr, w	Tryptamine	**1599**	[39, 516]
Erythrophleum chlorostachys	NQ	lf	Cinnamide	**1576–1579**	[539]
	NQ, NT	b	Diterpenoid alkaloid	**1580–1592**	[540–543]
Hovea longipes	CY	st	Pyridone	**1562**, **1561**, **1569**	[39, 535]
	NZ	lf, st	Triterpenoid	**1249**	[445]
Indigofera linnaei	NT	ap	Amino acid	**1596**	[545]
	NT	ap	Nitro compound	**1593–1595**	[544]
Ormosia ormondii	SEQ, NQ	b, st	Pyridone	**1566–1568**	[39, 536–538]
Senna coronilloides	SEQ	st	Chalcone	**1546**	[39]
	SEQ	st	Coumarin	**1547**	[39]
Sophora fraseri	SEQ	r	Benzofuran	**1531**	[519]
	SEQ	r	Isoflavonoid	**1524–1527**, **1530**	[519]
	SEQ	r	Pyridone	**1561**	[39]
Sophora tomentosa subsp. *australis*	CY	s, lf	Pyridone	**1563–1565**, **1561**	[39]

[a] SEQ = Southeast Queensland, NQ = North Queensland, CY = Cape York, Queensland, NNSW = Northern New South Wales, NSW = New South Wales, NT = Northern Territory, NZ = New Zealand, cult. = cultivated, [b] natural location shown for those species studied from cultivated material, [c] lf = leaves, s = seed, ap = aerial parts, r = roots, fl = flowesr, b = bark, hw = heartwood, w = wood, st = stems

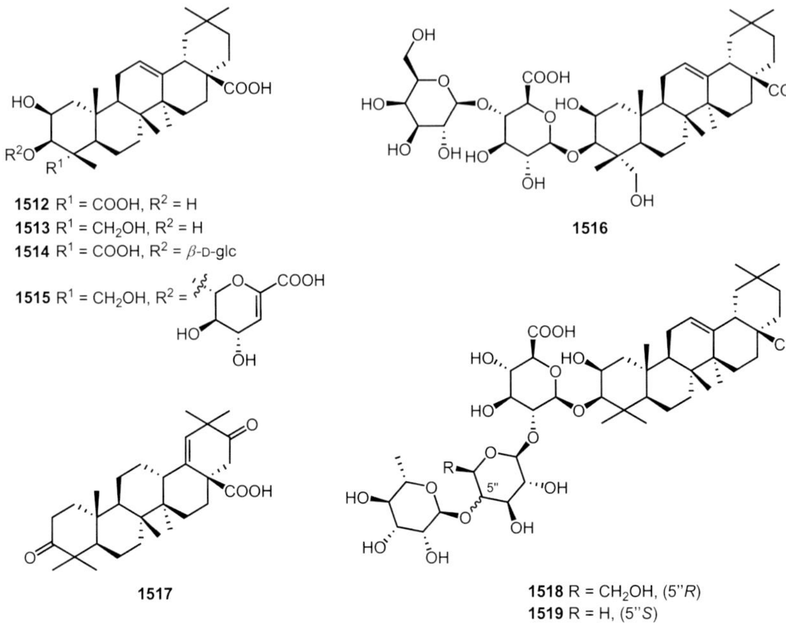

Fig. 117 Structures of compounds **1512–1519** reported from Fabaceae species

(**1537**), astragalin (**1538**), bayin (**1539**), 7,8,4′-trihydroxyflavonol (**1540**), (+)-2,3-*cis*-3,3′,4′,7,8-pentahydroxyflavan (4$\alpha \rightarrow$ 6)-isomelacacidin (**1541**), (+)-2,3-*trans*-3,3′,4′,7,8-pentahydroxyflavan (4$\beta \rightarrow$ 6)-isomelacacidin (**1542**), [4-*O*-4]-bis(2,3-*cis*-3,4-*trans*-3,3′,4′,7,8-pentahydroxyflavan (**1543**), and 2,3-*cis*-3,4-*trans*-3,3′,4′,7,8-pentahydroxyflavan-[4-0-4]-2,3-*cis*-3,4-*cis*-3,3′,4′,7,8-pentahydroxyflavan (**1544**)), a number with 7,8-dioxygenation, have been found in *C. australe, Acacia* and *Bauhinia* species [39, 517, 520–524]. Benzoic acid derivatives and chalcones (methyl gallate (**1545**), 3-(3,4-dihydroxyphenyl)-1-(3-β-D-glucopyranosyl-2,4-dihydroxyphenyl)-1-propanone (**1546**), methylumbelliferone (**1547**), and 4,2′,4′,6′-tetrahydroxychalcone 2′-[*O*-rhamnosyl-(1 $\rightarrow$ 4)-side] (**1548**)) have been reported from *Bauhinia*, *Senna*, and *Acacia* species [39, 525].

The rich diversity of alkaloids from rainforest Fabaceae species is predominantly from the subfamily Papilionoideae. A variety of indolizidines, pyrrolizidines, and related structures (castanospermine (**1549**), 6-*epi*-castanospermine (**1550**), 7-deoxy-6-*epi*-castanospermine (**1551**), 6,7-di-*epi*-castanospermine (**1552**), 3,8-di-*epi*-alexine (**1553**), australine (**1554**), 1-*epi*-australine (**1555**), 1,7α-diepialexine (**1556**), 1,7α-di-*epi*-alexine (**1557**), 1-*epi*-alexine (**1558**), *N*-hydroxyethyl)-2-hydroxymethyl)-3-hydroxypyrrolidine (**1559**), and fagomine (**1560**)) have been reported from *C. australe* seeds [526–533]. Castanospermine (**1549**) shows antiviral activity but was too toxic for HIV drug development. The seeds of *C. australe* have been a valuable source of starch for first nations rainforest dwellers, but the toxic alkaloids found in the seeds meant that they needed to be processed by pounding into

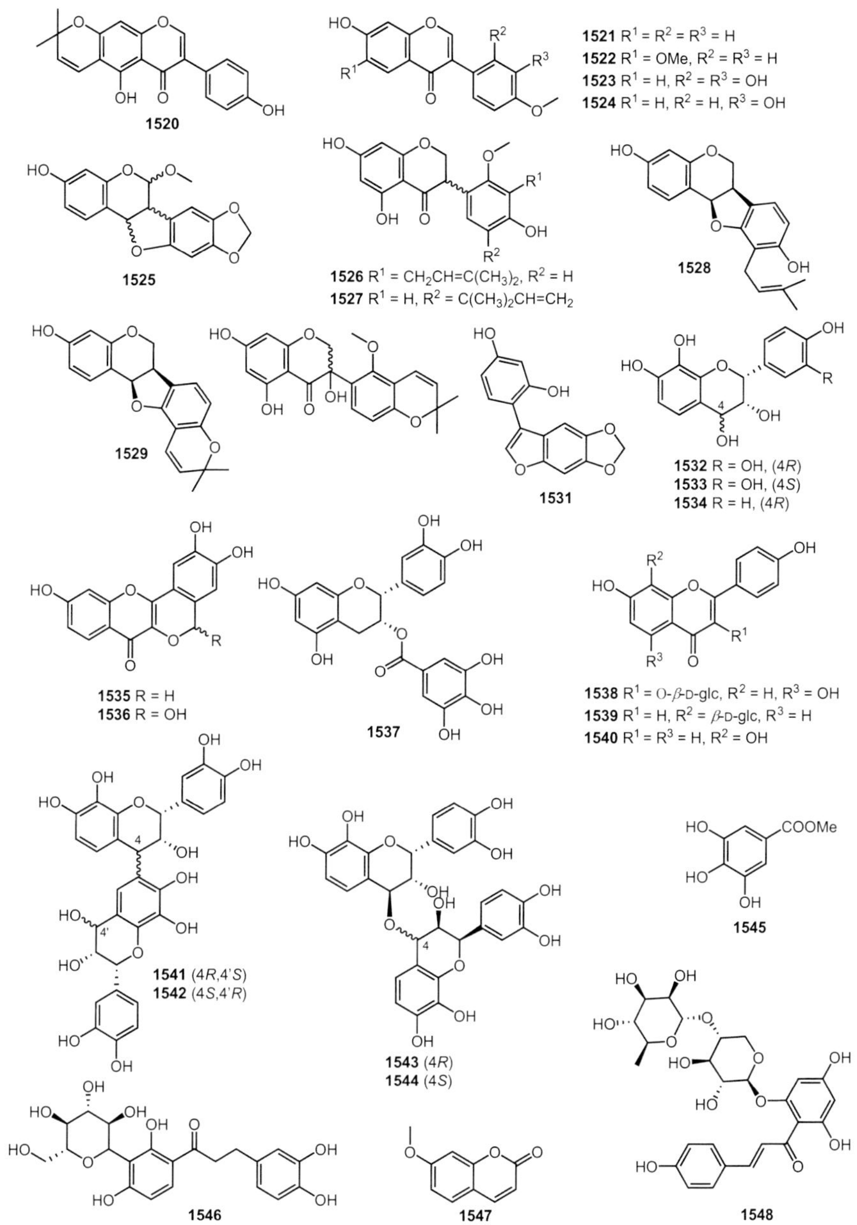

Fig. 118 Structures of compounds **1520–1548** reported from Fabaceae species

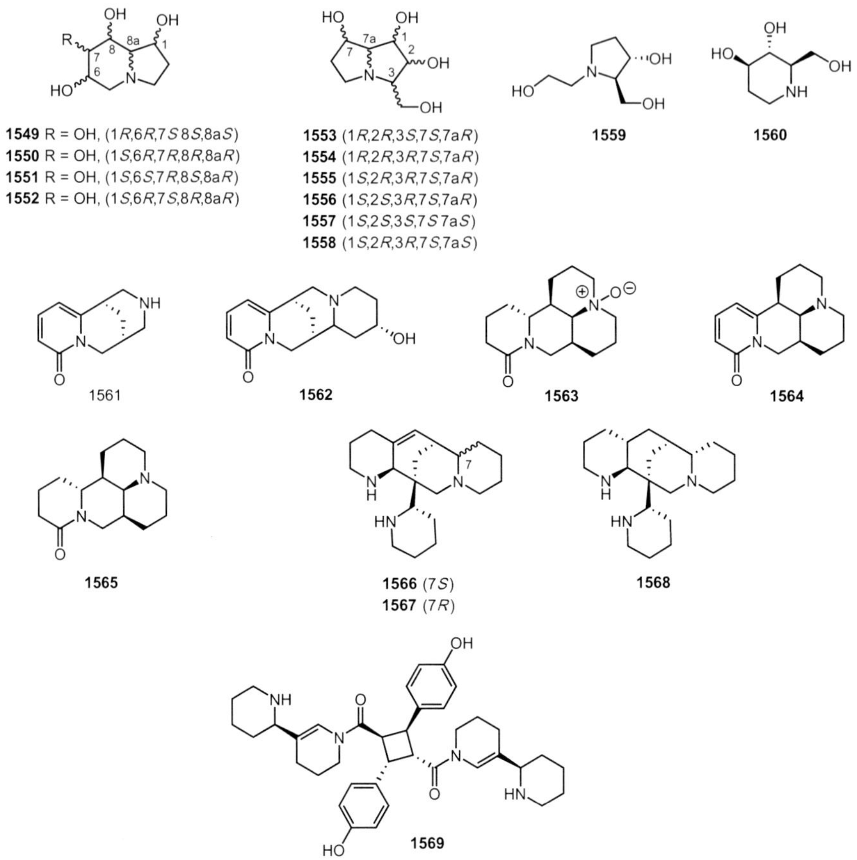

Fig. 119 Structures of compounds **1549–1569** reported from Fabaceae species

a meal and then washed repeatedly to remove the water-soluble alkaloids before they could be cooked and eaten [534]. *Sophora* and *Hovea* species both contain the nicotinic acetyl choline receptor agonist cytisine (**1561**), and these genera also contain other pyridone alkaloids of more complex structures such as the tetracyclic baptifoline (**1562**), and the matrine derivatives (oxymatrine (**1563**), sophoramine (**1564**), and matrine (**1565**)) [39, 535]. The bark of *Ormosia ormondii*, a taxon closely related to *Hovea* species, contains the more complex cytisine derivatives podopetaline (**1566**), 6-*epi*-podopetaline (**1567**), ormosanine (**1568**), and the cyclobutyl alkaloid, hoveine (**1569**) [39, 536–538].

Six *Erythrina* alkaloids, 11β-methoxyglucoerysovine (**1570**), erysotrine (**1571**), erysovine (**1572**), erythraline (**1573**), erythrinine (**1574**), and 11-methoxyerythraline (**1575**), have been found in *Erythrina vespertilio*, *E. numerosa*, or *E. variegata* [39, 516]. Several cinnamamides and ethylaminocinnamates (β-dimethylaminoethyl cinnamate (**1576**), *N*-2-hydroxyethyl-*N*-methyl cinnamamide

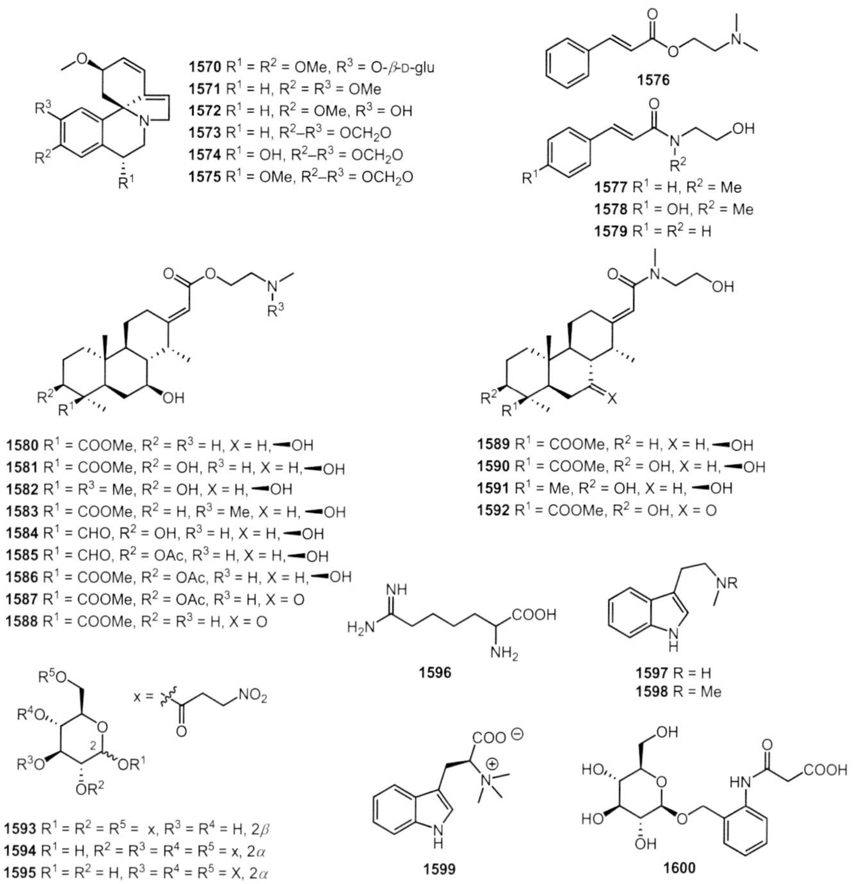

Fig. 120 Structures of compounds **1570–1600** reported from Fabaceae species

(**1577**), *N*-2-hydroxyethyl-*N*-methyl-(*E*)-*p*-hydroxycinnamamide (**1578**), and *N*-2-hydroxyethylcinnamamide (**1579**)) [539] and diterpenoid amines and amides (norcassamidine (**1580**), norerythrostachamine (**1581**), cassaidine (**1582**), cassamidine (**1583**), norerythrostachaldine (**1584**), norerythrostachaldine 3*β*-acetate (**1585**), 3*β*-acetoxynorerythrostachamine (**1586**), 3*β*-acetoxynorerythrophlamine (**1587**), norerythrophlamine (**1588**), norcassamidide (**1589**), norerythrostachamide (**1590**), norcassaidide (**1591**), and norerythrophlamide (**1592**)) have been isolated from the tropical tree *Erythrophleum chlorostachus* [540–543]. These compounds are highly poisonous and consumption of leaves by livestock has caused many deaths. The aerial parts of *Indigofera linnaei* contains three 3-nitropropionate esters of glucose, karakin (**1593**), 2,3,4,6-tetra-*O*-[3-nitropropanoyl]-*α*-D-glucopyranose (**1594**), and 3,4,6-tri-*O*-[3-nitropropanoyl]-*α*-D-glucopyranose (**1595**)) [544], along with an analogue of arginine, indospicine (**1596**), in which the amine of the guanidine is replaced with a methylene [545]. Three tryptophan derivatives have been isolated from *Acacia maidenii* (*N*-methyl tryptamine (**1597**) and *N,N*-dimethyl tryptamine (**1598**)) and *Eyrthrina vespertilio* (hypaphorine, (**1599**)) [516, 546]. The amino acid **993** was found in *Acacia maidenii* and the aniline amide vespertillioside (**1600**) was found in the fruits of *E. vespertilio* [516, 521].

Fagales

The Fagales has a clear divide between families of Northern and Southern Hemispheric distribution, with the Nothofagaceae and Casuarinaceae sharing a Gondwanan origin (Fig. 121) [547]. The center of extant species distribution of the Casuarinaceae is Australia [548]. Long-distance dispersal via ocean currents has resulted in littoral rainforest species from the genus *Casuarina* being found in Africa, Pacific islands, South and Central America, and part of Malesia. Long-distance dispersal is also thought to have resulted in *Gymnosperma* species colonizing New Guinea, New Caledonia, and Malesia.

The only compounds reported from Australian rainforest Casuarinaceae species are oleanolic acid derivatives (3-*O*-(*E*)-coumaroyl β-amyrin (**1601**), 3-*O*-(*Z*)-coumaroyl β-amyrin (**1602**), 3-*O*-dihydrocoumaroyl β-amyrin (**1603**), 3-*O*-(*E*)-coumaroyl erythrodiol (**1604**), 3-*O*-(*Z*)-coumaroyl erythrodiol (**1605**), 3-*O*-(*Z*)-coumaroyl oleanolic acid (**1606**), 3-*O*-(*E*)-coumaroyl oleanolic acid (**1607**), erythrodiol (**1608**), oleanolic acid (**1609**), and casuariene (**1610**)), benzoic and cinnamic acid derivatives (**542**, **592**, 3,5-dimethoxygallic acid (**1611**), 3,4-dihydroxybenoic acid (**1612**), 3-methoxy-4-hydroxybenzoic acid (**1613**) and *p*-hydroxycinnamic acid (**1614**)) [549], and macrocyclic diarylheptanoids (casuaritriol B (**1615**), adanin (**1616**), ostryopsitriol (**1617**), and casuaritriol A (**1618**)) (Fig. 122). Although the diarylheptanoids were screened for glycosidase and acetylcholine esterase activities, none were deemed active (Table 31) [550].

Rosales

Species in the Rosales have a cosmopolitan distribution. Six of its nine families have species that inhabit Australian rainforests (Fig. 123). Only Australian rainforest species from three of these familes have been investigated chemically (Table 32).

Rhamnaceae

Within the Rhamnaceae (Plate 26), several genera including *Alphitonia*, *Ventilago* and *Emmenosperma* are endemic to continental Australia. A number of *Alphitonia* species and *Emmenosperma alphitonioides* contain a series of betulinic acid derivatives (betulinic acid (**802**), alphitonic acid (**1619**), betulin (**1620**), (*Z*)-coumaroyl alphitolic acid (**1621**), (*E*)-coumaroyl alphitolic acid (**1622**), ceanothic acid (**1623**),

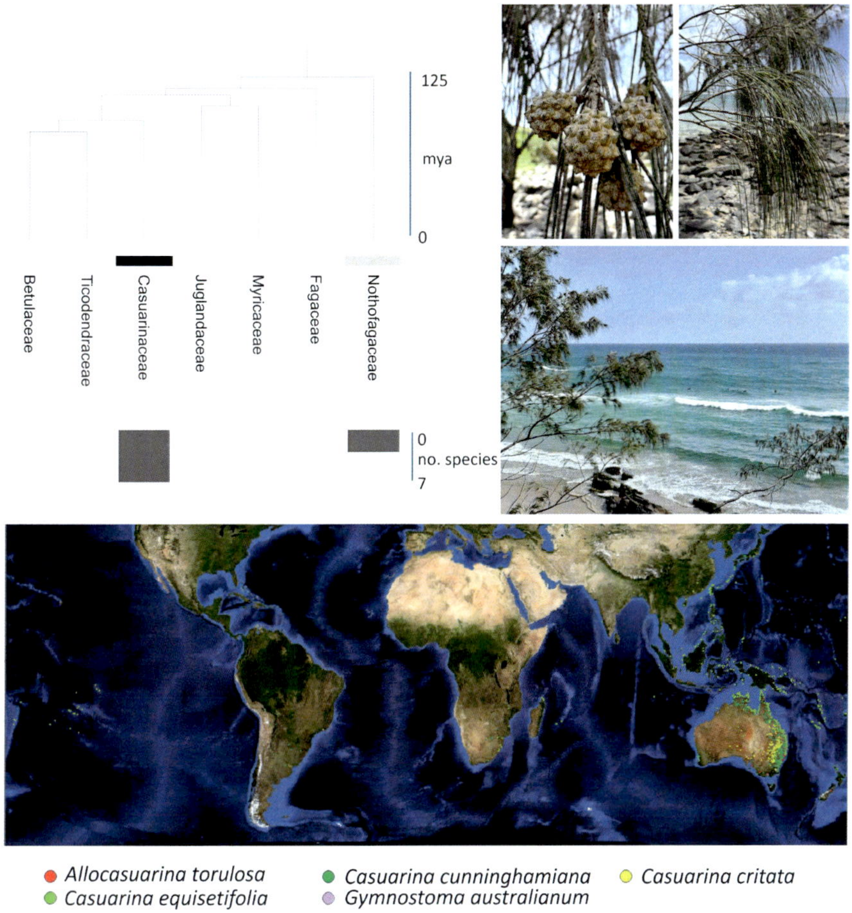

Fig. 121 Dated molecular phylogeny of the Fagales [7] showing those families that have species present in Australian rainforests (grey) and studied (black), and the number of species occurring in Australian rainforests. Photographs of *Casuarina equisetifolia,* a species found in littoral rainforest and along beaches. The map shows the herbarium records for Casuarinaceae species found in Australian rainforests [5, 43]. (Photographs AR Carroll)

alphitexolide (**1624**), ebelin lactone (**1625**), and jingullic acid (**1626**)), of which some have an A ring that has undergone an oxidative ring contraction [299, 551–555]. The benzofurans, alphitonin (**1627**) and maesopsin (**1628**), have also been reported from *Alphitonia* species [551]. *Ventilago viminalis* produces a naphthoquinone, ventilagone (**1629**), in addition to a series of anthraquinones, ventimalin (**1630**), viminalin (**1631**), chrysophanol (**1632**), islandicin (**1633**), and helminthosporin (**1634**) (Fig. 124) [556].

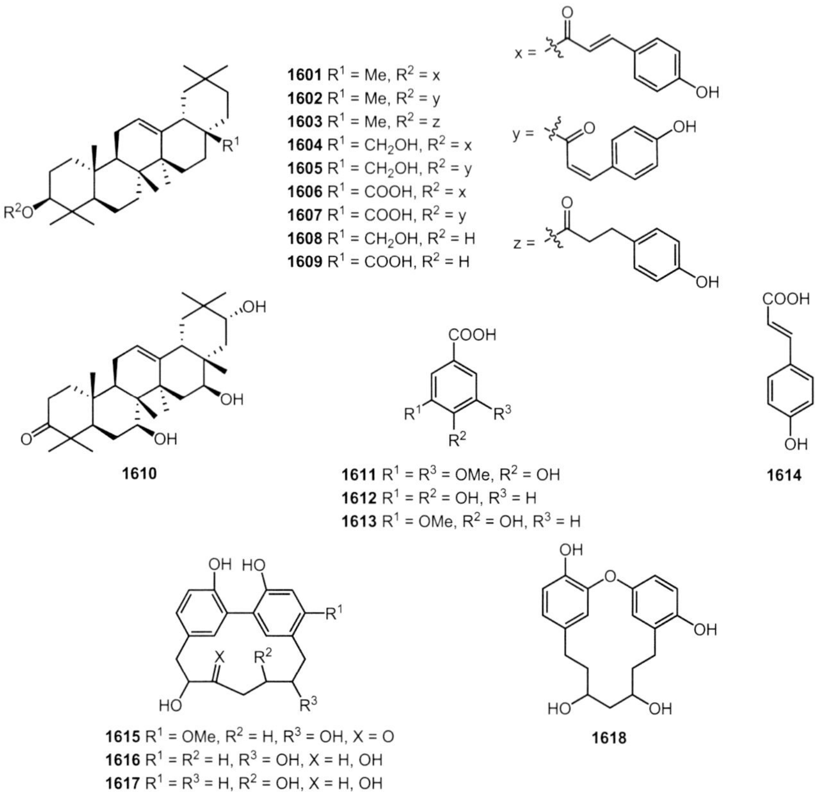

Fig. 122 Structures of compounds **1601–1618** reported from *Casuarina equisetifolia*

Table 31 Distribution of natural products isolated from the Casuarinaceae

Species	Collection location[a,b]	Plant part(s)[c]	Compound class	Compound(s)	Refs.
Casuarina equisetifolia	NQ (cult. (Japan))	lf	Benzoic acid	**542**, **592**	[549]
	NQ (cult. (Japan))	lf	Cinnamc acid	**1614**	[549]
	NQ (cult. (China))	lf, tw	Dibenzoheptane	**1615–1618**	[550]
	NQ (cult. (Japan))	lf	Gallic acids	**1611–1613**	[549]
	NQ (cult. (Japan))	lf, tw	Triterpenoid	**1601–1610**	[549, 550]

[a] NQ = North Queensland, cult. = cultivated, [b] natural location shown for those species studied from cultivated material, [c] lf = leaves, tw = twigs

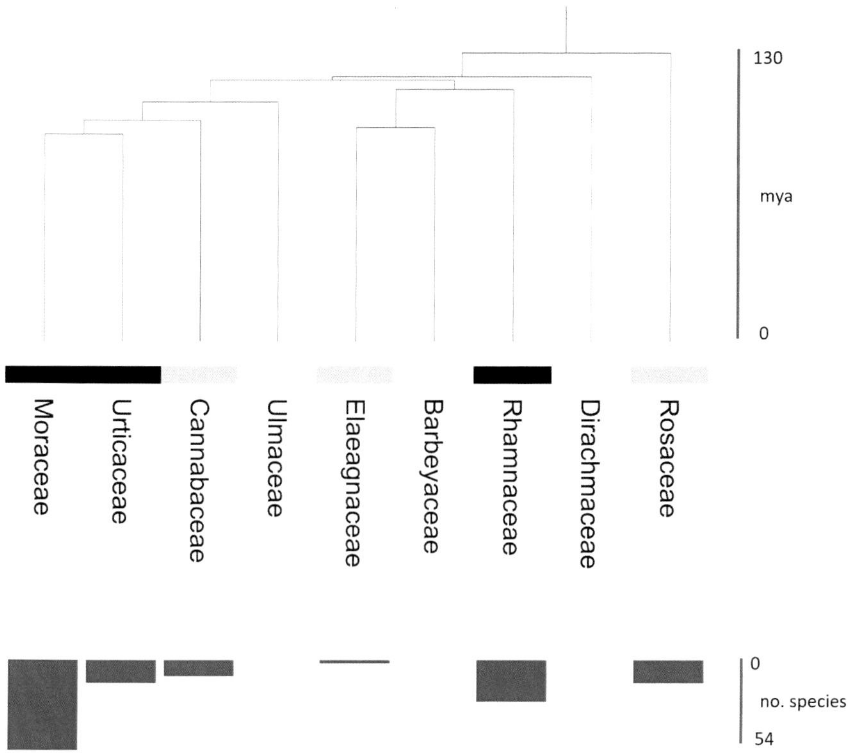

Fig. 123 Dated molecular phylogeny of the Rosales [7], showing those families that have species present in Australian rainforests (grey) and studied (black), and the number of species occurring in Australian rainforests

Urticaceae

The Australian subtropical and tropical rainforests contain several *Dendrocnide* species from the nettle family Urticaceae, which have leaves able to cause stings that are excruciatingly painful and long-lasting. *Dendrocnide* species are trees and shrubs having a tropical Asian origin and migrated into Australia during the late Miocene [557]. Two Australian species, *Dendrocnide moroides*, a shrub, and the giant stinging tree *D. excelsa*, a tall tree, both possess large heart-shaped leaves. The underside of the leaves and the petioles are covered with hairs called trichomes that act like tiny hypodermic needles. The trichomes contain a venomous solution that can be injected into the skin simply by brushing against the foliage (Fig. 126). The stings from these species are reputed to be the most painful of any plant worldwide. Initial investigation suggested that histamine, acetylcholine and/or serotonin might be the causative principles, but this has been disproven. An unusual cyclic peptide, moroidin (**1635**), was reported in the 1980's but this also did not fully account for the pain-inducing effects

Table 32 Distribution of natural products isolated from the Rosales

Species	Collection location[a]	Plant part(s)[b]	Compound class	Compound(s)	Refs.
Rhamnaceae					
Alphitonia excelsa	SEQ	w	Benzofuran	**1627**	[551]
	SEQ	w, b	Triterpenoid	**802**, **1619**, **1623**, **1624**	[551, 552]
Alphitonia petriei	SEQ	w	Benzofuran	**1627**	[551]
	SEQ, NQ	w, lf, st	Triterpenoid	**802**, **1619**, **1621–1623**	[551, 553]
Alphitonia whitei	NQ	w	Benzofuran	**1628**	[551]
	NQ	w	Triterpenoid	**802**, **1619**, **1620**, **1623**	[551]
Emmenosperma alphitonioides	SEQ, NNSW	w	Triterpenoid	**778**, **1623**, **1625**, **1626**	[299, 554, 555]
Ventilago viminalis	NSW	r	Anthraquinone	**1630–1634**	[556]
	NSW	r	Naphthoquinone	**1629**	[556]
Urticaceae					
Dendrocnide moroides	SEQ	lf	Peptide	**1635**, **1639**	[558, 559]
Dendrocnide excelsa	SEQ	lf	Peptide	**1636–1638**	[559]
Boehmeria platyphylla	NNSW	lf	Dibenzoisoquinoline	**391**, **1640**	[563]
	NNSW	lf	Piperidine	**1641**	[563]
Moraceae					
Ficus septica	NQ	lf	Dibenzoisoquinoline	**1183**, **1184**, **1642**	[564]
Ficus macrophyllaformacolumnaris	SEQ	lf	Triterpenoid	**1643**	[565]
Ficus pantoniana	NQ	lf	Flavonoid alkaloid	**1644**, **1645**	[566]

[a] SEQ = Southeast Queensland, NQ = North Queensland, NNSW = Northern New South Wales, NSW = New South Wales, [b] lf = leaves, r = roots, b = bark, w = wood, st = stems

Plate 26 Flowers and fruits of *Alphitonia excelsa*. (Photographs AR Carroll)

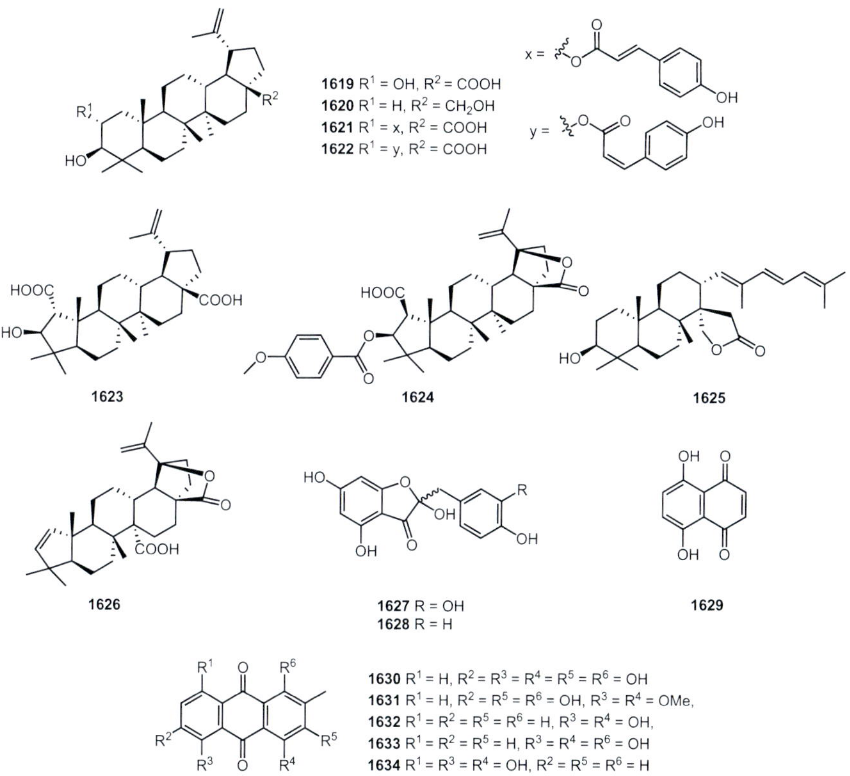

Fig. 124 Structures of compounds **1619–1634** reported from Rhamnaceae species

[558]. In 2020, four ~4 kda inhibitory cysteine knot (ICK) peptides named gympietides (ExTxA-C (**1636**), (**1637**), and (**1638**), and MoTxA (**1639**), each containing 36 amino acid residues and incorporating three disulfide bonds between their six cysteine residues, were identified as the causative principles (Figs. 125 and 126) [559]. These ICK peptides possess a unique amino acid sequence, are highly stable,

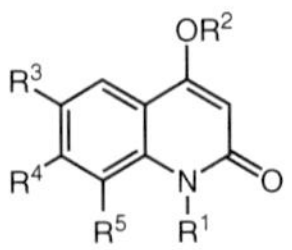

2016 $R^1 = R^3 = R^4 = R^5 = H$, $R^2 = Me$
2017 $R^1 = R^3 = R^4 = H$, $R^2 = Me$, $R^5 = OMe$
2018 $R^1 = R^2 = Me$, $R^3 = R^4 = R^5 = H$
2019 $R^1 = R^4 = H$, $R^2 = Me$, $R^3 = R^5 = OMe$
2020 $R^1 = R^2 = Me$, $R^3 = R^4 = H$, $R^5 = OMe$
2021 $R^1 = R^5 = H$, $R^2 = Me$, $R^3 = R^4 = OMe$
2022 $R^1 = R^3 = R^4 = R^5 = H$, $R^2 = CH_2CHC(CH_3)_2$

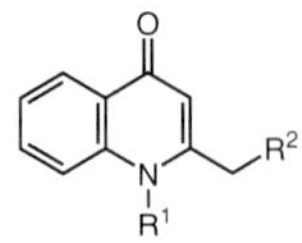

2023 $R^1 = H$, $R^2 = n$-But
2024 $R^1 = H$, $R^2 = Et$
2025 $R^1 = Me$, $R^2 = H$

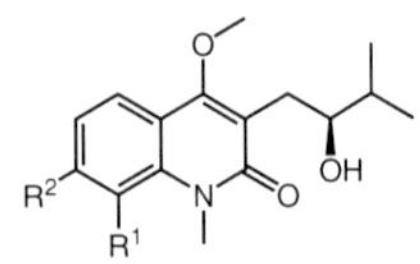

2028 $R^1–R^2 = OCH_2O$
2029 $R^1 = OMe$, $R^2 = H$

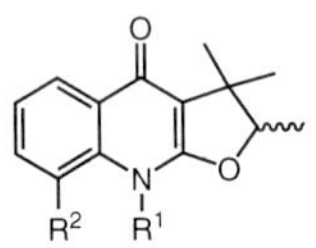

2026 $R^1 = Me$, $R^2 = H$
2027 $R^1 = H$, $R^2 = OMe$

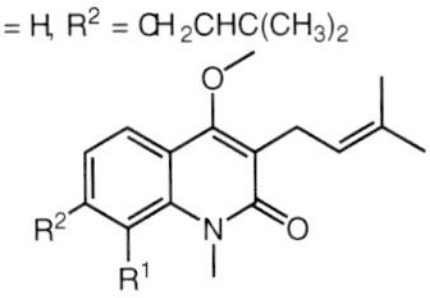

2030 $R^1–R^2 = OCH_2O$
2031 $R^1 = OMe$, $R^2 = H$

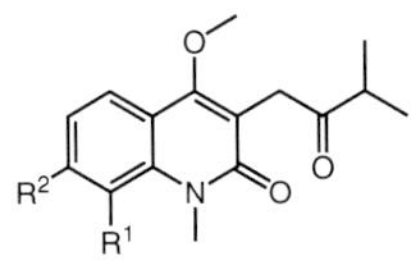

2032 $R^1–R^2 = OCH_2O$
2033 $R^1 = OMe$, $R^2 = H$

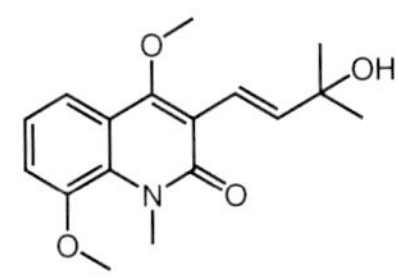

2034

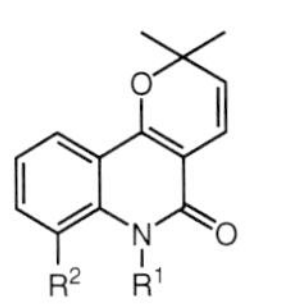

2035 $R^1 = R^2 = H$
2036 $R^1 = Me$, $R^2 = H$
2037 $R^1 = H$, $R^2 = OMe$
2038 $R^1 = CH_2OAc$, $R^2 = H$
2039 $R^1 = Me$, $R^2 = OMe$

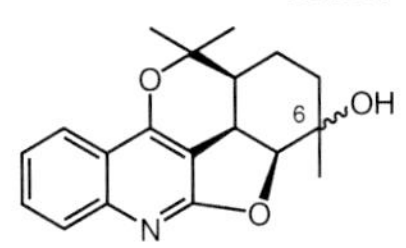

2040 (6*R*)
2041 (6*S*)

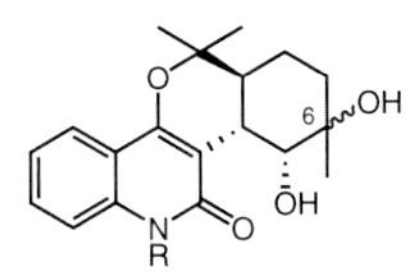

2042 R = Me, (6*S*)
2043 R = CH_2OAc, (6*S*)
2044 R = Me, (6*R*)

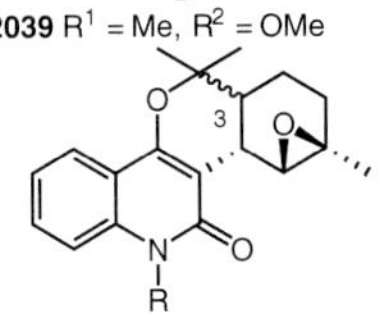

2045 R = Me, (3*S*)
2046 R = CH_2OAc, (3*S*)
2047 R = CH_2OAc, (3*R*)
2048 R = H, (3*S*)

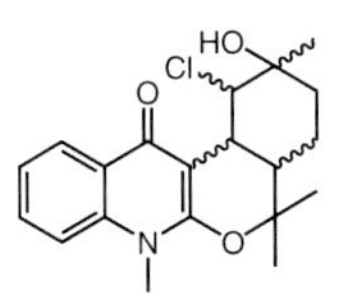

2049

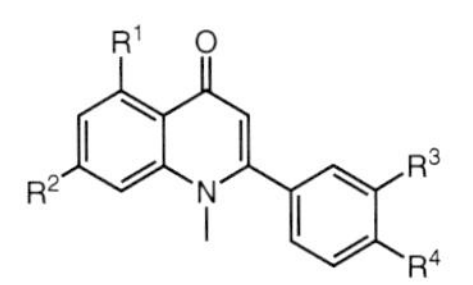

2050 $R^1 = R^3 = R^4 = H$, $R^2 = OMe$
2051 $R^1 = OH$, $R^2 = R^3 = R^4 = H$
2052 $R^1 = OH$, $R^2 = H$, $R^3–R^4 = OCH_2O$

Fig. 125 Structures of peptides **1635–1639** reported from *Dendrocnide* species

and, surprisingly, their quaternary structures are remarkably like peptides identified in cone snail venoms. Three of the peptides were found in *D. excelsa* leaves (**1636–1638**) and a fourth (**1639**) in *D. moroides* leaves. There are eight backbone positions that can have different amino acid residues and the *D. moroides* peptide MoTxA collectively has six different amino acid residues compared to ExTxA, whereas each of the ExTx congeners differ from each other by only one or two residue substitutions. The gympietide structures were confirmed through total synthesis and the natural and synthetic materials were shown to induce identical long-lasting pain response in mice through modulating inactivation of voltage-gated sodium channels (Na_v) in sensory neurons [560]. The mode action of ExTxA involves interaction with TMEM233, a

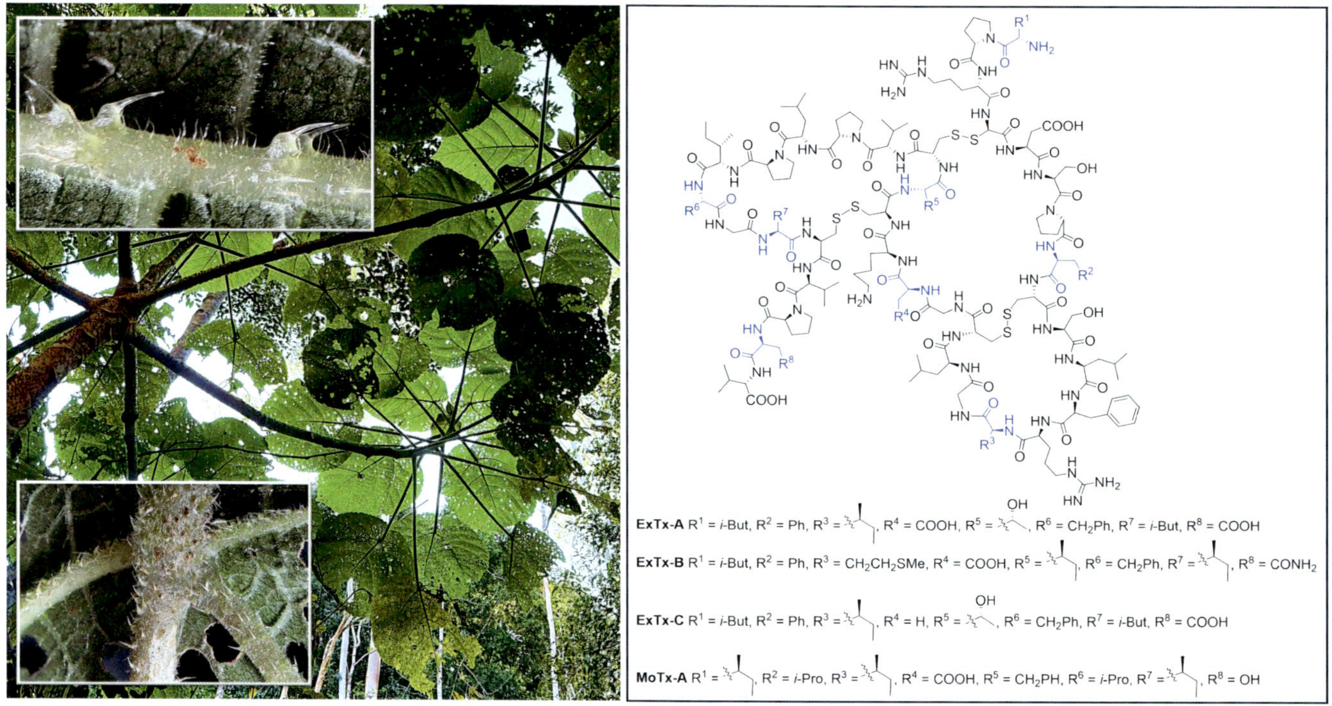

Fig. 126 Photographs of the giant stingingtree *Dendrocnideexcelsa*. The insets show the trichomes (top) and trichomes and hairs (bottom). Full peptide structures of compounds **1636–1639** are depicted to show the amino acid substitution positions in blue. (Photographs AR Carroll)

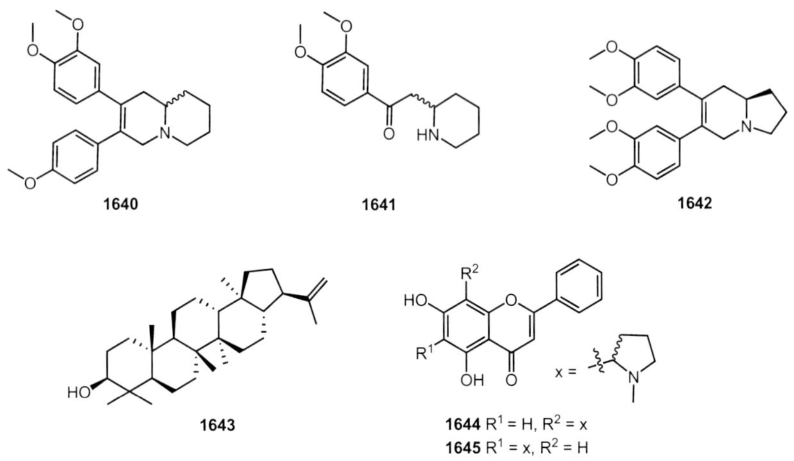

Fig. 127 Structures of compounds **1640–1645** reported from *Boehmeria platyphylla* and *Ficus* species)

new transmembrane protein in the dapanin family that is essential for modulating gating properties of Na_V1.7. These insights hold promise for the development of novel pain treatments. MoTxA is the most potent pain-inducing molecule, and this is in keeping with anecdotal reports that stings from *D. moroides* are more painful than those of *D. excelsa*. Using MALDI MS imaging of the petioles showed that the peptides are concentrated in the trichomes directly adjacent to the hollow silica needles [559].

Boehmeria platyphylla from northern New South Wales contains three alkaloids, including cryptopleurine (**391**) (isolated as a partial racemate), *seco*-phenanthroquinolizidine base (**1640**), and 3,4-dimethoxy-ω-(2′-piperidyl)acetophenone (**1641**) (Fig. 127). Crytopleurine was previously isolated from *Cryptocarya* species (Lauraceae), while **1641** is the methoxy ether of **389** that was purified earlier from *Cryptocarya pleurosperma*. Cryptopleurine was inactive against a variety of other cancer cell lines, but was found to be highly specific and extremely cytotoxic against Eagle's 9 KB carcinoma of the nasopharynx, exhibiting an ED_{50} value range of between 7.8×10^{-4} and 2.6×10^{-5} g/cm^3 [561–563].

Moraceae

In Australian rainforests, 48 fig (*Ficus*) species dominate the Rosales family Moraceae (57 total species). Only three of these species have been investigated, with the leaves of *F. septica* containing the indolizine alkaloids, (+)-tylocrebrine (**1183**), (−)-tylophorine (**1184**) and septicine (**1642**), which are the same compounds reported from *Tylophora* species and related to alkaloids that occur in *Cryptocarya* and *Boehmeria* species [564]. A hopene triterpenoid, moretenol (**1643**), was found in the leaves of *F. macrocarpa* [565], and two pyrrolidine-substituted flavones, ficine (**1644**) and isoficine (**1645**), were isolated from the leaves *F. pantoniana* collected in northern Queensland (Fig. 127) [566].

Malvids

The remaining orders of flowering plants are classified as Malvids. There are eight orders and 85 families within this group and species from 35 families are found in Australian rainforest. Several of these families are particularly species-rich (Euphorbiaceae, Phyllanthaceae, Sapindaceae, Rutaceae, and Malvaceae).

Celastrales

Of the three families that constitute the Celastrales, only Celastraceae species are represented in Australian rainforests. This family has a cosmopolitan distribution, but in Australia the genus *Denhamia* (endemic to Australia and New Caledonia) is the most species rich. Four of the 15 Australian rainforest *Denhamia* species have been investigated and three of these have proven to be a rich source of dihydro-β-agarofuran type sesquiterpenoids (Table 33).

Together with *Celastrus subspicata*, these *Denhamia* species have yielded 47 agarofurans (Figs. 128 and 129) (agarofuran 1 (**1646**), agarofuran 2 (**1647**), celastrofurans A–C (**1648–1650**), celastrofurans E–G (**1651–1653**), (*1S*,4*R*,5*S*,7*R*,9*S*,10*S*)-9-benzoyloxy-1-furoyloxydihydro-β-agarofuran (**1654**), (1*R*,2*R*,4*R*,5*S*,7*R*,9*S*,10*R*)-2-acetoxy-9-benzoyloxy-1-furoyloxydihydro-β-agarofuran (**1655**), bilocularins D–I (**1656–1661**), 1*α*,2*α*,6*β*,15-tetraacetoxy-9*β*-benzoyloxydihydro-β-agarofuran (**1662**), bilocularin A (**1663**), bilocularin C (**1664**), celastrine A (**1665**), 1,6,8-triacetoxy-9-benzoyloxydihydro-agarofuran (**1666**), 1,6-diacetoxy-9-benzoyloxy-8-hydroxydihydro-agarofuran (**1667**), 1,6-diacetoxy-9-benzoyloxydihydro-agarofuran (**1668**), Ejap-2 (**1669**), denhaminols O–R (**1670–1673**), denhaminol G (**1674**), denhaminols A–F (**1675–1680**), denhaminol H (**1681**), denhaminols K–N (**1682–1685**), wilforsinine F (**1686**), 1*α*,2*α*,6*β*,8*α*,15-pentaacetoxy-9*α*-benzoyloxydihydro-β-agarofuran (**1687**), celastrofuran D (**1688**), bilocularin B (**1689**), denhaminol I (**1690**), denhaminol J (**1691**), and 1*α*,2*α*,6*β*,15-tetraacetoxy-9*α*-benzoyloxy-8-oxodihydrobagarofuran (**1692**) [567–574]. Several of the agarofurans show moderate leucine uptake inhibition or P-glycoprotein inhibition.

Denhamia pittosporoides subsp. *angustifolia* and *D. oscura* contain four intensely orange-colored triterpenoid quinone methides, pristimerin (**1693**), celastrol (**1694**), tingenone (**1695**), and iguesterin (**1696**) [569, 575], along with β-amyrin triterpenoids friedelin (**1697**), multiflorenol (**1698**), and glutinol (**1699**) [575]. Also present are a series of abietane diterpenoids, ferruginol (**1700**), sugiol (**1701**), dispermol

Table 33 Distribution of natural products isolated from the Celastraceae

Species	Collection location[a]	Plant part(s)[b]	Compound class	Compounds	Refs.
Celastrus subspicata	SEQ, NNSW	s, lf	Sesquiterpenoid	**1646–1655**, **1688**	[567, 568]
Denhamia bilocularis	SEQ	r, lf	Sesquiterpenoid	**1656–1669**, **1689**	[569, 570]
Denhamia bilocularis	SEQ	R	Triterpenoid	**1693**, **1694**	[569]
Denhamia celastroides	NQ, SEQ	fr, lf	Sesquiterpenoid	**1670–1685**	[571–573]
Denhamia obscura	NT	r, b, s, lf	Diterpenoid	**1700–1706**	[575]
Denhamia obscura	NT	r, b, seed, lf	Triterpenoid	**1695–1699**	[575]
Denhamia pittosporoides subsp. *angustifolia*	SEQ	Lf	Sesquiterpenoid	**1662**, **1686**, **1687**, **1690–1692**	[574]
Elaeodendron australe var. *integrifolium*	SEQ	Wp	Steroid	**1707–1710**	[576]
Siphonodon australis	SEQ	Lf	Furanone	**1711**, **1712**	[577]

[a] SEQ = Southeast Queensland, NQ = North Queensland, NNSW = Northern New South Wales, NT = Northern Territory, [b] lf = leaves, s = seeds, r = roots, fr = fruits, wp = whole plant, b = bark

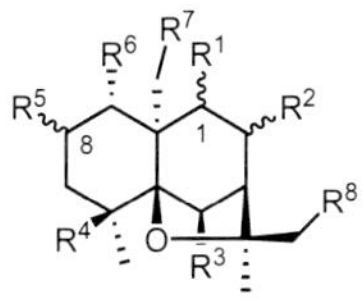

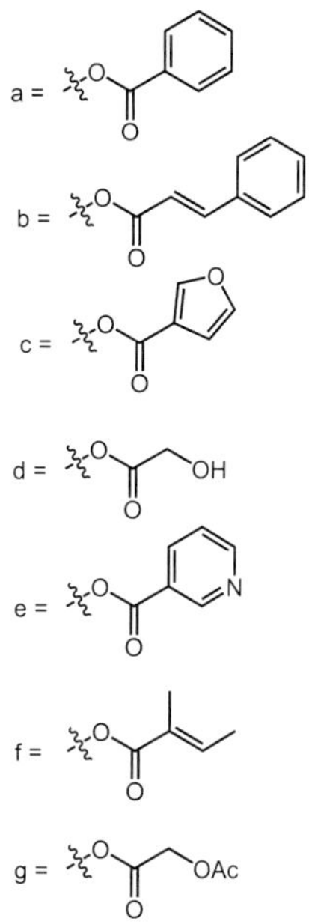

1646 $R^1 = R^3$ = a, $R^2 = R^4$ =OH, $R^5 = R^7 = R^8$ = H, R^6 = OAc, 1α, 2α
1647 $R^1 = R^3$ = a, $R^2 = R^6$ = OAc, R^4 = OH, $R^5 = R^7 = R^8$ = H, 1α, 2α
1648 $R^1 = R^6$ = a, $R^2 = R^3 = R^4 = R^5 = R^7 = R^8$ = H, 1β
1649 $R^1 = R^6$ = a, $R^2 = R^3 = R^4 = R^7 = R^8$ = H, R^5 = OAc, 1β, 8β
1650 $R^1 = R^6$ = a, R^2 = OH, $R^3 = R^4 = R^7 = R^8$ = H, R^5 = OAc, 1β, 2α, 8β
1651 R^1 = a, R^2 = OH, $R^3 = R^4 = R^7 = R^8$ = H, R^5 = OAc, R^6 = c, 1α, 2α, 8β
1652 $R^1 = R^6$ = c, R^2 = OH, $R^3 = R^4 = R^7 = R^8$ = H, R^5 = OAc, 1α, 2α, 8β
1653 R^1 = a, $R^2 = R^5$ = OAc, $R^3 = R^4 = R^7 = R^8$ = H, R^6 = c, 1α, 2α, 8β
1654 R^1 = a, R^2 = H, $R^3 = R^4 = R^5 = R^7 = R^8$ = H, R^6 = c, 1α
1655 R^1 = a, $R^2 = R^3 = R^4 = R^7 = R^8$ = H, R^5 = OAc, R^6 = c, 1α, 8β
1656 R^1 = b, $R^2 = R^8$ = H, $R^3 = R^5$ = OAc, R^4 = OH, R^6 = d, R^7 = a, 1β, 8α
1657 $R^1 = R^7$ = b, $R^2 = R^8$ = H, $R^3 = R^5$ = OAc, R^4 = OH, R^6 = d, 1β, 8α
1658 R^1 = b, $R^2 = R^8$ = H, $R^3 = R^5$ = OAc, $R^4 = R^7$ = OH, R^6 = d, 1β, 8α
1659 R^1 = b, $R^2 = R^8$ = H, $R^3 = R^5 = R^6 = R^7$ = OAc, R^4 = OH, 1β, 8α
1660 R^1 = b, $R^2 = R^8$ = H, $R^3 = R^6$ = OAc, $R^4 = R^5 = R^7$ = OH, 1β, 8α
1661 R^1 = b, $R^2 = R^7 = R^8$ = H, R^3 = e, $R^4 = R^5$ = OH, R^6 = OAc, 1β, 8α
1662 R^1 = a, $R^2 = R^4 = R^8$ = H, $R^3 = R^5 = R^6 = R^7$ = OAc, 1β, 8α
1663 R^1 = a, R^2 = OH, $R^3 = R^6 = R^7$ = OAc, $R^4 = R^5 = R^8$ = H,1β, 2α
1664 R^1 = a, $R^2 = R^6 = R^7$ = OAc, R^3 = OH, $R^4 = R^5 = R^8$ = H, 1β, 2α
1665 R^1 = a, $R^2 = R^3 = R^6 = R^7$ = OAc, $R^4 = R^5 = R^8$ = H, 1β, 2α
1666 R^1 = a, $R^2 = R^3 = R^6$ = OAc, $R^4 = R^5 = R^7 = R^8$ = H, 1β, 2α
1667 R^1 = a, R^2 = OH, $R^3 = R^6$ = OAc, $R^4 = R^5 = R^7 = R^8$ = H, 1β, 2α
1668 R^1 = a, $R^2 = R^4 = R^5 = R^7 = R^8$ = H, $R^3 = R^6$ = OAc, 1β
1669 R^1 = a, $R^2 = R^4 = R^5 = R^8$ = H, $R^3 = R^6 = R^7$ = OAc, 1β
1670 R^1 = b, $R^2 = R^5 = R^7$ = H, $R^3 = R^4$ = OH, R^6 = g, R^8 = f, 1β
1671 R^1 = b, $R^2 = R^5 = R^7$ = H, R^3 = OAc, R^4 = OH, R^6 = g, R^8 = f, 1β
1672 R^1 = b, $R^2 = R^3 = R^4$ = OH, $R^5 = R^7$ = H, R^6 = g, R^8 = f, 1β, 2β
1673 R^1 = b, $R^2 = R^3 = R^4$ = OH, $R^5 = R^7$ = H, R^6 = g, R^8 = f, 1β, 2β
1674 R^1 = b, $R^2 = R^4$ = OH, R^3 = OAc, $R^5 = R^7$ = H, R^6 = g, R^8 = f, 1β, 2β
1675 R^1 = b, $R^2 = R^5 = R^8$ = H, $R^3 = R^6 = R^7$ = OAc, R^4 = OH, 1β
1676 R^1 = b, $R^2 = R^4$ = OH, $R^3 = R^6 = R^7$ = OAc, $R^5 = R^8$ = H, 1β, 2β
1677 R^1 = b, $R^2 = R^5 = R^8$ = H, $R^3 = R^4$ = OH, $R^6 = R^7$ = OAc, 1β
1678 R^1 = b, $R^2 = R^5$ = H, $R^3 = R^4$ = OH, $R^6 = R^7$ = OAc, R^8 = a, 1β
1679 R^1 = b, $R^2 = R^5$ = H, $R^3 = R^6 = R^7$ = OAc, R^4 = OH, R^8 = a, 1β
1680 R^1 = b, $R^2 = R^4$ = OH, R^3 = OAc, $R^5 = R^7$ = H, R^6 = g, R^8 = a, 1β, 2β
1681 R^1 = b, $R^2 = R^5$ = H, $R^3 = R^7$ = OAc, R^4 = OH, R^6 = g, R^8 = f, 1β
1682 R^1 = b, $R^2 = R^5$ = H, $R^3 = R^6 = R^7$ = OAc, R^4 = OH, R^8 = f, 1β
1683 R^1 = b, $R^2 = R^5$ = H, $R^3 = R^7$ = OAc, R^4 = OH, R^6 = d, R^8 = f, 1β
1684 R^1 = b, $R^2 = R^5$ = H, $R^3 = R^6 = R^7$ = OAc, $R^4 = R^8$ = OH,1β
1685 R^1 = b, R^2 = OAc, $R^3 = R^4$ = OH, $R^5 = R^7 = R^8$ = H, R^6 = g, 1β, 2β
1686 R^1 = a, $R^2 = R^3 = R^7 = R^5$ = OAc, $R^4 = R^8$ = H, R^6 = OH, 1β, 2α, 8α
1687 R^1 = a, $R^2 = R^3 = R^5 = R^6 = R^7$ = OAc, $R^4 = R^8$ = H, 1β, 2α, 8α

Fig. 128 Structures of agarofuran sesquiterpenoids **1646**–**1687** reported from Celastraceae species

(**1702**), 12-hydroxytotarol (**1703**), maytenoquinone (**1704**), obscurol (**1705**), and 13-methoxy-sempervir-6-ene (**1706**) (Fig. 130) [575]. Four digitoxigenin steroids (digitoxigenin (**1707**), digitoxigenin-3-*O*-*β*-D-glucoside (**1708**), digitoxigenin-3-*O*-*β*-D-quinovoside (**1709**), and digitoxigenin-3-*O*-*β*-D-xyloside (**1710**), which show potent

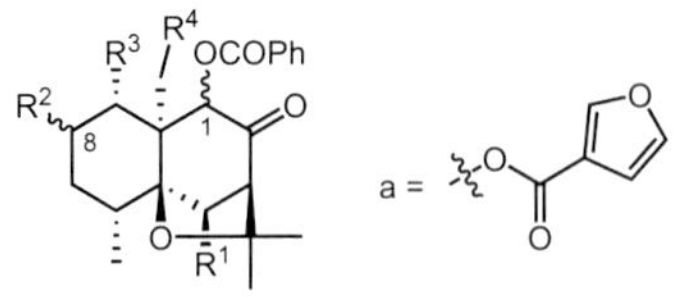

1688 R¹ = R⁴ = H, R² = OAc, R³ = a, 1β, 8β
1689 R¹ = R³ = R⁴ = OAc, R² = H, 1β
1690 R¹ = R³ = R⁴ = OAc, R² = OH,1α, 8α
1691 R¹ = OH, R² = R³ = R⁴ = OAc, 1α, 8α
1692 R¹ = R² = R³ = R⁴ = OAc, 1α, 8α

Fig. 129 Structures of agarofuran sesquiterpenoids **1688–1692** reported from Celastraceae species

IL-2 expression inhibition, have been reported from *Elaeodendron australe* var. *integrifolium* [576]. *Siphonodon australis* contains two simple lactones, siphonoside (**1711**) and siphonodin (**1712**) [577].

Malpighiales

The Malpighiales is a large order containing 37 families, of which 16 have species that occur in Australian rainforests. Just over half of these families have at least one Australian rainforest species that has been investigated chemically (Fig. 131, Table 34).

Ochnaceae and Clusiaceae

The Ochnaceae has only two species in Australian rainforests and the leaves of the north Queensland endemic understorey tree *Brackenridgea australiana* contain the triterpenoid **767** that shows DNA polymerase β inhibition [296]. This compound was also isolated from the bark of the Proteaceae tree *Bleasdalea bleasdalei* from the same region. *Calophyllum australianum* (Clusiaceae) contains the coumarin 5-methoxy-7,8-dimethyl-10-(3-methylbut-2-en-1-yl)-4-phenyl-7,8-dihydro-2*H*,6*H*-pyrano[3,2-*g*]chromene-2,6-dione (**1713**) [578], while the only study on Australian plant material from the widespread species *Calophyllum sil* reported two xanthones, euxanthone (**1714**) and jacareubin (**1715**) (Fig. 132) [579].

Euphorbiaceae

The Euphorbiaceae has a cosmopolitan tropical distribution with close to 7,000 species in 218 genera [580]. The Australian rainforest Euphorbiaceae includes 101 species across 32 genera (Plate 27). Constituents from 13 species have been reported. These studies have almost exclusively resulted in diterpenoids of various classes being isolated (Table 34).

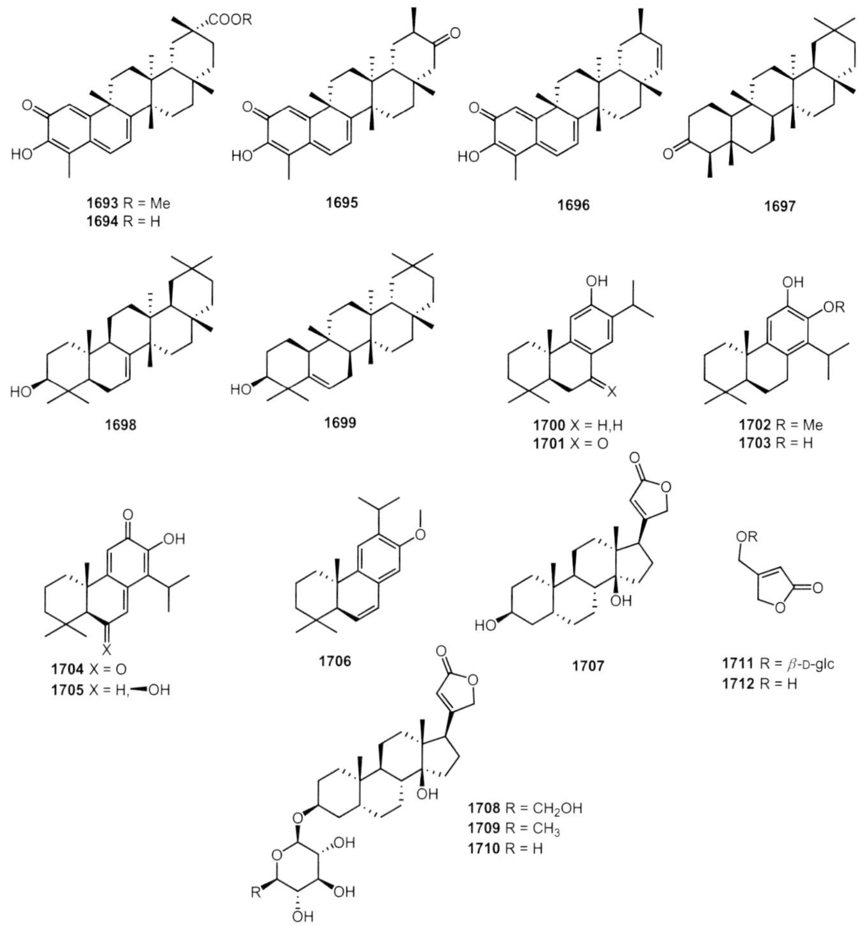

Fig. 130 Structures of compounds **1693–1712** reported from Celastraceae species

The most significant of these investigations is that on the north Queensland endemic *Fontainea picrosperma*. This genus is restricted to Queensland and New South Wales, with one species in New Caledonia. Extracts from the kernels of *F. picrosperma* showed potent cytotoxicity and this eventually led to the isolation of a series of phorbol esters as the active constituents (EBC-46 (**1716**), EBC-186 (**1717**), EBC-146 (**1718**), EBC-147 (**1719**), EBC-148 (**1720**), EBC-47 (**1721**), EBC-344 (**1722**), EBC-83 (**1723**), EBC-170 (**1724**), EBC-177 (**1725**), EBC-59 (**1726**), EBC-211 (**1727**), EBC-167 (**1728**), EBC-161 (**1729**), EBC-158 (**1730**), EBC-172 (**1731**), and EBC-188 (**1732**)) (Fig. 133). These compounds have a C-6/C-7 epoxide that is essential for cytotoxicity [581]. They impart their cytotoxic effect via activation of protein kinase C isoforms. The major variation within the series is the ester group attached to C-12, with a chain length greater than three carbons eliciting more

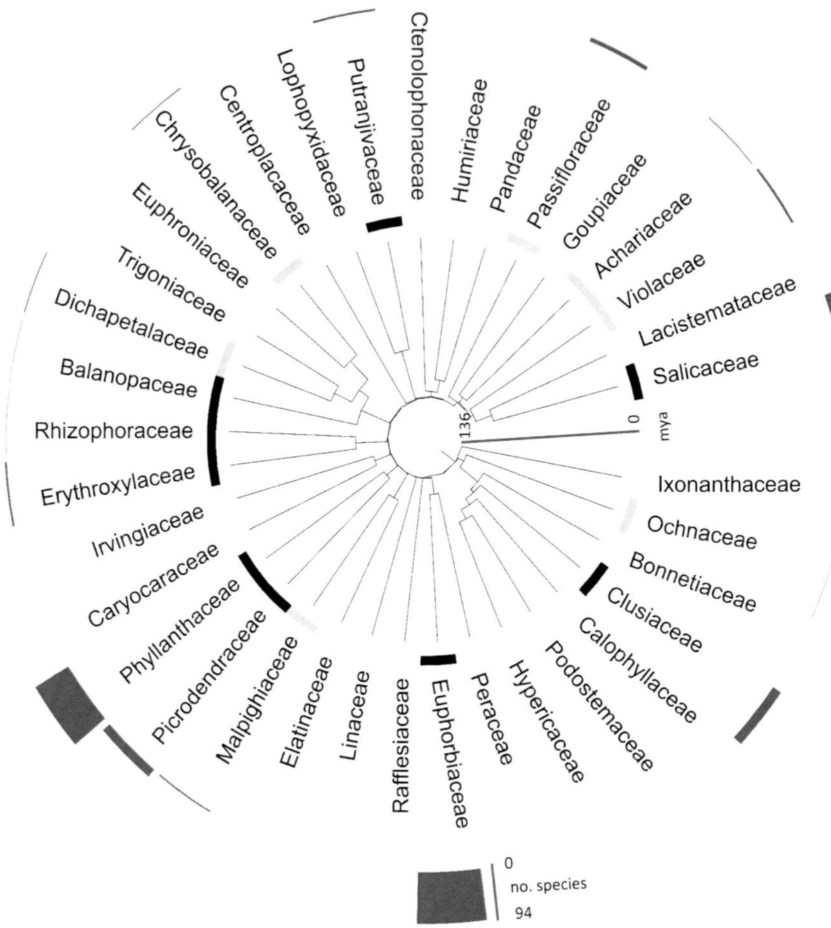

Fig. 131 Dated molecular phylogeny of the Malpighiales [7] showing those families that have species present in Australian rainforests (grey) and studied (black), and the number of species occurring in Australian rainforests

potent activation. Compounds with the epoxide shifted from C-6-/C-7 to C-5/C-6 or when ring opened are orders of magnitude less active. These findings showed that subtle changes to the structure of phorbol esters can shift them from being tumor promoters to tumor suppressors. One of the natural products, EBC-46 (**1716**), is now marketed as a veterinary anticancer drug tigilanol tiglate (Stelfonta®) and is undergoing human clinical trials for neck and head cancers [582]. Another phorbol ester, 12-deoxyphorhol 13-(3*E*,5*E*-decadienoate) (**1733**), lacking oxidation at C-5, C-6, C-7, and C-12 has been reported from the leaves and stems of *Excoecaria agallocha* when collected in the Kimberley region of northern Western Australia. This compound is a potent inhibitor of HIV replication [583].

Table 34 Distribution of natural products isolated from the Ochnaceae, Clusiaceae, and Euphorbiaceae

Species	Collection location[a]	Plant parts[b]	Compound class	Compounds	Refs.
Ochnaceae					
Brackenridgea australiana	SEQ	lf	Triterpenoid	**767**	[296]
Clusiaceae					
Calophyllum australianum	CY	b	Coumarin	**1713**	[578]
Calophyllum sil	NWA	w	Xanthone	**1714**, **1715**	[579]
Euphorbiaceae					
Alchornea ilicifolia	NQ	lf	Guanidine	**1786**	[39]
Alchornea rugosa	VIET	lf	Flavonoid alkaloid	**1787–1792**	[599]
Croton arnhemicus	NQ	st	Aporphine	**636**	[39]
Croton insularis	NQ	lf	Diterpenoid	**1734–1779**	[584–596]
Croton verreauxii	SEQ	ap	Diterpenoid	**1776**, **1777**	[595]
Endospermum myrmecophilum	NQ	lf	Piperidine	**1781–1783**	[598]
	NQ	lf	Pyrrolidine	**1784**	[598]
Excoecaria agallocha	NWA	lf, st	Diterpenoid	**1733**	[583]
Fontainea picrosperma	NQ	kern	Diterpenoid	**1716–1733**	[581]
Mallotus claoxyloides	NSW	s	Fatty acid	**1780**	[597]
Mallotus discolor	SEQ	s	Fatty acid	**1780**	[597]
Mallotus repandus	NQ	st	Pyridone	**1785**	[39]
Omphalea queenslandiae	NQ	lf	Piperidine	**1781–1783**	[598]
	NQ	lf	Pyrrolidine	**1784**	[598]

[a] SEQ = Southeast Queensland, NQ = North Queensland, CY = Cape York, NSW = New South Wales, NT = Northern Territory, NWA = North Western Australia, VIET = Vietnam, [b] lf = leaves, s = seeds, kern = kernels, ap = aerial parts, b = bark, st = stems

Croton insularis is a source of six pimaranes (Fig. 134), pimarane 5 (**1734**), keto pimarane 7 (**1735**), EBC-327 (**1736**), EBC-345 (**1737**), EBC-325 (**1738**), and EBC-326 (**1739**) and three cleistanthanes, EBC-316 (**1740**), spruceanol (**1741**), and sonderianol (**1742**), diterpenoids that are mostly weakly cytotoxic, apart from **1735** that showed moderate cytotoxicity towards melanoma (MM96L) and leukemia (LK562)

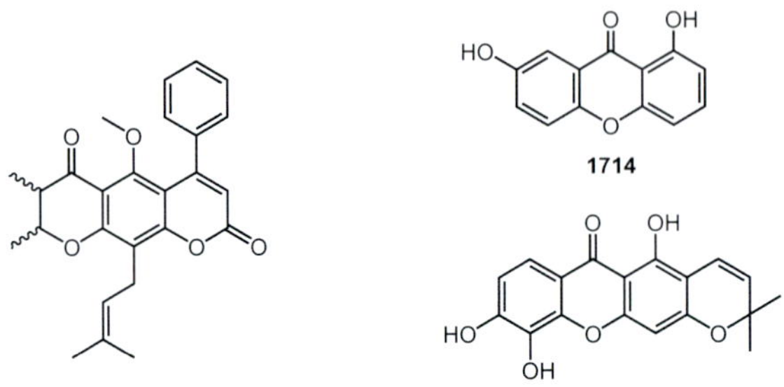

Fig. 132 Structures of compounds **1713–1715** reported from *Calophyllum* species

Plate 27 Rainforest Euphorbiaceae species; leaves and fruits of *Alchornea ilicifolia*, (top left and middle left) *Mallotus philippensis* (fruits top right and flowers bottom left), flower spikes of *Homolanthus populifolius* (bottom middle), and flowers of *Mallotus megadontus* (bottom right). (Photographs AR Carroll)

cells [584, 585]. Five crotofolane diterpenoids (EBC-233 (**1743**), EBC-300 (**1744**), EBC-240 (**1745**), EBC-241 (**1746**), and crotofolin C (**1747**)) were isolated from the leaves of *C. insularis* but were weak to inactive against a panel of six cancer cell lines [586]. The leaves of *C. insularis* also contained a series of nearly 20 casbane diterpenoid macrocycles and four oxidatively ring opened derivatives; (EBC-217 (**1748**),

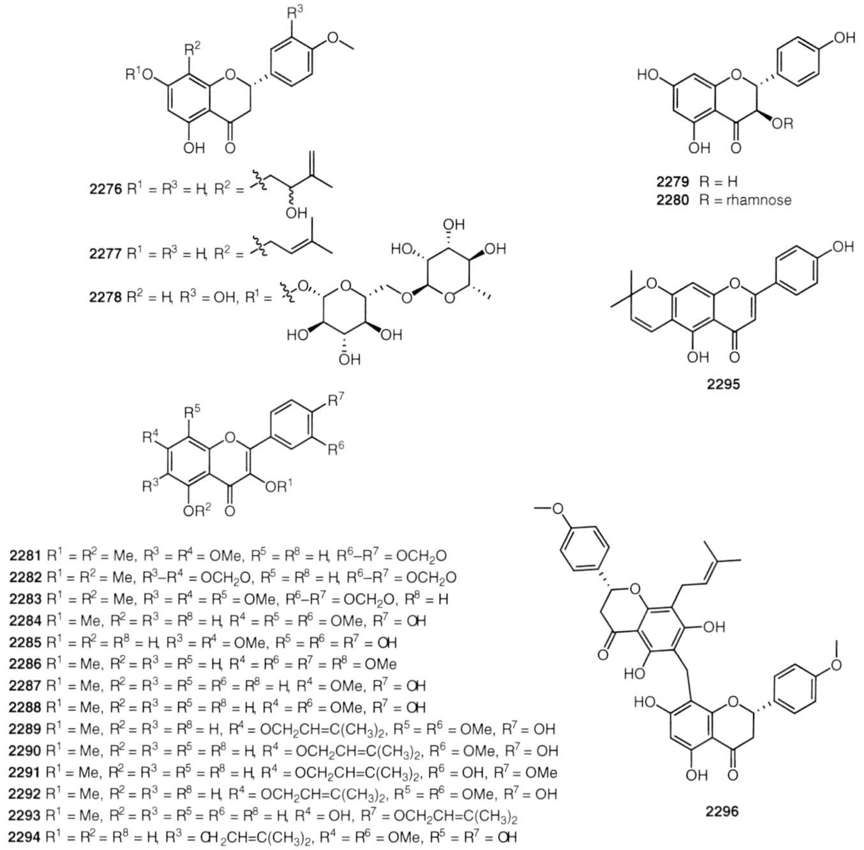

Fig. 133 Structures of phorbol diterpenoids **1716–1733** reported from Euphorbiaceae species

EBC-365 (**1749**), EBC-320 (**1750**), EBC-218 (**1751**), EBC-220 (**1752**), EBC-343 (**1753**), EBC-304 (**1754**), EBC-373 (**1755**), EBC-131 (**1756**), EBC-180 (**1757**), EBC-181 (**1758**), EBC-357 (**1759**), EBC-361 (**1760**), EBC-182 (**1761**), EBC-342 (**1762**), EBC-219 (**1763**), EBC-318 (**1764**), EBC-339 (**1765**), EBC-328 (**1766**), EBC-363 (**1767**), EBC-358 (**1768**), EBC-324 (**1769**), and EBC-329 (**1770**)) [587–593].

Nine halimane diterpenoids (Fig. 135) (EBC-204 (**1771**), EBC-205 (**1772**), EBC-201 (**1773**), EBC-202 (**1774**), EBC-203 (**1775**), croverin (**1776**), dihydrocroverin (**1777**), EBC-232 (**1778**), and EBC-323 (**1779**)) have been reported from *Croton insularis* and *C. verreauxii* and **1779** was weakly cytotoxic [594–596]. The seeds of *Mallotus claoxyloides* and *M. discolor* both contain the unsaturated fatty acid kamelolenic acid (**1780**) [597].

The leaves of *Endospermum myrmecophilum* from Mission Beach, and *Omphalea queenslandiae* from Babinda in north Queensland, both contain polyhydroxypiperidines (2,6-dideoxy-2,6-imino-*O*-glycero-L-gulo-heptitol (**1781**), 1,5-dideoxy-l,5-imino-D-glucitol (**1782**), and 1,5-dideoxy-l,5-imino-D-mannitol (**1783**)) and

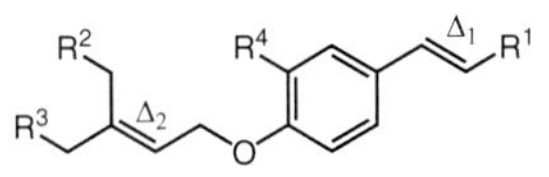

2297 R^1 = COOMe, $R^2 = R^3 = R^4$ = H
2298 R^1 = CH_2OH, $R^2 = R^3 = R^4$ = H, Δ_1 saturated
2299 R^1 = CH_2OH, R^2 = OH, $R^3 = R^4$ = H Δ_1, Δ_2 saturated
2300 R^1 = COOMe, $R^2 = R^4$ = H, R^3 = $CH_2CH{=}C(CH_3)_2$,
2301 R^1 = COOMe, R^2 = H, R^3 = $CH_2CH{=}C(CH_3)_2$, R^4 = OMe

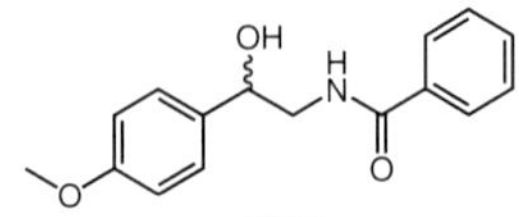

2305

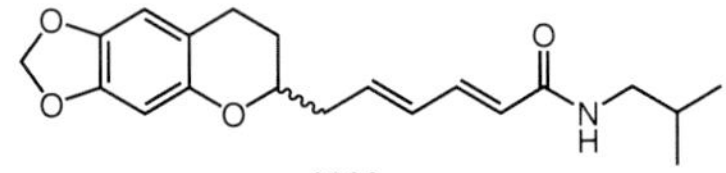

2306

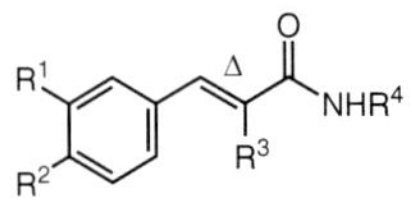

2302 $R^1 = R^3$ = H, R^2 = OH, R^4 = 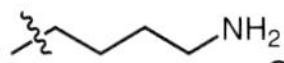

2303 $R^1 = R^3$ = H, R^2 = OH, R^4 = 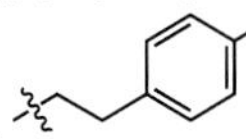

2304 $R^1 = R^2$ = OMe, R^3 = H, Δ saturated, R^4 = $CH_2C(CH_3)_2$

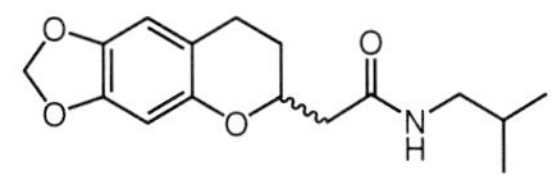

2307

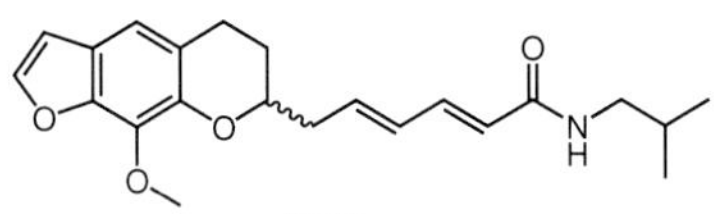

2308

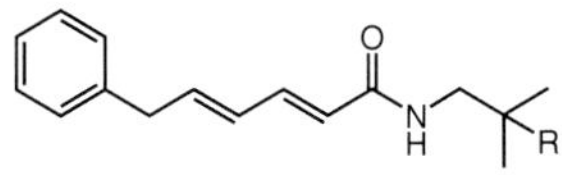

2309 R = OH
2310 R = H

Fig. 134 Structures of diterpenoids **1734–1770** reported from Euphorbiaceae species

(2*R*,5*R*)-dihydroxymethyl-(3*R*,4*R*)-dihydroxypyrrolidine (**1784**) that show glycosidase inhibitory activities [598]. *Mallotus repandus* produces the pyridone mallorepine (**1785**) [39]. *Croton arnhemicus* stems biosynthesize the aporphine alkaloid magnoflorine (**636**) [39].

Alchornea ilicifolia contains the guanidine alkaloid alchornine (**1786**) [39], while *A. rugosa* collected in Vietnam affords a series of six guanidine derivatives of an isoprenylated flavonoid (rugonidine A (**1787**), rugonidine B (**1788**), rugonidine C (**1789**), rugonidine D (**1790**), rugonidine E (**1791**), and rugonidine F (**1792**)) that show antidiabetes-related activities [599].

Picrodendraceae and Phyllanthaceae

The Picrodendraceae and Phyllanthaceae are sister families that are thought to have originated in the late Cretaceous (69 mya) [600]. These families were originally included within a much larger Euphorbiaceae but were split in the 1990’s. Picrodendraceae is represented in Australian rainforests by 16 species but only two of these have been studied (Plate 28, Table 35).

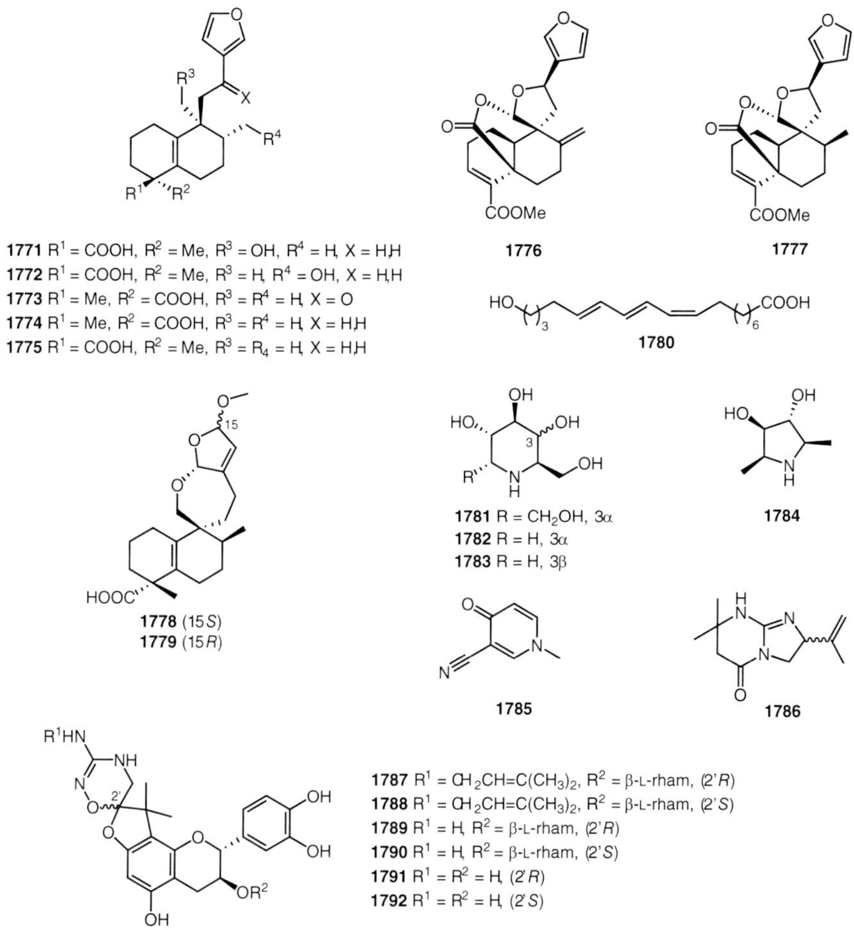

Fig. 135 Structures of compounds **1771–1792** reported from Euphorbiaceae species

Petalostigma pubescens from Victoria yielded the weakly cytotoxic pimarane diterpenoids petalostigmones A–C (**1793–1795**), pubescenone (**1796**) and (−)-sonderianol (**1797**) [601] and the leaves of *Austrobuxus swainii* from southeast Queensland were found to contain a series of non-cytotoxic picrotoxane diterpenoids (austrobuxusin N (**1798**), austrobuxusin A (**1799**), austrobuxusin B (**1800**), austrobuxusin C (**1801**), and austrobuxusin D (**1802**)) (Fig. 136) [602].

The Phyllanthaceae has 74 species in Australia rainforests with six having been investigated chemically (Fig. 137). These studies have yielded gallic acid esters of flavonoids ((2*R*,3*R*)-dihydromyricetin-4′-*O*-(3″-*O*-methyl)-gallate (**1803**), (2*R*,3*R*)-dihydromyricetin-3′-*O*-(3″-*O*-methyl)-gallate (**1804**), (2*R*,3*R*)-dihydromyricetin-4′-*O*-gallate (**1805**), (2*R*,3*R*)-dihydromyricetin-3′-*O*-gallate (**1806**), and (2*R*,3*R*)-dihydromyricetin (**1807**)), from *Glochidion sumatranum* collected in northern

Plate 28 Rainforest Picrodendraceae and Phyllanthaceae species; flowers and fruits of *Petalostigma pubescens* (top row), and flowers and fruits of *Glochidion ferdinandi* (bottom row). (Photographs AR Carroll)

Table 35 Distribution of natural products isolated from the Picrodendraceae and Phyllanthaceae

Species	Collection location[a]	Plant part(s)[b]	Compound class	Compound(s)	Refs.
Picrodendraceae					
Austrobuxus swainii	SEQ	lf	Diterpenoid	**1798–1802**	[602]
Petalostigma pubescens	Vic	hw	Diterpenoid	**1793–1797**	[601]
Phyllanthaceae					
Flueggea leucopyrus	NQ	b	Securinine alkaloid	**1814**	[39]
Flueggea virosa subsp. *melanthesoides*	NQ, CY	fl	Securinine alkaloid	**1815**, **1818**	[39]
Glochidion sumatranum	NQ	lf, st	Flavonoid	**1803–1807**	[603]
Margaritaria dubium-traceyi	CY	lf	Securinine alkaloid	**1815–1817**	[39]
Poranthera corymbosa	NNSW	wp	Azaphenalene alkaloid	**1808–1813**	[604]

[a] SEQ = Southeast Queensland, NQ = North Queensland, CY = Cape York, NNSW = Northern New Southeast Wales, Vic = Victoria, [b] lf = leaf, b = bark, hw = heartwood, st = stem

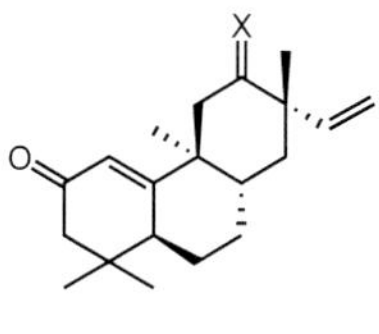

1793 X = H, ◂OH
1794 X = O

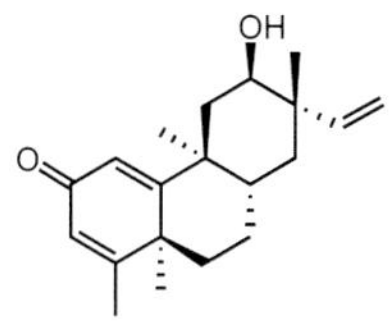

1795

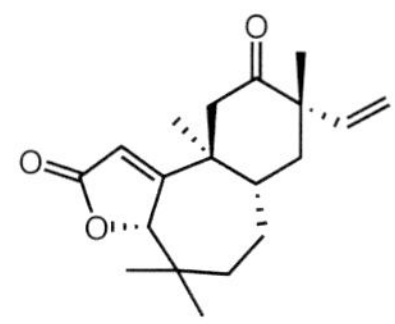

1796

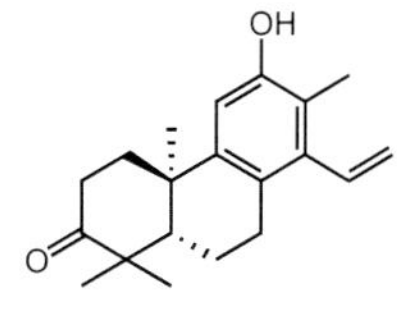

1797

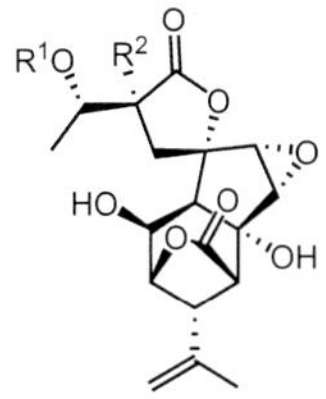

1798 R^1 = β-D-glc, R^2 = H
1799 R^1 = galloyl, R^2 = H
1800 R^1 = Me, R^2 = OH
1801 R^1 = Me, R^2 = H

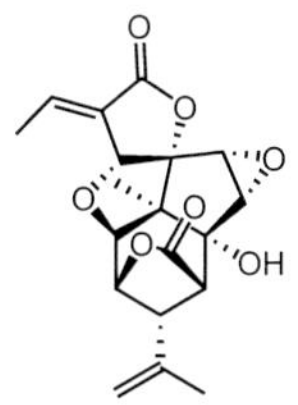

1802

Fig. 136 Structures of compounds **1793–1802** reported from *Petalostigma pubescens* and *Austrobuxus swainii*

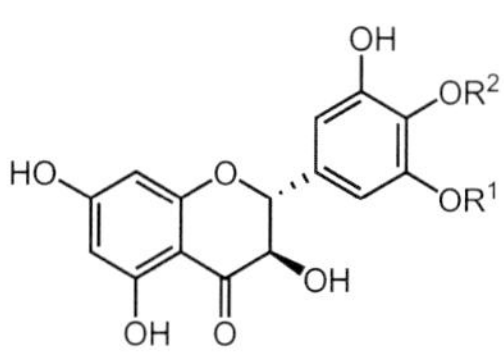

1803 R^1 = H, R^2 = x
1804 R^1 = x, R^2 = H
1805 R^1 = H, R^2 = y
1806 R^1 = y, R^2 = H
1807 R^1 = y, R^2 = H

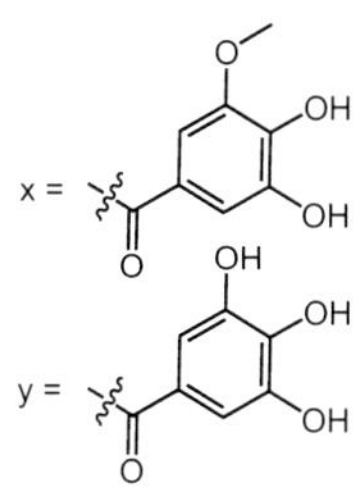

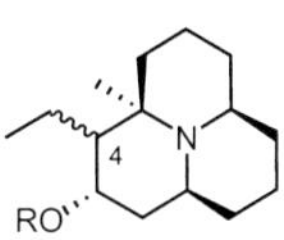

1808 R = H, 4β
1809 R = Ac, 4β
1810 R = Ac, 4α

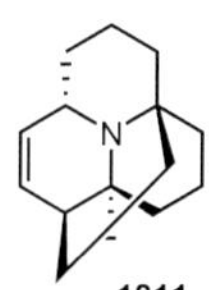

1811

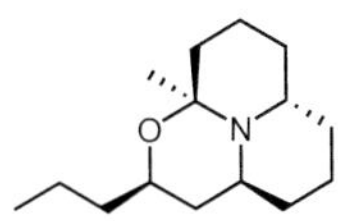

1812

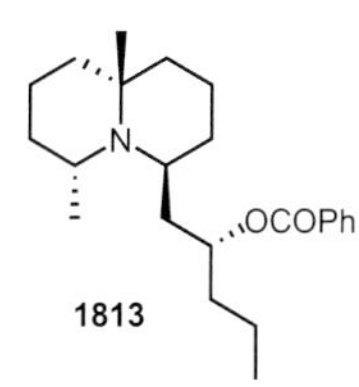

1813

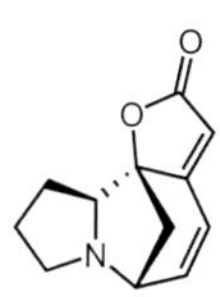

1814

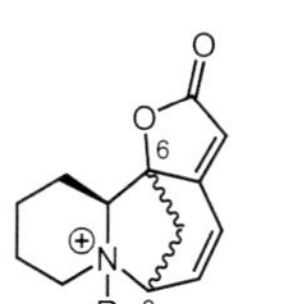

1815 6β, 8β, R = H
1816 6α, 8α, R = H
1817 6α, 8α, R = O^-

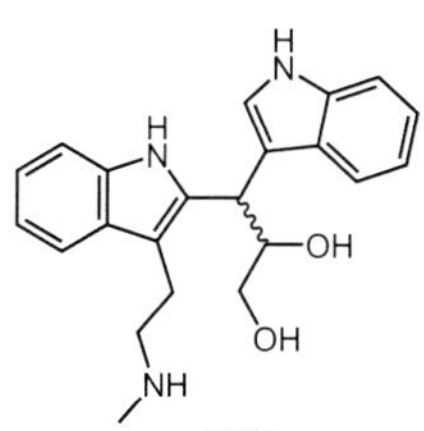

1818

Fig. 137 Structures of compounds **1803–1818** reported from Phyllanthaceae species

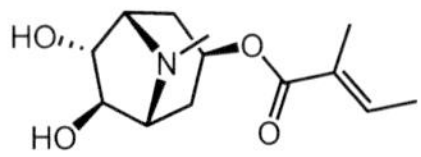

1819

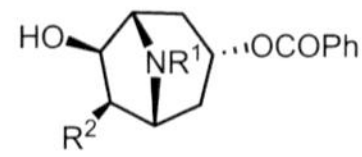

1820 R^1 = Me, R^2 = OH
1821 R^1 = R^2 = H

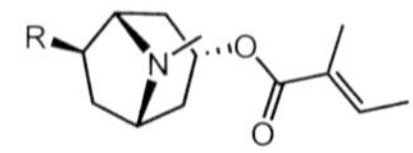

1822 R = OH
1823 R = H

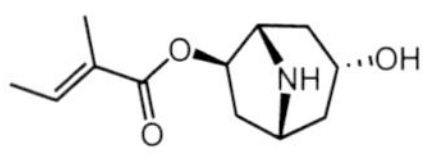

1824

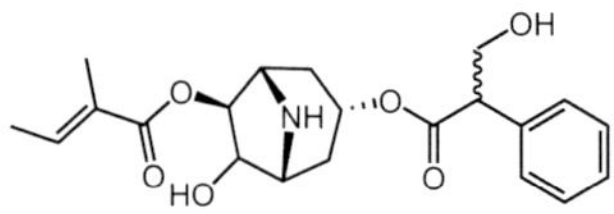

1825

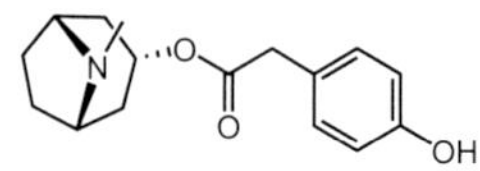

1826

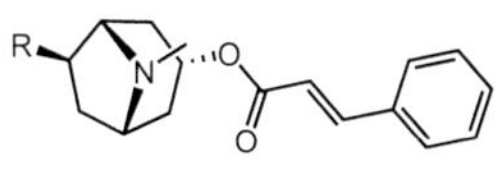

1827 R = OH
1828 R = H

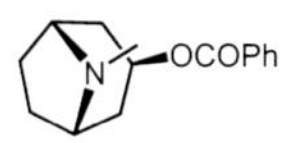

1829

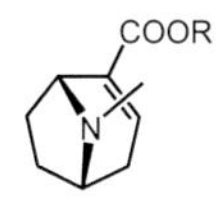

1830 R = H
1831 R = Me

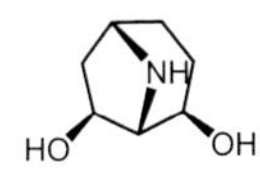

1832

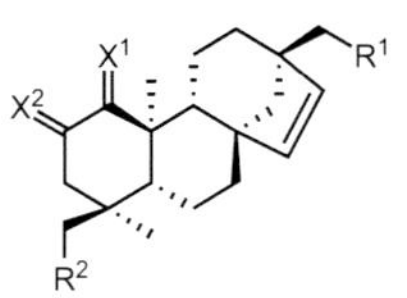

1833 R^1 = R^2 = H, X^1 = X^2 = H, H
1834 R^1 = R^2 = H, X^1 = O, X^2 = H, H
1835 R^1 = OH, R^2 = H, X^1 = X^2 = H, H
1836 R^1 = H, R^2 = OH, X^1 = X^2 = H, H
1837 R^1 = H, R^2 = OH, X^1 = O, X^2 = H, H
1838 R^1 = R^2 = H, X^1 = O, X^2 = H, βOH
1839 R^1 = R^2 = H, X^1 = H, βOH, X^2 = O
1840 R^1 = OH, R^2 = H, X^1 = O, X^2 = H, H
1841 R^1 = OH, R^2 = H, X^1 = O, X^2 = H, βOH
1842 R^1 = OH, R^2 = H, X^1 = H, βOH, X^2 = O
1843 R^1 = H, R^2 = OH, X^1 = O, X^2 = H, βOH
1844 R^1 = H, R^2 = OH, X^1 = H, βOH, X^2 = O
1845 R^1 = R^2 = H, X^1 = H, βOH, X^2 = H, H

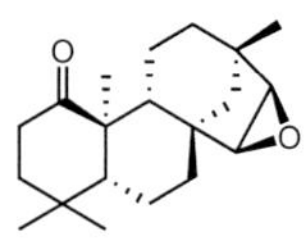

1846

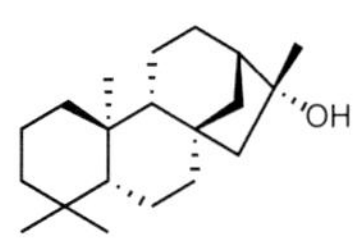

1847

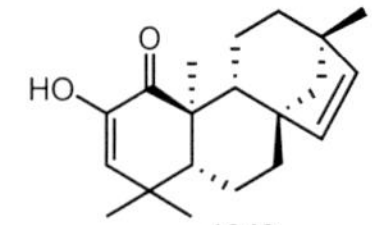

1848

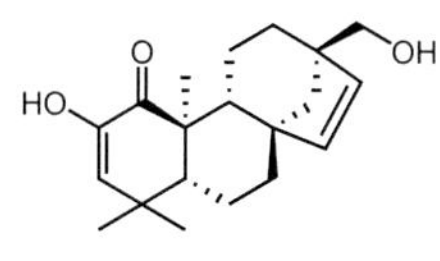

1849

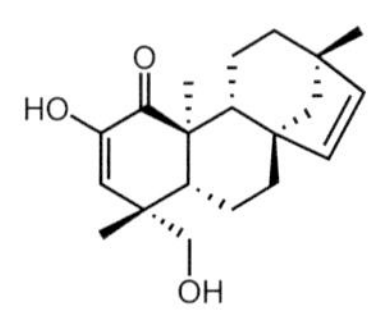

1850

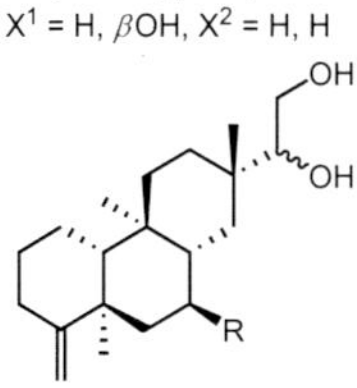

1851 R = H
1852 R = OH

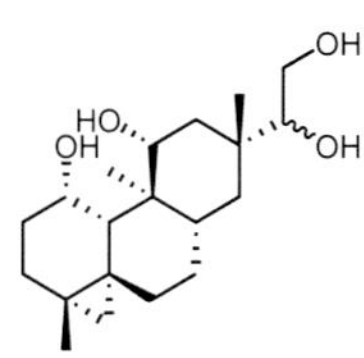

1853

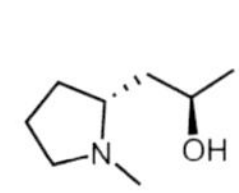

1854

Fig. 138 Structures of compounds **1819–1854** reported from *Erythroxylum* species and *Carallia brachiata*

Queensland [603] and the unprecedented reduced 9β-azaphenalenes from *Poranthera corymbosa* (poranthericine (**1808**), *O*-acetylporanthericine (**1809**), porantherilidine (**1810**), porantherine (**1811**) and porantheridine (**1812**)) [604]. This species also contains the quinolizidine alkaloid porantheriline (**1813**). Securinine-type alkaloids (norsecurinine (**1814**), securinine (**1815**), allosecurinine (**1816**) and allosecurinine-*N*-oxide (**1817**)) have been reported from *Flueggea leucopyrus*, *F. virosa* subsp. *melanthesoides*, and *Margaritaria dubium-traceyi*. A bis-indole fluevirine E (**1818**) was also found in *F. virosa* subsp. *melanthesoides* [39].

Erythroxylaceae and Rhizophoraceae

The Erythroxylaceae is a pantropical family mostly found in Central and South America, Central and Southern Africa, Madagascar, southeast Asia and Australia. This highlights the Gondwanan heritage of this family, originating in Africa with subsequent species dispersal and diversification into the Americas and the Indo-Pacific 50 mya [605]. *Erythroxylum coca* is the source of cocaine, and many other species are known to produce tropane alkaloids. In Australia, the four endemic *Erythroxylum* species are controlled plants, so that it is illegal to possess them without a license. Only two species have been investigated chemically (Table 36).

Erythroxylum australe and *E.* sp. Brewer LA (B. Hyland 13373) contain a variety of tropane esters (Fig. 138) (**649**, meteloidine (**1819**), 6β,7β-dihydroxytropan-3α-yl benzoate (**1820**), 3α-benzoyloxynortropan-6β-ol (**1821**), 6β-hydroxytropan-3α-yl tiglate (**1822**), tigloyltropine (**1823**), 3α-hydroxynortropan-6β-yl tiglate (**1824**), (+)-7-hydroxy-6-tigloyloxynortropan-3-yl 2-hydroxy-3-phenylpropionate (**1825**), 3α-(4-hydroxyphenylacetoxy)tropane (**1826**), 3-cinnamoyloxytropane-6,7-diol (**1827**), 3α-(*E*)-cinnamoyloxytropane (**1828**), **1829**, ecgonidine (**1830**), methyl ecgonidine (**1831**), and erycibelline (**1832**)) [39, 606–608]. Another study of the wood from *E. australe* showed that it contained over 20 kauranes and related diterpenoid structures (*ent*-beyer-15-ene (**1833**), *ent*-beyer-15-en-1-one (**1834**), *ent*-beyer-15-en-17-ol (**1835**), *ent*-beyer-15-en-19-ol (**1836**), *ent*-19-hydroxybeyer-15-en-1-one (**1837**), *ent*-2β-hydroxybeyer-15-en-l-one (**1838**), *ent*-lβ-hydroxybeyer-15-en-2-one (**1839**), *ent*-17-hydroxybeyer-15-en-l-one (**1840**), *ent*-2β-17-dihydroxybeyer-15-en-l-one (**1841**), *ent*-lβ,l7-dihydroxybeyer-l5-en-2-one (**1842**), *ent*-2β,19-dihydroxybeyer-15-en-l-one (**1843**), *ent*-lβ,19-dihydroxybeyer-15-en-2-one (**1844**), *ent*-beyer-15-en-1β-ol (**1845**), *ent*-15,16-epoxy-beyer-1-one (**1846**), *ent*-kauran-16β-ol (**1847**), *ent*-2-hydroxybeyer-2,15-dien-l-one (**1848**), *ent*-2,17-dihydroxybeyer-2,15-dien-l-one (**1849**), *ent*-2,19-dihydroxybeyer-2,15-dien-l-one (**1850**), erythroxydiol Y (**1851**), *ent*-dolabra-4(18)-en-7β,15ξ,16-triol (**1852**), and 1β,11β-dihydroxyerythroxydiol X (**1853**)) [609].

The Rhizophoraceae is sister to the Erythroxylaceae and mainly contains mangrove species. Mangrove and rainforest habitats meld in some locations. *Carallia brachiata*, known colloquially as freshwater mangrove, is the only Australian rainforest species in the family and it produces the pyrrolidine alkaloid (+)-hygroline (**1854**) [610].

Table 36 Distribution of natural products isolated from the Erythroxylaceae, Rhizophoraceae, Balanopaceae, Putranjivaceae, Achariaceae, and Salicaceae

Species	Collection location[a]	Plant part(s)[b]	Compound class	Compound(s)	Refs.
Erythroxylaceae					
Erythroxylum australe	SEQ	w	Diterpenoid	**1833–1853**	[609]
	NQ, CQ	rb, lf, b, r, st	Tropane	**649**, **1819–1832**	[39, 606–608]
Erythroxylum sp. Brewer LA (B. Hyland 13373)	NQ	lf	Tropane	**1821**, **1829**	[608]
Rhizophoraceae					
Carallia brachiata	NQ	lf	Pyrrolidine	**1854**	[610]
Balanopaceae					
Balanops australiana	NQ	b	Triterpenoid	**802**, **1855–1863**	[611, 612]
Putranjivaceae					
Drypetes acuminata	NQ	b	Steroid	**1864**, **1865**	[613]
	NQ	b	Triterpenoid	**1855**	[613]
Achariaceae					
Baileyoxylon lanceolatum	NQ	lf	Cyanohydrin	**1866**	[614]
Ryparosa kurrangii	NQ	lf	Cyanohydrin	**1866**	[615]
Salicaceae					
Casearia costulata	NQ	lf	Phenol	**1878**	[618]
Casearia grayi	NQ	lf, st	Diterpenoid	**1867–1870**, **1874–1877**	[617]
Casearia grewiifolia var. *gelonioides*	NQ	st	Diterpenoid	**1868**, **1869**, **1871–1873**	[616]
Casearia sp. Mission Beach (B.P. Hyland 773)	NQ	st	Shikimate	**1886**	[618]
Homalium brachybotrys	NQ	lf, st	Benzofuran	**1882–1885**	[619]
	NQ	lf, st	Phenol	**1879**, **1880**	[619]
Scolopia braunii	NNSW	lf	Phenol	**1881**	[618]

[a] SEQ = Southeast Queensland, NQ = North Queensland, CQ = central Queensland, NNSW = Northern New South Wales, [b] lf = leaves, r = roots, rb = root bark, b = bark, w = wood, st = stems

Balanopaceae, Putranjivaceae, Achariaceae, and Salicaceae

The remaining four Malpighiales families (Balanopaceae, Putranjivaceae, Achariaceae, and Salicaceae) have only 25 species that occur in Australian rainforests. The Balanopaceae is a small family containing nine species in a single genus (*Balanops*). Species are found in New Caledonia, Fiji, and Vanuatu, and there is a single species occurring in Queensland. The bark of *B. australiana* contains nine β-amyrin

Plate 29 Both pictures: flowers of *Drypetes deplanchei* (a close relative of *D. acuminata*). (Photographs AR Carroll)

triterpenoids, friedelin (**1855**), cerin (**1856**), zeylanol (**1857**), 21α-hydroxyfriedelan-3-one (**1858**), canophyllol (**1859**), friedelan-3β-ol (**1860**), friedelan-3α-ol (**1861**), friedelane-3α,21α-diol (**1862**), and friedelane-3α,28,29-triol (**1863**) (Fig. 139) [611, 612]. The Old-World family Putranjivaceae has four *Drypetes* species in Australian rainforests (Plate 29). The north Queensland species *D. acuminata* contains the β-amyrin triterpenoid friedelin (**1855**) and the two steroids aglaitriol (**1864**) and aglaitriol 3-caffeate (**1865**) [613].

The sole compound reported from Achariaceae (a family with only two Australian rainforest species), is a cyanogenic glycoside, gynocardin (**1866**), from the leaves of both *Baileyoxylon lanceolatum* and *Ryparosa kurrangii* [614, 615]. The cosmopolitan family Salicaceae has 18 Australian rainforest species of which five have been investigated chemically. A number of clerodane diterpenoids ((*rel*)-2(*R*)-hydroxy-(18*R*,19*R*)-diacetoxy-18,19-epoxy-(4*S*,5*S*,8*S*,9*S*,10*R*)-clerodan-13(16),14-dien-6-one, (**1867**), (*rel*)-(2*R*,18*S*,19*R*)-triacetoxy-18,19-epoxy-(4*S*,5*S*,9*S*,10*R*)-clerodan-13(16),14-dien-6-one (**1868**), (*rel*)-(2*R*,18*S*,19*R*)-triacetoxy-18,19-epoxy-(4*S*,5*R*,8*S*,9*S*,10*R*)-clerodan-13(16),14-diene (**1869**), (*rel*)-(18*S*,19*R*)-diacetoxy-18,19-epoxy-(2*R*)-hydroxy-(4*S*,5*R*,8*S*,9*S*,10*R*)-clerodan-13(16),14-diene (**1870**), (*rel*)-(18*S*,19*R*)-diacetoxy-18,19-epoxy-(2*R*)-2ξ-methylbutanoyl-(4*S*,5*S*,9*S*,10*R*)-clerodan-13(16),14-dien-6-one (**1871**), (*rel*)-(18*S*,19*R*)-diacetoxy-18,19-epoxy-(2*R*)-isobutanoyl-(4*S*,5*R*,9*S*,10*R*)-clerodan-13(16),14-diene (**1872**), (*rel*)-(18*S*,19*R*)-diacetoxy-18,19-epoxy-(2*R*)-2ξ-methylbutanoyl-(4*S*,5*R*,9*S*,10*R*)-clerodan-13(16),14-diene (**1873**), (*rel*)-(2*R*,18*S*,19*R*)-triacetoxy-18,19-epoxy-(4*S*,5*S*,8*R*,9*S*,10*R*)-clerodan-12(*E*),14-dien-6-one (**1874**), (*rel*)-(2*R*)-hydroxy-(18*S*,19*R*)-diacetoxy-18,19-epoxy-(4*S*,5*S*,8*S*,9*S*,10*R*)-clerodan-(12*E*),14-dien-6-one (**1875**), (*rel*)-(18*S*,19*R*)-diacetoxy-18,19-epoxy-(2*R*)-hydroxy-(4*S*,5*R*,8*S*,9*S*,10*R*)-clerodan-(12*E*),14-diene (**1876**), and 16-hydroxy-halima-5(10),(13*E*)-dien-15-oic acid (**1877**)) were found in *Casearia grewiifolia* and *C. grayi* [616, 617]. Also, the glycosylated phenols idesin (**1878**), salirepin (**1879**), poliothrysoside (**1880**), and (*rel*)-2-*O*-(6-benzoylglucopyranosyl)-7-*O*-(1α,2α,6α-trihydroxy-5-oxo-cyclohex-3-enoyl)-4-hydroxybenzyl alcohol (**1881**) were reported from *C. costulata*, *Homalium brachybotrys*, and *Scolopia braunii*

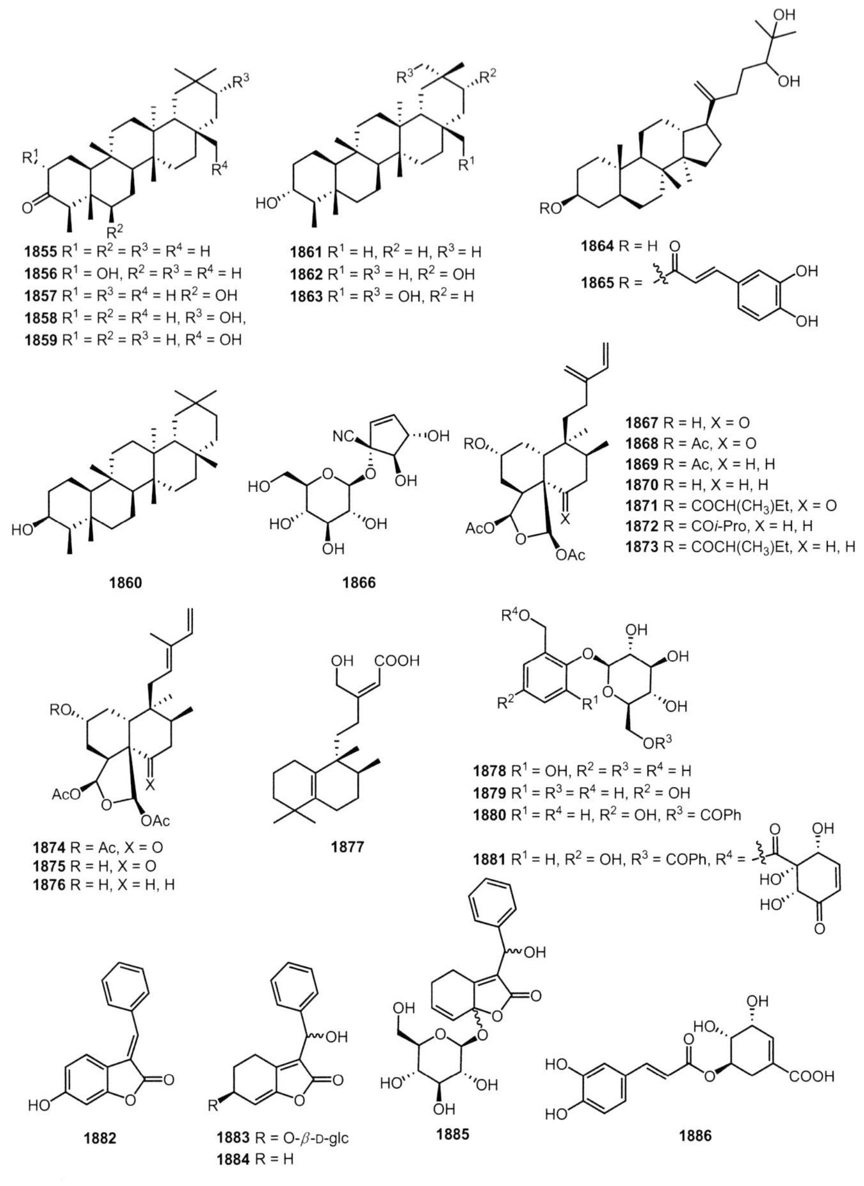

Fig. 139 Structures of compounds **1855–1886** reported from Balanopaceae, Achariaceae, Putranjivaceae, and Salicaceae species

[618, 619], with benzyl furanones (10-deoxy-4,5,6-dehydrocochinolide (**1882**), cochinolide-6-*O*-*β*-glucopyranoside (**1883**), 6-deoxycochinolide (**1884**), and isocochinolide-7*a*-*β*-glucopyranoside (**1885**)) obtained from *H. brachybotrys* [619], and 5-*O*-caffeoylshikimic acid (**1886**) was found in *Casearia* sp. Mission Beach (B.P. Hyland 773) [618].

Oxalidales

The Oxalidales is an order with 10 families, of which four are found in Australian rainforests. However, only two families, Cunoniaceae and Elaeocarpaceae, have been investigated chemically (Fig. 140, Table 37).

Cunoniaceae

Of 33 Australian rainforest Cunoniaceae, some species are commercially important. For example, the leatherwood (*Eucryphia lucida*) is a dominant tree in Tasmanian rainforests and the honey derived from it is highly prized. *Davidsonia pruriens* (Davidson plum (Plate 30)) produces an edible fruit that is used to produce preserves and cordials. Several other species are valued for their cabinet timber. Only the bark of *Ceratopetalum apetalum* and *Davidsonia pruriens* have been studied (Fig. 141) and triterpenoids related to β-amyrin and ursolic acid (**1861**), including methyl 3-oxo-bauer-7-en-28-oate (**1887**), 3-oxo-bauer-7-en-28-oic acid (**1888**), and 3*β*-hydroxy-bauer-7-en-28-oic acid (**1889**) have been reported [620, 621].

Elaeocarpaceae

The Elaeocarpaceae is a predominantly Southern Hemispheric family, with Australasia holding the greatest extant taxonomic diversity [622]. The family has a Gondwanan origin with vicariance and long-distance dispersal leading to the current geographic distribution patterns (Fig. 140) [622]. The genus *Aristotelia* has one species in South America, two in New Zealand and two in Australia. The two species of *Peripentadenia* being restricted to north Queensland, and the *Aceratium* species are Australian (5) or New Guinean (13) while *Elaeocarpus* and *Sloanea* are speciose with close to 500 and 100 species, respectively. *Elaeocarpus* species are distributed across southeast Asia, Pacific islands, Malesia, and Australia and *Sloanea* species are found across Asia, Malesia, Central and South America, Madagascar, New Caledonia, and Australia. Australian rainforests contain 31 *Elaeocarpus* and five *Sloanea* species. The basal lineages of the family include the sister clades of *Aristotelia/Vallea* and *Sloanea*, with subsequent divergence of *Peripentadenia*/*Chrinodendron* (South American) from the rest of the genera in the late Cretaceous [622].

The *Aristotelia* species are a major source of monoterpenoid indole alkaloids that are different structurally to those found in the Gentianales (Fig. 142). The two

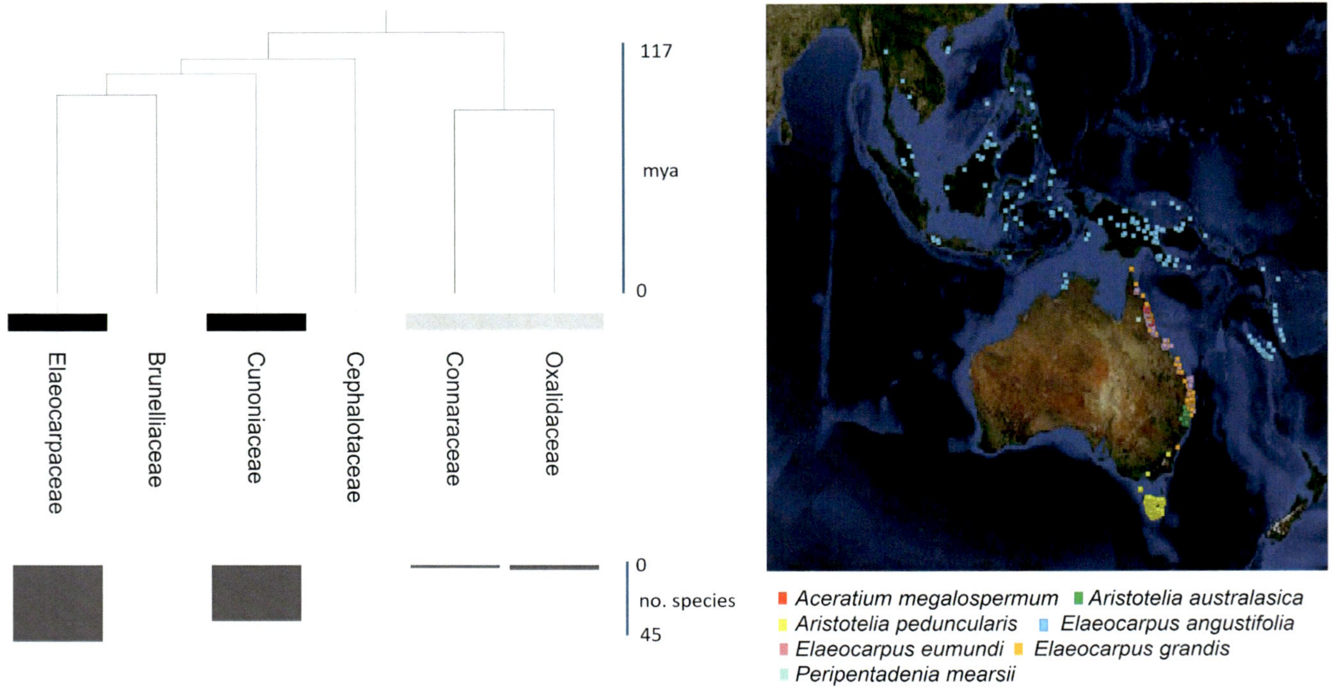

Fig. 140 Dated molecular phylogeny of the Oxalidales [7], showing those families that have species present in Australian rainforests (grey) and studied (black), and the number of species occurring in Australian rainforests. The map depicts collection locations for herbarium records for the Elaeocarpaceae species that have been studied [5, 43]

Table 37 Distribution of natural products isolated from the Oxalidales

Species	Collection location[a]	Plant part(s)[b]	Compound class	Compound(s)	Refs.
Cunoniaceae					
Ceratopetalum apetalum	NSW	b	Triterpenoid	**1860**	[620]
Davidsonia pruriens	NQ	b	Triterpenoid	**1887–1889**	[621]
Elaeocarpaceae					
Aceratium megalospermum	NQ	lf	Tetralin glycoside	**1965**	[648]
Aristotelia australasica	NNSW	lf, st	Terpenol indole alkaloid	**1890–1905**, **1917–1919**	[623–627]
Aristotelia peduncularis	Tas	r, lf, st	Terpenoid indole alkaloid	**1906–1916**	[628–634]
Elaeocarpus eumundi	NQ	lf	Stilbene	**1963**, **1964**	[647]
Elaeocarpus grandis	SEQ, NQ	lf	Indolizidine	**1931–1938**	[642, 643]
Elaeocarpus sericopetalus	NQ	lf	Cyanohydrin	**1966**	[649]
Peripentadenia mearsii	NQ	b, lf, s	Pyrrolidine	**1922–1930**	[39, 638–641]
	NQ	lf	Tropane	**1829**, **1920**, **1921**	[637]
Elaeocarpus angustifolia	PNG, S China	lf	Indolizidine	**1933**, **1939–1959**	[644–646]

[a] SEQ = Southeast Queensland, NQ = North Queensland, NNSW = Northern New South Wales, NSW = New South Wales, Tas = Tasmania, PNG = Papua New Guinea, S China = Southern China, [b] lf = leaves, s = seeds, r = roots, b = bark, st = stems

Australian species *A. peduncularis* from Tasmania and *A. australasica* from northern New South Wales have yielded over 25 alkaloids in this class (aristone (**1890**), 11-*epi*-aristoteline (incorrect) (**1891**), corrected to *allo*-aristoteline (**1892**), dehydro-9,10-aristotellne (**1893**), aristolasicone (incorrect) (**1894**), corrected to aristolasicone (**1895**), aristolasicol (**1896**), aristocarbinol (**1897**), aristolasicolone (**1898**), aristotelinone (**1899**), *epi*-3 aristoserratenine (**1900**), aristoserratenine (**1901**), aristolasol (**1902**), aristolasene (**1903**), 17-hydroxyhobartine (**1904**), hobartidiol (**1905**), peduncularine (incorrect) (**1906**), corrected to peduncularine (**1907**), sorellin (**1908**), hobartin (**1909**), tasmanin (**1910**), aristoserratin (**1911**), aristoteline (**1912**), pedunculaistin (**1913**), triabunnin (**1914**), aristolarin (**1915**), isopeduncularine (**1916**), bisaristone A (**1917**), bisaristone B (**1918**), and aristoaristone (**1919**)) [623–634]. These alkaloids are derived from the conjugation of tryptamine with terpineol and are thus uniquely different from the iridoid/trypamine monoterpenoid indole alkaloids found

Plate 30 Rainforest Oxalidales species: fruits and flowers of *Davidsonia pruriens* (left), tree and fruits of *Elaeocarpus grandis* (middle and top right) and flowers of *E. eumundii* (bottom right). (Photographs AR Carroll)

in the Gentianales. The alkaloids include those such as aristoserratenine (**1901**) and 3-*epi*-aristoserratenine (**1900**), in which the terpenoid is spiro-linked to the indole C-3 carbon and these molecules can rearrange to generate aristoteline- or *allo*-aristotelin-type structures, with the latter containing an indole with the ethyl amino side chain now linked to C-2 rather than C-3. Three indole alkaloid dimers, **1917**, **1918**, and **1919** have also been reported [623, 635]. In these compounds, the indole nitrogen in one monomer is conjugated to an aryl carbon of the second indole. Both the New Zealand and Chilean species also contain many of the same and related alkaloids, and some have been shown to be potent non-competitive inhibitors of nicotinic acetylcholine receptors, with higher selectivity for the h$\alpha3\beta4$ subtype when compared with $\alpha4\beta2$ and h$\alpha7$ AChRs [636]. This suggests a role in treatments for addiction and depression. None of the later diverging genera within the family contain monoterpenoid indole alkaloids, thus providing another example of early diverging lineages housing chemical diversity that is lost or dormant in younger lineages.

The north Queensland rainforest tree *Peripentadenia mearsii* yielded the three tropane alkaloids **1829**, 2α-benzyloxy-3β-hydroxytropane (**1920**), and 6β-hydroxy-3α-acetoxyhydroxytropane (**1921**) (Fig. 143) [637]. However, only a single collection of *P. mearsii* contained these alkaloids and several attempts to recollect more material from the species in the same location failed to yield tropanes. This suggests a

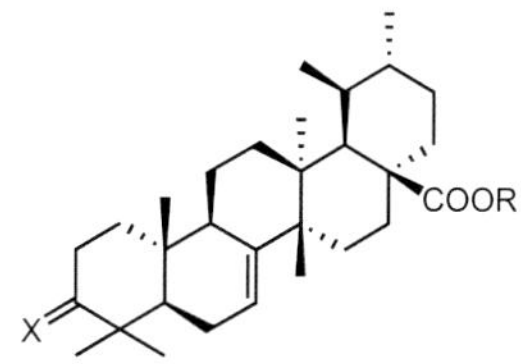

Fig. 141 Structures of compounds **1887–1889** reported from *Ceratopetalum apetalum* and *Davidsonia prunens*

misidentification (or mislabeling) of the original collection sample even though the botanist involved was quite familiar with this species. Subsequent investigations of the species have yielded a series of 1-amido-3-aminopropane derivatives, namely, peripentadenine (**1922**), di-*nor*-peripentadenine (**1923**), peripentamine (**1924**), anhydroperipentamine (**1925**), peripentonine A (**1926**), peripentonine B (**1927**), peripentonine C (**1928**), and mearsamine (**1929**) and an unusual isoquinuclidine alkaloid, mearsine (**1930**) [39, 638–641]. All of these structures contain a diamino-propane with the other amino group forming an amide bonded to either a butyric or hexanoic acid. The *Peripentadenia* alkaloids have shown affinity towards δ-opioid receptors.

Elaeocarpus grandis is a large and fast-growing tree that is endemic to rainforests from central New South Wales to north Queensland and New Guinea. It produces masses of thin-skinned bright blue fruits that contain a hard nut (Plate 30). A closely related species, *E. angustifolia*, occurs in the Northern Territory and is also widespread from India to southeast Asia, and southern China through to New Caledonia. *Elaeocarpus angustifolia* is also synonymous with *E. sphaericus* and *E. ganitrus*. Collections of *E. grandis* from southeast Queensland have yielded eight indolizidine-type alkaloids (Fig. 143) (grandisine A (**1931**), grandisine B (**1932**), isoelaeocarpiline (**1933**), grandisine C (**1934**), grandisine F (**1935**), grandisine G (**1936**), grandisine D (**1937**), and grandisine E (**1938**)) with unprecedented structures. Several of these compounds also have shown activity at the δ-opioid receptor sites [642, 643].

Collections of *E. angustifolia* from New Guinea (reported under the synonym *E. sphaericus*) and China have yielded 21 indolizidine and related alkaloids (Fig. 144) (**1933**, (+)-3-oxoelaeocarpin (**1939**), (−)-3-oxoelaeocarpin (**1940**), 3-oxoisoelaeocarpin (**1941**), elaeocarfoline A (**1942**), (−)-elaeocarfoline B (**1943**), (+)-elaeocarfoline B (**1944**), 2-[3-(3-hydroxypropylamino)propyl]-5-methylchroman-4-one (**1945**), 8,9-dehydroelaeocarpine (**1946**), 9-*epi*-elaeocarpine-(*Z*)-*N*-oxide (**1947**), 9-elaeocarpine (*Z*)-*N*-oxide (**1948**), elaeocarpine (*E*)-*N*-oxide (**1949**), isoelaeocarpine (**1950**), elaeocarpine (**1951**), 1-(2-hydroxy-6-methylphenyl)-2-(1-(3-hydroxypropyl)pyrrolidin-2-yl)ethan-1-one (**1952**), (+)-*epi*-isoelaeocarpiline (**1953**), (+)-elaeocarpiline (**1954**), (−)-*epi*-elaeocarpiline (**1955**), (+)-*epi*-alloelaeocarpiline (**1956**), (−)-alloelaeocarpiline (**1957**), and (+)-pseudo-*epi*-isoelaeocarpiline (**1958**)) [644–646]. Some of these compounds weakly inhibited the enzyme butyrylcholine esterase. *Elaeocarpus angustifolia* from New Guinea has also yielded the pentacyclic indole alkaloid elaeocarpidine (**1959**). The complexity

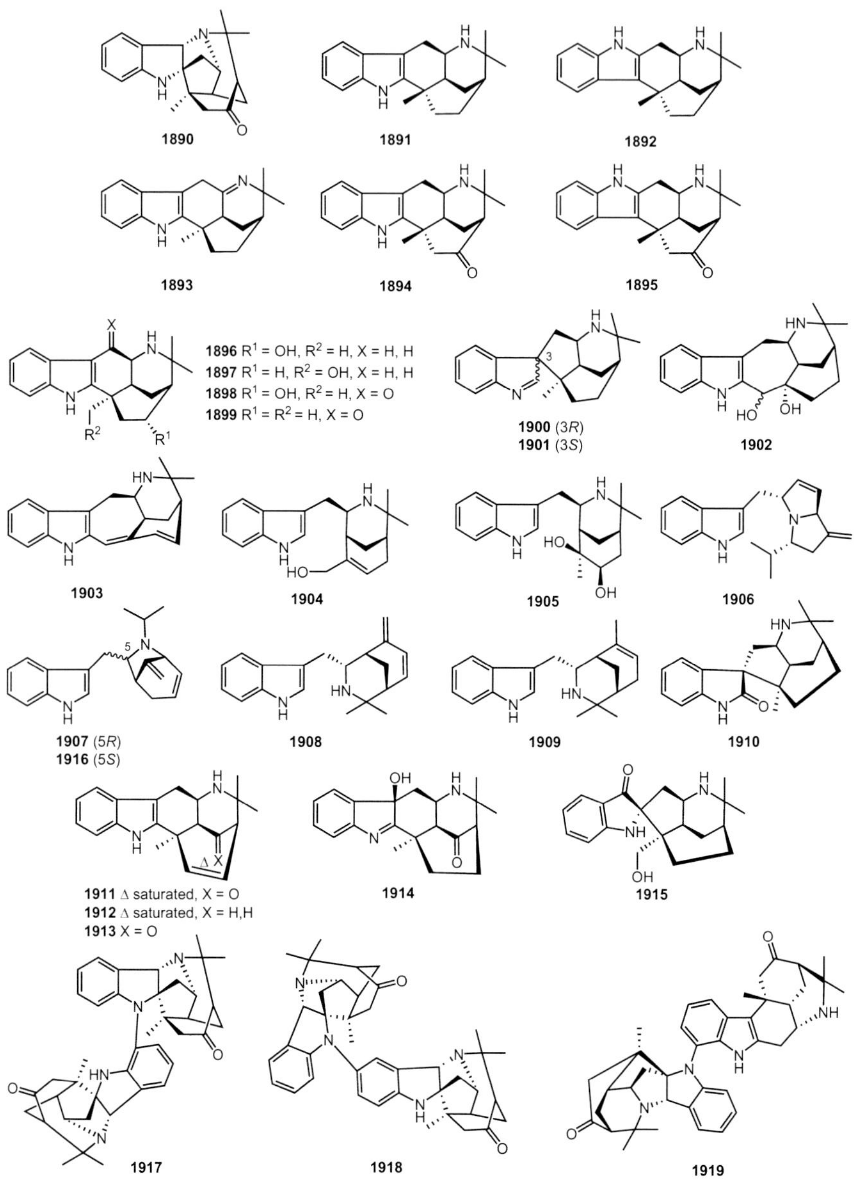

Fig. 142 Structures of alkaloids **1890–1919** reported from *Aristotelia* species

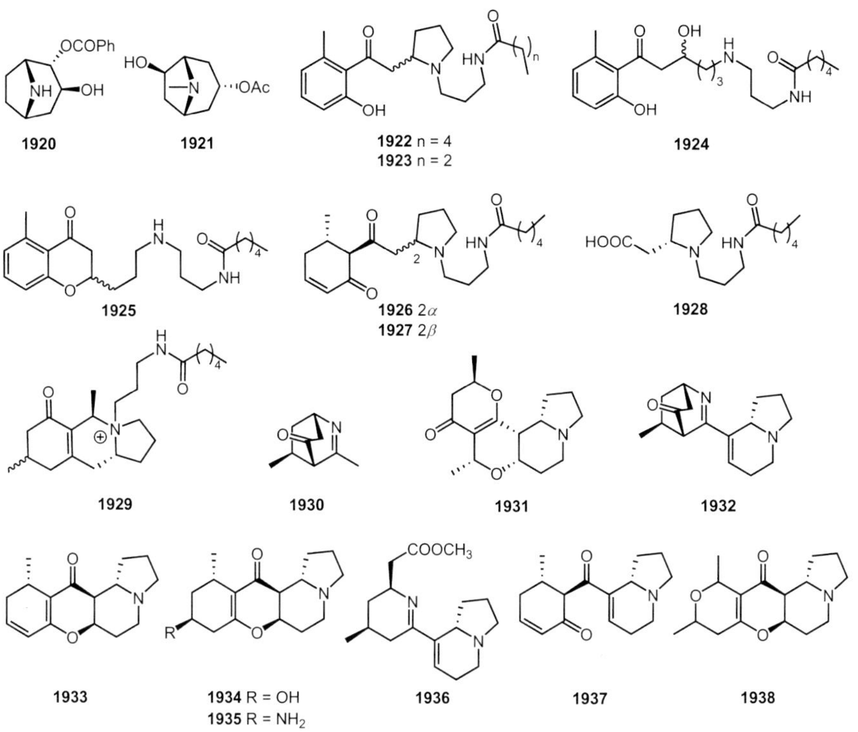

Fig. 143 Structures of alkaloids **1920–1938** reported from *Peripentadenia mearsii* and *Elaeocarpus grandis*

of the alkaloid mixtures found in *E. grandis* and *E. angustifolia* has meant that their isolation procedures and structure elucidation have been challenging. Some of the plant materials contain individual alkaloids as racemates while other investigations have yielded enantiomerically pure alkaloids. The elaeocarpine and elaeocarpiline congeners have three or four stereogenic centers, respectively, and all possible diastereomers have been reported [646].

The *Peripentadenia* and *Elaeocarpus* alkaloids are related biogenetically. A possible biogenesis involves five Claisen condensations of malonate to generate intermediate **1960** (Scheme 4). A pyrrolidinium intermediate (**1961**) derived from arginine and methionine provides the core alkaloid substructure. Conjugation of the pyrrolinium intermediate with tryptamine generates eleaocarpidine (**1959**), while conjugation of **1961** with the polyketide intermediate (**1960**) followed by elimination of CO_2 generates a precursor (**1962**) to all other *Elaeocarpus* and *Peripentadenia* alkaloids. A retro-aldol reaction followed by alternate aldol condensations generate the pyrano derivatives instead of the aryl/cyclohexyl derivatives.

The leaves of a north Queensland sample of *Elaeocarpus eumundi* (Plate 30) were found to contain two stilbene glycosides, 6′-*O*-galloyldihydropieceid (**1963**) and dihydropieceid (**1964**), showing anti-inflammatory activity [647], and the leaves

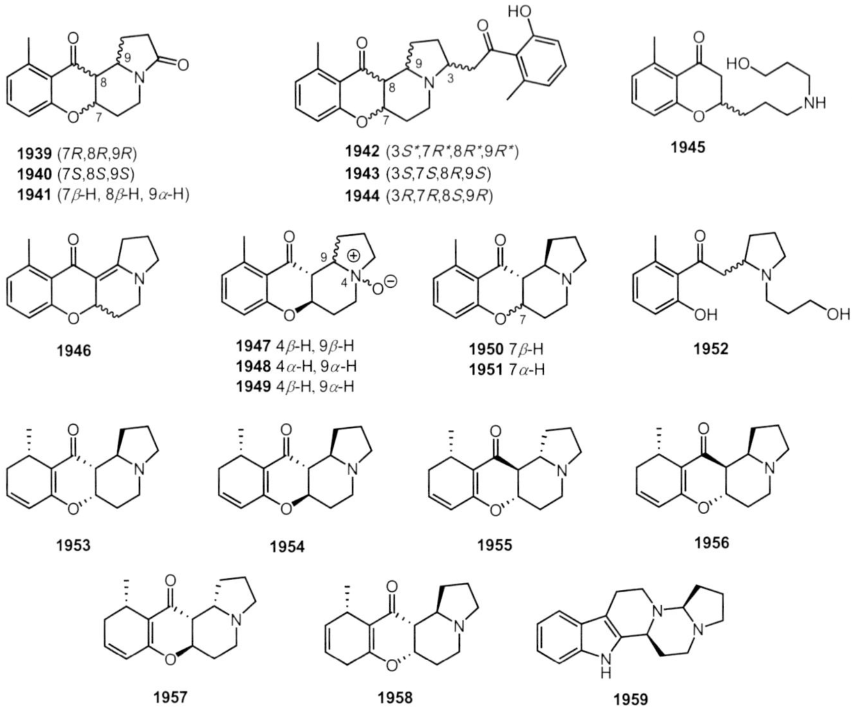

Fig. 144 Structures of alkaloids **1939–1959** reported from *Elaeocarpus angustifolia*

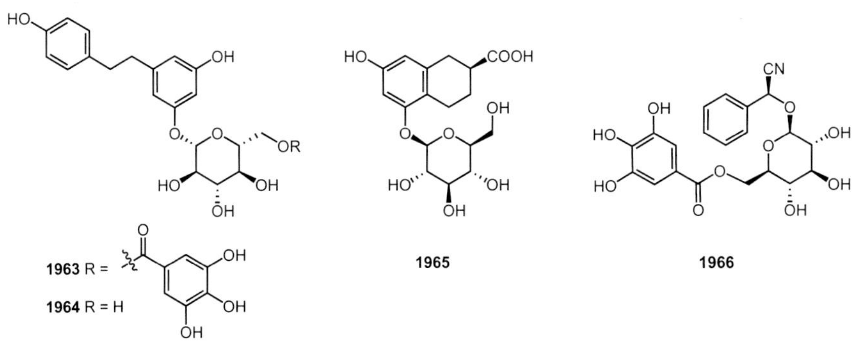

Fig. 145 Structures of compounds **1963–1966** reported from *E. eumundi* and *Aceratium megalospermum*

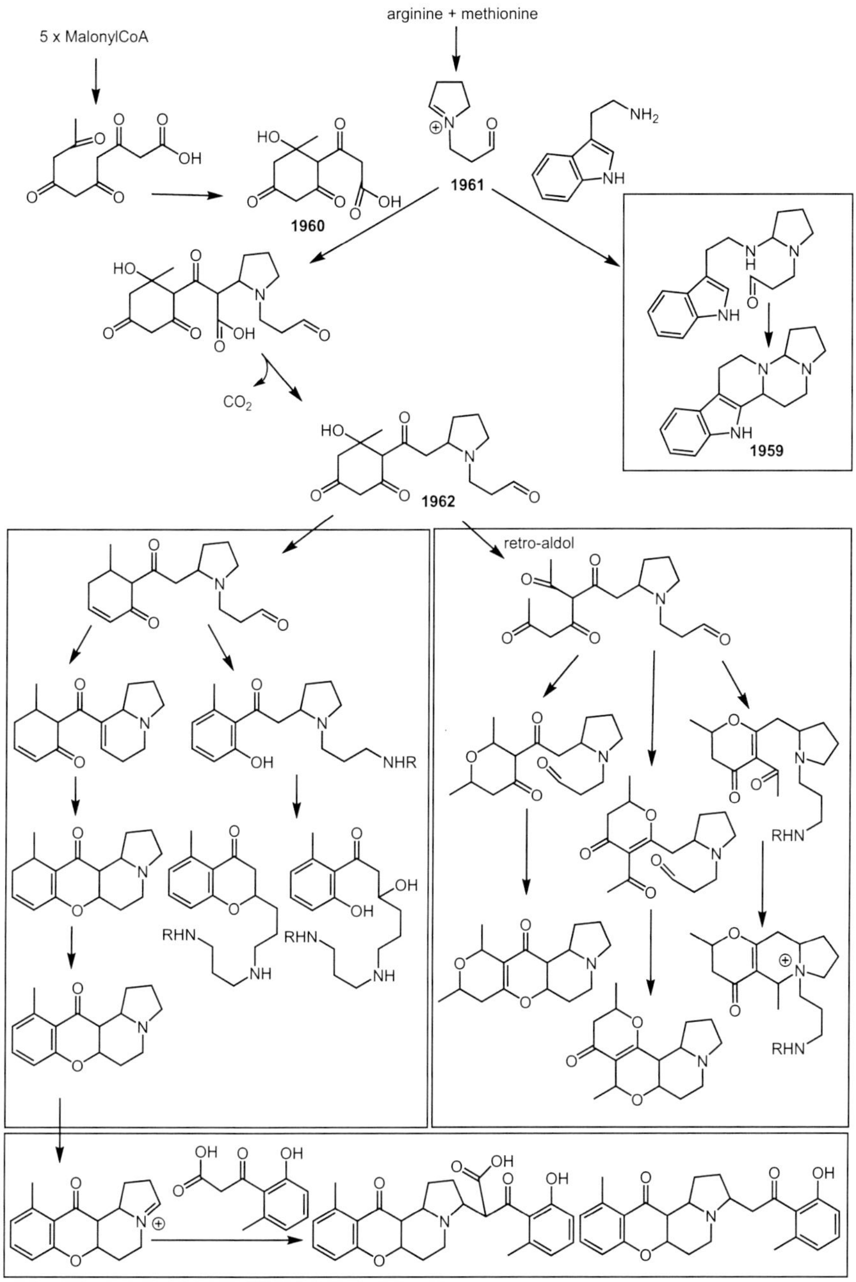

Scheme 4 Proposed biogenesis of the *Elaeocarpus* and *Peripentadenia* alkaloids

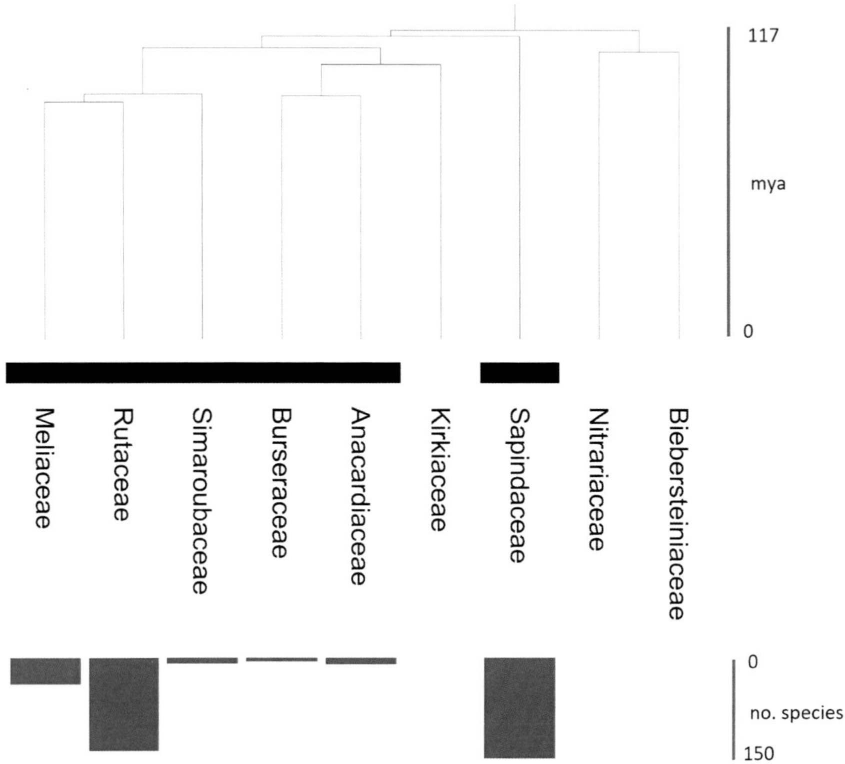

Fig. 146 Dated molecular phylogeny of the Sapindales [7], showing those families that have species present in Australian rainforests (grey) and studied (black), and the number of species occurring in Australian rainforests

of *Aceratium megalospermum*, also from north Queensland produce the weakly cytostatic tetralin glycoside, aceratioside (**1965**) (Fig. 145) [648]. These three compounds share a polyketide derivation with the *Elaeocarpus*/*Peripentadenia* alkaloids. *Elaeocarpus sericopetalus* leaves contain the cyanogenic glycoside, 6′-*O*-galloylsambunigrin (**1966**) [649].

The observation of shared biosynthetic origins of compounds from *Aceratium*, *Elaeococarpus*, and *Peripentadenia* provides an expectation that when more detailed investigations of plants from the family are undertaken, polyketide-derived molecules are likely to predominate.

Sapindales

The Sapindales contains nine families including six that have species in Australian rainforests. The Sapindaceae (154 species), Meliaceae (41 species), and Rutaceae

Plate 31 Rainforest Sapindaceae species; fruits and flowers of *Harpullia pendula* (top three left photographs), and flowers of *Alectryon coriaceus* (right top and bottom), *Dodonaea lanceolata* whole plant (with fruit—inset), female (middle center) and male flowers (bottom center), and *Mischocarpus anodontus* fruits (bottom middle right). (Photographs AR Carroll)

(143) are particularly species-rich compared to the other three families (Anacardiaceae, Burseraceae, and Simaroubaceae) that each contain less than 12 species (Fig. 146). Many natural products have been reported from species in this grouping and their chemistry is diverse. However, most studies have focused on species from the Rutaceae, with a fewer number of studies performed (only nine species) on the equally speciose Sapindaceae (Table 38).

Sapindaceae

The Sapindaceae has a Eurasian origin with dispersal into the Southern Hemisphere occurring in the late Paleocene [650]. The Australian taxa (Plate 31) most likely migrated from southeast Asia once the Australian tectonic plate drifted north after the break up of Gondwana [650].

The limited investigations conducted on Australian rainforest Sapinaceae species (Fig. 147) have yielded a series of labdane and clerodane diterpenoids (7,8-baiyunol (**1967**), 17-hydroxy-13-methyl-labda-7,(13*Z*)-diene-15-oic acid (**1968**), 13-methyl-17-oxolabda-7,(13*Z*)-diene-15-oic acid (**1969**), 13,17-epoxy-13-methyl-15-*oxo*-labda-7-ene (**1970**), 15,16-epoxy-(8*R*)-(benzoyloxy)methylcleroda-3,13(16),14-trien-18-oic acid (**1971**), polyandric acid A (**1972**), 15,16-epoxy-(8*R*)-(benzoyloxy)methyl-2-oxocleroda-3,13(16),14-trien-18-oic acid (**1973**), and 15,16-epoxy-(2*R*)-benzoyloxycleroda-3,13(16),14-trien-18-oic acid (**1974**)) from

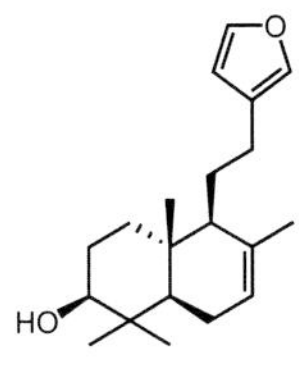

1967

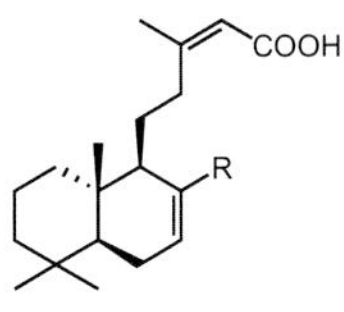

1968 R = CH_2OH
1969 R = CHO

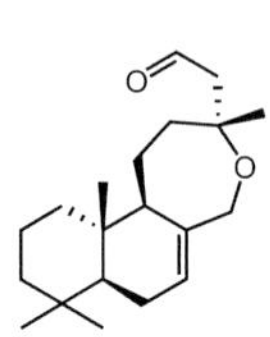

1970

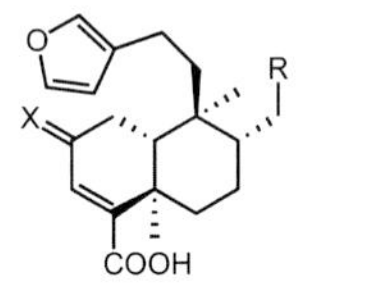

1971 R = OCOPh, X = H, H
1972 R = OCOPh, X = H, ····OH
1973 R = OCOPh, X = O
1974 R = H, X = H, ····OCOPh

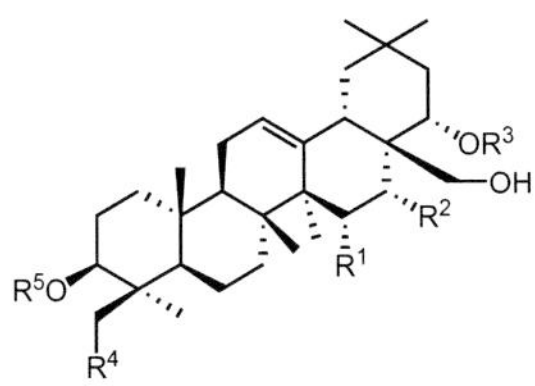

a =

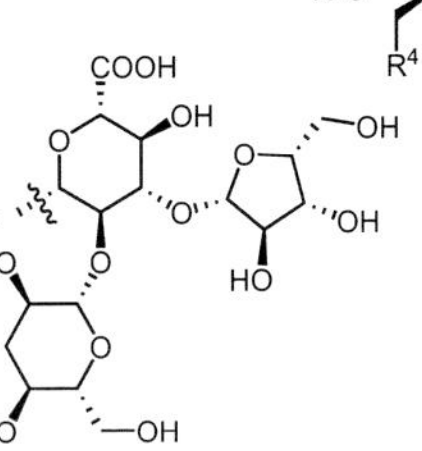

1975 $R^1 = R^2$ = OH, $R^3 = R^4 = R^5$ = H
1976 $R^1 = R^3 = R^4 = R^5$ = H, R^2 = OH
1977 $R^1 = R^2 = R^3 = R^4 = R^5$ = H
1978 $R^1 = R^2$ = OH, $R^4 = R^5$ = H, R^3 =
1979 $R^1 = R^3 = R^4 = R^5$ = H, R^2 =
1980 $R^1 = R^4 = R^5$ = H, R^2 = OH, R^3 =
1981 R^1 = H, $R^2 = R^4$ = OH, R^5 = a, R^3 =
1982 $R^1 = R^3$ = H, R^4 = OH, R^5 = a, R^2 =

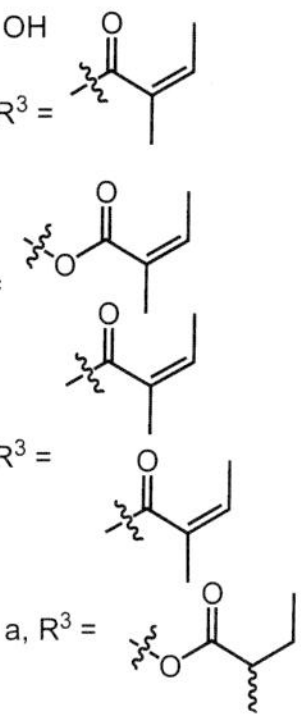

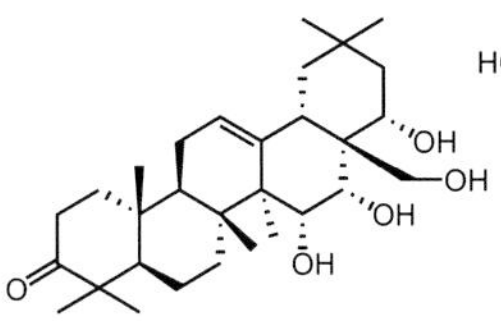

1983

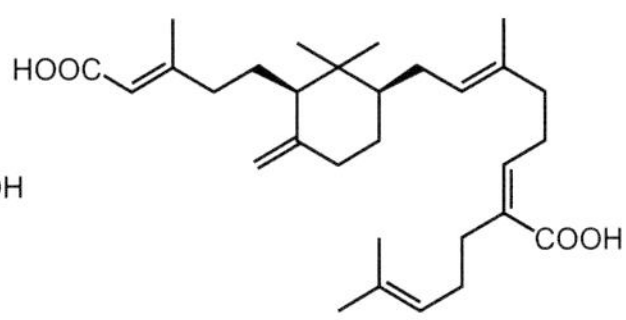

1984

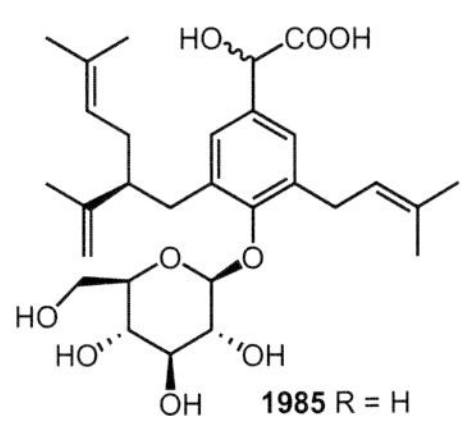

1985 R = H
1986 R = Me

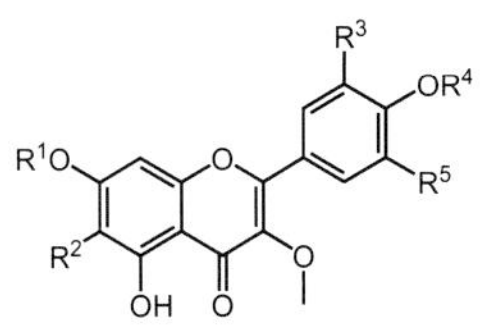

1987 $R^1 = R^2 = R^4 = R^5$ = H, R^3 = $CH_2CH{=}C(CH_3)_2$
1988 $R^1 = R^2 = R^5$ = H, R^3 = $CH_2CH{=}C(CH_3)_2$, R^4 = Me
1989 $R^1 = R^2 = R^4$ = H, $R^3 = R^5$ = $CH_2CH{=}C(CH_3)_2$
1990 $R^1 = R^4$ = H, R^2 = OMe, $R^3 = R^5$ = $CH_2CH{=}C(CH_3)_2$
1991 $R^1 = R^5$ = H, R^2 = OMe, R^3 = $CH_2CH{=}C(CH_3)_2$, R^4 = Me
1992 R^1 = Me, $R^2 = R^3 = R^4 = R^5$ = H

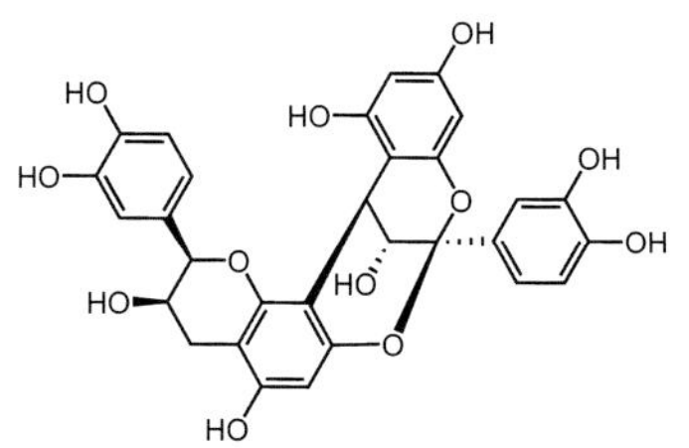

1993

Fig. 147 Structures of compounds **1967–1993** reported from Sapindaceae species

Table 38 Distribution of natural products isolated from the Sapindales (excluding Rutaceae and Meliaceae)

Species	Natural location[a,b]	Plant part(s)[c]	Compound class	Compound(s)	Refs.
Sapindaceae					
Alectryon oleifolius subsp. *elongatus*	NWA	lf	Bisflavonoid	**1993**	[659]
Dodonaea polyandra	CY	lf, lf res	Diterpenoid	**1967–1974**	[651, 652]
	CY	lf, st	Flavonoid	**1987–1992**	[658]
Harpullia pendula	SEQ	b	Cinnamate	**760**	[653]
	SEQ (cult. (Egypt))	lf	Isoprenyl phenol	**1985**, **1986**	[657]
	SEQ	b	Triterpenoid	**1975–1980**, **1983**	[653, 654]
	SEQ (cult. (Egypt))	s	Triterpenoid	**1981**, **1982**	[655]
Mischocarpus pyriformis subsp. *pyriformis*	NQ	b	Triterpenoid	**1984**	[656]
Anacardiaceae					
Pleiogynium timoriense	TON	b	5-Alkylphenol	**1994–1996**	[660]
	NNSW	fr	Anthocyanin	**1998**	[662]
Rhus taitensis	PNG	lf, tw	Triterpenoid	**1999–2002**	[663, 664]
Semecarpus australiensis	NQ	s	5-Alkylphenol	**1997**	[661]
Burseraceae					
Canarium muelleri	NQ	lf	Triterpenoid	**2003**	[665]
Simaroubaceae					
Samadera bidwillii	NT	wp	Acridone	**2007**, **2008**	[667]
	NT	wp	Quinolone	**2006**	[667]
	NT	wp	Lignan	**2009–2015**	[667]
	NT	wp	Triterpenoid	**2004**, **2005**	[667]

[a] SEQ = Southeast Queensland, NQ = North Queensland, CY = Cape York, Queensland, NNSW = Northern New South Wales, NT = Northern Territory, NWA = North Western Australia, PNG = Papua New Guinea, TON = Tonga, cult. = cultivated, [b] natural location shown for those species studied from cultivated material, [c] lf = leaves, lf res = leaf resin, s = seeds, tw = twigs, wp = whole plant, b = bark, st = stems

Dodonaea polyandra from Cape York in the far north of Queensland [651, 652]. Two compounds, **1972** and **1974**, showed anti-inflammatory activities.

Polyhydroxylated triterpenoid glycosides and aglycones related to β-amyrin (A1-barrigenol (**1975**), camelliagenin A (**1976**), 22*α*-hydroxyerythrodiol (**1977**), A1-barrigenol 22-angelate (**1978**), camelliagenin A 16-angelate (**1979**), camelliagenin A 22-angelate (**1980**), pendulaoside A (**1981**), pendulaoside B (**1982**), and 15*α*,16*α*,22*α*,28-tetrahydroxyolean-12-en-3-one (**1983**)) are constituents of the

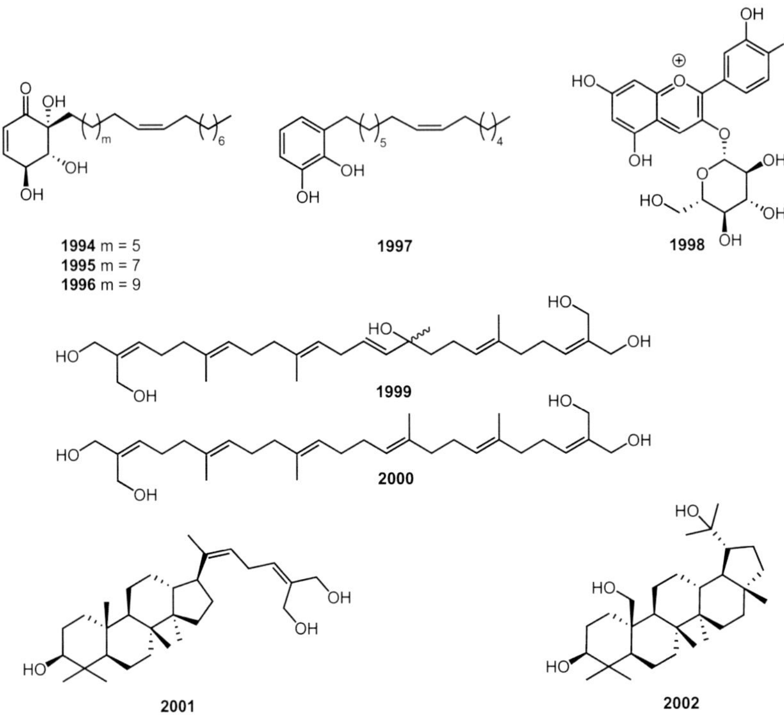

Fig. 148 Structures of compounds **1994–2002** reported from Anacardiaceae species

bark and seeds of *Harpullia pendula* [653–655]. These compounds are closely comparable to triterpenoids found in the Pittosporaceae and Lecythidaceae. Pendulaosides A (**1981**) and B (**1982**) are weakly cytotoxic. A linear triterpenoid dicarboxylic acid, mispyric acid (**1984**), from the stem bark of *Mischocarpus pyriformis* subsp. *pyriformis* (endemic to the subtropical and tropical rainforests of New South Wales and Queensland), and studied in the U.S.A. showed weak DNA polymerase inhibition [656]. The leaves of *H. pendula* cultivated in Egypt produced two compounds, harpulliasides A (**1985**) and B (**1986**) that are glycosides of 4-hydroxymandelic acid substituted by three isoprene units [657]. Isoprenylated flavones (5,7,4′-trihydroxy-3′-(3-methylbut-2-enyl)-3-methoxyflavone (**1987**), 5,7-dihydroxy-3′-(3-methylbut-2-enyl)-3,4′-dimethoxyflavone (**1988**), 5,7,4′-trihydroxy-3′,5′-(3-methylbut-2-enyl)-3-methoxyflavone (**1989**), 5,7,4′-trihydroxy-3′,5′-(3-methylbut-2-enyl)-3,6-dimethoxyflavone (**1990**), viscosol (**1991**), and 5,4′-dihydroxy-3,7-dimethoxyflavone (**1992**) were also present in the leaves of *D. polyandra* [658]. A bis-flavonoid, procyanidin A2 (**1993**), with anthelminthic activity, was found in the leaves of the Western Australian *Alectryon oleifolius*. Procyanidin A2 significantly inhibited the development of the free-living stages of cyathostomin nematodes that are parasites of horses, as well as inhibiting the motility of infective larval nematodes [659].

Anacardiaceae

Species in the Anacardiaceae are well known to cause dermatitis and this is attributed to their propensity to produce alkyl phenols. The bark of *Pleiogynium timoriensis* from Tonga and the seeds of *Semecarpus australiensis* from north Queensland contain pleiogenones A–C (**1994–1996**), derivatives of urushiol (**1997**) (Fig. 148) [660, 661]. The pleiogenones show high n*M* cytotoxicity towards A2780 ovarian cancer cells. *Pleiogynium timoriensis* fruits collected in northern New South Wales contain the anthocyanin cyanogen-3-glucoside (**1998**) [662]. The leaves and twigs of *Rhus taitensis* from Papua New Guinea accumulate triterpenoids that inhibit *Mycobacterium tuberculosis* and these include two hydroxylated linear triterpenoids, 1,10,24,25,30-pentahydroxysqualene (**1999**) and 1,10,25,30-tetrahydroxysqualene (**2000**), and two tetracyclic triterpenoids, dammar-20(22),24-diene-3,26,27-triol (**2001**) and lupane-3,20,25-triol (**2002**) [663, 664].

Burseraceae and Simaroubaceae

Only one natural product has been reported from the Burseraceae. *Canarium muelleri* leaves collected in Queensland contain the A-ring oxidized lupane triterpenoid, canaric acid (**2003**) (Fig. 149) [665].

The Simaroubaceae has a pantropical distribution and only six of 109 species are found in Australian rainforests. The family is sister to the Rutaceae and Malvaceae [666]. The only Australian rainforest species studied, *Samadera bidwillii* from the Northern Territory, contains compounds typically associated with Rutaceae species. These include the limonoid triterpenoid limonin (**2004**), the quassinoid triterpenoid, 2′-acetoxyglaucarubin (**2005**), the quinolone, 2-(10′-acetoxyundecanyl)-1-acetoxymethyl-4-quinolone (**2006**), two acridones, 1-methoxy-10-methyl-acridan-9-one (**2007**) and 1,8-dihydroxyacridan-9-one (**2008**), and the tetrahydrofuran lignans (−)-*epi*-sesamin (**2009**), neofargesin (**2010**), (−)-*epi*-eudesmin (**2011**), (−)-eudesmin (**2012**), (−)-sesamin (**2013**), fargesin (**2014**), and (−)-kobusin (**2015**) [667].

Rutaceae

By far the most extensively studied family within Australian rainforests is the Rutaceae. Prof. Peter Waterman and his colleagues from Strathclyde University (UK) (and later Southern Cross University, Lismore, New South Wales) have contributed significantly to our knowledge of the chemistry of Australian rainforest Rutaceae species. Over 400 compounds from 23 genera and 57 species have been reported. This

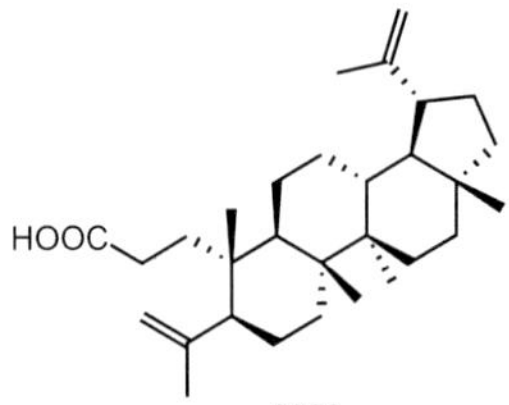

2003

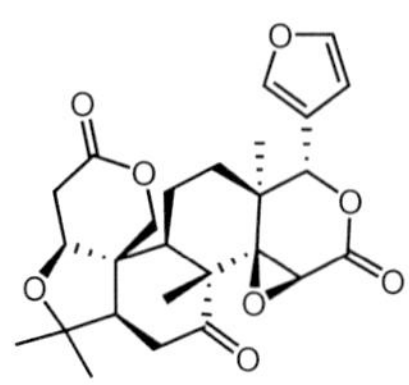

2004

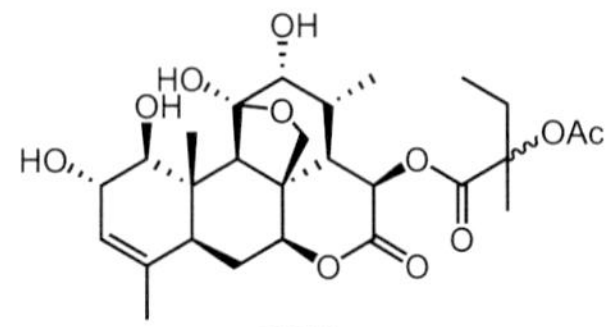

2005

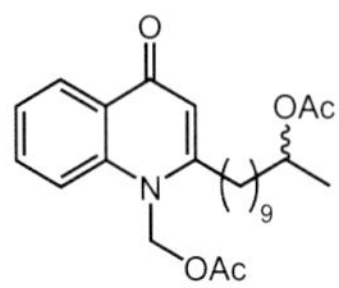

2006

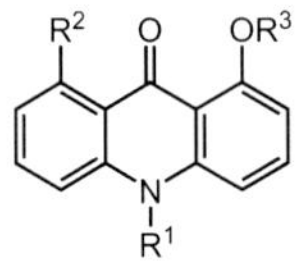

2007 R^1 = Me, R^2 = H, R^3 = Me
2008 R^1 = R^3 = H, R^2 = OH

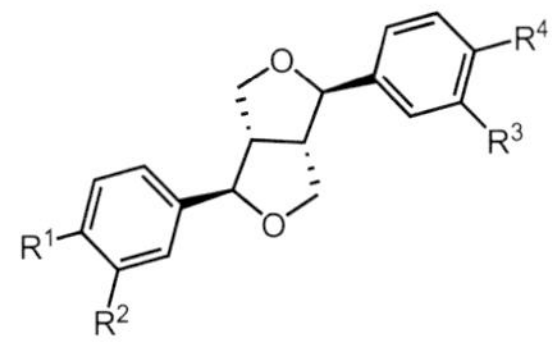

2013 R^1–R^2 = R^3–R^4 = OCH_2O
2014 R^1 = R^2 = R^3 = R^4 = OMe
2015 R^1 = R^2 = OMe, R^3–R^4 = OCH_2O

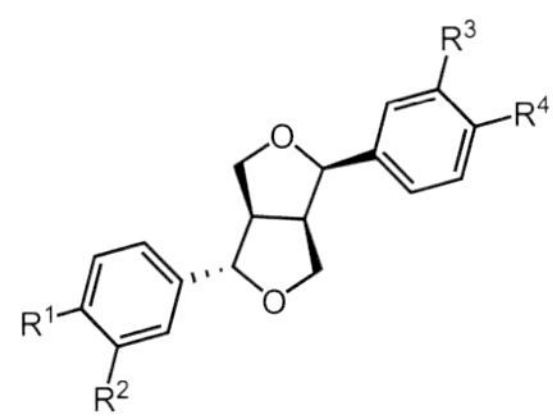

2009 R^1–R^2 = R^3–R^4 = OCH_2O
2010 R^1 = R^2 = R^3 = R^4 = OMe
2011 R^1–R^2 = OCH_2O, R^3 = R^4 = OMe
2012 R^1 = R^2 = OMe, R^3–R^4 = OCH_2O

Fig. 149 Structures of compounds **2003–2015** reported from *Canarium muelleri and Samadera bidwillis*

is comprehensive given that the total genus and species diversity within Australian rainforest Rutaceae is represented by 34 genera and 143 species. The occurrence of benzyl isoquinoline alkaloids in *Zanthoxylum* and closely related genera led Waterman and his associates to propose a close phylogenetic relationship between the proto-Rutaceae with the basal eudicot families in the Ranunculales [668]. However, subsequent molecular evidence did not support this hypothesis, suggesting convergent evolution is at play [669]. Other molecular evidence suggests that genes for key benzyl isoquinoline alkaloid enzymes are widespread but mostly dormant in most angiosperms [10].

The Rutaceae is a pantropical family and contains approximately 160 genera and 2,000 species. It is divided into six subfamilies with the Cneoroideae sister to all other families, and the largest subfamily Zanthoxyloideae (containing 110 genera) sister to the remaining four subfamilies (Fig. 150) [669]. Australia is a center of diversity and likely ancestral home of most genera in the subfamily Zanthoxyloideae, with most of the species in Clades C4b-f (containing close to one third (48) of all Rutaceae

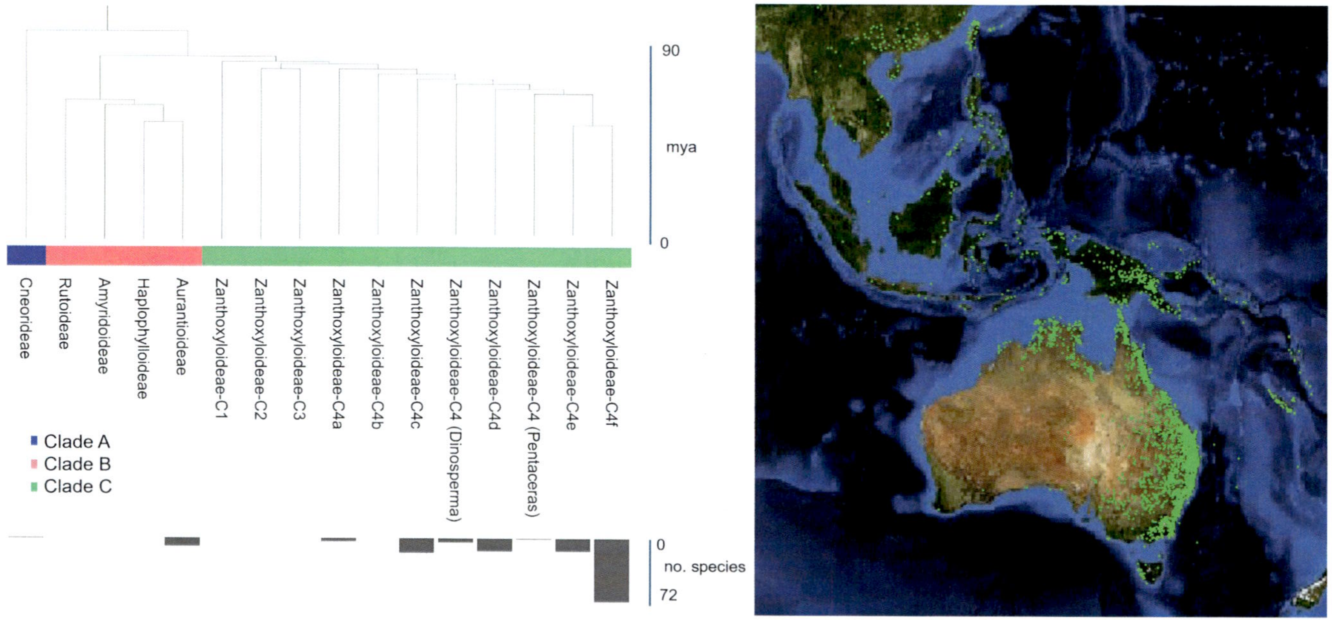

Fig. 150 Dated molecular phylogeny of the Rutaceae [669] showing tribes within Clades A–C and the number of species within each group that occur in Australian rainforests. The genera that occur in Australia's rainforest Clades are listed as follows: Cneoroideae: *Harrisonia*; Aurantioideae: *Glycosmis, Micromelum, Clausena, Murraya, Luvunga, Citrus, Microcitrus*; Zanthoxyloideae: C4a—*Zanthoxylum*; C4c—*Flindersia, Lunasia, Coatesia, Geijera*; C4d—*Bosistoa, Bouchardatia, Acradenia*; C4e—*Halfordia, Correa, Leionema, Eriostemon, Phebalium, Nematolepis;* C4f—*Boronia, Zieria, Brombya, Euodia, Melicope, Medicosma, Acronychia, Pitaviaster, Sarcomelicope*. The map shows the collection locations for herbarium records for Australian rainforest Rutaceae species [5, 43]

Plate 32 *Flinderisa* species: *F. brayleyana* (top left), *F. schottiana* (top, middle, and right), close-up of flowers of *F. brayleyana* (bottom left) fruits and flowers of *F. brayleyana* (bottom middle left) and flowers and opened seed pod of *F. australis* (bottom right). (Photographs AR Carroll)

genera) endemic to Australia or New Caledonia (Fig. 150). The other clades in the Zanthoxyloideae are C1 (four genera, mainly Japan), C2 (10 genera, mainly South Africa) C3 (24 genera, mainly North, Central and South America) and C4a (five genera, occurring across Asia, Africa, the Americas, and Australia).

The natural product chemistry of the Rutaceae is dominated by coumarins, alkaloids of various structural classes, flavonoids, and triterpenoids (Table 39). The biosynthetic precursor to most of the alkaloids is anthranilic acid with polyketide and isoprenyl elaboration contributing to their structural diversity. Given the extensive sampling of Australian rainforest Rutaceae species, the chemical diversity found reflects almost the entirety of the chemical diversity of the Rutaceae worldwide.

Thirty-seven congeners of quinoline/quinolone alkaloids, with the phenyl ring either unsubstituted, mono or dioxygenated at either C-7, C-8, and/or C-9 have been reported from the Australian rainforest Rutaceae (Fig. 151). These comprise 4-methoxy-2-quinolone (**2016**), 4-methoxy-1-methyl-2(1*H*)-quinolinone (**2017**), 4,6,8-trimethoxy-2(1*H*)-quinolinone (**2018**), halfordamine (**2019**), 4,8-dimethoxy-1-methyl-2(1*H*)-quinolinone (**2020**), 4,8-dimethoxy-2(1*H*)-quinolinone (**2021**), 4,6,7-trimethoxy-2(1*H*)-quinolinone (**2022**), 2-*n*-pentyl-4(1*H*)-quinolinone (**2023**), 2-*n*-propyl-4(1*H*)-quinolinone (**2024**), 1,2-dimethylquinol-4-one (**2025**), ifflaiamine (**2026**), *N*-demethyl-8-methoxyifflaiamine (**2027**), 7-[(2*R*)-2-hydroxy-3-methylbutyl]-6-methoxy-9-methyl-1,3-dioxolo[4,5-*h*]quinolin-8(9*H*)-one (**2028**), 3-[(2*R*)-2-hydroxy-3-methylbutyl]-4,8-dimethoxy-1-methyl-2(1*H*)-quinolinone (**2029**), pteleprenine (**2030**), *O*-methylglycosolone (**2031**), lunidonine (**2032**),

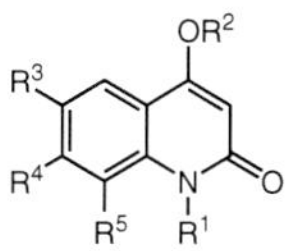

2016 $R^1 = R^3 = R^4 = R^5 = H$, $R^2 = Me$
2017 $R^1 = R^3 = R^4 = H$, $R^2 = Me$, $R^5 = OMe$
2018 $R^1 = R^2 = Me$, $R^3 = R^4 = R^5 = H$
2019 $R^1 = R^4 = H$, $R^2 = Me$, $R^3 = R^5 = OMe$
2020 $R^1 = R^2 = Me$, $R^3 = R^4 = H$, $R^5 = OMe$
2021 $R^1 = R^5 = H$, $R^2 = Me$, $R^3 = R^4 = OMe$
2022 $R^1 = R^3 = R^4 = R^5 = H$, $R^2 = CH_2CHC(CH_3)_2$

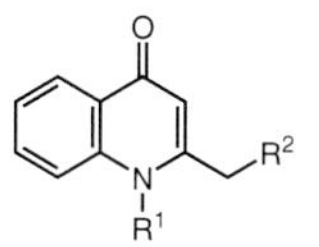

2023 $R^1 = H$, $R^2 = n$-But
2024 $R^1 = H$, $R^2 = Et$
2025 $R^1 = Me$, $R^2 = H$

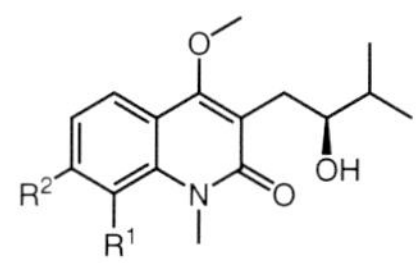

2028 $R^1–R^2 = OCH_2O$
2029 $R^1 = OMe$, $R^2 = H$

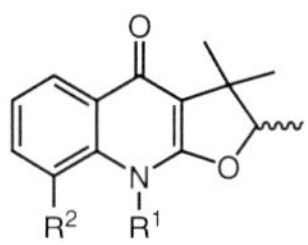

2026 $R^1 = Me$, $R^2 = H$
2027 $R^1 = H$, $R^2 = OMe$

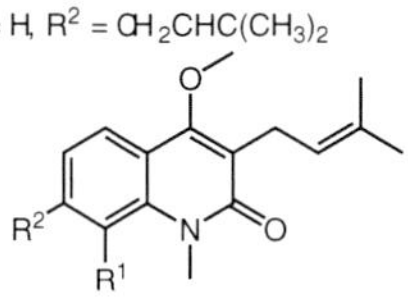

2030 $R^1–R^2 = OCH_2O$
2031 $R^1 = OMe$, $R^2 = H$

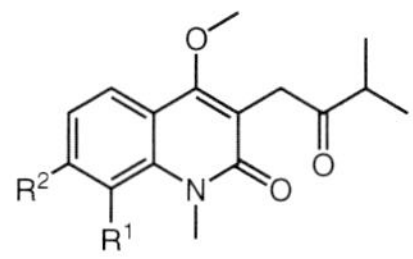

2032 $R^1–R^2 = OCH_2O$
2033 $R^1 = OMe$, $R^2 = H$

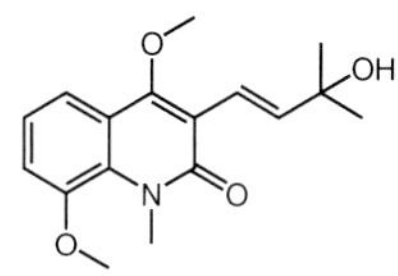

2034

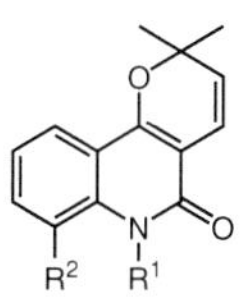

2035 $R^1 = R^2 = H$
2036 $R^1 = Me$, $R^2 = H$
2037 $R^1 = H$, $R^2 = OMe$
2038 $R^1 = CH_2OAc$, $R^2 = H$
2039 $R^1 = Me$, $R^2 = OMe$

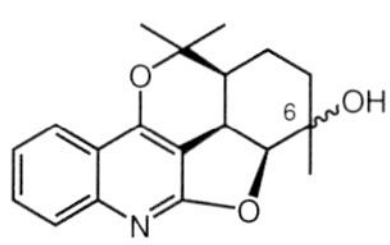

2040 (6*R*)
2041 (6*S*)

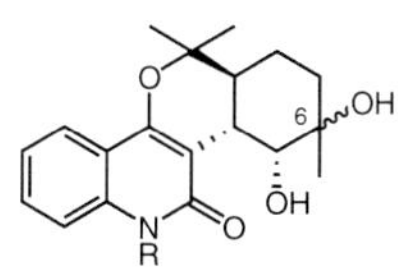

2042 R = Me, (6*S*)
2043 $R = CH_2OAc$, (6*S*)
2044 R = Me, (6*R*)

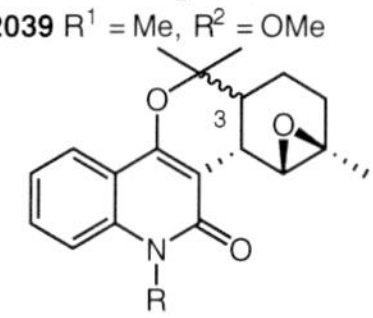

2045 R = Me, (3*S*)
2046 $R = CH_2OAc$, (3*S*)
2047 $R = CH_2OAc$, (3*R*)
2048 R = H, (3*S*)

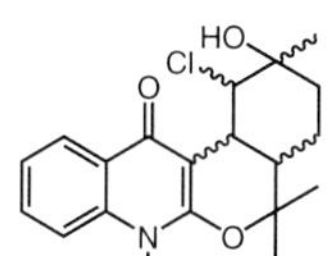

2049

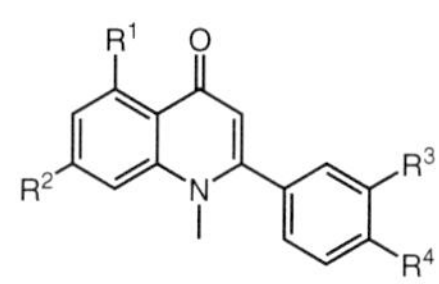

2050 $R^1 = R^3 = R^4 = H$, $R^2 = OMe$
2051 $R^1 = OH$, $R^2 = R^3 = R^4 = H$
2052 $R^1 = OH$, $R^2 = H$, $R^3–R^4 = OCH_2O$

Fig. 151 Structures of quinoline alkaloids **2016–2052** reported from Rutaceae species

orixalone A (**2033**), glycocitlone C (**2034**), flindersine (**2035**), *N*-methylflindersine (**2036**), 8-methoxyflindersine (**2037**), *N*-acetoxymethylflindersine (**2038**), 8-methoxy-*N*-methylflindersine (**2039**), furoerioaustralasine (**2040**), furoerzoaustralasine (**2041**), *trans*-deacetoxyaustralasine hydrate (**2042**), *trans*-australasine hydrate (**2043**), *trans*-10-*epi*-deacetoxyaustralasine hydrate (**2044**), *trans*-deacetoxyarioaustrasine (**2045**), *trans*-erioaustralasine (**2046**), *cis*-erioaustralasine (**2047**), synthetic product (**2048**), and (1*R*,2*R*,4*αS*,12*βS*)-1-chloro-1,2,3,4,4*α*,5,7,12*β*-octahydro-2-hydroxy-2,5,5,7-tetramethyl-12*H*-[2]benzopyrano[3,4-*b*]quinolin-12-one (**2049**) [39, 670–680]. Three phenyl quinolones, 7-methoxy-1-methyl-2-phenyl-4-quinolone (**2050**), 5-hydroxy-1-methyl-2-phenyl-4-quinolone (**2051**), and

Table 39 Distribution of natural products isolated from the Rutaceae

Species	Collection location[a,b]	Plant part(s)[c]	Compound class	Compound(s)	Refs.
Acradenia frankliniae	ACT, Tas	lf	Acylphloroglucinol	**2242**, **2243**, **2254**, **2259**, **2263**	[743, 744]
	ACT	lf	Triterpenoid	**2343**	[743]
Acronychia acidula	NQ	lf, st	Acridone	**2103**, **2104**	[709]
Acronychia crassipetala	NQ	fr	Acylphloroglucinol	**2244**, **2246–2250**	[745, 746]
Acronychia pubescens	NNSW	r	Acylphloroglucinol	**2245**, **2251**, **2264**	[747]
Acronychia vestita	SEQ	b	Isoprenylphloroglucinol	**2268**	[742]
Boronia alulata	NQ	ap	Quinolone	**2023**, **2024**	[674]
Boronia lanceolata	NT	ap	Acridone	**2095–2101**	[707]
	NT	ap	Coumarin	**2183**, **2184**, **2187**	[725]
	NT	ap	Pyranocoumarin	**2233**, **2235–2238**	[725]
Bosistoa floydii	NNSW	lf	Flavonoid	**2286**, **2288**	[721]
	NNSW	lf	Xanthine	**2164**	[721]
Bosistoa medicinalis	NQ	lf, st	Flavonoid	**2276**, **2277**, **2287**, **2293**, **2296**	[754]
	NQ	lf, st	Triterpenoid	**801**, **1860**, **2341**, **2342**	[754]
Bosistoa pentacocca	SEQ	lf, st	Isoprenylphloroglucinol	**2265**, **2266**, **2269–2271**	[752]
Bosistoa transversas	SEQ	ap	Acridone	**2102**, **2118–2124**	[708]
Bouchardatia neurococca	SEQ	ap	β-Carboline	**2152–2155**	[683]
	SEQ	ap	Furanoquinoline	**2077**	[683]
	SEQ	ap	Triterpenoid	**2351–2353**	[683]
	SEQ	ap	Monoterpenoid	**2355**	[683]
Brombya platynema	NQ	b	Phenylalkyl polyketide	**2311–2316**	[760, 761]

(continued)

Table 39 (continued)

Species	Collection location[a,b]	Plant part(s)[c]	Compound class	Compound(s)	Refs.
Brombya smithii	NQ	ap	Coumarin	**2182**, **2185**, **2186**, **2188**, **2207**, **2208**	[684]
	NQ	ap	Furanoquinoline	**2081**, **2082**	[684]
	NQ	ap	Phenylalkyl polyketide	**2317**, **2318**	[684]
	NQ	ap	Pyranocoumarin	**2231**	[684]
Citrus garrawayi	NQ	fr	Furanocoumarin	**2213**, **2221–2225**, **2227–2230**	[736]
	NQ	fr	Triterpenoid	**2344**	[736]
Clausena brevistyla	NQ	lf	Amide	**2305**	[758]
Coatesia paniculata	SEQ	b, lf	Isoprenylphloroglucinol	**2267**, **2272**	[737]
	SEQ	b, lf	Pyranocoumarin	**2232**	[737]
Dinosperma erythrococcum	NQ	ap	Amide	**2306–2310**	[759]
Dinosperma melanophloium	SEQ	ap	Amide	**162**, **2304**	[738]
	SEQ	ap	Furanocoumarin	**2214–2216**	[738]
	SEQ	ap	Pyranocoumarin	**2239**	[738]
Dinosperma stipitatum	SEQ	lf, b, st	Acyphloroglucinol	**2251–2254**, **2256**, **2257**	[748]
Eriostemon banksii	CY	ap	Cinnamate	**2297**	[677]
	CY	ap	Quinolone	**2040**, **2041**, **2046**, **2047**	[677, 679]
Flindersia acuminata	NQ	ap	Flavonoid	**2278**	[685]
	CY	ap	Furanoquinoline	**2077**, **2086**	[685]
	NQ	b	Isoprenylindole	**2138, 2145–2147**	[717]

(continued)

Table 39 (continued)

Species	Collection location[a,b]	Plant part(s)[c]	Compound class	Compound(s)	Refs.
Flindersia australis	SEQ (cult. (Sri Lanka))	b	Cinnamate	**2298**, **2299**	[756]
	SEQ (cult. (Sri Lanka))	b	Flavonoid	**2119**, **2279**, **2280**	[755]
	NNSW	w	Quinolone	**2035**	[678]
Flindersia bennettii	SEQ	r, lf, b, st	Coumarin	**2194**, **2198**	[39, 686]
	SEQ	lf, b, st	Furanocoumarin	**2216**	[686]
	SEQ	lf, b, st	Furanoquinoline	**2085–2088**	[39, 686]
	SEQ	lf	Indole	**2163**	[39]
	SEQ	lf, b, st, r	Pyranocoumarin	**2240**	[686]
	SEQ	r	Quinolone	**2018**, **2035**	[39]
Flindersia bourjotiana	NQ	b	Flavonoid	**2278**	[687]
	NQ	r, b	Furanoquinoline	**2081**, **2085–2087**	[39, 687]
	NQ	lf	Triterpenoid	**2344–2347**	[764]
Flindersia brassii	CY	lf	Lignan	**2334**	[763]
	CY	lf	Phenylalkyl polyketide	**2319–2327**	[762]
Flindersia brayleyana	NQ (cult. (Brazil))	b	Coumarin	**2189**, **2196**, **2197**	[726]
	NQ (cult. (Brazil))	b	Pyranocoumarin	**2241**	[726]

(continued)

Table 39 (continued)

Species	Collection location[a,b]	Plant part(s)[c]	Compound class	Compound(s)	Refs.
Flindersia collina	SEQ	b	Coumarin	**2199**	[727]
	SEQ	lf, b, st	Furanoquinoline	**2080**, **2081**, **2085**	[39, 688]
Flindersia ifflana	NQ	w	Quinolone	**2026**	[676]
	NQ	w	Triterpenoid	**2349**	[766]
Flindersia laevicarpa var. *laevicarpa*	NQ	lf	β-Carboline	**2156–2159**	[719]
	NQ	w	Flavonoid	**2278**	[689]
	NQ	w	Furanoquinoline	**2087**	[689]
	NQ	b	Triterpenoid	**2354**	[767]
	NQ	b	Monoterpenoid	**2356**	[719]
	NQ	w, b	Phenylalkyl polyketide	**2328–2333**	[689, 719]
Flindersia pimenteliana	NQ	lf	Flavonoid	**804**, **2295**	[715]
	PNG	b	Furanoquinoline	**2077**, **2087**	[690]
	NQ	lf, b	Isoprenylindole	**2139–2144**, **2148–2151**	[715, 716]
	PNG	b	Pyranocoumarin	**2233**, **2240**	[690]
	PNG	b	Triterpenoid	**1620**	[690]
Flindersia schottiana	SEQ	lf	Acridone	**2126**	[39]
	NQ	lf, b	Coumarin	**2198**	[691]
	NQ, SEQ	lf, b	Furanoquinoline	**2077**, **2081**, **2083–2087**	[39, 691]
	NQ	lf, b	Pyranocoumarin	**2240**	[691]
Flindersia xanthoxyla	NNSW	lf, b	Flavonoid	**2278**	[692]
	NNSW	b	Furanoquinoline	**2085**, **2086**	[692]

(continued)

Table 39 (continued)

Species	Collection location[a,b]	Plant part(s)[c]	Compound class	Compound(s)	Refs.
	SEQ	lf, st	Isoprenylindole	**2138**	[39]
Geijera parviflora	NNSW	lf	Anthranilic acid	**2129–2133**	[713]
	SEQ	lf, fr	Coumarin	**2190**, **2200–2005**, **2209–2211**	[676, 728–730]
	NNSW	lf	Flavonoid	**2285**	[673]
	SEQ	lf	Monoterpenoid	**2357**, **2358**	[768]
	NNSW	lf, fr	Quinolone	**2022**, **2035**, **2037**, **2038**	[673, 729]
Geijera salicifolia	PNG, SEQ	lf, b	Coumarin	**2182**, **2195**, **2201**	[731, 732]
	PNG	lf	Furanoquinoline	**2061**, **2062**, **2087**, **2088**	[693]
Halfordia kendack	CY	lf	Coumarin	**2212**	[39]
	SEQ	b, ap	Furanocoumarin	**2217–2220**, **2226**	[739–741]
	SEQ, CY	b	Furanoquinoline	**2077**, **2079**, **2081**	[39, 694]
	SEQ, CY	b	Oxazole	**2134**, **2136**, **2137**	[39, 694]
	CY	b	Pyranocoumarin	**2234**	[39]
	SEQ, NQ, CY	ap, lf, st	Quinolone	**2019**, **2042–2046**, **2048**, **2049**	[39, 672, 680]
Halfordia scleroxyla	SEQ	b	Furanoquinoline	**2089**	[694]
	SEQ	b	Oxazole	**2134–2137**	[694, 714]
	SEQ	b	Quinolone	**2019**	[694]
Lunasia amara	CY, INDO	b	Furanoquinoline	**2055**, **2056**, **2058**, **2065–2067**, **2070**, **2077**, **2087**, **2088**	[670, 681, 695, 696]
	CY, INDO	b, w, lf	Quinolone	**2016–2018**, **2020**, **2021**, **2027–2034**, **2036**, **2039**, **2050–2052**	[670, 671, 681, 682]
Medicosma cunninghamii	SEQ	b	Furanoquinoline	**2053**, **2077**, **2090**	[697, 698]

(continued)

Table 39 (continued)

Species	Collection location[a,b]	Plant part(s)[c]	Compound class	Compound(s)	Refs.
Medicosma fareana	NQ	lf	Acridone	**2106–2109**, **2112**	[699]
	NQ	lf	Benzaldehyde	**2362**	[699]
	NQ	lf	Furanoquinoline	**2060**, **2087**, **2091**	[699]
	NQ	lf	Phenyl ethyl alcohol	**2361**	[699]
Medicosma sessiliflora	NQ	lf	Acylphloroglucinol	**2260–2262**	[751]
Melicope affinis	NQ	r	Furanoquinoline	**2094**	[39]
Melicope broadbentiana	NQ	w, b	Flavonoid	**2281–2283**	[39, 750]
	NQ	r	Furanoquinoline	**2093**	[39]
	NQ	lf	Isoprenylphloro-glucinol	**2274**, **2275**	[750]
Melicope elleryana	CQ, SEQ	lf	Acylphloroglucinol	**2255**, **2258**	[749]
	SEQ	fr	Flavonoid	**2284**, **2288–2291**	[749]
	PNG	lf	Furanoquinoline	**2072**, **2077**, **2087**	[705]
	NQ	lf	Isoprenylphloroglucinol	**2273**	[753]
Melicope hayesii	SEQ	lf, st	Amide	**2303**	[733]
	SEQ	lf, st	Coumarin	**2182**, **2191**, **2206**	[733]
	SEQ	lf, st	Lignan	**2336–2338**	[733]
	SEQ	lf, st	Pyrrolidine	**2364**	[733]
Melicope micrococca	NNSW	lf, st	Acridone	**2105**	[712]
	SEQ	st	Amide	**2302**	[39]
	NNSW	lf, st	Flavonoid	**527**, **2292**, **2294**	[712]
	SEQ	st	Furanoquinoline	**2094**	[39]

(continued)

Table 39 (continued)

Species	Collection location[a,b]	Plant part(s)[c]	Compound class	Compound(s)	Refs.
	NNSW	lf, st	Inositol	**2363**	[712]
	NNSW	lf, st	Lignan	**2335**, **2338**	[712]
	NNSW	lf, st	Steroid	**2350**	[712]
Melicope vitiflora	NQ	lf	Coumarin	**2192**, **2193**	[734]
Melicope xanthoxyloides	CQ, NQ	st, b	Acridone	**2103**, **2104**, **2109**, **2111**	[39, 700]
	NQ	b	Anthranilic acid	**2128**	[39]
	CY, NQ	lf, st, b	Furanoquinoline	**2066**, **2068**, **2069**, **2071**, **2078**, **2081**, **2085**, **2086**	[39, 700]
Micromelum minutum	NQ	lf	Coumarin	**2198**, **2203**	[735]
Pentaceras australis	NNSW	lf, st	Anthraquinone dimer	**2340**	[710]
	SEQ	b	β-Carboline	**2160–2162**	[720]
	NNSW	lf, st	Coumarin	**2198**	[710]
	NNSW	lf, st	Pyranocoumarin	**2240**	[710]
Phebalium squamulosum subsp. *verrucosum*	NSW	lf, st	Sesquiterpenoid	**2360**	[770]
	ACT	ap	Triterpenoid	**2348**	[765]
Pitaviaster haplophyllus	NQ	r, st, b, lf	Acridone	**2105**	[701, 706]
	NQ	r, st, b, lf	Furanoquinoline	**2057**, **2059**, **2063**, **2064**, **2073–2076**, **2081**, **2090**	[701, 706]
Sarcomelicope simplicifolia subsp. *simplicifolia*	SEQ	b, lf	Acridone	**2103**, **2107–2110**, **2112–2117**, **2125–2127**	[39, 675, 702, 703, 711]
	SEQ	b	Cinnamate	**2300**, **2301**	[757]
	SEQ	lf, st, b	Furanoquinoline	**2054**, **2081**, **2087**, **2089**, **2091**	[39, 702, 703]
	SEQ	b	Quinolone	**2025**	[675]
	SEQ	b	Triterpenoid	**801**	[757]

(continued)

Table 39 (continued)

Species	Collection location[a,b]	Plant part(s)[c]	Compound class	Compound(s)	Refs.
Zanthoxylum brachyacanthum	NQ, SEQ	wp, b	Aporphine	**634**, **636**	[39, 723]
	CQ	st	β-Carboline	**2152**	[39]
	SEQ, CQ, NQ	b, st, wp	Benzophenanthridine	**2168**, **2169**, **2171**, **2174**	[39, 723]
	SEQ	b	Berberine	**2180**	[723]
	CQ	b	BIA	**2179**	[39]
Zanthoxylum nitidum	NQ	lf, st	Aporphine	**187**	[722]
	NQ	lf, st	Benzophenanthridine	**2165**–**2167**, **2170**, **2171**, **2176**, **2177**	[722]
	NQ	lf, st	Flavonoid	**2278**	[722]
	NQ	lf, st	Lignan	**2013**, **2339**	[722]
	NQ	lf, st	Triterpenoid	**590**	[722]
Zanthoxylum ovalifolium	NQ	B	β-Carboline	**2161**	[704]
	NQ	B	Benzophenanthridine	**2171**–**2173**, **2175**	[39, 704]
	NQ	B	Furanoquinoline	**2092**	[704]
Zanthoxylum rhetsa	CY	wp, b	Benzophenanthridine	**2169**, **2171**, **2178**	[39]
	CY	B	Berberine	**2181**	[39]
	CY	B	BIA	**2179**	[39]
Zieria smithii	SEQ	wp	Monoterpenoid	**2359**	[769]

[a] SEQ = Southeast Queensland, NQ = North Queensland, CQ = Central Queensland, CY = Cape York, Queensland, NNSW = Northern New South Wales, NSW = New South Wales, ACT = Australian Capital Territory, Tas = Tasmania, NT = Northern Territory, PNG = Papua New Guinea, INDO = Indonesia, cult. = cultivated, [b] natural location shown for those species studied from cultivated material, [c] lf = leave, ap = aerial parts, r = roots, fr = fruits, wp = whole plant, b = bark, w = wood, st = stems

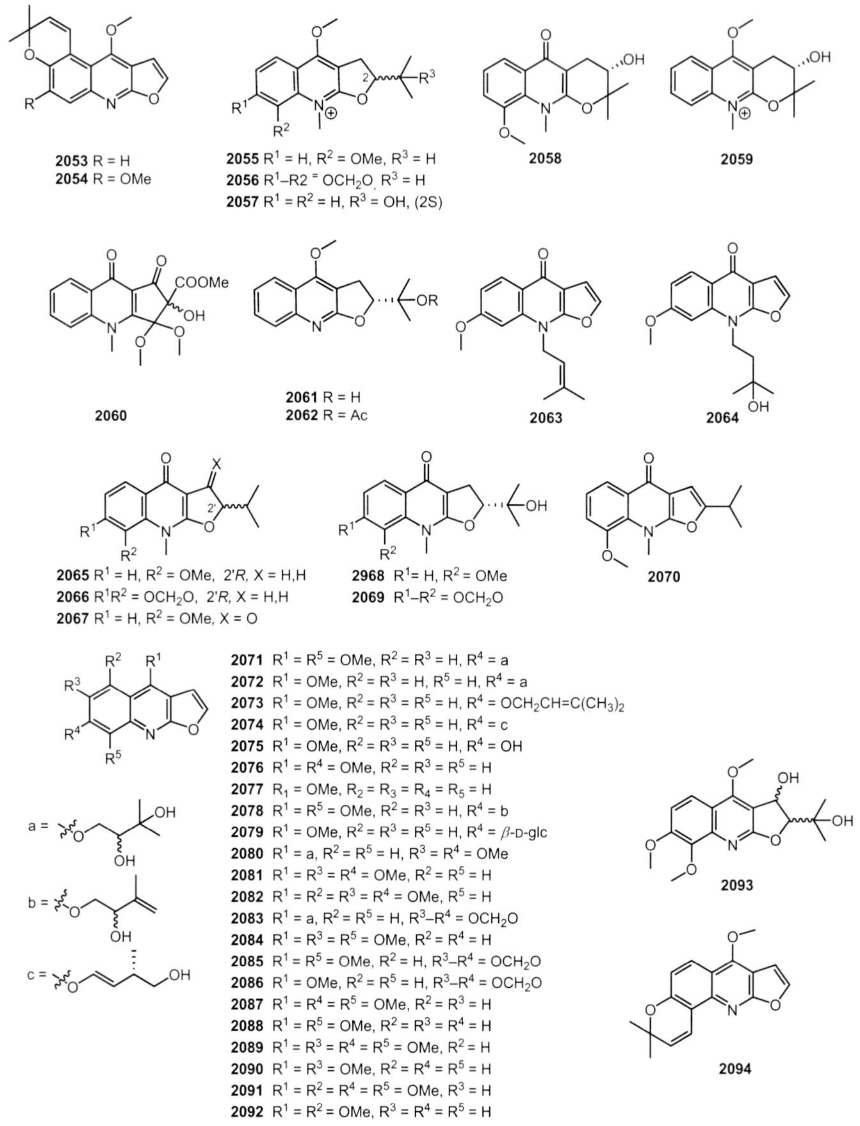

Fig. 152 Structures of furanoquinoline alkaloids **2053–2094** reported from Rutaceae species

5-hydroxygraveroline (**2052**), have been reported from *Lunasia amara* collected in Cape York, in the far north of Queensland or from Japan [142, 681, 682]. The heterocyclic ring in the quinolones is often substituted with an isoprene or monoterpenoid unit attached at C-3. The quinolones are restricted to *Boronia, Eriostemon, Flindersia* (Plate 32), *Geijera, Halfordia, Lunasia,* and *Sarcomelicope* species (all clade 4C, Zanthoxyloideae). Six of the quinolones have methyl, propyl, pentyl or phenyl side

Plate 33 Pink flowers of *Melicope elleryana* (top row), *M. rubra* fruit (bottom left), *Halfordia kendack* leaves (bottom middle) and *Bosistoa transversa* vegetation (bottom right). (Photographs AR Carroll)

chains attached to C-2. *Eriostemon* and *Halfordia* are the only genera that have the monoterpenoid, terpineol, conjugated to the quinolone at C-3.

The isoprenes attached at C-3 of the quinolones can undergo further cyclization with the oxygen at C-2 to generate a furan ring and, in most cases, this leads to elimination of the isopropyl sidechain to generate a furanoquinoline (Fig. 152). Furanoquinoline alkaloids are unique to the Rutaceae and Australian rainforest species (*Medicosma*, *Lunasia*, *Flindersia*, *Geijera*, *Melicope*, *Halfordia*, *Brombya*, *Sarcomelicope*, *Pitaviastia*, and *Zanthoxylum*), which have yielded 42 representatives of this class (medicosmine (**2053**), acronidine (**2054**), lunasine (**2055**), (−)-*O*-methylluninium base (**2056**), *N*-methylplatydesminium (**2057**), (−)-lunacrinol (**2058**), tabouensinium (**2059**), fareanine (**2060**), platydesmine (**2061**), platydesmine acetate (**2062**), acrophylline (**2063**), acrophyllidine (**2064**), lunacrine (**2065**), lunine (**2066**), 3-oxolunacrine (**2067**), balfourodine (**2068**), hydroxylunine (**2069**), 2,3-dehydrolunacrine (**2070**), evoxine (**2071**), evellerine (**2072**), melineurine (**2073**), leptanoine D (**2074**), 7-hydroxydictamnine (**2075**), evolitrine (**2076**), dictamnine (**2077**), evodine (**2078**), 4-methoxyfuro[2,3-*b*]quinolin-7-yl β-D-mannopyranoside (**2079**), 1-[(6,7-dimethoxyfuro[2,3-*b*]quinolin-4-yl)oxy]-3-methyl-2,3-butanediol (**2080**), kokusaginine (**2081**), 5,6,7-trimethoxydictamnine (**2082**), maculosine (**2083**), maculosidine (**2084**), flindersiamine (**2085**), maculine (**2086**), skimmianine (**2087**), fagarine (**2088**), halfordinine (**2089**), pteleine (**2090**),

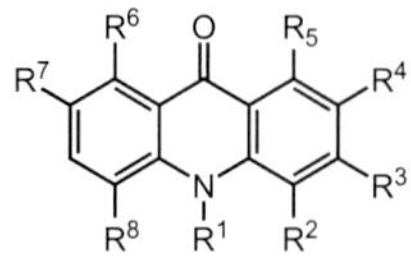

2095 $R^1 = R^2 = R^3 = R^4 = R^6 = R^7 = R^8 = H, R^5 = OH$
2096 $R^1 = Me, R^2 = R^3 = R^4 = R^6 = R^7 = R^8 = H, R^5 = OH$
2097 $R^1 = R^2 = R^3 = R^4 = R^6 = R^8 = H, R^5 = R^7 = OH$
2098 $R^1 = Me, R^2 = R^4 = R^6 = R^8 = H, R^3 = R^5 = R^7 = OH$
2099 $R^1 = R^2 = R^3 = R^4 = R^7 = R^8 = H, R^5 = R^6 = OH$
2100 $R^1 = Me, R^2 = R^3 = R^4 = R^7 = R^8 = H, R^5 = R^6 = OH$
2101 $R^1 = Me, R^2 = R^3 = R^4 = R^7 = R^8 = H, R^5 = R^6 = OH$
2102 $R^1 = Me, R^2 = R^4 = R^7 = H, R^3 = OMe, R^5 = R^6 = R^8 = OH$
2103 $R^1 = R^6 = R^7 = R^8 = H, R^2 = R^5 = OMe, R^3–R^4 = OCH_2O$
2104 $R^1 = Me, R^2 = R^5 = OMe, R^3–R^4 = OCH_2O, R^6 = R^7 = R^8 = H$
2105 $R^1 = Me, R^2 = R^6 = R^7 = R^8 = H, R^3 = R^4 = OMe, R^5 = OH$
2106 $R^1 = Me, R^2 = R^3 = R^5 = OMe, R^4 = R^6 = R^7 = R^8 = H$
2107 $R^1 = Me, R^2–R^3 = OCH_2O, R^4 = R^5 = OMe, R^6 = R^7 = R^8 = H$
2108 $R^1 = Me, R^2 = R^3 = R^4 = R^5 = OMe, R^6 = R^7 = R^8 = H$
2109 $R^1 = Me, R^2 = R^5 = OMe, R^3–R^4 = OCH_2O, R^6 = R^7 = R^8 = H$
2110 $R^1 = Me, R^2 = OMe, R^3–R^4 = OCH_2O, R^5 = OH, R^6 = R^7 = R^8 = H$
2111 $R^1 = Me, R^2 = R^6 = R^7 = R^8 = H, R^3–R^4 = OCH_2O, R^5 = OMe$
2112 $R^1 = Me, R^2 = R^3 = R^4 = OMe, R^5 = OH, R^6 = R^7 = R^8 = H$
2113 $R^1 = Me, R^2 = R^4 = R^6 = R^7 = R^8 = H, R^3 = R^5 = OMe$
2114 $R^1 = Me, R^2–R^3 = OCH_2O, R^4 = OMe, R^5 = OH, R^6 = R^7 = R^8 = H$
2115 $R^1 = Me, R^2 = R^4 = OMe, R^3 = R^5 = R_6 = R^7 = R^8 = H$
2116 $R^1 = Me, R^2 = R^4 = R^5 = OMe, R^3 = R^6 = R^7 = R^8 = H$
2117 $R^1 = Me, R^2 = R^3 = R^4 = OMe, R^5 = R^6 = R^7 = R^8 = H$

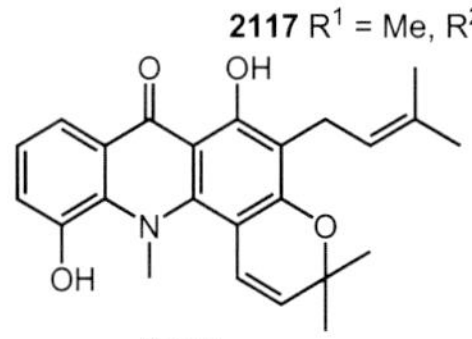

2118

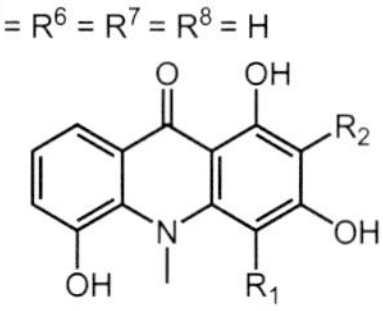

2119 $R^1 = H, R^2 = a$
2120 $R^1 = R^2 = a$
2121 $R^1 = a, R^2 = b$
2122 $R^1 = H, R^2 = b$

a =, b =

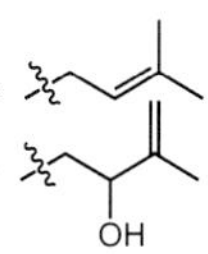

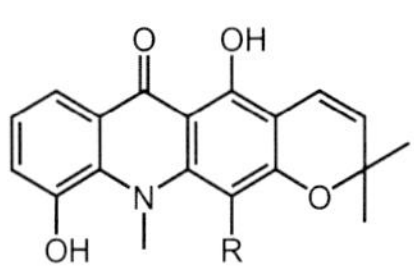

2123 R = b
2124 R = H

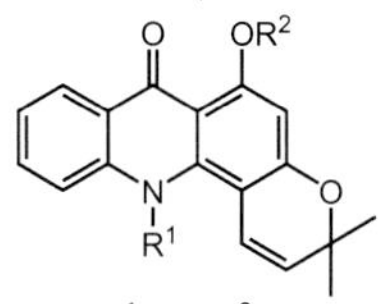

2125 $R^1 = H, R^2 = Me$
2126 $R^1 = R^2 = Me$
2127 $R^1 = Me, R^2 = H$

Fig. 153 Structures of acridone alkaloids **2095–2127** reported from Rutaceae species

acronycidine (**2091**), 5-methoxydictamnine (**2092**), 7,8-dimethoxymyrtopsine (**2093**), and dutadrupine (**2094**)) [39, 670, 683–706].

Acridones are also commonly encountered (Fig. 153), and 33 congeners (1-hydroxy-9(10*H*)-acridinone (**2095**), 1-hydroxy-10-methyl-9(10*H*)-acridinone

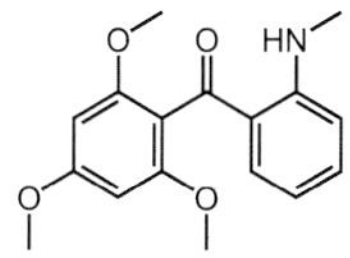

2128

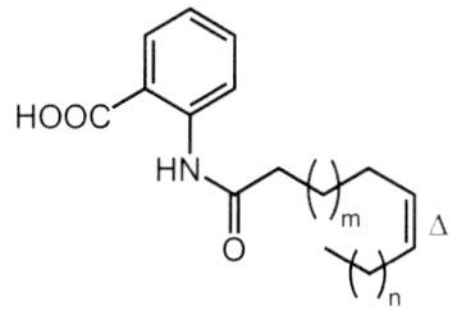

2129 m = 7, n = 3
2130 m = 5, n = 5
2131 m = 3, n = 7
2132 m = 7, n = 3, Δ saturated

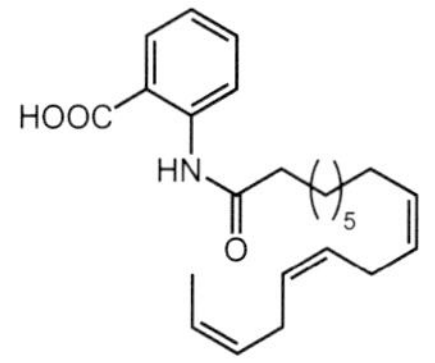

2133

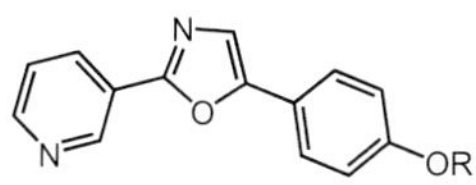

2134 R = H

2135 R = 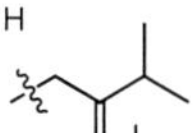

2136 R = 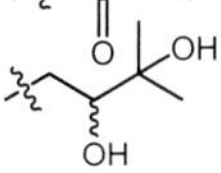

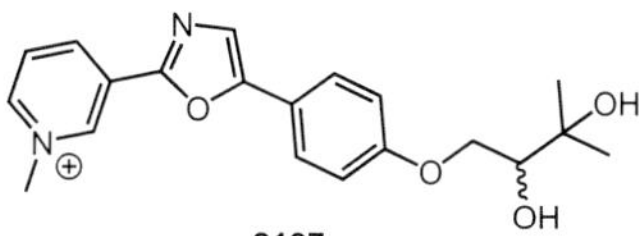

2137

Fig. 154 Structures of anthranilic acid-derived compounds **2128**–**2133** and oxazoles **2134**–**2137** reported from Rutaceae species

(**2096**), 1,7-dihydroxy-9(10*H*)-acridinone (**2097**), 1,3-dihydroxy-10-methyl-9(10*H*)-acridinone (**2098**), 1,8-dihydroxy-9(10*H*)-acridinone (**2099**), 1,8-dihydroxy-10-methyl-9(10*H*)-acridinone (**2100**), 1,3,8-trihydroxy-10-methyl-9(10*H*)-acridinone (**2101**), citrusamine (**2102**), xanthevodine (**2103**), melicopidin (**2104**), arborinine (**2105**), 1,3,4-trimethoxy-10-methyl acridone (**2106**), melicopine (**2107**), melicopicine (**2108**), melicopidine (**2109**), normelicopidine (**2110**), evoxanthine (**2111**), melicopicine (**2112**), 2,4-dimethoxy-10-methylacridone (**2113**), normelicopine (**2114**), 1,3-dimethoxy-10-methylacridone (**2115**), 1,3,4-trimethoxy-10-methyl acridone (**2116**), 1,2,3-trimethoxy-10-methyl acridone (**2117**), *N*-methylataphyllinin (**2118**), junosine (**2119**), *N*-methylataphylline (**2120**), bosistine (**2121**), bosistidine (**2122**), 4-(2-hydroxy-3-methylbut-3-enyl)-yukocitrine (**2123**), yukocitrine (**2124**), *des*-*N*-methyl acronycine (**2125**), acronycine (**2126**), and noracronycine (**2127**)), have been reported from Australian rainforest species (*Boronia*, *Bosistoa*, *Acronychia*, *Pitaviastia*, *Melicope, Medicosma*, and *Sarcomelicope* (Plate 33)) [675, 699–703, 706–712]. Hydroxylation on either phenyl ring occurs, although in most cases one of the phenyl groups is unsubstituted, in keeping with an anthranilic acid biogenesis. Ten acridones contain one or two isoprene groups attached at C-2 and/or C-4.

A 2-aminobenzophenone, tecleanone (**2128**), has been found in *Melicope xanthoxyloides* [39]. Long-chain fatty acid amides of anthranilic acid include 11′-hexadecenoylanthranilic acid (**2129**), 9′-hexadecenoylanthranilic acid (**2130**), 7′-hexadecenoylanthranilic acid (**2131**), hexadecanoylanthranilic acid (**2132**), and 9,12,15-octadecatrienoylanthranilic acid (**2133**). Amides have been reported also

Plate 34 *Flindersia pimenteliana* with purple flowers (left, unusual in the genus *Flindersia*), close-up of flowers of *F. brayleyana*, (top middle) *F. schottiana* (top right), *F. pimenteliana* (bottom middle) and *F. australis* (bottom right). (Photographs AR Carroll)

from *Geijera parviflora* (Fig. 154) [713]. Four 2-pyridyl-5-phenyloxazoles, halfordinone (**2134**), halfordinol (**2135**), halfordine (**2136**), and *N*-methylhalfordinium chloride (**2137**), have been reported from *Halfordia kendack* and *H. scleroxyla* [39, 694, 714].

Flindersia (a genus of 17 species restricted to continental Australia, and one species in New Caledonia, Clade C4c (Plate 34)), has yielded potently antiplasmodial isoprenylated bisindoles of the borreverine class (Fig. 155) (isoborreverine (**2138**), dimethylisoborreverine (**2139**), 10,10-dimethoxydimethylisoborreverine (**2140**), 10-methoxydimethylisoborreverine (**2141**), 10′-methoxydimethylisoborreverine (**2142**), 4-methylborreverine (**2143**), and borreverine (**2144**)) [39, 715–717]. Borreverine was originally isolated from the unrelated Rubiaceae species, *Spermacoce verticillata* (syn. *Borreria verticillata*), suggesting a polyphyletic origin for this alkaloid structural class [718]. Since then, seven congeners have been reported from *Flindersia* species. Three members of a related class, flinderoles A–C (**2145–2147**), have been reported from Australian and Papua New Guinean species [717]. Other isoprenylated indoles have also been found in *Flindersia* species, including the glycosylated pimentelamines A–C (**2148–2150**) and 2-isoprenyl-*N*,*N*-dimethyltryptamine (**2151**) [715].

β-Carboline alkaloids have been reported, with rutaecarpine (**2152**), 1-hydroxyrutaecarpine (**2153**), 1,2-dihydroxyrutaecarpine (**2154**), and bouchardatine (**2155**) found in *Bouchardatia neurococca* [683], and *N*-[2-(2-acetylindol-3-yl)ethyl]acetamide (**2156**), harmalan (**2157**), acetylharmalan (**2158**), and acetylharmalan dimer (**2159**) isolated from *Flindersia laevicarpa* var. *laevicarpa* from north Queensland [719].

Pentaceras and *Zanthoxylum* species (both from Clade C4d) have also yielded β-carboline alkaloids, including 5-methoxycanthinone (**2160**), 6-canthinone (**2161**) and 4-methylthiocanthin-6-one (**2162**) [704, 720]. The final indole derivative found in Australian rainforest Rutaceae species is *N*,*N*-dimethyl-5-hydroxy-tryptamine (**2163**), from *Flindersia bennettii* collected in southeast Queensland [39]. The

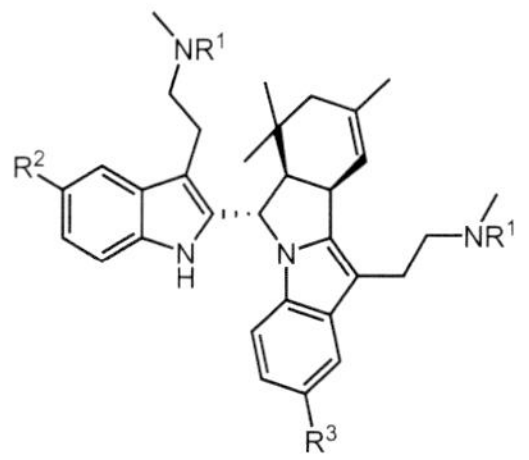

2138 $R^1 = R^2 = R^3 = H$
2139 $R^1 = Me$, $R^2 = R^3 = H$
2140 $R^1 = Me$, $R^2 = R^3 = OMe$
2141 $R^1 = Me$, $R^2 = OMe$, $R^3 = H$
2142 $R^1 = Me$, $R^2 = H$, $R^3 = OMe$

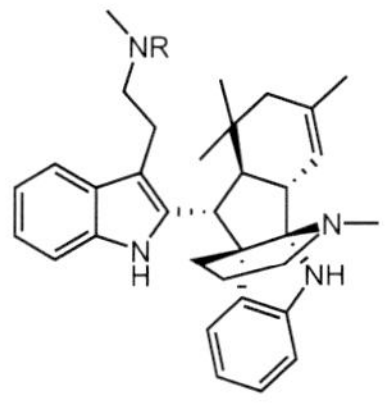

2143 R = Me
2144 R = H

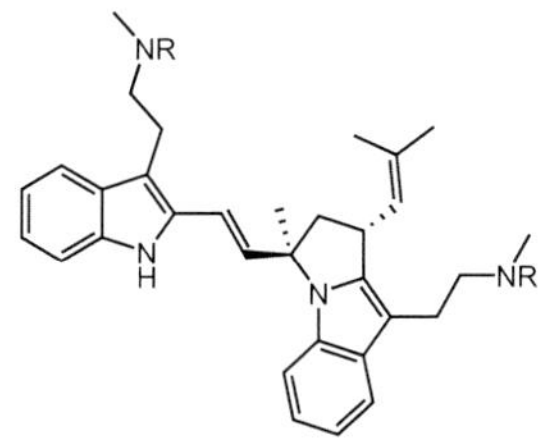

2145 R = H
2146 R = Me

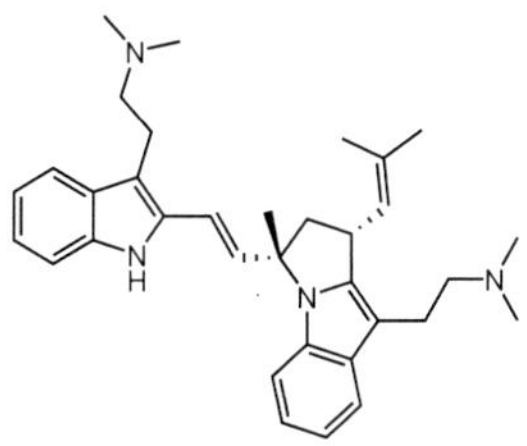

2147

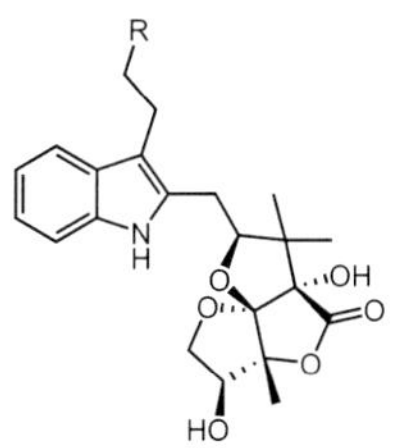

2148 R = $N(CH_3)_2$
2149 R = $NHCH_3$
2150 R = $N^+(CH_3)_2OH$

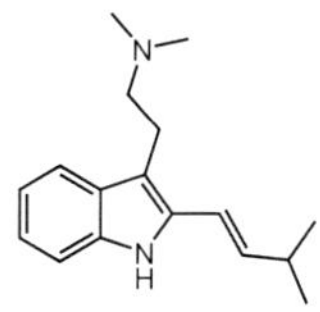

2151

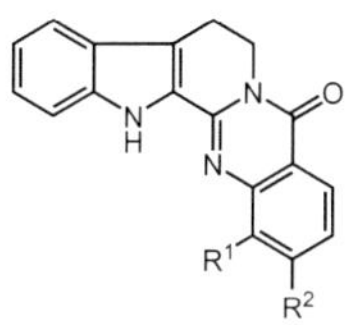

2152 $R^1 = R^2 = H$
2153 $R^1 = OH$, $R^2 = H$
2154 $R^1 = R^2 = OH$

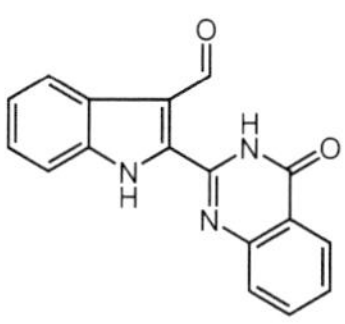

2155

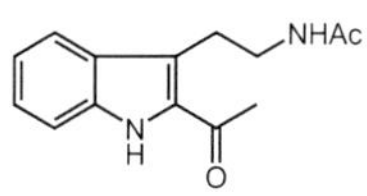

2156

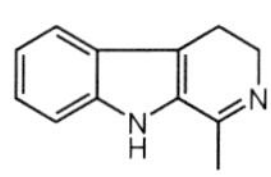

2157

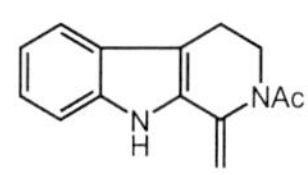

2158

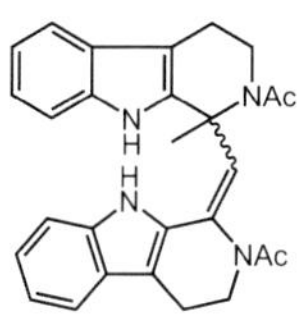

2159

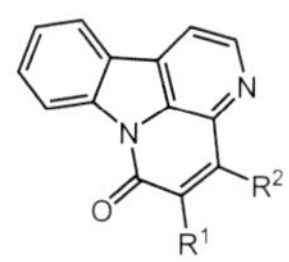

2160 $R^1 = OMe$, $R^2 = H$
2161 $R^1 = R^2 = H$
2162 $R^1 = H$, $R^2 = SMe$

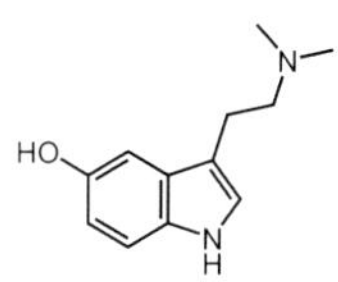

2163

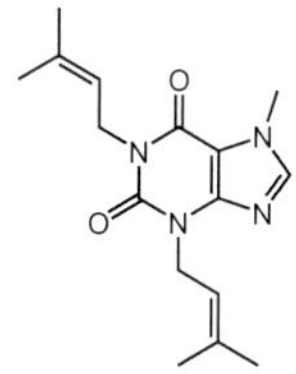

2164

Fig. 155 Structures of indole alkaloids **2134–2163** and xanthine **2164** reported from Rutaceae species

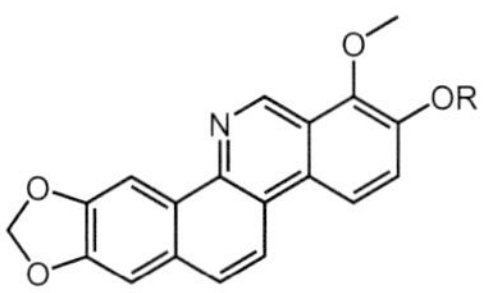

2165 R = H
2166 R = Me
2167 R = Ac

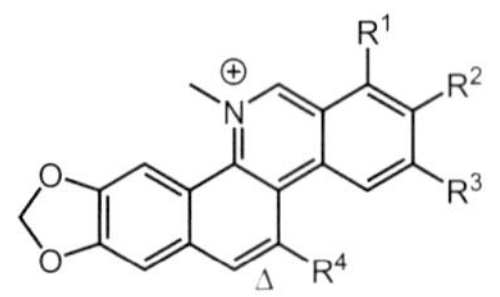

2169 $R^1 = R^2 = OMe, R^3 = R^4 = H$
2170 $R^1 = OH, R^2 = R^4 = OMe, R^3 = H$, Δ saturated
2171 $R^1 = R^4 = H, R^2 = R^3 = OMe$
2172 $R^1 = R^4 = H, R^2 = OH, R^3 = OMe$
2173 $R^1 = R^4 = H, R^2 = OMe, R^3 = OH$
2174 $R^1 = OMe, R^2 = OH, R^3 = R^4 = H$
2175 $R^1 = R^4 = H, R^2–R^3 = OCH_2O$

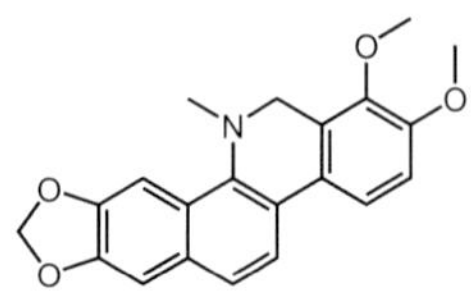

2176

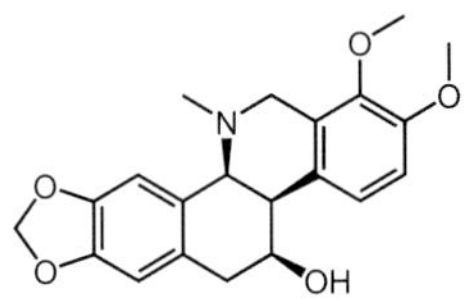

2168

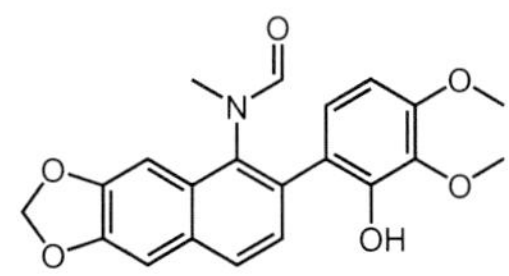

2177

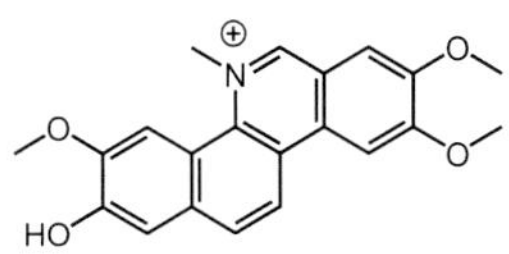

2178

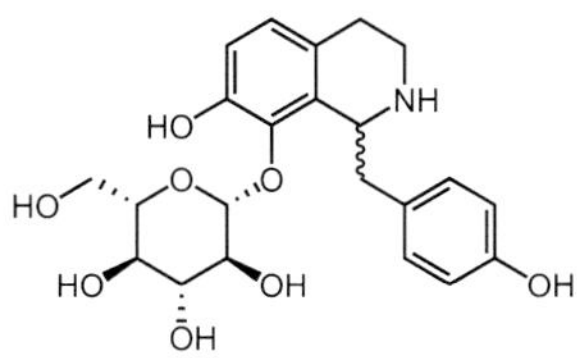

2179

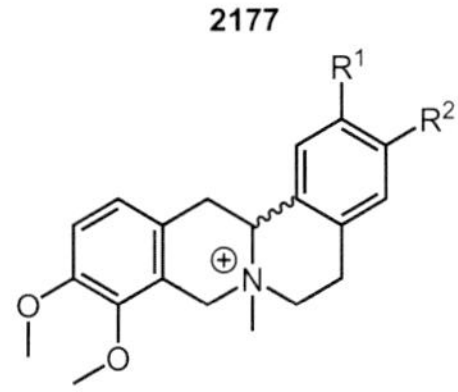

2180 $R^1–R^2 = OCH_2O$
2181 $R^1 = OMe, R^2 = OH$

Fig. 156 Structures of BIAs **2165–2181** reported from *Zanthoxylum* species

isoprenylated xanthine 1,3-di-(3-methylbut-2-enyl)-7-methylxanthine (**2164**) was reported from *Bosistoa medicinalis* from northern New South Wales [721].

Benzophenanthridine alkaloids (Fig. 156) (decarine (**2165**), norchelerythrine (**2166**), decarine acetate (**2167**), homo-chelidonine (**2168**), chelerythrine (**2169**), 6-methoxy-5,6-dihydrofagaridine (**2170**), nitidine (**2171**), terihanine (**2172**), isoterihanine (**2173**), fagaridine (**2174**), avicine (**2175**), dihydrochelerythrine (**2176**), arnottianamide (**2177**), and fagaronine (**2178**)) have been reported exclusively from *Zanthoxylum* species [39, 704, 722, 723]. In addition, three aporphine alkaloids, **187**, **634**, and **636**, a benzylisoquinoline alkaloid ((1*R*)-1,2,3,4-tetrahydro-7-hydroxy-1-[(4-hydroxyphenyl)methyl]-8-isoquinolinyl-*β*-D-glucopyranoside (**2179**), and two berberine alkaloids, cadanine methiodide (**2180**) and β-*N*-methylcorypalmine (**2181**), are also present in *Zanthoxylum* species [39, 722, 723].

Flindersia species are generally accepted to have evolved in rainforests, but four of these species are now restricted to semi-arid and arid areas of Australia. A study showed that alkaloid diversity increased along an aridity gradient and the radiation in alkaloid diversity and structure classes along this gradient is likely to have been driven by now extinct megafauna herbivory and water scarcity [724].

Coumarins that are widespread throughout all subfamilies in the Rutaceae are also commonly encountered in Australian rainforest species (Fig. 157). These compounds are always oxygenated at C-7 and mostly contain an isoprenyl or geranyl side chain either directly attached at C-6 and/or C-8 or ether linked

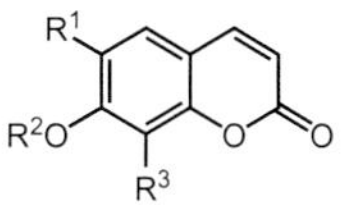

2182 $R^1 = R^2 = R^3 = H$
2183 $R^1 = R^2 = H$, $R^3 = c$
2184 $R^1 = b$, $R^2 = R^3 = H$, (2′*S*)
2185 $R^1 = b$, $R^2 = R^3 = H$,
2186 $R^1 = R^2 = H$, $R^3 = a$
2187 $R^1 = a$, $R^2 = R^3 = H$
2188 $R^1 = a$, $R^2 = Me$, $R^3 = H$
2189 $R^1 = OMe$, $R^2 = R^3 = a$
2190 $R^1 = d$, $R^2 = Me$, $R^3 = H$
2191 $R^1 = R^3 = H$, $R^2 = a$
2192 $R^1 = R^3 = H$, $R^2 = f$
2193 $R^1 = R^3 = H$, $R^2 = e$
2194 $R^1 = H$, $R^2 = Me$, $R^3 = g$
2195 $R^1 = h$, $R^2 = Me$, $R^3 = H$
2196 $R^1 = OMe$, $R^2 = a$, $R^3 = i$
2197 $R^1 = OMe$, $R^2 = a$, $R^3 = j$
2198 $R^1 = H$, $R^2 = Me$, $R^3 = a$
2199 $R^1 = H$, $R^2 = c$, $R^3 = OMe$
2200 $R^1 = R^3 = H$, $R^2 = c$
2201 $R^1 = R^3 = H$, $R^2 = k$
2202 $R^1 = OMe$, $R^2 = k$, $R^3 = H$
2203 $R^1 = l$, $R^2 = Me$, $R^3 = H$
2204 $R^1 = R^3 = H$, $R^2 = m$
2205 $R^1 = R^3 = H$, $R^2 = n$
2206 $R^1 = R^3 = H$, $R^2 = o$
2207 $R^1 = p$, $R^2 = R^3 = H$
2208 $R^1 = q$, $R^2 = R^3 = H$

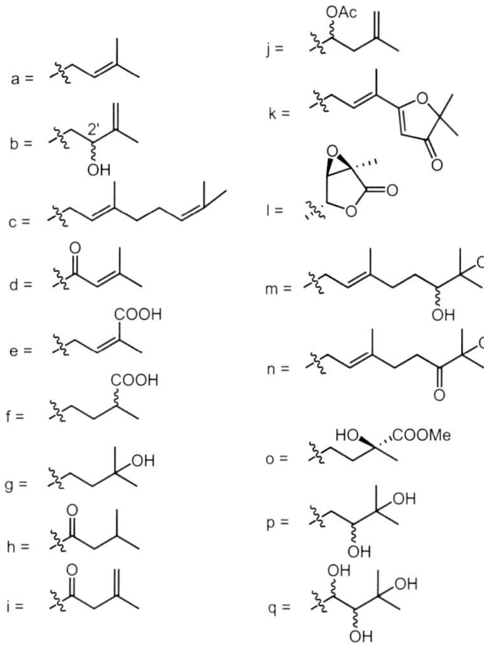

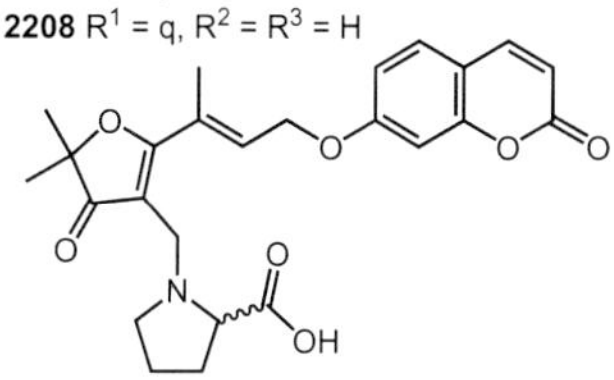

2209

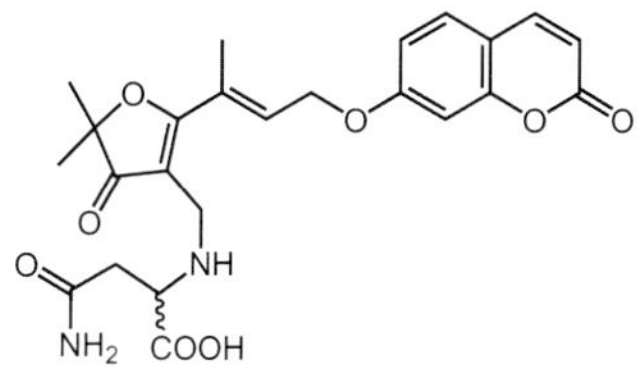

2210

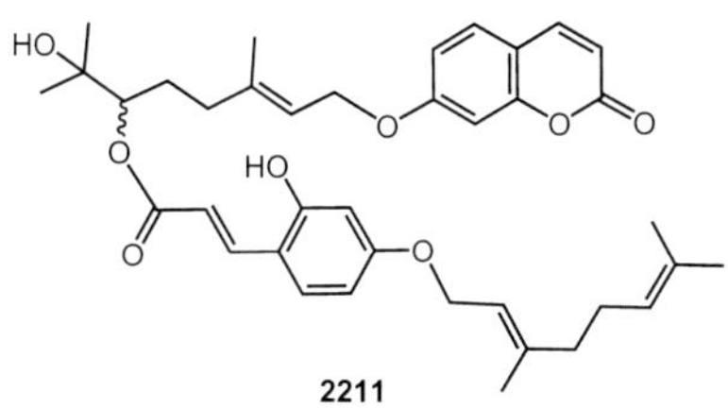

2211

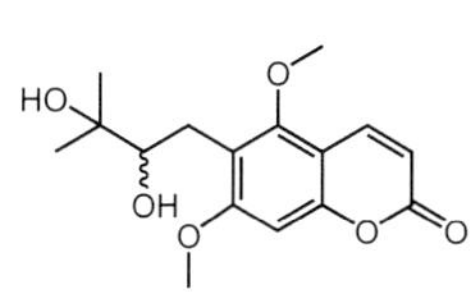

2212

Fig. 157 Structures of coumarins **2182–2212** reported from Rutaceae species

Plate 35 *Micromelum minutum* shrub (left), close-up of its flowers (middle), female *Papilio aegeus* (Papilionidae), a butterfly that uses *M. minutum* as a food plant for its caterpillars (right), pupa on the stem of *M. minutum* (inset). (Photographs AR Carroll)

at O-7. There are a large number of variants of this class reported from Australian rainforest Rutaceae from the genera *Boronia*, *Brombya*, *Flindersia*, *Geijera*, *Melicope*, and *Micromelum* (Plate 35). These compounds include umbelliferone (**2182**), 7-hydroxy-8-geranylcoumarin (**2183**), 7-hydroxy-6-2*R*-hydroxy-3-methylbut-3-enyl)coumarin (**2184**), demethyltamarin (**2185**), osthenol (**2186**), demethylsuberosin (**2187**), suberosin (**2188**), brayleyanin (**2189**), dehydrogeijerin (**2190**), 7-(3-methylbut-2-enoxy)umbelliferone (**2191**), 7-(3′,3′-dimethylallyloxy)-coumarin (**2192**), 7-(3′-carboxy-2-butenoxy)-coumarin (**2193**), herniarin (**2194**), geijerin (**2195**), 6-methoxy-8-(3-methyl-2-oxobut-3-en-1-yl)-7-((3-methylbut-2-en-1-yl)oxy)-2*H*-chromen-2-one (**2196**), 1-(6-methoxy-7-((3-methylbut-2-en-1-yl)oxy)-2-oxo-2*H*-chromen-8-yl)-3-methylbut-3-en-2-yl acetate (**2197**), osthol (**2198**), collinin (**2199**), 7-geranyloxycoumarin (**2200**), geiparvarin (**2201**), 6-methoxylgeiparvarin (**2202**), micromelum (**2203**), marmin (**2204**), 6′-dehydromarmin (**2205**), (+)-7-(3-methyl-4-carboxybutanoxy)umbelliferone methyl ester (**2206**), peucedanol (**2207**), 6-(1′,2′,3-trihydroxy-3-methylbutyl)-7-hydroxy-2*H*-1-benzopyran-2-one (**2208**), parvifloranine B (**2209**), parvifloranine A (**2210**), 2-hydroxy-2,6-dimethyl-8-((2-oxo-2*H*-chromen-7-yl)oxy)oct-(6*E*)-en-3-yl (*E*)-3-(4-((3,7-dimethylocta-(2*E*),6-dien-1-yl)oxy)-2-hydroxyphenyl)acrylate (**2211**), and toddalolactone (**2212**) [39, 673, 684, 725–735].

Analogous to the quinolones, further cyclization of an isoprene unit attached at C-6 through an ether linkage to O-7 in the coumarins generates a furan that often eliminates the isopropyl group. There are 18 furanocoumarins reported from Australian rainforest *Boronia*, *Citrus*, *Coatesia*, *Dinosperma*, *Flindersia*, *Halfordia*, and *Pentaceras* species (Fig. 158), namely, garracoumarin D (**2213**), bergapten (**2214**), xanthotoxin (**2215**), isopimpinellin (**2216**), halkendin (**2217**), halfordinin (**2218**), isohalfordin (**2219**), halfordin (**2220**), garracoumarin C (**2221**), garracoumarin A (**2222**), garracoumarin B (**2223**), garracoumarin D

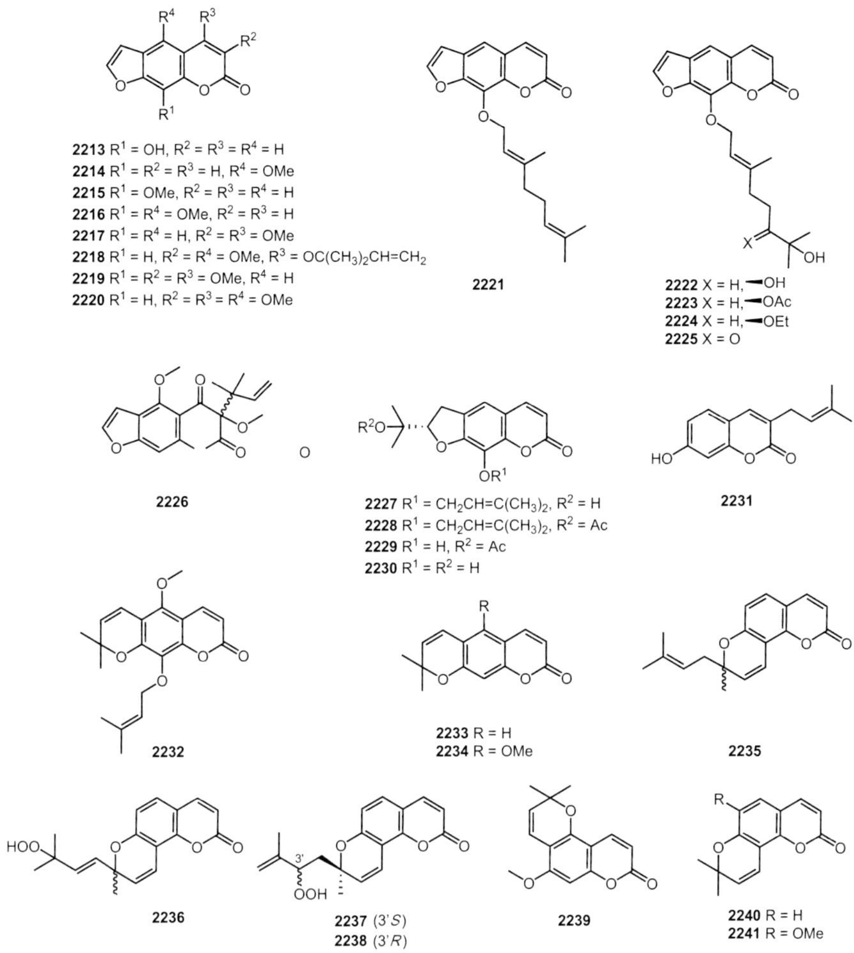

Fig. 158 Structures of coumarins **2213–2241** reported from Rutaceae species

(**2224**), garracoumarin C (**2225**), 3,5-dimethoxy-3-(1,1-dimethylprop-2-enyl)-3,4-dihydropsoralen-4-one (**2226**), garracoumarin B (**2227**), garracoumarin A (**2228**), garracoumarin E (**2229**), and garracoumarin E (**2230**). Pyranocoumarins are also commonly encountered, embracing 3-(3-methylbut-2-enyl)umbelliferone (**2231**), 8-*O*-(3-methylbut-2-enyl)xanthoxyletin (**2232**), xanthyletin (**2233**), xanthoxyletin (**2234**), 5-deoxyprotobruceol-1 (**2235**), 5-deoxyprotobruceol-II hydroperoxide (**2236**), 5-deoxyprotobruceol-III-hydroperoxide 1 (**2237**), 5-deoxyprotobruceol-III-hydroperoxide 2 (**2238**), alloxanthoxyletin (**2239**), seselin (**2240**), and braylin (**2241**) [686, 690, 691, 710, 725, 726, 735–741].

Plate 36 Flowers of *Acronychia* species: *A. imperforata* (top and bottom left), *A. littoralis* (top middle and right), and *Zieria smitthii* (bottom middle and right). (Photographs AR Carroll)

Phloroglucinol acetophenones (Fig. 159) (34 natural products: xanthoxylin (**2242**), methyl xanthoxylin (**2243**), 6-demethylacronylin (**2244**), acronyculatin A (**2245**), methyl-acronylin (**2246**), acronyol B (**2247**), acronyol A (**2248**), crassipetalone A (**2249**), crassipetalonol A (**2250**), acronylin (**2251**), furostipitol (**2252**), evodionol (**2253**), *O*-methylevodionol (**2254**), evodione (**2255**), 3-hydroxydihydropyranostipitol ethyl ether (**2256**), 3,4-dihydroxydihydropyranostipitol (**2257**), alloevodione (**2258**), *O*-methyl-*allo*-evodionol (**2259**), sessiliflorene (**2260**), sessiliflorol A (**2261**), sessiliflorol B (**2262**), franklinone (**2263**), acrotrione (**2264**), pentacoccol (**2265**), pentacoccol-5-methyl ether (**2266**), eriostoic acid (**2267**), acrovestone (**2268**), 5-acetoxy-3,4-dihydro-3ξ,7-dihydroxy-6-(3,7-dimethyloct-2,5-dienyl)-1,2-dimethyl-[2*H*]-benzopyran (**2269**), 2-(1-hydroxyisopropyl)furanopentacoccol (**2270**), furanopentacoccol (**2271**), eriostemoic acid (**2272**), muellitol (**2273**), melicopol (**2274**), and methylmelicopol (**2275**)), substituted by one or two isoprene units, have been reported from *Acronychia* (Plate 36), *Melicope*, *Medicosma* (Clade 4f), and *Bosistoa*, *Coatesia*, *Dinosperma*, and *Acradenia* (Clade 4d) species [737, 742–753]. Two of these are dimers. These compounds are polyketides derived from condensation of four malonates, followed by isoprenylation.

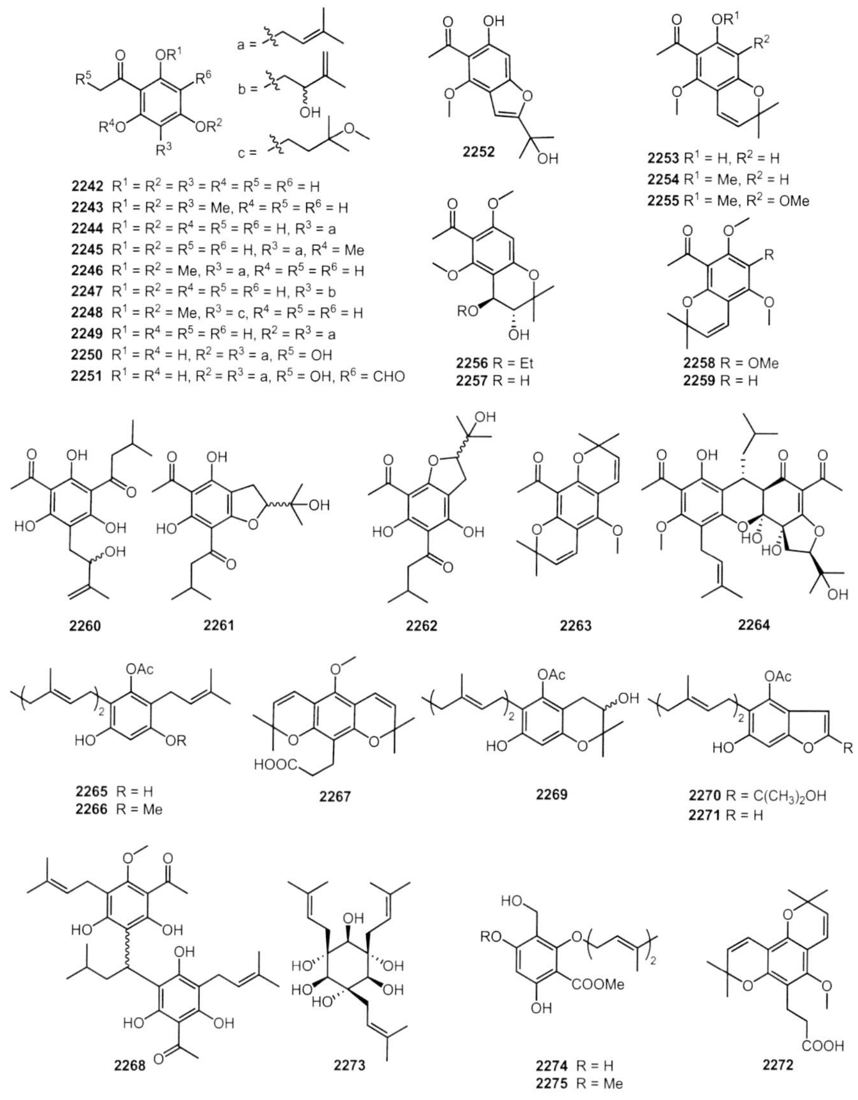

Fig. 159 Structures of phloroglucinol derivatives **2242–2275** reported from Rutaceae species

Flavonoids (23 natural products (Fig. 160), predominately isoprenylated: **527**, **588**, 5,7-dihydroxy-4′-methoxy-8-(2-hydroxy-3-methylbut-3-enyl)flavanone (**2276**), 5,7-dihydroxy-4′-methoxy-8-(3-methylbut-2-enyl)flavanone (**2277**), hesperidin (**2278**), dihydrokaempferol (**2279**), engeletin (**2280**), melisimplexin (**2281**), meliternatin (**2282**), melibentin (**2283**), ternatin (**2284**), 3,5,8,4′-tetrahydroxy-6,7-dimethoxyflavone (**2285**), 5-hydroxy-3,3,4,5,7-pentamethoxyflavone (**2286**),

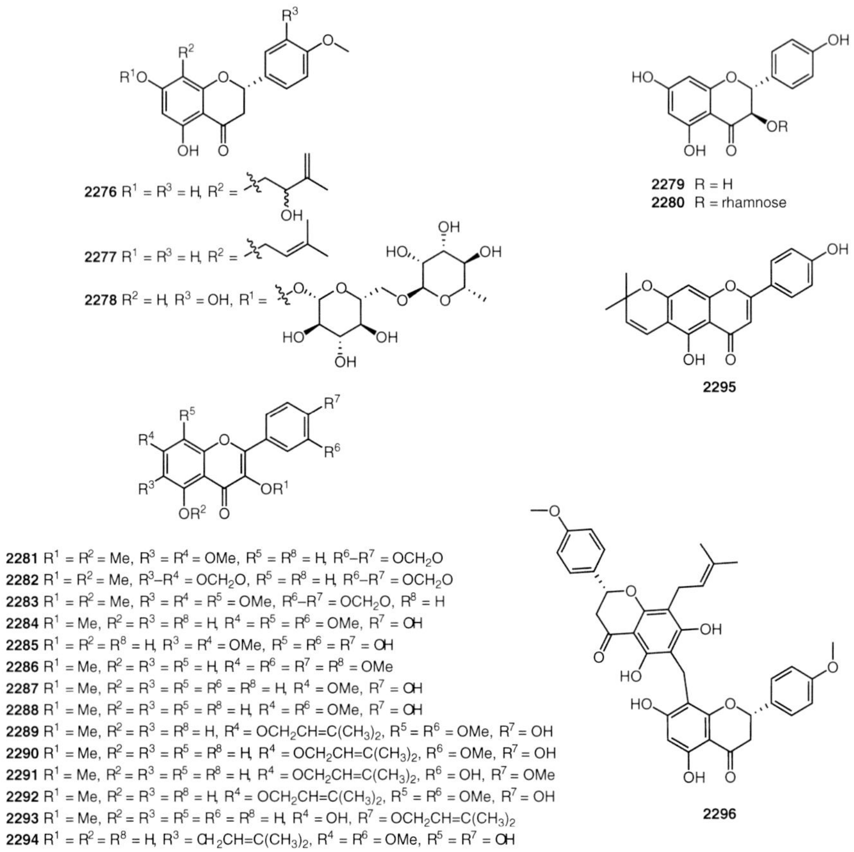

Fig. 160 Structures of flavonoids **2276–2296** reported from Rutaceae species

kumatakenin (**2287**), pachypodol (**2288**), 4,5-dihydroxy-3,3,8-trimethoxy-7-(3-methylbut-2-enyloxy)flavone (**2289**), 4,5-dihydroxy-3,3-dimethoxy-7-(3-methylbut-2-enyloxy)flavone (**2290**), 3,5-dihydroxy-3,4-dimethoxy-7-(3-methylbut-2-enyloxy)flavone (**2291**), 3,5,4′-trihydroxy-8,3′-dimethoxy-7-(3-methyl-but-2-enoxy)flavone (**2292**), 5,7-dihydroxy-3-methoxy-4′-(3-methylbut-2-enyloxy)flavone (**2293**), 3,5,8,4′-tetrahydroxy-7,3′-dimethoxy-6-(3-methyl-but-2-enyl)flavone (**2294**), carparchromene (**2295**), and bosistoabiflavanone dimer (**2296**)) have been isolated from *Bosistoa*, *Melicope*, *Flindersia*, and *Zanthoxylum* species [673, 685, 687, 689, 712, 715, 721, 722, 749, 750, 754, 755].

Five isoprenyl aryl ethers of methyl cinnamates or phenylpropanols (methyl (*E*)-4-(3′-methyl-2′-eny1oxy)cinnamate (**2297**), 4′-(3″-methylbut-2″-enyloxy)-3-phenylpropanol (**2298**), 4′-(3″-methyl-4″-hydroxybutyloxy)-3-phenylpropanol (**2299**), methyl-*p*-geranyloxy-*trans*-cinnamate (**2300**), and methyl *p*-geranyloxyferulate. (**2301**)) from *Eriostemon*, *Flindersia*, and *Sarcomelicope* have been reported (Fig. 161) [677, 756, 757]. *Melicope* and

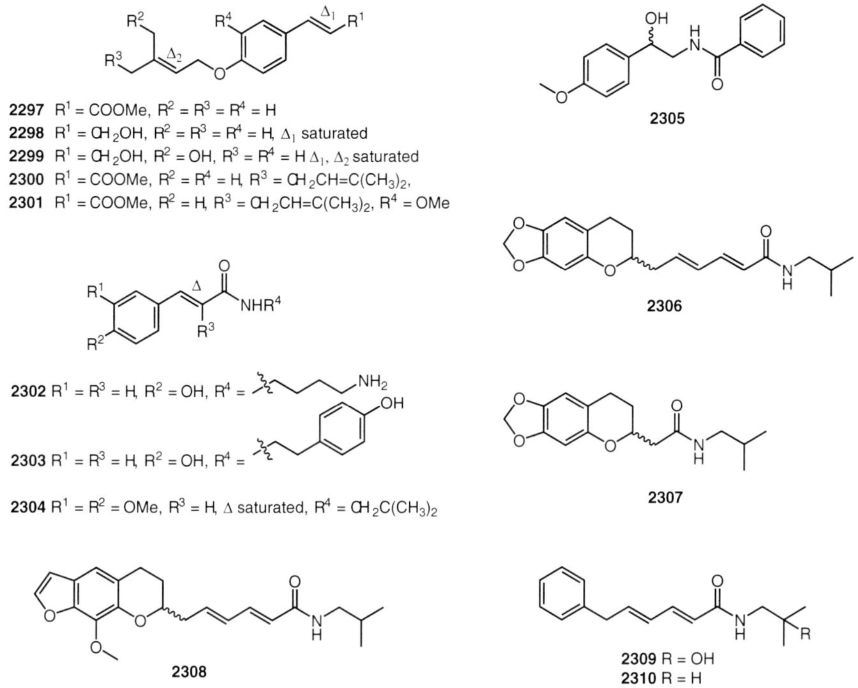

Fig. 161 Structures of compounds **2297–2310** reported from Rutaceae species

Dinosperma species contain the four cinnanamides, *N*-*trans*-*p*-coumaroylputrescine (**2302**), *N*-*p*-coumaroyltyramine (**2303**), 3,4-dimethoxy-*N*-(2′-methylpropyl)-phenylpropylamide (**2304**) and **162**, as also found in *Piper* species [39, 733, 738]. A related compound, *N*-benzoyl[2-hydroxy-2-(4′-methoxyphenyl)]ethylamine (**2305**) is the only natural product reported from *Clausena brevistyla* collected in Australia [758]. The coumarins and cinnanamides are shikimate-derived and another series of Rutaceae compounds is generated when the cinnamate carbonyl undergoes a further Claisen condensation with additional malonate groups via a polyketide biosynthetic route. These include a series of amides (erythrococcamides A–C (**2306–2308**), *N*-(2-hydroxy-2-methylpropyl)-6-phenyl-(2*E*,4*E*)-hexadienamide (**2309**) and *N*-(2-methylpropyl)-6-phenyl-(2*E*,4*E*)-hexadienamide (**2310**)), isolated from *Dinosperma erythrococca* that are remarkably like the pungent amides reported from certain Piperaceae species [759]. Interestingly, *D. erythrococca* is called colloquially tingle tongue, because chewing the bark causes a tingling sensation in the mouth.

Six related polyketides, brombyins I–VI (**2311–2316**), have been reported from *Brombya platynema* collected at Boonjie in north Queensland (Fig. 162) [760, 761]. These compounds are aryl decalins, reminiscent of polyketides reported from *Endiandra* and *Beilschmiedia* species in the Lauraceae. (*E*)-Piperonylprop-2-enal (**2317**) and 7-(3,4-methylenedioxyphenyl)octene-2-one (**2318**) from *B. smithii*

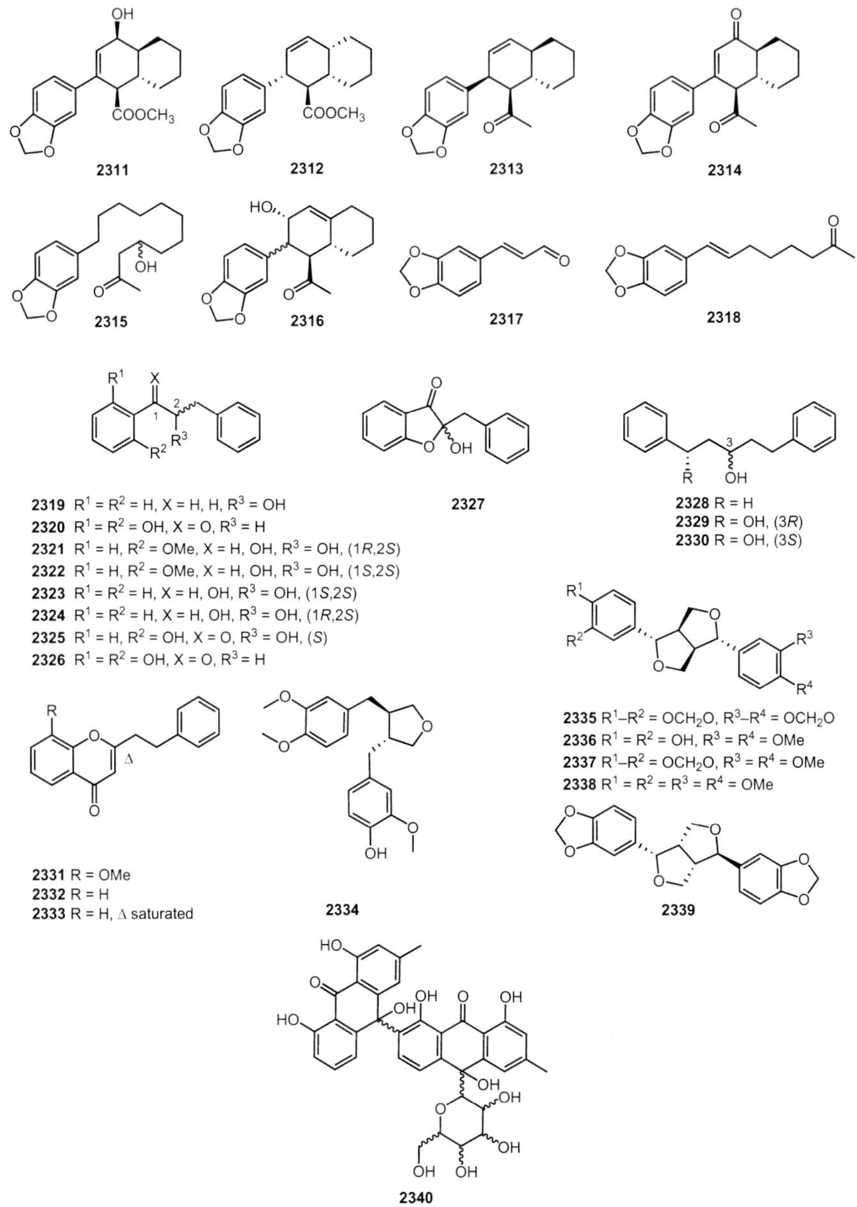

Fig. 162 Structures of compounds **2311–2340** reported from Rutaceae species

from Cedar Bay in north Queensland are also biosynthetically related [684]. Dihydrochalcones ((*R*)-(+)-1,3-diphenylpropan-1-ol (**2319**), 1-(2′-hydroxyphenyl)-3-phenylpropan-1-one (**2320**), *erythro*-1-(2′-methoxyphenyl)-3-phenylpropane-1,2-diol (**2321**), *threo*-1-(2′-methoxyphenyl)-3-phenylpropane-1,2-diol (**2322**), *threo*-1,3-diphenylpropane-1,2-diol (**2323**), *erythro*-1,3-diphenylpropane-l,2-diol (**2324**), 2-hydroxy-1-(2-hydroxyphenyl)-3-phenylpropan-1-one (**2325**), 1-(2,6-dihydroxyphenyl)-3-phenylpropan-1-one (**2326**) and 2-benzyl-2-hydroxybenzfuran-3(2*H*)-one (**2327**) and 1,5-diphenylpentanes (1,5-diphenylpentan-3-ol (**2328**), (−)-*erythro*-1,5-diphenylpentan-1,3-diol (**2329**), (+)-*threo*-1,5-diphenylpentan-1,3-diol (**2330**), 8-methoxyflindersiachromone (**2331**), flindersiachromone (**2332**), and 2,3-dihydroflindersiachromone (**2333**)), also likely to be of polyketide origin, have been reported from *Flindersia brassii* and *F. laevicarpa* var. *laevicarpa* from north Queensland [689, 719, 762]. The tetrahydro-furanolignans, brassilignan (**2334**), (+)-sesamin (**2335**), lignan 1 (**2336**), (+)-kobusin (**2337**), (+)-eudesmin/ (+) pinoresinol dimethyl ether (**2338**), and (−)-asarinin (**2339**) have been reported from *Zanthoxylum*, *Flindersia*, and *Melicope* species [712, 722, 733, 763]. The anthraquinone dimer ramosin (**2340**) has been isolated from *Pentaceras australis* [710].

β-Amyrin-type triterpenoids (**590**, **1860**, multiflorenol (**2341**), D-friedo-olean-14-ene-3,7a-diol (**2342**), and crusgallin (**2343**)), tirucallane triterpenoids (bourjotinolone A (**2344**), bourjotinolone B (**2345**), bourjotinolone C (**2346**), bourjotone (**2347**), and phebaloparvilactone (**2348**)), α-amyrin-type triterpenoids (**801** and ifflaionic acid (**2349**)), a lupane-type triterpenoid, **1620**, and a sterol glycoside, daucosterol (**2350**) have been reported from *Bosistoa*, *Zanthoxylum*, *Acradenia*, *Flindersia*, *Sarcomelicope*, *Phebalium*, and *Melicope* species [690, 712, 722, 736, 743, 754, 757, 764–766]. The limonoids veprisone (**2351**), 23-oxo-21-hydroxy-21,23-dihydroveprisone (**2352**), 21-oxo-23-hydroxy-21,23-dihydroveprisone (**2353**) and flindercarpin-2 (**2354**) have been found to occur in *Bouchardatia* and *Flindersia* species (Fig. 163) [683, 767]. Five monoterpenoids, (−)-boscialin (**2355**), (−)-nerolidol (**2356**), pregeijerene (**2357**), geijerene (**2358**) and filifolone (**2359**), a sesquiterpenoid, squamulosone (**2360**) a phenylethylalcohol, fareanol (**2361**), *p*-hydroxybenzaldehyde (**2362**), an inositol, sequoyitol (**2363**), and *N*-methyl-2-pyrolidinone (**2364**) were also isolated from Rutaceae species (Fig. 164) [683, 699, 712, 719, 733, 768–770].

Murraya paniculata, *Glycosmis trifoliata* (syn. *G. pentaphylla*), and *Clausena excavata* are additional Rutaceae species found in Australian tropical and subtropical rainforests as well as throughout Malesia and southeast Asia. No studies have been undertaken on Australian-collected material of *M. paniculata* or *G. trifoliata* but all three have been extensively studied in Asia [771–773].

The bioactivities of acridone and furanoquinoline alkaloids have been extensively reviewed and many congeners in both classes have been reported to possess cytotoxic, antibacterial, antiviral, antiprotozoal, and anti-inflammatory effects [774, 775]. Several of these compounds were active against several unrelated targets, suggesting non-specific activity. Acronycine was investigated as a candidate to treat solid tumors,

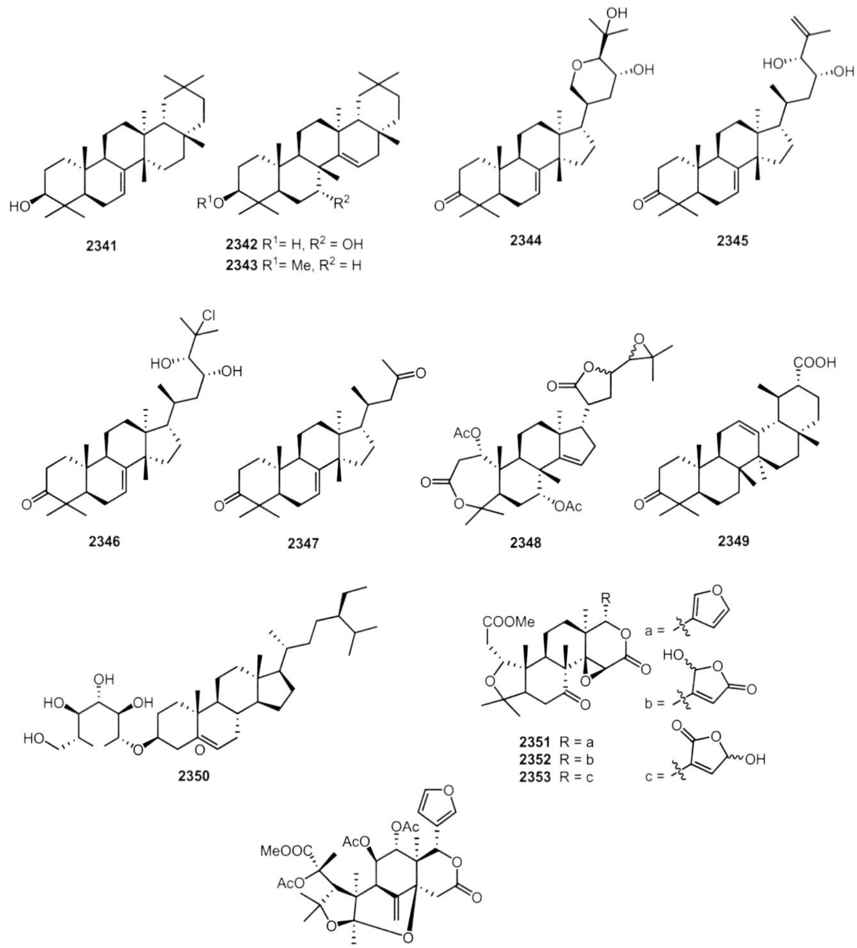

Fig. 163 Structures of triterpenoids **2341–2354** reported from Rutaceae species

but clinical trials performed on this compound were disappointing. The molecule acts as a topoisomerase-I and -II inhibitor [776].

Meliaceae

The Australian rainforest Meliaceae are represented almost exclusively by species in the Melioideae subfamily (Plate 37), but the widespread Asian species *Toona citiata* and *Melia azedarch* are also present. The chemistry of these two species has been studied extensively in Asia [777, 778]. The Meliaceae is of Gondwanan (African) origin with dispersal into the Americas and Asia [779]. Australian species are likely

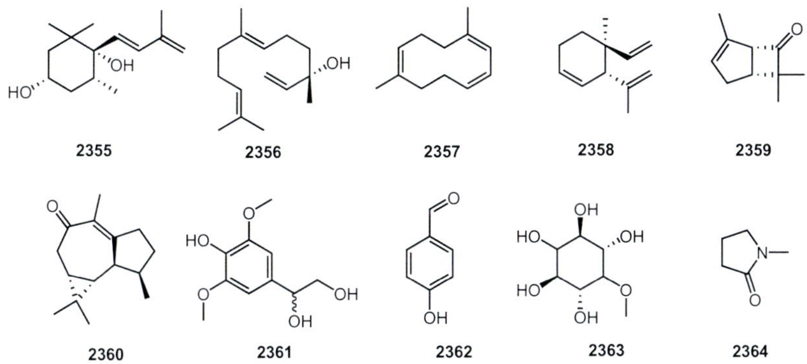

Fig. 164 Structures of compounds **2355–2364** reported from Rutaceae species

Plate 37 Rainforest Meliaceae: large leaves (>30 cm in length) of *Dysoxylum rufum* (left) and *Synoum glandulosum* subsp. *glandulosum* flowers (right). (Photographs AR Carroll)

to be recent emigrants from Asia. There have been limited chemical investigations of Australian rainforest Meliaceae species (Table 40).

Of the eight species investigated, tirucallane triterpenoids were found to predominate (Fig. 165) (3-oxo-tirucalla-7, 24-dien-21-al (**2365**), 3-oxo-7,(24*Z*)-tirucalladien-26-oic acid (**2366**), 3-oxo-7,(24*E*)-tirucalladien-26-oic acid (**2367**), 3*α*-hydroxy-7,(24*Z*)-tircalladien-26-oic acid (**2368**), dysoxylic acid A (**2369**), dysoxylic acid B (**2370**), 3-isobutyl-7-deactylglabretal (**2371**), 7-deacetylglabretal 3-acetate (**2372**), 7-deacetylglabretal-3*α*-tiglate (**2373**), (24*S*),25-dihydroxytirucall-7-en-3-one (**2374**), cabraleadiol 3-acetate (**2375**), cabraleone (**2376**), aglinin C (**2377**), aglinin D (**2378**), 21,(24*R*)-epoxy-25-hydroxytirucall-7-en-3-one (**2379**), 6*α*-acetoxydeoxyhavanensin (**2380**), and 28-deoxonimbolide (**2381**)) [780–784]. In

Table 40 Distribution of natural products isolated from the Meliaceae

Species	Collection location[a]	Plant part(s)[b]	Compound class	Compound(s)	Refs.
Aglaia australiensis	NQ	lf	Cinnamate	**2386**	[787]
Aglaia ferruginaea	NQ	b	Rocaglate	**2384**, **2385**	[785, 786]
	MAL	b	Steroid	**2382**, **2383**	[784]
	NQ, MAL	hw, b	Triterpenoid	**2372**, **2373**, **2375–2378**	[783, 784]
Dysoxylum pettigrewianum	NQ	b	Triterpenoid	**2366–2370**	[780]
Dysoxylum schiffneria	NQ	lf	Sesquiterpenoid	**2387–2389**	[788]
Owenia acidula	CQ	fr	Triterpenoid	**2371**, **2380**	[781]
Owenia cepiodora	CQ	lf, b	Triterpenoid	**2365**, **2374**, **2379**, **2381**	[782]
Owenia venosa	CQ	fr	Triterpenoid	**2371**, **2380**	[781]

[a] NQ = North Queensland, MAL = Malaysia, [b] lf = leaves, fr = fruits, b = bark, hw = heartwood

addition, two steroids (aglatomins A (**2382**) and B (**2383**), the flavagline rocaglaol (**2384**), initially assigned structure **2385** and given the name ferrugin, and a putricin diamide, aglairubine (**2386**), have been reported [784–787]. The sesquiterpenoids 8-hydroxycalamenene (**2387**), schifnerone A (**2388**), and schifnerone B (**2389**) were found in *Dysoxylum schiffneria* [788].

Malvales

The Malvales contains ten families, but only three have species that occur in the Australian rainforests. The Malvaceae, with 96 species, is the most diverse family, but the Thymelaeaceae with 18 species and Bixaceae with one species are both poorly represented (Fig. 166).

Chemical investigations are very limited (Table 41). *Pimelea ligustrina* and *P. linifolia* (Thymelaeaceae) have yielded daphnane orthoesters (gnidimacrin (**2390**), linimacrin D (**2391**), pimelea factor P3 (**2392**), and linimacrin C (**2393**)) (Fig. 167) [789]. *Argyrodendron peralatum* (Malvaceae) contains *N*-cinnamoylhistamine (**2394**) that is present in many other angiosperms [790] and *Cochlospermum gillivraei* (Bixaceae) contains five widespread flavonoids, **2395**, naringenin (**2396**), (+)-afzelechin (**2397**), prunin (**2398**), and cosmosiin (**2399**) [791]. The Illawarra flame tree, *Brachychiton acerifolius* (Malvaceae), is a stunning deciduous tree that produces a flush of bright red flowers during late spring and summer. The tree creates a stark contrast to the surrounding dense green foliage in the rainforest when it is flowering (Plate

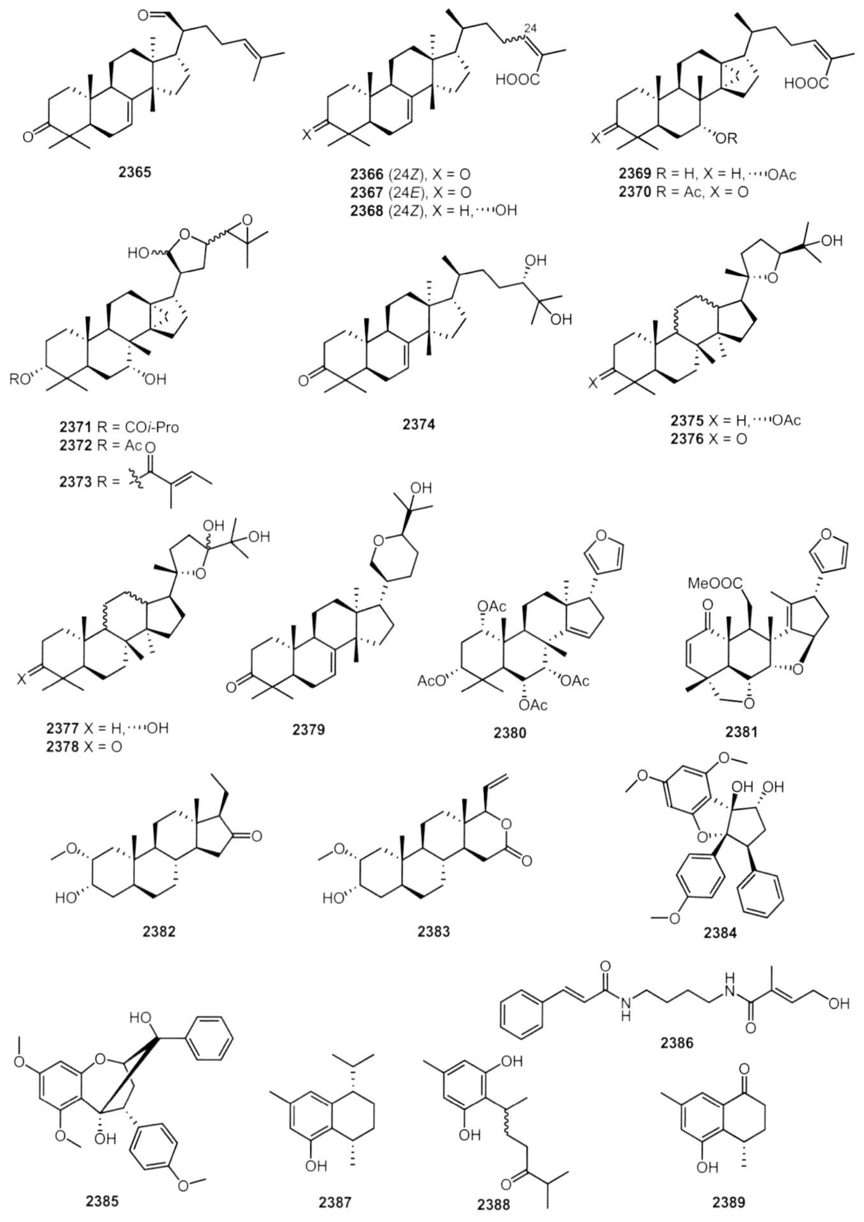

Fig. 165 Structures of compounds **2365–2389** reported from Meliaceae species

38). The red pigment in the flowers is pelargonidin 3-(6-coumaroylglucoside)-5-(6-acetylglucoside) (**2400**). Five additional flavonoids, **588**, naringenin 7-*O*-*β*-glucuronide (**2401**), naringenin 7-*O*-*β*-glucuronide methyl ester (**2402**), quercetin (**2403**), and kaempferol 3-rutinoside (**2404**) were isolated from the flowers [792].

Brassicales

The last order in the Malvids is the Brassicales, but this too is poorly represented in Australian rainforests. Only three families (Akaniaceae (one species), Capparaceae (18 species), Gyrostemonaceae (one species)) out of 19 families are present. *Codonocarpus attenuatus* (Gyrostemonaceae) is the only species studied previously, and it has yielded two macrocyclic spermidines, codonocarpine (**2405**) and *N*-methyl codonocarpine (**2406**) (Fig. 168) [793].

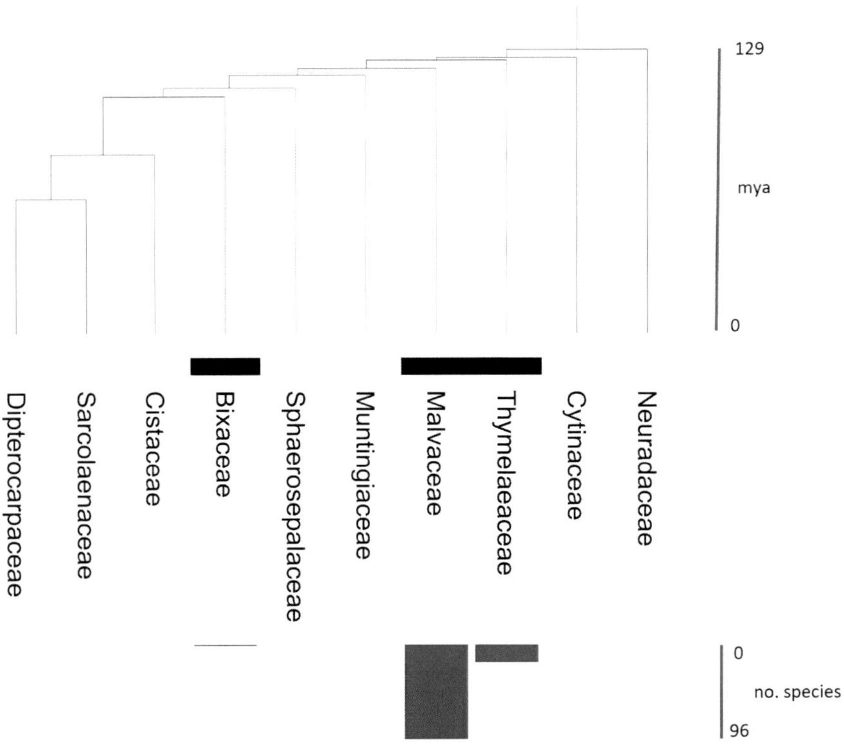

Fig. 166 Dated molecular phylogeny of the Malvales showing those families that have species present in Australian rainforests (grey) and studied (black) and the number of species occurring in Australian rainforests

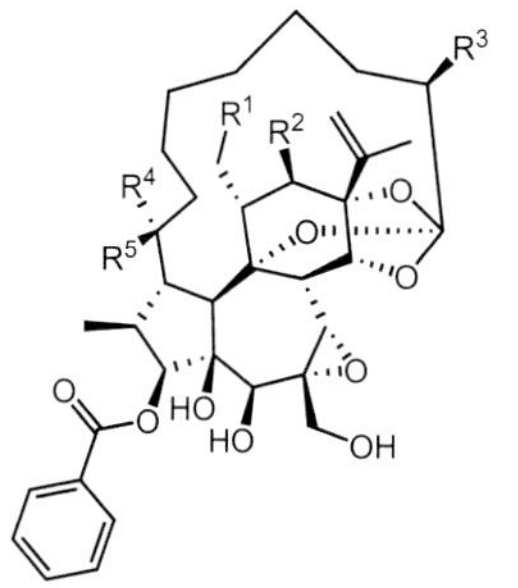

2390 R^1 = OCOPh, R^2 = R^5 = H, R^3 = OH, R^4 = Me
2391 R^1 = R^3 = R^5 = H, R^2 = OAc, R^4 = Me
2392 R^1 = R^2 = R^3 = R^5 = H, R^4 = Me
2393 R^1 = R^2 = R^4 = H, R^3 = OH, R^5 = Me

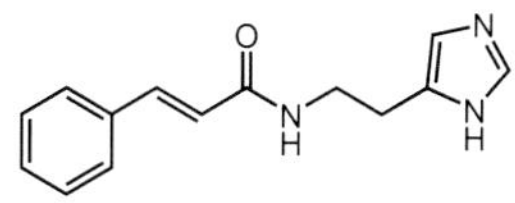

2394

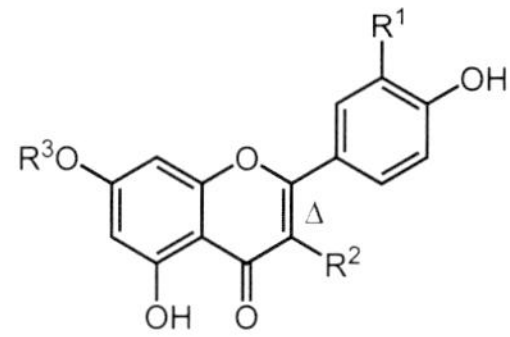

2395 R^1 = R^2 = R^3 = H
2396 R^1 = R^2 = R^3 = H, Δ saturated
2399 R^1 = R^2 = H, R^3 = β-D-glc
2403 R^1 = R^2 = OH, R^3 = H
2404 R^1 = R^3 = H, R^2 = rutinoside

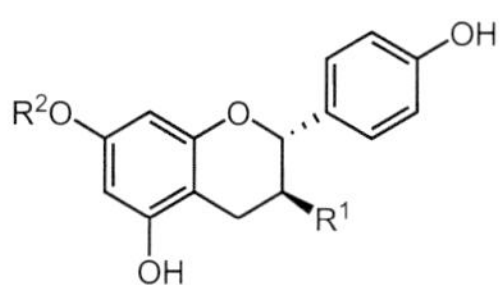

2397 R^1 = OH, R^2 = H
2398 R^1 = H, R^2 = β-D-glc

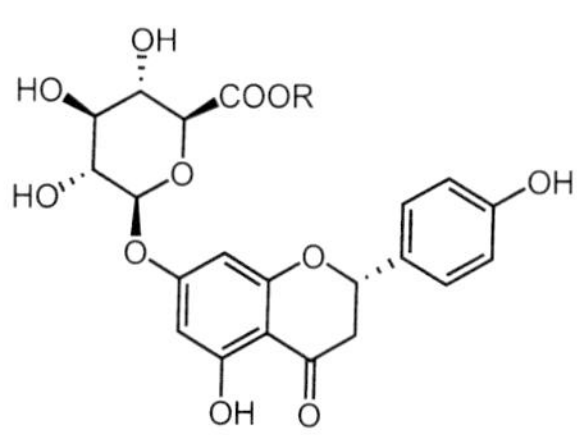

2401 R = H
2402 R = Me

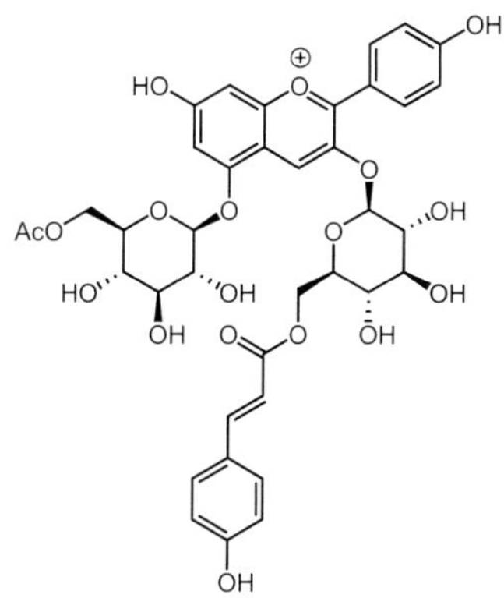

2400

Fig. 167 Structures of compounds **2390–2404** reported from Malvales and Brassicales species

5 Australian Rainforest Plant Chemistry Within a Chemical Diversity Context

A broad analysis of the natural product chemistry reported from Australian rainforest seed plants shows that alkaloids represent 40% of all compounds reported, while terpenoids account for 24% and phloroglucinols 13%. Although the present author has not quantified the occurrence of Australian rainforest seed plant-derived natural products from other organisms in other parts of the world in the process of generating the entire Australian rainforest natural product database, many examples of Australian rainforest-derived natural products occurred that were also found from other sources.

Table 41 Distribution of natural products isolated from the Thymelaeaceae, Malvaceae, Bixaceae, and Gyrostemonaceae

Species	Collection location[a]	Plant part(s)[b]	Compound class	Compound(s)	Refs.
Thymelaeaceae					
Pimelea ligustrina subsp. *hypericina*	NSW	lf, st	Daphnane	**2390**	[789]
Pimelea linifolia	NSW	lf, st	Daphnane	**2391–2393**	[789]
Malvaceae					
Argyrodendron peralatum	NQ	lf, st	Cinnamate	**2394**	[790]
Brachychiton acerifolius	NSW	fl	Flavonoid	**588**, **2400–2404**	[792]
Bixaceae					
Cochlospermum gillivraei	NQ	b	Flavonoid	**2395–2399**	[791]
Gyrostemonaceae					
Codonocarpus attenuatus	SEQ	b	Spermidine	**2405**, **2406**	[793]

[a] SEQ = Southeast Queensland, NQ = North Queensland, NSW = New South Wales, [b] lf = leaves, st = stems, fl = flowesr, b = bark

Plate 38 Flame tree *Brachychiton acerifolius* (left) and flowers (top right) and *Argyrodendron trifoliolatum* (bottom right). (Photographs AR Carroll)

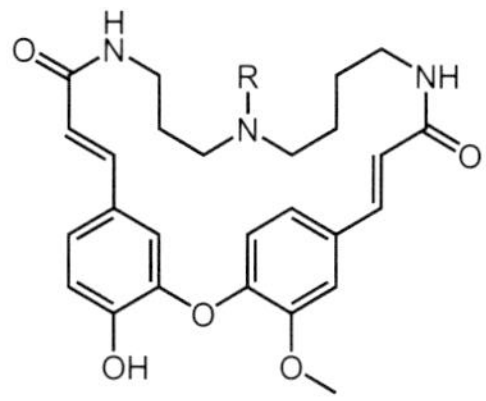

2405 R = H
2406 R = Me

Fig. 168 Structures of compounds **2405** and **2406**

The age, diversification and speciation of flowering plant genera has been studied. This has shown that the age of genera follows a latitudinal gradient, with the oldest genera being more concentrated in the wet and warm tropics and these decrease with latitude [5]. In contrast, mean speciation and net diversification rates increase with latitude [5]. The biosynthesis of plant natural products is genetically encoded, with key enzymes required to manufacture specific structure classes. Mutations of these genes is then linked to diversification of the molecules within structural classes [794]. Since the flowering plants radiated in the Cretaceous around 145 mya to generate the extant orders and later, families, the genes encoding natural product structure class biosynthesis would have already been established prior to the breakup of the supercontinents. Ecology and physiology then likely influenced the expression of specific natural products within lineages. In some cases, this led to specific structure classes becoming dormant or lost entirely through time [10]. Secologanin-derived monoterpenoid indole alkaloids are found exclusively in the order Gentianales. Within the family Apocynaceae these alkaloids are only found in the subfamily Rauvolfiodeae. This is an example of an early diverging group (Apocynaceae crown age 104 mya) that has retained the capacity to biosynthesize these alkaloids, while later diverging groups (58 mya) within the Apocynaceae have switched to the expression of triterpenoids and steroids. The diversification of structures within the monoterpenoid indole alkaloids coincided with diversification and speciation within the Rauvolfiodeae [369]. For example, the ellipticine alkaloids are restricted to *Strychnos* species, the basal Apocynaceae tribe Aspidospermateae (*Aspidosperma* spp.), and the tribe Vinceae (*Ochrosia* spp.) in the subfamily Rauvolfiodeae. *Strychnos* species occur in the Loganiaceae, a family that split with the Apocynaceae 104 mya. The presence of ellipticine alkaloids within these species reflects their ancient origins. This is an example where an ancient lineage within an order possesses unique chemistry. It is reasonable to assume that other ancient lineages would likely contain ancient structural class lineages as well. The antiquity of Australia's rainforest may therefore explain why they represent a repository of ancient natural product lineages. Irrespective of how and when Australia's extant rainforest plants arrived and diversified (either through vicariance or long-distance dispersal) the flora contains a high proportion of tropical and thus ancient lineages and lineages that have become extinct elsewhere. Many families have basal groups in Australia's rainforest and these basal lineages often have higher diversities of natural product structural classes (or different structural classes), when compared to their more modern relatives that have speciated (or diversified) in arid and temperate regions.

Self-organizing maps (SOMs) can be a powerful tool to visualize chemical diversity [795]. They represent chemical diversity as vectors separated from each other by dissimilarity (defined by molecular fragments) on a 2D vector similarity landscape analogous to a topographic map. Valleys contain molecules that are like each other, and these are separated from dissimilar structures by hills and mountain ranges. The right and left, and top and bottom edges also form a continuous connection, and one can imagine that the map is a doughnut cut along the lateral and longitudinal axes and laid flat. The free cheminformatics software package DataWarrior can be used to generate SOMs [796]. The software generates a variety of different molecular descriptors and using the fragFP molecular descriptor (a descriptor of 512 predefined structure fragments that is used to generate a fingerprint for each molecule) is useful for identifying molecules that are biosynthetically related (i.e., structures that share the same carbon framework). The natural product chemistry reported from Australia's rainforest spermatophytes (2,378 compounds) combined with world plant chemical diversity (105,550 compounds) obtained from the DEREP-NP database was mapped using a 200 $\times$ 200 cell SOM [32]. All structures were classified using NP-classifier (an online informatics tool that is part of GNPS suite hosted by the Scripps Institution of Oceanography, San Diego, CA, USA) [797] and color-coding the SOM by natural product structure class highlights the discriminating power of SOMs to segregate structures based on molecular features (Fig. 169).

Australian rainforest natural product chemical diversity is intrinsically linked to phylogeny, suggesting an evolutionary link. Therefore, one should expect structual classes to be segregated taxonomically, and this is clearly evident in Fig. 169. Analysis of very broadly classified natural product structure classes in a 2020 paper suggested that most plant natural product structure classes result from independent and frequent evolution [798]. The natural product structural classes chosen in this study (tannins, alkaloids, steroids, terpenoids, flavonoids, phenolic acids, quinones and phenylpropanoids) are too broadly grouped, and their general conclusions may be correct for some structural classes but not for others. If the chemical diversity within each broad natural product group is clustered into sub-classes (e.g., *ent*-kaurene diterpenoid, monoterpenoid indole alkaloid, deoxylignan, etc.) natural product chemical diversity data more likely points to monophyletic origins of many natural product structure classes within these broadly defined categories [798]. To make sense of phylogenetic trends associated with natural product structures, substructural features within broadly grouped structural classes such as flavonoids or terpenoids need to be considered. When the SOM in Fig. 169 is color-coded by plant order, the demarcation of structures at this taxonomic level is clear. The resultant SOM (Fig. 170) shows examples of the limited (but phylogenetically grouped) distributions of various structure classes, and this supports the theory that plant natural product chemical diversity is intrinsically linked to plant evolution and thus phylogeny [799].

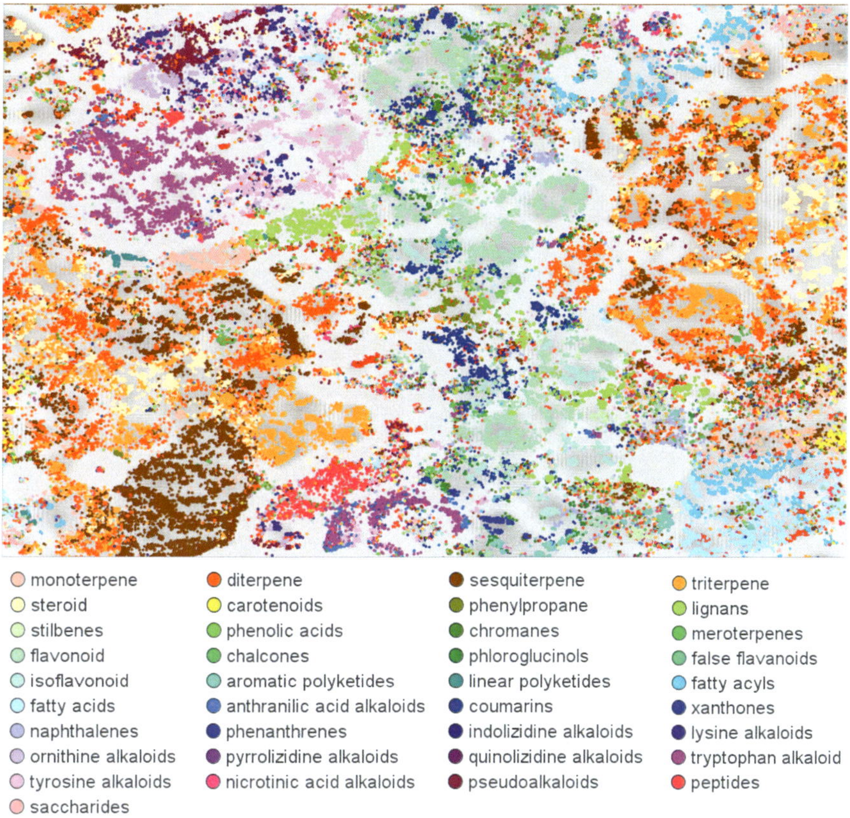

Fig. 169 Self organizing map (200 × 200 neurons) of all plant natural products found in DEREP-NP (105,550). The natural products are color coded using a variety of structure classes adapted from NP-classifier

Secologanin monoterpenoid indole alkaloids from the Gentianales, benzyl isoquinoline alkaloids from the Magnoliales, Laurales, and Ranunculales, 9,9′-deoxylignans from the Amborellales, Nymphaeales, Austrobaileyales, and Magnoliid angiosperms, β-triketone and acylphloroglucinols in the Myrtales and acridone and quinoline alkaloids from the Sapindales (overwhelmingly in Rutaceae species) are examples of these demarcations. When Australian rainforest derived NPs are color coded differently to all other plant NPs, their widespread chemical diversity is apparent, and most natural product structure classes are reported in Australian rainforest species studied to date (Fig. 171). This is not surprising given that Australia's rainforests contain a diversity of species spread across most plant Orders and over half of plant diversity at the family level. Only ~10% of all Australian rainforest species have been investigated for their natural product constituents and in most cases only from leaf, bark and/or stem. Most studies have been on plant material collected from a single location and, as many of the maps in the previous section show, some

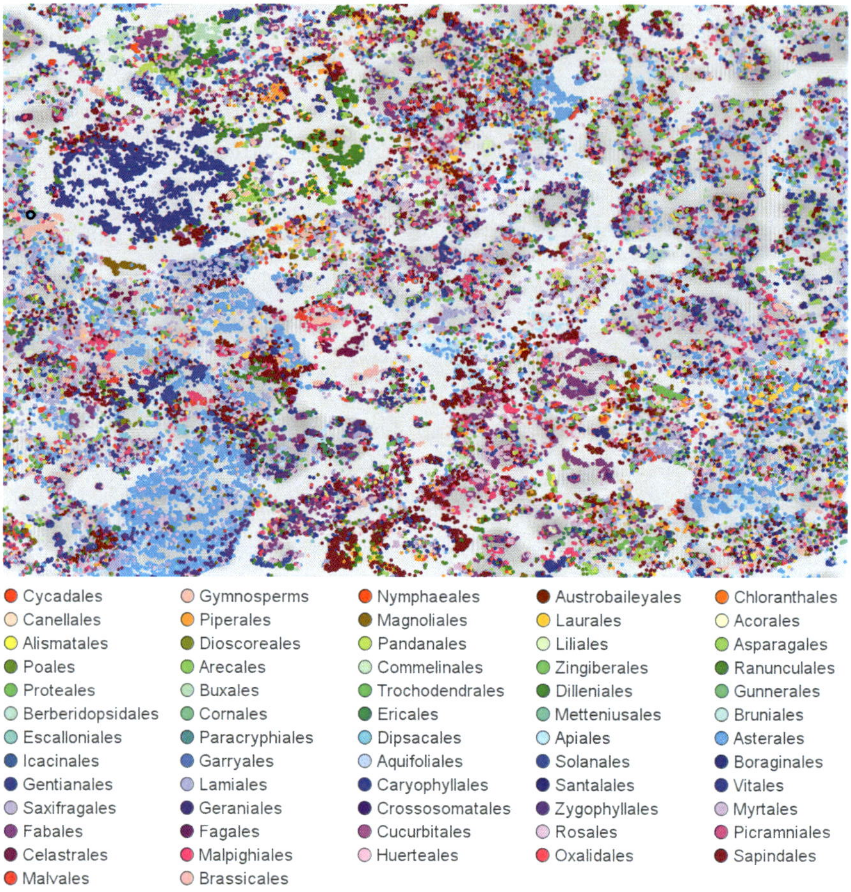

Fig. 170 Self organizing map (200 × 200 neurons) of all plant natural products found in DEREP-NP (105,550). The natural products are color coded by angiosperm order

species have a broad distribution across the Australian rainforests while others are distributed further north into New Guinea and Malesia, or east into New Caledonia and the Pacific islands. There is clearly much more chemistry to be revealed from studying Australian rainforest species.

The unique or rarely encountered chemistry found in Australian rainforest species includes (i) the *Galbulimima* alkaloids, (ii) the *Anopterus* alkaloids, (iii) the Proteaceae tropanes, (iv) the Elaeocarpaceae monoterpenoid indole alkaloids and the indolizidine alkaloids, (v) the *Flindersia* isoprenylated bis-indoles, (vi) the *Eupomatia* lignans and naphthyridine alkaloids, (vii) the *Fontainea* phorbols, (vii) the unusual diversity of β-triketones from Myrtaceae species, (viii) the Atherospermataceae polycyclic bis-BIAs, (ix) homoerythrinan and homo-aporphine alkaloids of the Colchicaceae, (x) the chirally unique tris-pyrroloindoline of *Idiospermum*, and (xi) the gympietide ICK peptides from *Dendrocnide* species. The Asteraceae are

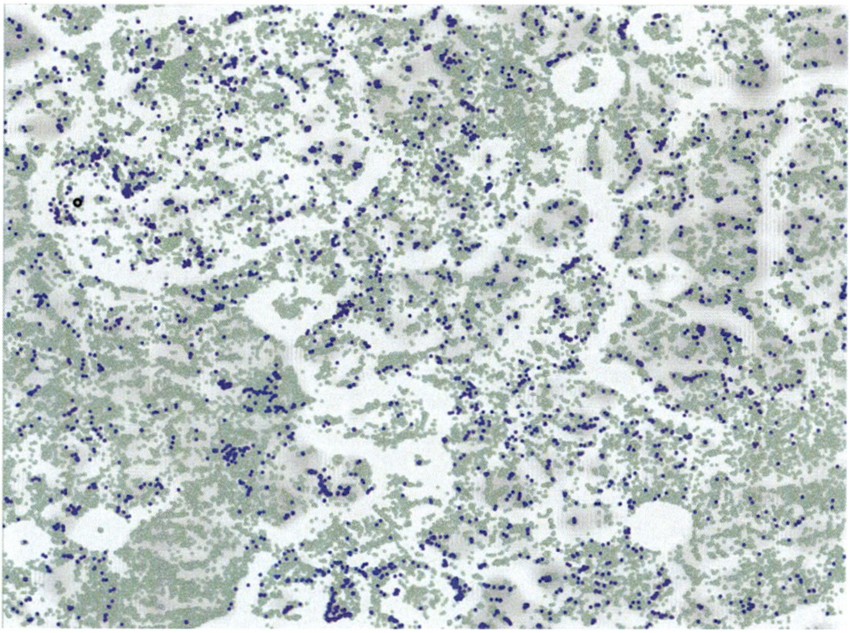

Fig. 171 Self organizing map (200 × 200 neurons) showing Australian rainforest natural products (2,378 compounds, dark blue) versus all spermatophyte natural product chemical diversity (105,550 compounds, light green)

underrepresented in Australian rainforests and since these are the predominant source of sesquiterpenoids it follows that this compound class is also underrepresented.

Indole alkaloids have a disjunct distribution across the angiosperms and are not found in the gymnosperms. The north Queensland rainforest tree *Idiospermum australiense*, an ancestral relic from the Calycanthaceae dating back over 90 million years, is one of only a small number of Magnoliid species to contain indole alkaloids and most other species in the clade (apart from the much younger Calycanthaceae siblings *Calycanthus* and *Chimonanthus*) are from the Annonaceae that produce isoprenylated indoles [800]. *Idiospermum australiense* is the most ancient of the indole alkaloid-producing angiosperms and variants of its dimeric and trimeric pyrrolidinoindoline structures (albeit with different relative and absolute configurations) found in unrelated Rubiaceae species (including the North Queensland *Eumachia*) likely represent products of convergent evolution. The *Flindersia* indole alkaloids derived from 2-isoprenyl tryptamine share features in common only with a small number of alkaloids from the Rubiaceae [801] and marine bryozoans from the genus *Flustra* [802]. These disjunct distributions imply paraphyletic origins. Members of the Rutaceae tribe Aurantioideae (including *Clausena excavata*, *Murraya paniculata*, and *Glycosmis trifoliata* (syn. *G. pentaphylla*), found in Australian rainforests) contain carbazole alkaloids. These compounds are also derived from 2-isoprenylindole but have undergone an intramolecular cyclization with subsequent elimination of ethyl amine to generate the carbazole moiety (Scheme 5). Carbazoles are only found in natural products from the Apocynaceae, Ranunculaceae, and Rutaceae. The shared presence of 2-isoprenyl-indoles in the Aurantioideae and Zanthoxyloideae that diverged around 85 mya indicates that their common ancestors had the capacity to make 2-isoprenyl-indoles and therefore these indoles also have an ancient origin. Related isoprenylindoles lacking a C-3 ethyl amino group have also been reported from Annonaceae species [800] but this disjunct taxonomic distribution suggests convergent evolution (Fig. 172).

The Apocynaceae are the most dominant source of indole alkaloids in Australian rainforests (and indeed throughout the seed plants) and the rare ellipticine and unprecedented actinophyllic acid, structures from Australian species provide some distinction from species found elsewhere. The quite different monoterpenoid indole alkaloids from *Aristotelia* species (Elaeocarpaceae), with vicariant distributions only in Australian, New Zealand and Chilean rainforests, suggests they represent another group of ancient structures.

The benzyl isoquinoline alkaloids are found predominantly in Magnoliid and basal Eudicots. In Australian rainforests, the Magnoliales (Eupomatiaceae, two species, and Annonaceae) and the Laurales (Lauraceae, mostly from the basal tribe Cryptocaryeae, and Atherospermataceae, 65% of all species) have yielded both ubiquitous and unusual benzyl isoquinoline alkaloid structures. The extradiol-oxidized aporphines from Australian *Eupomatia*, *Goniothalamus*, *Cananga*, and *Mitrephora* again appear mostly to be associated with the early diverging genera within the Order, while *Aristolochia* species contain characteristic nitro-phenanthrenes (aristolochic acids). The bis-BIAs from within Australian rainforest Atherospermataceae, Hernandiaceae, and Menispermaceae species, and elsewhere in the Piperales, Magnoliales, Laurales, and Ranunculales are unique within the angiosperms.

The homoerythrinan, homoaporphine, and colchicine alkaloids from the Colchicaceae are mostly associated with the Northern Hemisphere *Colchicum* lilies, but their occurrence in the near basal Tripladenieae tribe from Australian rainforests point to the ancestry of this alkaloid class being in Australia [69].

Although the chemistry of the diverse and speciose Proteaceae has not received much attention and Australia's rainforests harbor the highest proportion of genus-level diversity, structurally unique tropane alkaloids have been found in taxonomically disparate species from the tropics, subtropics, temperate rainforests and alpine areas. Outside of Australia, these alkaloids have only been found in two New Caledonian rainforest Proteaceae species, *Eucarpha strobilina* and *E. deplanchei*, which are closely related to Australia's tropane alkaloid producing *Triunia* species. Why is this the case? A 2024 paper proposes a northwest African origin for the Proteaceae 132 mya followed by migration to northeastern South America over the next 30 million years, during which time the three extant subfamilies Persoonioideae, Proteoideae, Grevilleoideae, and the isolated genus *Carnarvonia* (now only found in north Queensland rainforests) diverged. Southward migration of these subfamilies into Antarctica then saw two routes of migration into Australia between 100 and 45 mya via dry and fire-prone northern Antarctica into southwestern Australian or via warm and wet coniferous rainforests in southern Antarctica into Tasmania and up

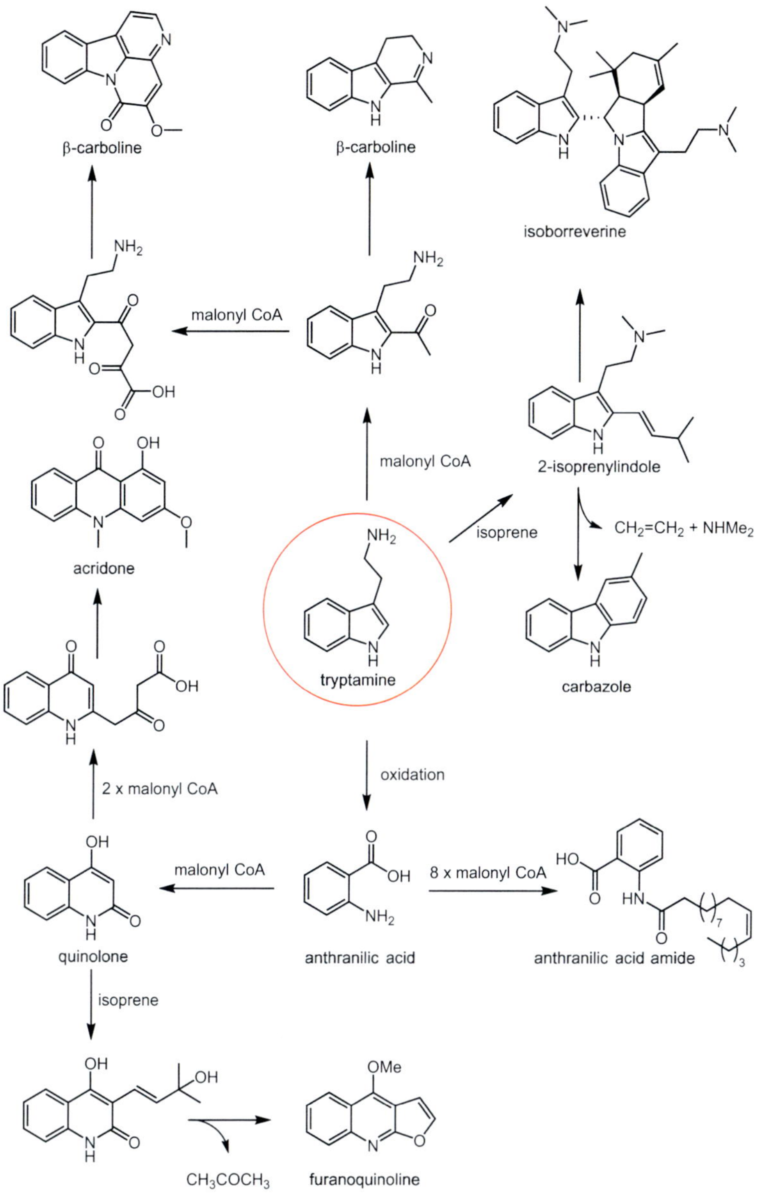

Scheme 5 Biosynthetic origins of Rutaceae alkaloids from tryptamine. The structures with bold generic names represent the major classes of alkaloids found in Australian rainforest Rutaceae species

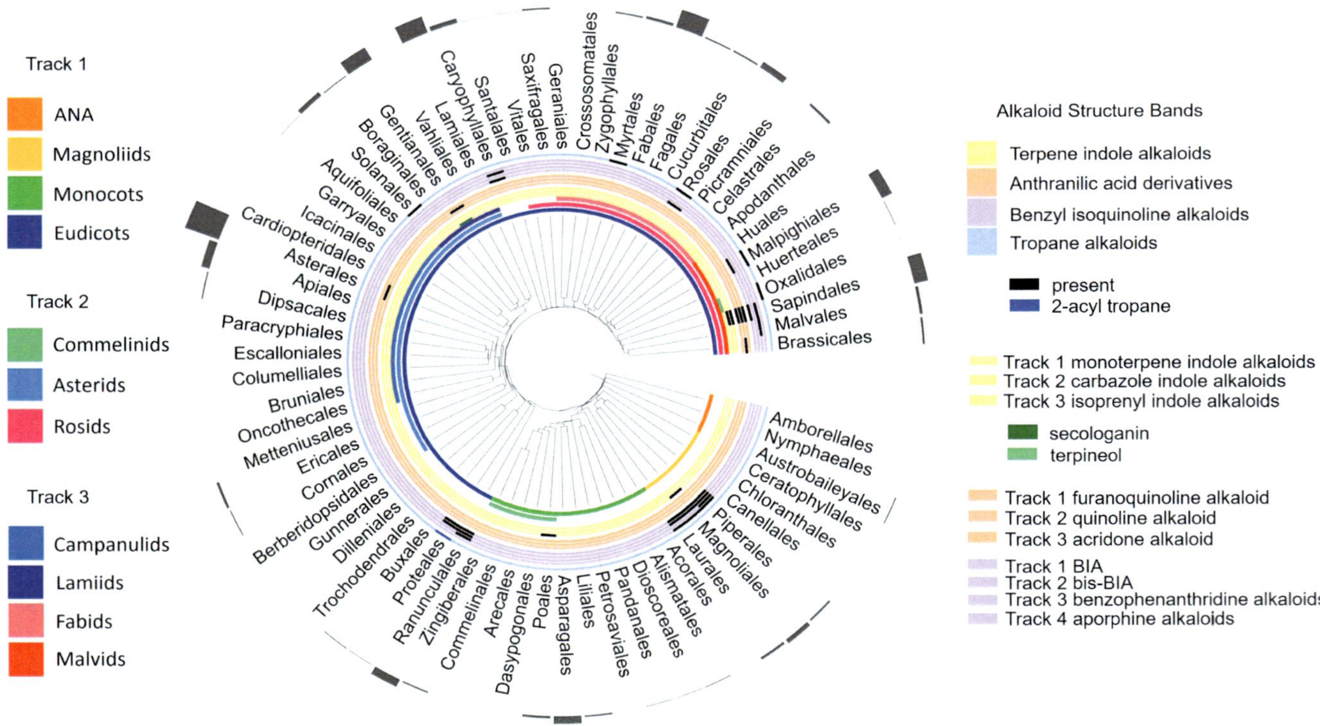

Fig. 172 Angiosperm phylogenetic tree showing orders that have been reported to contain indole alkaloids (pale yellow band—tracks 1–3), anthranilic acid-derived polyketides (pale orange band tracks 1–3), subclasses of BIAs (pink band tracks 1–4) and tropane alkaloids (blue band track 1). The outer track is a bar chart depicting the proportion of worldwide species in each order

the east coast of Australia (Fig. 2) [262]. The rainforest Proteaceae species diversified during their southern Antarctic migration while the sclerophyllous species diversified and speciated during their northern Antarctic migration. Species from the three subfamilies migrated into Australia via both routes leading to the diversity that now dominates in Australia. These migrations were followed by extinctions along their migration routes, and this has resulted in Australia now representing a museum for the Proteaceae. Long-distance dispersal from Australia and diversification led to the *Eucarpha* species in New Caledonia. The tropane alkaloids have only been reported from species that arrived and/or diversified in Australia via southern Antarctica. This implies that the climatic, edaphic and/or biotic conditions in the rainforest environment were conducive to alkaloid production. Other rainforest Proteaceae species both in Australia and in South America may therefore also yield 2-acyl tropanes. The phylogenetic distribution of tropane across the three subfamilies also points to an ancient origin.

Lignans that lack 9,9′-oxygenation are almost exclusively found in Amborellales, Nymphaeales, Austrobaileyales, and Magnoliid species while the more oxidized lignans have a more widespread distribution across both gymnosperms and angiosperms (Fig. 173). In Australian rainforests these reduced lignans occur in *Austrobaileya* (Austrobaileyaceae), *Trimenia* (Trimeniaceae), *Eupomatia* (Eupomatiaceae), *Galbulimima* (Himantandraceae), and *Endiandra* (Lauraceae) species. Related structures are found in the Piperaceae, Aristolochiaceae, Annonaceae, Sauraceae, Schizandraceae, Magnoliaceae, and Myristicaceae from South and Central America, Africa, and Asia [803]. These molecules are derived from phenyl propene precursors and the enzymes responsible for their synthesis and oxidative coupling are unknown but likely conserved within these ancient lineages [803]. The lignans in gymnosperms and eudicots by comparison result from dimerization of coniferyl alcohol derived from tyrosine [803].

The interrelatedness of what appear to be structurally disparate molecules can provide some insight regarding common precursors that are likely to be involved in the biosynthesis of compounds from taxonomically related species. For example, species from the Magnoliid families Winteraceae (*Zygogynum acsmithii*), Lauraceae (*Lindera* spp., *Clinostemon mahuba*), and Annonaceae (*Meiogyne cylindrocarpa*) contain C-21 butenolides that are potential precursors to the *Galbulimima* alkaloids found in the Australian Magnoliid *Galbulimima baccata* (Himantandraceae) [146, 162, 804, 805]. The compounds are also related to the acetogenins that are almost exclusive but widely distributed in the Annonaceae [806]. The butenolides appear to be derived from linear polyunsaturated fatty acids that have had a 2-propanol moiety added to generate the lactone [806]. Since the Magnoliid orders and families began to diverge from each other ~122 mya, the common polyketide biosynthetic genes required to make these compounds needed to have arisen prior to this divergence, suggesting that these molecules are likely have an ancient origin [7].

In general, flavonoids are widespread across Australian rainforest species, but other polyketide-derived NPs have a more segregated phylogenetic distribution. The biosynthesis of polyketides in plants is exclusively via type III polyketide synthases (PKS) starting from a carboxylic acid starter unit, elongation with malonyl-CoA

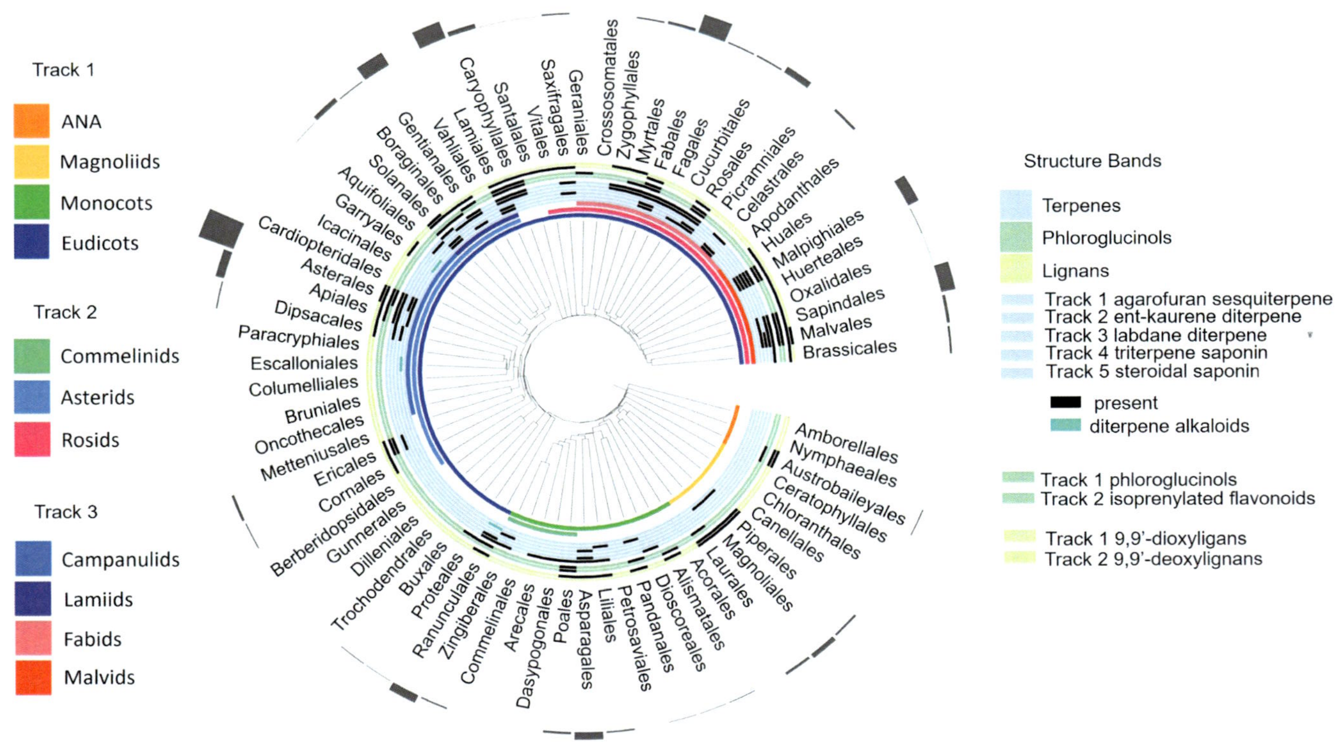

Fig. 173 Angiosperm phylogenetic tree showing orders that have been reported to contain terpenoids (blue band—tracks 1–5), phloroglucinols and isoprenylflavonoids (green band—tracks 1 and 2) and lignans (yellow band—tracks 1 and 2). The outer ring is a bar chart depicting the proportion of worldwide species in each order

extender units and finally cyclization via several different routes. The starter unit carboxylic acid specificity, extension count, and cyclization routes are associated with the structural differences in the type III PKS and this is linked phylogenetically to variations in key amino acid residues [807]. Starter units include the aliphatic acids acetic, malonic, valeric, isovaleric, isobutyric, and 2-methylbutyric acids (with the latter four from branched-chain amino acids) and the aromatic acids, benzoic, cinnamic, and anthranilic acids and their more oxidized phenolic analogues [808]. Some PKS's have been shown to accommodate long-chain dicarboxylic acid starter units that ultimately lead to the biosynthesis of the bis-5-alkylresorcinols that are almost exclusive to Proteaceae species [809]. The PKS biosynthesis pathway can generate a range of structural classes including the flavonoids, other phloroglucinols, alkyl resorcinols, pyrones, and the quinolones and acridones found exclusively in the Sapindales families Simaroubacece and especially Rutaceae. The flavonoids are widespread across gymnosperms and angiosperms and so few obvious phylogenetic relationships can be inferred. Other polyketides with aryl starter units are: (i) the endiandric acids (**402–406**) and related compounds from the Lauraceae; (ii) the aryl tetralones (**140**) from the Winteraceae; (iii) brombyins (**2311–2316**) from the Rutaceae; (iv) pipercallosine and related amides **158**, **164**, and **165** from the Piperaceae, and the erythrococcamides (**2306–2310**) from the Rutaceae; (v) the aryl-substituted tropanes from the Proteaceae (**688** and **689**); (vi) benzophenone derivatives of the Myrtaceae, and (vii) the α-pyrones from the Annonaceae and Lauraceae (Fig. 174). These compounds differ in the number of elongation steps used and the ultimate cyclization route.

Although it is tempting to draw taxonomic relationships between species that produce similar polyketide compounds, genomic analysis suggests that convergent evolution has given rise to structurally distinct type-III PKS's across the angiosperms that can make structurally related molecules [807]. The Magnoliid polyketides are therefore more likely to be derived from a common PKS but the structurally related polyketides from the Rutaceae likely result from a different PKS. This may be exemplified by the brombyins from the Rutaceae that are remarkably like the cyclostachines isolated from the stems of *Piper trichostachyon* (Piperaceae) from India [810]. There are still insights to be gained from analysis of polyketide structures obtained from different plant families. The Australian Myrtaceae is a major source of phloroglucinols that are structurally distinct from those reported from other families. The starter specificity is predominantly 2-methyl isobutyrate, isovalerate, isobutyrate and benzoate and this gives rise to a great diversity of both phloroglucinols and β-triketones involving mixed biosynthesis with monoterpenoids or sesquiterpenoids, further diversifying the natural product structures. These natural products are concentrated in the early diverging groups within the Myrtaceae that have their ancestral origins in Australia. The tropane alkaloids from the Proteaceae also owe their unique structural diversity to polyketide precursors that show a similar plasticity in starter unit specificity to that found in the Myrtaceae. They obviously differ in that mostly two (but occasionally three) malonyl-CoA elongation steps are followed by conjugation to pyrroline, leading to the tropanes, while exclusively three malonyl-CoA elongations generate the Myrtaceae phloroglucinols. As noted above, the occurrence

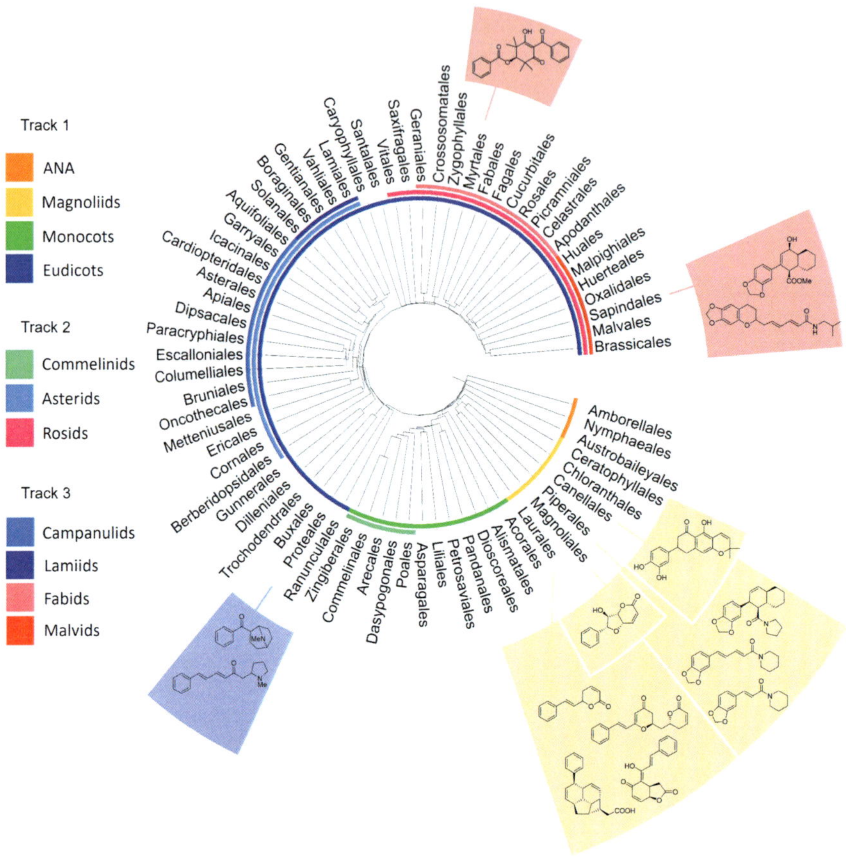

Fig. 174 Angiosperm phylogenetic tree showing Australian rainforest orders that have been reported to contain polyketides with a benzoic acid starter unit (excluding flavonoids)

of 2-acyltropanes exclusively in eastern Australasian Proteaceae suggests an isolated and ancient biosynthetic legacy. The alkyl resorcinols and bis 5-alkylresorcinols are not restricted to Australian rainforest species but are also found in species that likely migrated to Western Australia via north Antarctica and in species in southern Africa. This suggests that the compound class has its origins in the early divergence of the family prior to its migration to both Australia and South Africa. The very restricted occurrence of these types of molecules points to their uniqueness as Gondwanan relicts from their ancestral homeland in north Africa and South America. The Elaeocarpaceae polyketide alkaloids (*Elaeocarpus* and *Peripentadenia* alkaloids) are derived from malonate followed by four malonyl-CoA elongation steps and subsequent conjugation to 3-aminopropylpyrroline. These alkaloids are characteristic of the Australasian Elaeocarpaceae, suggesting an independent evolution of a type III PKS in Australia.

The Rutaceae contains a remarkable diversity of polyketides, including isoprenylated flavonoids and simple acetophenones (derived from four malonyl-CoA units) that predominately are further conjugated to isoprenes, but the polyketides that derive from an anthranilic acid starter unit are almost exclusive to the family and give rise to three structural classes: acridones, quinoline/quinolones, and polyunsaturated acyl-anthranilic acids. The acridone biosynthesis is catalyzed by acridone synthase and requires three malonyl-CoA elongations from anthranilic acid and cyclization to generate an aromatic moiety that condenses with the aniline nitrogen whereas the quinolone biosynthesis is catalyzed by quinolone synthase and only involves one malonyl-CoA condensation with anthranilic acid [808]. The structural diversity of the Rutaceae quinolones results from further conjugation with one or more isoprene groups and cytochrome P450 oxidation of the aromatic ring. The acridones and quinolines are found in Simaroubaceae and Rutaceae species but not in other taxa, indicating that the PKSs acridone synthase and quinolone synthase are enzymes inherited from ancestors before the families split 90 mya. The acridones and quinolones are distributed widely within the Rutaceae but, given that the Australian rainforests contain high genus and species diversity, the acridone/quinolone chemical diversity within these species matches with the entire worldwide chemical diversity.

The chemistry of the Rutaceae is bewildering, as noted by Waterman who suggested that the family has close affinities with the basal Eudicots [668]. However, this hypothesis is refuted by molecular evidence [669]. Examples of this dichotomy are: (i) benzyl isoquinoline alkaloids in common with the Ranunculales and Magnoliids; (ii) 2-isoprenyl-indoles in common with the Annonaceae and Rubiaceae; (iii) alkyl-phenyl carboxylic acids and amides in common with the Lauraceae and Piperaceae, and (iv) food plant associations with Papilionidae butterflies that otherwise rely on Magnoliid species [811]. The genes encoding the enzymes responsible for making the benzyl isoquinoline alkaloids are likely to have evolved in some of the earliest angiosperms and are widespread (albeit dormant in many clades) and their presence in *Zanthoxylum* in the Rutaceae species may suggest that the ecological conditions required to maintain their expression have been conserved. The polyketide structural diversity on the other hand reflects a diversification of PKSs within the family since genomic studies suggest that the PKS's have repeatedly evolved in different groups of plants under ecological pressures in keeping with theories of convergent evolution [808]. A major feature of Rutaceae chemistry is the dominance of isoprenylation. Within Australian rainforest Rutaceae, 25 of 34 quinolones, all 37 furanoquinolines, 10 of 32 acridones, 12 of 24 indole alkaloids, 61 of 62 coumarins, 10 of 23 flavonoids, and 32 of 34 acyl phloroglucinols are conjugated to either an isoprene or a monoterpenoid. The only other Australian rainforest families to contain isoprenyl conjugates are the Proteaceae (three naphthoquinone conjugates) Fabaceae (nine isoflavonoid conjugates), Euphorbiaceae (seven guanidine conjugates) and the Clusiaceae (two phloroglucinol conjugates). This disjunct family distribution is supported by analysis of natural products in the DEREP-NP database, which shows that the Fabaceae is the major source of isoprenylated natural products (1,092 natural products that are almost exclusively isoflavonoid or flavonoid conjugates), with Clusiaceae (320 natural products that are predominantly phloroglucinol conjugates), Moraceae (290 flavonoid conjugates), and Rutaceae (383 natural products across a variety of structure classes), accounting for the vast majority of the remaining isoprenyl conjugates. Prenyl transferases are enzymes that catalyze the transfer of a prenyl group to a natural product and these are widely distributed through living organisms [812]. They are also used to modulate protein function [812]. The dominance of prenylated NPs across a range of structural classes in the Rutaceae highlights a uniqueness within the angiosperms. Their presence in species across the breadth of the clades suggests an ancient origin within the family. Interestingly, the other Sapindales families, Sapindaceae, Simaroubaceae and Meliaceae have very few isoprenylated natural products reported (either isoprenylated flavonoids or indoles), suggesting that substrate plasticity leading to a diversification of the pathway within the Rutaceae occurred after its divergences 85 mya. Almost 50% of the isoprenylated natural products reported from Rutaceae species worldwide come from Australian rainforest species.

A second major producer of terpenoid conjugates are Myrtaceae species. In this case, the adducts are meroterpenoids, exclusively monoterpenoid or sesquiterpenoid conjugates of acyl phloroglucinols or acyl-2,2,4,4-tetramethyl-β-triketones. As discussed earlier, the starter units for the polyketide-derived acyl phloroglucinols in the Myrtaceae are branched chain carboxylic acids derived from amino acids. These meroterpenoids are unprecedented in the seed plants and their presence across ancient lineages of the Myrtaceae points to Australia being a center of origin and diversification of these structures. The sclerophyllous Myrtaceae flora (including *Eucalyptus*, *Corymbia*, *Angophora*, *Melaleuca*, and *Callistemon* species) have their origins in the Australian rainforests but now represent a dominant part of Australia's dry land habitats and are a major source of these phloroglucinols. Some of these taxa migrated north into Asia and the Mediterranean region, where they diversified and/or speciated. They are now represented by one or two species that contain meroterpenoids or acylphloroglucinols (for example, *Myrtus communis* (Mediterranean region), *Rhodomyrtus tormentosa*, and *Baeckea frutescens* (southeast Asia)).

Australian rainforest-derived sesquiterpenoids are concentrated in the Winteraceae (22 drimanes), Celastraceae (47 agarofuran), Cupressaceae (nine guaiane-, elemane-, or eudesmane-type) and Asteraceae (nine mostly guaiane), with Annonaceae (2), Lauraceae (4), Himantandraceae (1), Apocynaceae (3), Scrophulariaceae (5), and Solanaceae (2) contributing the remainder. There are undoubtedly more species that contain mono- and sesquiterpenoid-rich essential oils but the analysis of such information is outside the scope of this volume.

Altogether, 204 diterpenoids have been reported from only 14 families of Australian rainforest spermatophytes. The Euphorbiaceae has the most with 63 compounds (mainly phorbols and casbanes, but also the only sources of crotofolane- and halimane-structural types) followed by the Araucariaceae (29, a major source of labdanes), Erythroxylaceae (21 *ent*-kauranes), Escalionaceae (13 *ent*-kaurane alkaloids), Fabaceae (13) and Lamiaceae (13). Most of these diterpenoids are either labdane (40), *ent*-kaurane (34), abietane (26), clerodane (22), casbane (24), phorbol (22), or pimarane (15). The most notable occurrence here is the uniquely oxidized phorbols from *Fontainea* (Euphorbiaceae) and daphnanes from *Pimelea*

(Thymelaeaceae), but related structures have been reported from other Euphorbiaceae and Thymelaeaceae species elsewhere.

The *ent*-kaurene diterpenoid alkaloids have a patchy distribution across the phylogeny of angiosperms. They mostly occur in either species in Australian rainforests (*Anopterus* Escallionaceae, Campanulid), alpine meadows in the northern hemisphere (*Aconitum*, Ranunculaceae basal eudicot) or American dry shrubland areas (*Garrya*, Garryaceae, Lamiid) [813]. These alkaloids probably result from a late-stage amination of an oxidized *ent*-kaurene diterpenoid precursor and therefore the capacity to make these alkaloids is likely to be directly linked to the occurrence of *ent*-kaurene diterpenoids [813]. The highest numbers of *ent*-kaurene diterpenoids are reported from Lamiid's (mainly Lamiaceae) and Campanulids (mainly Asteraceae), but bryophyte, lycopod, gymnosperm, Magnoliid, Monocot, Ranunculaceae, Fabid, and Mavid species have also been reported to contain *ent*-kaurene diterpenoids [813]. The gibberellin plant growth regulators are derived from *ent*-kaurene, highlighting that the biosynthetic capacity to make these diterpenoids is found in all seed plants and that these compounds have an ancient origin [814]. The kaurene diterpenoid alkaloids in all cases have a nitrogen bridge between the two methyl carbons attached to C-3 and C-10 but the *Anopterus* alkaloids from Australian rainforests are unique because they contain an additional bond between C-14 and C-17.

Species from 29 families contain triterpenoids and, of the 241 structures reported, amyrin/lupeol (57) or betulinic/oleanolic/ursolic acid (143) types are dominant. Within the triterpenoid acids, 81 are glycosylated. Gymnosperms are known to produce amyrin/lupeol-type triterpenoids, but it is the basal eudicots where glycosylated betulinic/oleanolic/ursolic acid derivatives first appear with any great diversity. Across the Superrosids and Superasterids, their presence is patchy but over 5,000 compounds in this class have been reported worldwide. The Australian rainforest examples therefore represent neither an enhanced nor a depleted presence and the families where they are found are the same ones that have been reported to produce them elsewhere. These terpenoids therefore simply represent a common class of NPs that are found across many different habitats and continents.

6 Conclusion and Future Directions

In summary, Australian rainforest plants represent a structurally rich repository of natural product chemical diversity. There is still much to be learnt from studying the natural product chemistry of Australia's rainforest plants. Many species-rich families have barely been studied (Lauraceae, Orchidaceae, Proteaceae, Sapotaceae, Rubiaceae, Lamiaceae, Myrtaceae, Fabaceae, Euphorbiaceae, Phyllanthaceae, Sapindaceae, and Malvaceae) as have specialized plant parts (roots, flowers, and fruits) in those species that have been investigated. Methods to deconvolute and identify NPs found within complex mixtures remains a major hurdle that hampers rapid analysis of the rich repertoire of natural products likely to be found within unstudied (and under-studied) species. The present author believes that this still represents a "grand challenge" for natural product chemists. To illustrate this point, the NP database COCONUT contains just over 407,000 entries (with a very significant contribution of either natural product derivatives or synthetic-inspired analogues) [815]. This database has 97,000 unique molecular formulas but over 75% of the compounds (325,000) share only 15,000 molecular formulas (an average of 22 compounds per molecular formula). This makes MS-based metabolomic methods challenging to discriminate molecules solely based on mass spectrometric data. Nuclear magnetic resonance analysis can provide complimentary information that can greatly facilitate the identification of unique structures possessing the same molecular formula but the challenge is how to marry the MS data with NMR fingerprints for each of the individual compounds The power of using structural features derived from NMR fingerprints through application of covariance NMR methods [473], the separation of molecules using DOSY methods [90] and augmentation with computational methods and neural network analysis could help to provide a solution. The bioactivity of natural products reported from Australian rainforest species warrants much more study since most natural products have not been tested against any target and, in other cases, have only undergone screening against a limited number of disease targets. The precedent for the biomedical potential of Australian rainforest natural products is strong since drugs to treat cardiovascular disease, ophthalmic conditions and cancer have already been developed. The promise of future treatments for pain control, malaria (and other protozoan parasites), cancer, and inflammatory disorders are already evident. Much of Australia's rainforests are now protected in national parks, but there are still threats to their existence from land clearing, disease, invasive species, and climate change, and so it is imperative to protect and study this unique biodiversity while there is still have the opportunity to do so.

Acknowledgements I thank my daughter Caitlin Carroll for help with curating and formatting the references and my wife Claire Dickinson for proof reading. I thank Griffith University for providing me with a six-month sabbatical to research and write the chapter.

References

1. Sanmartín I (2012) Historical biogeography: evolution in time and space. Evol Educ Outreach 5:555
2. Carta A, Peruzzi L, Ramirez-Barahona S (2022) A global phylogenetic regionalization of vascular plants reveals a deep split between Gondwanan and Laurasian biotas. New Phytol 233:1494
3. Wang XQ, Ran JH (2014) Evolution and biogeography of gymnosperms. Mol Phylogenet Evol 75:24
4. Yang Y, Ferguson DK, Liu B, Mao KS, Gao LM, Zhang SZ, Wan T, Rushforth K, Zhang ZX (2022) Recent advances on phylogenomics of gymnosperms and a new classification. Plant Divers 44:340
5. Dimitrov D, Xu X, Su X, Shrestha N, Liu Y, Kennedy JD, Lyu L, Nogués-Bravo D, Rosindell J, Yang Y, Fjeldså J, Liu J, Schmid B, Fang J, Rahbek C, Wang Z (2023) Diversification of flowering plants in space and time. Nat Commun 4:7609

6. The Angiosperm Phylogeny Group (2016) An update of the Angiosperm Phylogeny Group classification for the orders and families of flowering plants: APG IV. Bot J Linn Soc 181:1
7. Zuntini AR, Carruthers T, Maurin O, Bailey PC, Leempoel K, Brewer GE, Epitawalage N, Françoso E, Gallego-Paramo B, McGinnie C, Negrão R, Roy SR, Simpson L, Toledo Romero E, Barber VMA, Botigué L, Clarkson JJ, Cowan RS, Dodsworth S, Johnson MG, Kim JT, Pokorny L, Wickett NJ, Antar GM, Debolt L, Gutierrez K, Hendriks KP, Hoewener A, Hu AQ, Joyce EM, Kikuchi IABS, Larridon I, Larson DA, De Lírio EJ, Liu JX, Malakasi P, Przelomska NAS, Shah T, Viruel J, Allnutt TR, Ameka GK, Andrew RL, Appelhans MS, Arista M, Ariza MJ, Arroyo J, Arthan W, Bachelier JB, Bailey CD, Barnes HF, Barrett MD, Barrett RL, Bayer RJ, Bayly MJ, Biffin E, Biggs N, Birch JL, Bogarín D, Borosova R, Bowles AMC, Boyce PC, Bramley GLC, Briggs M, Broadhurst L, Brown GK, Bruhl JJ, Bruneau A, Buerki S, Burns E, Byrne M, Cable S, Calladine A, Callmander MW, Cano Á, Cantrill DJ, Cardinal-Mcteague WM, Carlsen MM, Carruthers AJA, de Castro MA, Chase MW, Chatrou LW, Cheek M, Chen S, Christenhusz MJM, Christin PA, Clements MA, Coffey SC, Conran JG, Cornejo X, Couvreur TLP, Cowie ID, Csiba L, Darbyshire I, Davidse G, Davies NMJ, Davis AP, Van Dijk KJ, Downie SR, Duretto MF, Duvall MR, Edwards SL, Eggli U, Erkens RHJ, Escudero M, de la Estrella M, Fabriani F, Fay MF, Ferreira PDL, Ficinski SZ, Fowler RM, Frisby S, Fu L, Fulcher T, Galbany-Casals M, Gardner EM, German DA, Giaretta A, Gibernau M, Gillespie LJ, González CC, Goyder DJ, Graham SW, Grall A, Green L, Gunn BF, Gutiérrez DG, Hackel J, Haevermans T, Haigh A, Hall JC, Hall T, Harrison MJ, Hatt SA, Hidalgo O, Hodkinson TR, Holmes GD, Hopkins HCF, Jackson CJ, James SA, Jobson RW, Kadereit G, Kahandawala IM, Kainulainen K, Kato M, Kellogg EA, King GJ, Klejevskaja B, Klitgaard BB, Klopper RR, Knapp S, Koch MA, Leebens-Mack JH, Lens F, Leon CJ, Léveillé-Bourret É, Lewis GP, Li DZ, Li L, Liede-Schumann S, Livshultz T, Lorence D, Lu M, Lu-Irving P, Luber J, Lucas EJ, Luján M, Lum M, Macfarlane TD, Magdalena C, Mansano VF, Masters LE, Mayo SJ, McColl K, McDonnell AJ, McDougall AE, McLay TGB, McPherson H, Meneses RI, Merckx VSFT, Michelangeli FA, Mitchell JD, Monro AK, Moore MJ, Mueller TL, Mummenhoff K, Munzinger J, Muriel P, Murphy DJ, Nargar K, Nauheimer L, Nge FJ, Nyffeler R, Orejuela A, Ortiz EM, Palazzesi L, Peixoto AL, Pell SK, Pellicer J, Penneys DS, Perez-Escobar OA, Persson C, Pignal M, Pillon Y, Pirani JR, Plunkett GM, Powell RF, Prance GT, Puglisi C, Qin M, Rabeler RK, Rees PEJ, Renner M, Roalson EH, Rodda M, Rogers ZS, Rokni S, Rutishauser R, De Salas MF, Schaefer H, Schley RJ, Schmidt-Lebuhn A, Shapcott A, Al-Shehbaz I, Shepherd KA, Simmons MP, Simões AO, Simões ARG, Siros M, Smidt EC, Smith JF, Snow N, Soltis DE, Soltis PS, Soreng RJ, Sothers CA, Starr JR, Stevens PF, Straub SCK, Struwe L, Taylor JM, Telford IRH, Thornhill AH, Tooth I, Trias-Blasi A, Udovicic F, Utteridge TMA, Del Valle JC, Verboom GA, Vonow HP, Vorontsova MS, de Vos JM, Al-Wattar N, Waycott M, Welker CAD, White AJ, Wieringa JJ, Williamson LT, Wilson TC, Wong SY, Woods LA, Woods R, Worboys S, Xanthos M, Yang Y, Zhang YX, Zhou MY, Zmarzty S, Zuloaga FO, Antonelli A, Bellot S, Crayn DM, Grace OM, Kersey PJ, Leitch IJ, Sauquet H, Smith SA, Eiserhardt WL, Forest F, Baker WJ (2024) Phylogenomics and the rise of the angiosperms. Nature 629:843
8. Chassagne F, Cabanac G, Hubert G, David B, Marti G (2019) The landscape of natural product diversity and their pharmacological relevance from a focus on the Dictionary of Natural Products®. Phytochem Rev 18:601
9. Hagel JM, Facchini PJ (2013) Benzylisoquinoline alkaloid metabolism: a century of discovery and a brave new world. Plant Cell Physiol 54:647
10. Liscombe DK, MacLeod BP, Loukanina N, Nandi OI, Facchini PJ (2005) Evidence for the monophyletic evolution of benzylisoquinoline alkaloid biosynthesis in angiosperms. Phytochemistry 66:1374
11. Marchese C (2015) Biodiversity hotspots: a shortcut for a more complicated concept. Glob Ecol Conserv 3:297
12. Mcloughlin S (2001) The breakup history of Gondwana and its impact on pre-Cenozoic floristic provincialism. Aust J Bot 49:271

13. Crisp MD, Laffan S, Linder HP, Monro A (2002) Endemism in the Australian flora. J Biogeogr 28:183
14. Fujioka T, Chappell J (2010) History of Australian aridity: chronology in the evolution of arid landscapes. Geol Soc Spec Pub 346:121
15. Sanderson R (2008) Re-writing the history of Australian tropical rainforests: 'Alien invasives' or 'ancient indigenes'? Environ Hist 14:165
16. Byrne M, Steane DA, Joseph L, Yeates DK, Jordan GJ, Crayn D, Aplin K, Cantrill DJ, Cook LG, Crisp MD, Keogh JS, Melville J, Moritz C, Porch N, Sniderman JMK, Sunnucks P, Weston PH (2011) Decline of a biome: evolution, contraction, fragmentation, extinction and invasion of the Australian mesic zone biota. J Biogeogr 38:1635
17. Metcalfe DJ, Ford AJ (2009) Floristics and plant biodiversity of the rainforests of the wet tropics. In: Stork NE, Turton SM (eds) Living in a dynamic tropical forest landscape. Blackwell Publishing, Malden, Oxford, UK, p 123
18. Kondo T, Crisp MD, Linde C, Bowman DMJS, Kawamura K, Kaneko S, Isagi Y (2012) Not an ancient relic: the endemic Livistona palms of arid central Australia could have been introduced by humans. Proc R Soc Lond B, Biol Sci 279:2652
19. Viscarra Rossel RA, Bui EN (2016) A new detailed map of total phosphorus stocks in Australian soil. Sci Total Environ 542:1040
20. Carr DJ, Carr SGM, Hyland BPM, Wilson PG, Ladiges PY (2002) *Stockwelia quadrifida* (Myrtaceae), a new Australian genus and species in the eucalypt group. Bot J Linn Soc 139:415
21. Roberts P, Buhrich A, Caetano-Andrade V, Cosgrove R, Fairbairn A, Florin SA, Vanwezer N, Boivin N, Hunter B, Mosquito D, Turpin G, Ferrier A (2021) Reimagining the relationship between Gondwanan forests and Aboriginal land management in Australia's "Wet Tropics". iScience 24:102190
22. Fletcher MS, Hall T, Alexandra AN (2021) The loss of an indigenous constructed landscape following British invasion of Australia: an insight into the deep human imprint on the Australian landscape. Ambio 50:138
23. Bradshaw CJA (2012) Little left to lose: deforestation and forest degradation in Australia since European colonization. J Plant Ecol 5:109
24. Chapman AD (2009) Numbers of living species in Australia and the world, 2nd edn. Australian Biodiversity Information Services, Toowoomba, Queensland
25. Australian Government, Department of Climate Change, Energy, the Envirnonment and Water, 2024 Australia's World Heritage List 2024. http://www.dcceew.gov.au/parks-heritage/heritage/places/world-heritage-list
26. Webb LJ, Tracey JG, Williams WT (1984) A floristic framework of Australian rainforests. Aust J Ecol 9:169
27. Forests of Australia (2018) dataset [Online]. Australian Bureau of Agricultural and Resource Economics and Sciences, Forests of Australia (2018), Australian Bureau of Agricultural and Resource Economics and Sciences, Canberra, December, CC BY 4.0. 1 February 2025
28. CSIRO (2020) Tropical Rainforest Plants ed 8. https://apps.lucidcentral.org/rainforest/text/intro/index.html
29. Harden G, Nicholson H, McDonald B, Nicholson N, Tame T, Williams J (2016) Rainforest plants of Australia interactive identification key (mobile app). Terania Rainforest Publishing, Terania Creek, New South Wales

30. Jaman SJ, Kantvilas G, Brown MJ (1991) Floristic and ecological studies in Tasmanian rainforest. Tasmanian NRCP Technical Report No. 3, the Tasmanian component of the National Rainforest Conservation Program Hobart, Tasmania
31. Zhou S, Renner SS, Wen J (2006) Molecular phylogeny and intra- and intercontinental biogeography of Calycanthaceae. Mol Phylogenet Evol 39:1
32. Zani CL, Carroll AR (2017) Database for rapid dereplication of known natural products using data from MS and fast NMR experiments. J Nat Prod 80:1758
33. Price JR, Lamberton JA, Culvenor CCJ (1992) The Australian phytochemical survey: historical aspects of the CSIRO search for new drugs in Australian plants. Hist Rec Aust Sci 9:1960
34. Collins DJ, Culvenor CCJ, Lamberton JA, Loder JW, Price JR (1990) Plants for medicines—a chemical and pharmacological survey of plants in the Australian region. CSIRO Publishing, Clayton, Victoria
35. Laird S, Monagie C, Johnston S (2008) Queensland biodiscovery collaboration: the Griffith University AstraZeneca Partnership for Natural Product Drug Discovery. In: United Nations University Institute of Advanced Studies (UNU-IAS). Yokohama, Japan
36. Unpublished data that is associated with plant collections provided by the Queensland herbarium to Griffith University
37. Quinn RJ, Carroll AR, Pham NB, Baron P, Palframan ME, Suraweera L, Pierens GK, Muresan S (2008) Developing a drug-like natural product library. J Nat Prod 71:464
38. Camp D, Davis RA, Campitelli M, Ebdon J, Quinn RJ (2011) Drug-like properties: guiding principles for the design of natural product libraries. J Nat Prod 75:72
39. Personal communication: known natural products and the species from which they have been obtained from are provided in the accompanying tables
40. Griffith University (2024) NatureBank Information Leaflet. http://www.griffith.edu.au/__data/assets/pdf_file/0024/720492/Nature-Bank-Flyer.pdf
41. Qbiotics (2024) Our Science Web page (2024). http://www.qbiotics.com/science/our-science
42. Carroll AR, Copp BR, Davis RA, Keyzers RA, Prinsep MR (2022) Marine natural products. Nat Prod Rep 39:1122
43. Atlas of Living Australia Occurrence. https://doi.org/10.26197/5d7f2348165a0. Accessed 30 April 2024
44. Riggs NV (1954) The occurrence of macrozamin in the seeds of cycads. Aust J Chem 7:123
45. Moretti A, Sabato S, Gigliano GS (1981) Distribution of macrozamin in Australasian cycads. Phytochemistry 20:1415
46. Jones W, Hill K, Allen J (1995) *Wollemia nobilis*, a new living Australian genus and species in the Araucariaceae. Telopea 6:173
47. Bates RB, Cai S, Cantor RS, Carducci MD, Irvine AK, Jiorle BV, Nakkiew P, Setzer WN, Trinh LN (2003) Agatholic acid. Acta Crystallogr E 59:o97
48. Carman RM, Marty RA (1966) Diterpenoids. IX. *Agathis microstachya* oleoresin. Aust J Chem 19:2403
49. Carman RM, Craig WJ, Shaw IM (1973) Diterpenoids. XXXI. Three new resin acids. Aust J Chem 26:209
50. Carman RM, Dennis N (1964) The diterpene acids of *Agathis robusta* oleoresin. Aust J Chem 17:390
51. Ebada SS, Talaat AN, Labib RM, Mándi A, Kurtán T, Müller WEG, Singab A, Proksch P (2017) Cytotoxic labdane diterpenes and bisflavonoid atropisomers from leaves of *Araucaria bidwillii*. Tetrahedron 73:3048
52. Caputo R (1974) New diterpenes from *Araucaria cunninghamii*. Phytochemistry 13:475
53. Venditti A, Frezza C, Sciubba F, Foddai S, Serafini M, Bianco A (2017) Terpenoids and more polar compounds from the male cones of *Wollemia nobilis*. Chem Biodivers 14:e1600333
54. Frezza C, Venditti A, Rossi G, Serafini I, Pitorri M, Ciccòla A, Foddai S, Bianco A, Serafini M (2017) Phytochemical study on the leaves of *Wollemia nobilis*. Biochem Syst Ecol 74:63

55. Venditti A, Frezza C, Vincenti F, Brodella A, Sciubba F, Montesano C, Franceschin M, Sergi M, Foddai S, Di Cocco ME, Curini R, Delfini M, Bianco A, Serafini M (2019) A *syn-ent*-labdadiene derivative with a rare spiro-β-lactone function from the male cones of *Wollemia nobilis*. Phytochemistry 158:91
56. Venditti A, Frezza C, Campanelli C, Foddai S, Bianco A, Serafini M (2017) Phytochemical analysis of the ethanolic extract of *Agathis robusta* (C. Moore ex F. Muell.) F.M. Bailey. Nat Prod Res 31:1604
57. Khan NU, Ilyas M, Rahman W, Okigawa M, Kawano N (1970) On the bisflavones in the leaves of *Araucaria bidwillii* Hooker. Tetrahedron Lett 11:2941
58. Khan NU, Ilyas M, Rahman W, Mashima T, Okigawa M, Kawano N (1972) Biflavones from the leaves of *Araucaria bidwillii* Hooker and *Agathis alba* Foxworthy (Araucariaceae). Tetrahedron 28:5689
59. Khan NU, Ansari WH, Usmani JN, Ilyas M, Rahman W (1971) Biflavonyls of the Araucariales. Phytochemistry 10:2129
60. Gleńsk M, Włodarczyk M, Stefanowicz P, Kucharska A (2013) Biflavonoids from the Wollemi Pine, *Wollemia nobilis* (Araucariaceae). Biochem Syst Ecol 46:18
61. Venditti A, Frezza C, Rossi G, Serafini I, Ciccola A, Sciubba F, Foddai S, Tomassini L, Bianco A, Serafini M (2019) A new bicyclic monoterpene glucoside and a new biflavone from the male reproduction organs of *Wollemia nobilis*. Fitoterapia 133:62
62. Chen J, Yang ML, Zeng J, Gao K (2013) Antimicrobial activity of *Araucaria cunninghamii* Sweet and the chemical constituents of its twigs and leaves. Phytochem Lett 6:41
63. Carman RM, Russell-Maynard JKL, Schumann RC (1985) Isoflavones from *Podocarpus amarus*. Aust J Chem 38:485
64. Crowden RK, Grubb MJ (1971) Anthocyanins from five species of the Podocarpaceae. Phytochemistry 10:2821
65. Galbraith MN, Horn DHS, Middleton EJ, Hackey RJ (1969) The structure of podecdysone B, a new phytoecdysone. J Chem Soc D 402
66. Galbraith MN, Horn DHS, Porter QN, Hackney RJ (1968) Structure of podecdysone A, a steroid with moulting hormone activity from the bark of *Podocarpus elatus* R. Br. J Chem Soc D 971
67. Baggaley KH, Erdtman H, Mclean NJ, Norin T, Eriksson G, Blinc R, Paušak S, Ehrenberg L, Dumanović J (1967) Chemistry of the order Podocarpales. I. Heartwood constituents of the huon pine (*Dacrydium franklinii* Hook. f.). Acta Chem Scand 21:2247
68. Panichanun S, Ralph C. Bick I (1984) *Athrotaxis* alkaloids. Part I. Alkaloids of *A. cupressoides*. Tetrahedron 40:2677
69. Panichanun S, Ralph C, Bick I (1984) *Athrotaxis* alkaloids. Part II. Alkaloids of *A. selaginoides* and *A. laxifolia*. Tetrahedron 40:2685
70. Larsson S, Ronsted N (2014) Reviewing Colchicaceae alkaloids—perspectives of evolution on medicinal chemistry. Curr Top Med Chem 14:274
71. Brophy JJ, Goldsack RJ, Fookes CJR, Rozefelds AC (2002) The chemistry of the Australian gymnosperms—part 4: the leaf oils of the genus *Athrotaxis* D. Don (Cupressaceae). J Essent Oil Res 14:109
72. Daniels P, Erdtman H, Nishimura K, Norin T, Kierkegaard P, Pilotti AM (1972) Athrotaxin, a C17-phenolic constituent from *Athrotaxis selaginoides* Don. J Chem Soc, Chem Commun 246
73. Sadgrove NJ, Jones GL (2014) Medicinal compounds, chemically and biologically characterised from extracts of Australian *Callitris endlicheri* and *C. glaucophylla* (Cupressaceae): used traditionally in Aboriginal and colonial pharmacopoeia. J Ethnopharmacol 153:872
74. Watanabe Y, Mitsunaga T, Yoshimura T (2005) Investigating antitermitic compounds from Australian white cypress heartwood (*Callitris glaucophylla* Thompson et Johnson) against *Coptotermes formosanus* Shiraki. J Essent Oil Res 17:346
75. Wanner J, Jirovetz L, Schmidt E (2015) *Callitris intratropica* R.T. Baker & H.G. Smith as a novel rich source of deoxypodophyllotoxin. Curr Bioact Compd 11:73

76. Oyedeji AO, Ekundayo O, Sonwa MM, Fricke C, König WA (1998) (−)-Eudesma-1,4(15),11-triene from the essential oil of *Callitris intratropica*. Phytochemistry 48:657
77. Carman RM, Lambert LK, Robinson WT, Vandonge N (1986) 3,10-Dihydroxydielmentha-5,11-diene-4,9-dione. A diterpenoid (bismonoterpenoid) with a novel carbon skeleton. Aust J Chem 39:1843
78. Gadek PA, Quinn CJ (1982) Amentoflavones from *Callitris* species. Phytochemistry 21:248
79. Metcalfe DJ, Ford AJ (2009) A re-evaluation of Queensland's wet tropics based on primitive plants. Pac Conserv Biol 15:80
80. Murphy ST, Ritchie E, Taylor WC (1975) Some constituents of *Austrobaileya scandens* (Austrobaileyaceae): structures of seven new lignans. Aust J Chem 28:81
81. Tran TD, Pham NB, Booth R, Forster PI, Quinn RJ (2016) Lignans from the Australian endemic plant *Austrobaileya scandens*. J Nat Prod 79:1514
82. Riggs NV, Stevens JD (1962) Some constituents of *Piptocalyx moorei* Oliv. Aust J Chem 15:305
83. Da Silva Filho AA, Albuquerque S, Silva ML, Eberlin MN, Tomazela DM, Bastos JK (2004) Tetrahydrofuran lignans from *Nectandra megapotamica* with trypanocidal activity. J Nat Prod 67:42
84. Lowry JB, McAlpine JB, Riggs NV (1975) The absolute stereochemistry of piptoside and dihydropiptosidin. Aust J Chem 28:109
85. Muller S, Salomo K, Salazar J, Naumann J, Jaramillo MA, Neinhuis C, Feild TS, Wanke S (2015) Intercontinental long-distance dispersal of Canellaceae from the New to the Old World revealed by a nuclear single copy gene and chloroplast loci. Mol Phylogenet Evol 84:205
86. Mathie K, Lainer J, Spreng S, Dawid C, Andersson DA, Bevan S, Hofmann T (2017) Structure-pungency relationships and TRP channel activation of drimane sesquiterpenes in Tasmanian Pepper (*Tasmannia lanceolata*). J Agric Food Chem 65:5700
87. Cock IE (2013) The phytochemistry and chemotherapeutic potential of *Tasmannia lanceolata* (Tasmanian pepper): a review. Pharmacogn Commun 3:13
88. Deans BJ, Just J, Tedone L, Kilah NL, Jordan GJ, Bissember AC, Smith JA (2020) Distinct drimane chemotypes in Tasmanian mountain pepper (*Tasmannia lanceolata*): differences in the profiles of pungent leaf phytochemicals associated with altitudinal cline. J Agric Food Chem 68:315
89. Jansen BJ, De Groot A (2004) Occurrence, biological activity and synthesis of drimane sesquiterpenoids. Nat Prod Rep 21:449
90. Kleks G, Holland DC, Porter J, Carroll AR (2021) Natural products dereplication by diffusion ordered NMR spectroscopy (DOSY). Chem Sci 12:10930
91. Porter JD (2022) A chemical investigation of Australian Winteraceae. PhD Thesis, Griffith University, Southport, Queensland, p 523
92. Thomas N, Bruhl JJ, Ford A, Weston PH, Ebach M (2014) Molecular dating of Winteraceae reveals a complex biogeographical history involving both ancient Gondwanan vicariance and long-distance dispersal. J Biogeogr 41:894
93. Allouche N, Morleo B, Thoison O, Dumontet V, Nosjean O, Gueritte F, Sevenet T, Litaudon M (2008) Biologically active tetralones from New Caledonian *Zygogynum* spp. Phytochemistry 69:1750
94. Isnard S, Prosperi J, Wanke S, Wagner ST, Samain M-S, Trueba S, Frenzke L, Neinhuis C, Rowe NP (2012) Growth form evolution in *Piperales* and its relevance for understanding Angiosperm diversification: an integrative approach combining plant architecture, anatomy, and biomechanics. Int J Plant Sci 173:610
95. Liu R, Zhang H-C (2020) Chemical constituents from *Aristolochia tagala* and their chemotaxonomic significance. Biochem Syst Ecol 90:104037
96. Moretti C, Rideau M, Chenieux JC, Viel C (1979) [Isolation of aristolochic acid from two Madagascan Aristolochiaceae. Determination of its toxicity on vegetal cells. Comparison with animal cells (author's transl)]. Planta Med 35:360
97. Mebs D, Schneider M (2002) Aristolochic acid content of South-East Asian troidine swallowtails (Lepidoptera: Papilionidae) and of *Aristolochia* plant species (Aristolochiaceae). Chemoecology 12:11

98. Braby MF, Trueman JWH, Eastwood R (2005) When and where did troidine butterflies (Lepidoptera: Papilionidae) evolve? Phylogenetic and biogeographic evidence suggests an origin in remnant Gondwana in the Late Cretaceous. Invertebr Syst 19:113
99. Loder JW, Moorhouse A, Russell GB (1969) Tumour inhibitory plants. Amides of *Piper novae-hollandiae* (Piperaceae). Aust J Chem 22:1531
100. Setzer WN, Setzer MC, Bates RB, Nakkiew P, Jackes BR, Chen L, McFerrin MB, Meehan EJ (1999) Antibacterial hydroxycinnamic esters from *Piper caninum* from Paluma, north Queensland, Australia. The crystal and molecular structure of (+)-bornyl coumarate. Planta Med 65:747
101. Thebpatiphat S, Pengprecha S, Ternai B (1988) Some constituents of the stems of *Piper interruptum* Opiz. ScienceAsia 14:225
102. Doyle JA, Endress PK (2024) Integrating Cretaceous fossils into the phylogeny of living angiosperms: fossil Magnoliales and their evolutionary implications. Int J Plant Sci 185:42
103. Bergstrom G, Groth I, Pellmyr O, Endress PK, Thien LB, Hubener A, Francke W (1991) Chemical basis of a highly specific mutualism—chiral esters attract pollinating beetles in Eupomatiaceae. Phytochemistry 30:3221
104. Bowden BF, Picker K, Ritchie E, Taylor WC (1975) Constituents of *Eupomatia* species. IV. Structure and synthesis of eupolauridine (EL base-1) and some observations on the structures of eupolauramine (EL base-2) and hydroxyeupolauramine (EL base-3). Aust J Chem 28:2681
105. Read RW, Taylor WC (1979) Constituents of *Eupomatia* species. V. The isolation of eupomatenoid-13 (a new neolignan), (±)-trans-dehydrodiisoeugenol, and other extractives from the bark of *Eupomatia laurina*. Aust J Chem 32:2317
106. Carroll AR, Taylor WC (1991) Constituents of *Eupomatia* species. XII. Isolation of constituents of the tubers and aerial parts of *Eupomatia bennettii* and determination of the structures of new alkaloids from the aerial parts of *E. bennettii* and minor alkaloids of *E. laurina*. Aust J Chem 44:1615
107. Bowden BF, Freeman HC, Jones RDG (1976) Molecular and crystal structure of eupolauramine, a novel type of alkaloid from *Eupomatia laurina*. J Chem Soc, Perkin Trans 2:658
108. Taylor WC (1984) Constituents of *Eupomatia* species. IX. NMR evidence for the structure of eupolauramine and hydroxyeupolauramine. Aust J Chem 37:1095
109. Khan MR, Kihara M, Omoloso AD (2008) Antimicrobial activity of the alkaloidal constituents of the root bark of *Eupomatia laurina*. Pharm Biol 41:277
110. Bugg TDH, Lin G (2001) Solving the riddle of the intradiol and extradiol catechol dioxygenases: how do enzymes control hydroperoxide rearrangements? Chem Commun: 941
111. Taylor WC (1985) Chapter 1, *Eupomatia* alkaloids. In: Brossi A (ed) The alkaloids: chemistry and pharmacology, vol 24. Academic Press, Orlando, San Diego, New York, London, Toronto, Montreal, Sydney, Tokyo, p 1
112. Shamma M, Guinaudeau H (1984) Biogenetic pathways for the aporphinoid alkaloids. Tetrahedron 40:4795
113. Lucio ASSC, da Silva Almeida JRG, da-Cunha EVL, Tavares JF, Barbosa Filho JM (2015) Alkaloids of the Annonaceae: occurrence and a compilation of their biological activities. In: Knölker H-J (ed) The alkaloids: chemistry and biology, vol 74. Elsevier, Amsterdam, p 233
114. Doyle JA, Le Thomas A (2007) Phylogeny and geographic history of Annonaceae. Géogr Phys Quat 51:353
115. Kitahara Y, Onikura H, Shibano Y, Watanabe S, Mikami Y, Kubo A (1997) Synthesis of eupomatidines 1, 2 and 3 and related compounds including iminoquinolinequinone structure. Tetrahedron 53:6001
116. Pan E, Cao S, Brodie PJ, Callmander MW, Randrianaivo R, Rakotonandrasana S, Rakotobe E, Rasamison VE, Tendyke K, Shen Y, Suh EM, Kingston DG (2011) Isolation and synthesis of antiproliferative eupolauridine alkaloids of *Ambavia gerrardii* from the Madagascar dry forest. J Nat Prod 74:1169
117. Thi N, Oanh T, Tung H, Hue T, Thanh LN, Thao T, Hang M (2017) Two new alkaloids from *Polyalthia nemoralis*. Nat Prod Commun 12:1025

118. Carroll AR, Taylor WC (1991) Constituents of *Eupomatia* species. XIII. The structures of new lignans from the tubers and aerial parts of *Eupomatia bennettii*. Aust J Chem 44:1627
119. McCredie RS, Ritchie E, Taylor WC (1969) The structure and synthesis of eupomatene. A lignan of novel type from *Eupomatia laurina* R.Br. Aust J Chem 22:1011
120. Bowden B, Ritchie E, Taylor W (1972) Constituents of *Eupomatia* Species. II. Isolation and structure determination of further eupomatenoid lignans from the bark of *Eupomatia laurina*. Aust J Chem 25:2659
121. Raju R, Cullen JK, Bruce ZC, Reddell P, Munch G (2021) Eupomatenes A–E: neolignans isolated from the leaves of Australian rainforest plant *Eupomatia laurina*. Fitoterapia 153:104972
122. Liu SQ, Yang YP, Hussain N, Jian YQ, Li B, Qiu YX, Yu HH, Wang HZ, Wang W (2023) Dibenzocyclooctadiene lignans from the family Schisandraceae: a review of phytochemistry, structure-activity relationship, and hepatoprotective effects. Pharmacol Res 195:106872
123. Bowden BF, Read RW, Taylor WC (1980) Constituents of *Eupomatia* species. VI. The structures of eupodienone-1, -2 and -3. Aust J Chem 33:1823
124. Read RW, Taylor WC (1981) Constituents of *Eupomatia* species. VIII. The structures of eupodienone-4, -5, -6 and -7. Aust J Chem 34:1125
125. Carroll AR, Taylor WC (1990) Constituents of *Eupomatia* species. XI. New eupodienone lignans from the flowers of *Eupomatia laurina*. Aust J Chem 43:1871
126. He D, Ding L, Xu H, Lei X, Xiao H, Zhou Y (2012) Gymnothelignans A–O: conformation and absolute configuration analyses of lignans bearing tetrahydrofuran from *Gymnotheca chinensis*. J Org Chem 77:8435
127. Xiao SJ, Lao YZ, Chen F, Xu HX, Ding LS, Zhou Y (2014) Two new lignans from *Gymnotheca chinensis* Decne. Phytochem Lett 8:38
128. Xiao SJ, Lei XX, Xia B, Xiao HP, He DH, Fang DM, Qi HY, Chen F, Ding LS, Zhou Y (2014) Two novel norlignans from *Gymnotheca chinensis*. Tetrahedron Lett 55:2869
129. Xiao SJ, Chen F, Ding LS, Zhou Y (2014) Two new eupodienone lignans from *Gymnotheca chinensis*. Chin Chem Lett 25:463
130. Xiao SJ, Fang DM, Xia B, Xu LF, Xu HX, Jiang Y, Chen F, Ding LS, Zhou Y (2014) Biphenyl lignans with a tetrahydrofuran moiety from *Gymnotheca chinensis* Decne. Chin J Org Chem 34:1677
131. Xiao SJ, He DH, Fang DM, Chen F, Ding LS, Zhou Y (2014) Further biphenyl lignans with a tetrahydrofuran moiety from *Gymnotheca chinensis*. Helv Chim Acta 97:499
132. Xiao SJ, Lei XX, Xia B, Xu DQ, Xiao HP, Xu HX, Chen F, Ding LS, Zhou Y (2014) Two novel polycyclic spiro lignans from *Gymnotheca involucrata*. Tetrahedron Lett 55:5949
133. Xiao S, Zhang MS, Lang LH, Fu SB, Chen F, Ding LS, Zhou Y (2016) Phytochemical investigation of *Gymnotheca chinensis*. Chem Nat Compd 52:979
134. Lu DW, Zhang X, Chen MH (2022) A new lignan from *Gymnotheca involucrata*. Chem Nat Compd 58:411
135. Carroll AR, Taylor WC (1991) Constituents of *Eupomatia* species. XIV. The structures of eupomatilone-1, -2, -3, -4, -5, -6 and -7 isolated from *Eupomatia bennettii*. Aust J Chem 44:1705
136. Carroll AR (1988) A chemical investigation of the Eupomatiaceae and related studies. PhD Thesis, University of Sydney, Sydney, New South Wales, p 248
137. Gurjar MK, Cherian J, Ramana CV (2004) Synthesis of the putative structure of eupomatilone-6. Org Lett 6:317
138. Coleman RS, Gurrala SR (2004) Total synthesis of eupomatilones 4 and 6: structurally rearranged and atropisomerically fluxional lignan natural products. Org Lett 6:4025
139. Picker K, Ritchie E, Taylor WC (1973) Constituents of *Eupomatia* species. III. New eupomatenoid lignans from the leaves and wood of *Eupomatia laurina*. Aust J Chem 26:1111
140. Su YCF, Saunders RMK (2009) Evolutionary divergence times in the Annonaceae: evidence of a late Miocene origin of *Pseuduvaria* in Sundaland with subsequent diversification in New Guinea. BMC Evol Biol 9:153

141. Johns SR, Lamberton JA, Li CS, Sioumis AA (1970) Alkaloids from *Pseuduvaria* species, *Schefferomitra subaequalis*, and *Polyalthia nitidissima* (Annonaceae): isolation of a new alkaloid shown to be 1,2,9,10-tetramethoxynoraporphine(norglaucine). Aust J Chem 23:423
142. Jossang A, Leboeuf M, Cabalion P, Cavé A (1983) Alkaloids from Annonaceae. Planta Med 49:20
143. Levrier C, Balastrier M, Beattie KD, Carroll AR, Martin F, Choomuenwai V, Davis RA (2013) Pyridocoumarin, aristolactam and aporphine alkaloids from the Australian rainforest plant *Goniothalamus australis*. Phytochemistry 86:121
144. Mueller D, Davis RA, Duffy S, Avery VM, Camp D, Quinn RJ (2009) Antimalarial activity of azafluorenone alkaloids from the Australian tree *Mitrephora diversifolia*. J Nat Prod 72:1538
145. Wu YC, Chang GY, Duh CY, Wang SK (1993) Cytotoxic alkaloids of *Annona montana*. Phytochemistry 33:497
146. Bousserouel H, Awang K, Guéritte F, Litaudon M (2012) Enyne- and enediyne-γ-lactones from the bark of *Meiogyne cylindrocarpa*. Phytochem Lett 5:29
147. Litaudon M, Bousserouel H, Awang K, Nosjean O, Martin MT, Dau ME, Hadi HA, Boutin JA, Sévenet T, Guéritte F (2009) A dimeric sesquiterpenoid from a Malaysian *Meiogyne* as a new inhibitor of Bcl-xL/BakBH3 domain peptide interaction. J Nat Prod 72:480
148. Chaudhary NK, Taylor WC, Mander LN, Karuso P (2021) Isolation and structure elucidation of additional alkaloids from the tropical rainforest tree *Galbulimima baccata*. J Nat Prod 84:2525
149. Hughes GK, Ritchie E (1954) The chemical constituents of *Himantandra* species. I. The lignins of *Himantandra baccata* Bail. and *H. belgraveana* F. Muell. Aust J Chem 7:104
150. Birch AJ, Mostyn KMC (1955) A new sesquiterpene alcohol from *Himantandra baccata* Bail. Aust J Chem 8:550
151. Brown FFC, Drummond R, Fogerty AC, Hughes GK, Pinhey JT, Ritchie E, Taylor WC (1956) Corrigenda—the chemical constituents of *Himantandra* species. II. The isolation of the alkaloids of *Himantandra baccata* Bail. and *H. belgraveana* F. Muell. Aust J Chem 9:283
152. Guise GB, Mander LN, Prager RH, Rasmussen M, Ritchie E, Taylor WC (1967) Corrigenda—the chemical constituents of *Galbulimima* species. VIII. The structures of the ester alkaloids. Aust J Chem 20:1029
153. Mander LN, Prager RH, Rasmussen M, Ritchie E, Taylor WC (1967) The chemical constituents of *Galbulimima* species. IX. The structures of himbadine and alkaloid G.B. 13. Aust J Chem 20:1473
154. Mander LN, Prager RH, Rasmussen M, Ritchie E (1967) The chemical constituents of *Galbulimima* species. X. The structure of himgaline. Aust J Chem 20:1705
155. Mander LN, Willis AC, Herlt AJ, Taylor WC (2009) The structures of three new *Galbulimima* alkaloids. Tetrahedron Lett 50:7089
156. Bradford TA, Willis AC, White JM, Herlt AJ, Taylor WC, Mander LN (2011) The structures of four new himbacine-like *Galbulimima* alkaloids. Tetrahedron Lett 52:188
157. Lan P, Herlt AJ, Willis AC, Taylor WC, Mander LN (2018) Structures of new alkaloids from rain forest trees *Galbulimima belgraveana* and *Galbulimima baccata* in Papua New Guinea, Indonesia, and Northern Australia. ACS Omega 3:1912
158. Behenna DC, Stockdill JL, Stoltz BM (2008) The biology and chemistry of the zoanthamine alkaloids. Angew Chem Int Ed Engl 47:2365
159. Chackalamannil S, Wang Y, Greenlee WJ, Hu Z, Xia Y, Ahn HS, Boykow G, Hsieh Y, Palamanda J, Agans-Fantuzzi J, Kurowski S, Graziano M, Chintala M (2008) Discovery of a novel, orally active himbacine-based thrombin receptor antagonist (SCH 530348) with potent antiplatelet activity. J Med Chem 51:3061
160. Knight E, Robinson E, Smoktunowicz N, Chambers RC, Aliev AE, Inglis GG, Chudasama V, Caddick S (2016) Synthesis of novel and potent vorapaxar analogues. Org Biomol Chem 14:3264
161. Oikawa H (2010) Diels–Alderases. In Liu H-W, Mander L (Eds) Comprehensive natural products II: chemistry and biology. Elsevier, Oxford, UK, p 277

162. Allouche N, Apel C, Martin MT, Dumontet V, Gueritte F, Litaudon M (2009) Cytotoxic sesquiterpenoids from Winteraceae of Caledonian rainforest. Phytochemistry 70:546
163. Renner SS (1999) Circumscription and phylogeny of the Laurales: evidence from molecular and morphological data. Am J Bot 86:1301
164. Kuke RK, Allan RD, Johnston GA, Mewett KN, Mitrovic AD, Duke CC, Hambley TW (1995) Idiospermuline, a trimeric pyrrolidinoindoline alkaloid from the seed of *Idiaospermum australiense*. J Nat Prod 58:1200
165. Calixto NO, Pinto MEF, Ramalho SD, Burger MCM, Bobey AF, Young MCM, Bolzani VS, Pinto AC (2016) The genus *Psychotria*: phytochemistry, chemotaxonomy, ethnopharmacology and biological properties. J Braz Chem Soc 27:1355
166. Soh KS, Lahey FN, Greenhalgh R (1966) The structure of hernandine. Tetrahedron Lett 7:5279
167. McKenzie AW, Price JR (1953) The alkaloids of *Gyrocarpus americanus* Jacq. Aust J Chem 6:180
168. Levrier C, Rockstroh A, Gabrielli B, Kavallaris M, Lehman M, Davis RA, Sadowski MC, Nelson CC (2018) Discovery of thalicthuberine as a novel antimitotic agent from nature that disrupts microtubule dynamics and induces apoptosis in prostate cancer cells. Cell Cycle 17:652
169. Chalandre MC, Bruneton J, Cabalion P, Guinaudeau H (1986) Étude des Hernandiacées. XII. Dimères aporphine–benzylisoquinoléine originaux isolés de *Hernandia peltata*. Can J Chem 64:123
170. Bick IRC, Geewananda YA, Gunawardana P, Patrick VA, White AH (1985) Crystallographic characterization of isosevanine, a new benzylisoquinoline alkaloid. Aust J Chem 38:1571
171. Gunawardana YAGP, Cordell GA, Bick IRC (1988) (–)-5-*exo*-Hydroxyborneol: isolation and NMR spectral assignments. J Nat Prod 51:142
172. Chanderbali AS, Van Der Werff H, Renner SS (2001) Phylogeny and historical biogeography of Lauraceae: evidence from the chloroplast and nuclear genomes. Ann Missouri Bot Gard 88:104
173. Clezy PS, Gellert E, Lau DYK, Nichol AW (1966) The alkaloids of *Beilschmiedia elliptica*. Aust J Chem 19:135
174. Cooke RG, Haynes HF (1954) The alkaloids of *Cryptocarya angulata* C.T. White and *C. triplinervis*. Aust J Chem 7:99
175. Jones SR, Lamberton JA (1966) *Cassytha* alkaloids. I. New aporphine alkaloids from *Cassytha filiformis* L. Aust J Chem 19:297
176. Johns SR, Lamberton JA, Sioumis AA (1966) *Cassytha* alkaloids. II. Alkaloids of *Cassytha pubescens* R. Br. Aust J Chem 19:2331
177. Ndi CP, Sykes MJ, Claudie DJ, Mckinnon RA, Semple SJ, Simpson BS (2016) Antiproliferative aporphine alkaloids from *Litsea glutinosa* and ethnopharmacological relevance to Kuuku I'yu traditional medicine. Aust J Chem 69:145
178. Lamberton J, Vashist V (1972) Alkaloids of *Litsea leefeana* and *Cryptocarya foveolata* (Lauraceae). Aust J Chem 25:2737
179. Tran TD, Pham NB, Fechner G, Quinn RJ (2010) Chemical investigation of drug-like compounds from the Australian tree, *Neolitsea dealbata*. Bioorg Med Chem Lett 20:5859
180. Ellis J, Gellert E, Summons R (1972) Alkaloids of the genus *Cinnamomum*. III. Identification of reticuline as the major bark alkaloid of *Cinnamomum pleurosperma*. Aust J Chem 25:1829
181. Gellert E (1959) The constituents of *Cryptocarya pleurosperma* White & Francis. I. Pleurospermine: a new alkaloid of the leaves. Aust J Chem 12:90
182. Johns SR, Lamberton JA, Sioumis AA, Willing RI (1970) New alkaloids from *Cryptocarya pleurosperma* (Lauraceae): the structures of cryptopleuridine and cryptopleurospermine. Aust J Chem 23:353
183. Gellert E, Riggs NV (1954) Crytopleurine: and alkaloid of *Cinnamomum pleurosperma* White & Francis. Aust J Chem 7:113
184. Bick IRC, Sinchai W (1981) Phenanthroindolizidine and phenanthroquinolizidine alkaloids. In: Brossi A (ed) The alkaloids: chemistry and pharmacology, vol 19. Academic Press

Amsterdam, Boston, Heidelberg, London, New York, Oxford, Paris, San Diego, San Francisco, Sydney, Tokyo, p 193
185. Banfield JE, Black DS, Collins DJ, Hyland BPM, Lee JJ, Pranowo SR (1994) Constituents of some species of *Beilschmiedia* and *Endiandra* (Lauraceae): new endiandric acid and benzopyran derivatives isolated from *B. oligandra*. Aust J Chem 47:587
186. Raju R, Gunawardena D, Reddell P, Munch G (2022) Cryptocaryoic acids A–C: new phenyl alkyl acids isolated from the leaves of Australian rainforest plant *Cryptocarya mackinnonia*. Fitoterapia 162:105266
187. Dong L, Schill H, Grange RL, Porzelle A, Johns JP, Parsons PG, Gordon VA, Reddell PW, Williams CM (2009) Anticancer agents from the Australian tropical rainforest: spiroacetals EBC-23, 24, 25, 72, 73, 75 and 76. Chemistry 15:11307
188. Davis RA, Demirkiran O, Sykes ML, Avery VM, Suraweera L, Fechner GA, Quinn RJ (2010) 7′,8′-Dihydroobolactone, a typanocidal alpha-pyrone from the rainforest tree *Cryptocarya obovata*. Bioorg Med Chem Lett 20:4057
189. Lenta BN, Chouna JR, Nkeng-Efouet PA, Sewald N (2015) Endiandric acid derivatives and other constituents of plants from the genera *Beilschmiedia* and *Endiandra* (Lauraceae). Biomolecules 5:910
190. Dong L, Gordon VA, Grange RL, Johns J, Parsons PG, Porzelle A, Reddell P, Schill H, Williams CM (2008) Structure and absolute stereochemistry of the anticancer agent EBC-23 from the Australian rainforest. J Am Chem Soc 130:15262
191. Brophy JJ, Forster PI, Goldsack RJ (2016) Coconut laurels: the leaf essential oils from four endemic Australian *Cryptocarya* species: *C. bellendenkerana*, *C. cocosoides*, *C. cunninghamii* and *C. lividula* (Lauraceae). Nat Prod Commun 11:255
192. Bhat ZS, Rather MA, Maqbool M, Lah HU, Yousuf SK, Ahmad Z (2017) α-pyrones: small molecules with versatile structural diversity reflected in multiple pharmacological activities—an update. Biomed Pharmacother 91:265
193. Davis RA, Carroll AR, Duffy S, Avery VM, Guymer GP, Forster PI, Quinn RJ (2007) Endiandrin A, a potent glucocorticoid receptor binder isolated from the Australian plant *Endiandra anthropophagorum*. J Nat Prod 70:1118
194. Davis RA, Barnes EC, Longden J, Avery VM, Healy PC (2009) Isolation, structure elucidation and cytotoxic evaluation of endiandrin B from the Australian rainforest plant *Endiandra anthropophagorum*. Bioorg Med Chem 17:1387
195. Ritchie E, Senior RG, Taylor WC (1969) Cycloneolitsin, a cyclopropanoid triterpene ether with a novel side-chain. Aust J Chem 22:2371
196. Wu X, Vogler B, Jackes BR, Setzer WN (2008) Terpenoids from *Neolitsea dealbata*. Nat Prod Commun 3:129
197. Bick IRC, Clezy PS, Crow ED (1956) The alkaloids of *Atherosperma moschatum* Labill. I. Isolation of the alkaloids and structure of berbamine. Aust J Chem 9:111
198. Bick IRC, Bremner JB, Leow HM, Wiriyachitra P (1972) Alkaloids of *Daphnandra* species. Part XI. Some dibenzo-p-dioxin-type biscoclaurine alkaloids from an unnamed species. The structure of micranthine. J Chem Soc, Perkin Trans 2884
199. Bick IRC, Sotheeswaran S (1978) Alkaloids of *Daphnandra apatela*. Aust J Chem 31:2077
200. Bick IRC, Leow HM (1978) Alkaloids of *Daphnandra johnsonii*. Aust J Chem 31:2539
201. Bick IRC, Bremner JB, Leow HM, Wiriyachitra P (1972) The structure and stereochemistry of micranthine. Tetrahedron Lett 13:33
202. Bick IRC, Taylor WI, Todd AR (1953) 140. Alkaloids of *Daphnandra* species. Part V. An examination of the alkaloidal content of *D. micrantha* and of *D. tenuipes*. Isolation of some minor *Daphnandra* alkaloids. J Chem Soc 695
203. Schiff LJP, Galal AM, Critchett CD, Bick RCI, Sotheeswaran S, Gao C-Y, Lin F-T, Duah FK, Fu EW, Wong L (1989) Nortenuipine-2′b-N-oxide, a new bisbenzylisoquinoline alkaloid from *Daphnandra dielsii*. Heterocycles 29:1689
204. Howard AS, Harleymason J, Bick IRC (1990) Alkaloids of *Daphnandra dielsii*. II. Daphnine. Isolation and characterization. Aust J Chem 43:1129

205. Guilhem J, Bick IRC (1981) Daphnine, a unique bisbenzylisoquinoline alkaloid: X-ray crystal structure of its dihydroiodide. J Chem Soc, Chem Commun 1007
206. Bick IC, Leo W, Richards M (1980) Alkaloids of *Doryphora aromatica*. Aust J Chem 33:225
207. Liu M, Han J, Feng Y, Guymer G, Forster PI, Quinn RJ (2021) Antimicrobial benzyltetrahydroisoquinoline-derived alkaloids from the leaves of *Doryphora aromatica*. J Nat Prod 84:676
208. Bick IRC, Harley-Mason J, Sheppard N, Vernengo MJ (1961) 367. Structural correlations in the nuclear magnetic resonance spectra of bisbenzylisoquinoline and aporphine alkaloids. J Chem Soc 1896
209. Carroll AR, Davis RA, Roberts E (Griffith University) (1998) Alkaloid compounds. Australian patent number 199863618
210. Ripper J, Jenkins I, Van Le P, Quinn R, Janssen C (AstraZeneca) (2009) 1,2,3,4-Tetrahydroisoquinoline derivatives, preparations thereof and uses thereof. U.S. patent application number 20090023728A1
211. Bick IRC, Douglas GK (1965) The structures of atheroline and moschatoline. Tetrahedron Lett 6:4655
212. Bick IR, Douglas GK (1965) Atheroline, a yellow phenolic alkaloid from *Atherosperma moschatum* Labill. Tetrahedron Lett 2399
213. Gharbo SA, Beal JL, Schlessinger RH, Cava MP, Svoboda GH (1965) A phytochemical study of *Doryphora sassafras*. I. Isolation of eight crystalline alkaloids from the leaves. Lloydia 28:237
214. Chen CR, Beal JL, Doskotch RW, Mitscher LA, Svoboda GH (1974) A phytochemical study of *Doryphora sassafras*. II. Isolation of eleven crystalline alkaloids from the bark. Lloydia 37:493
215. Buchanan MS, Davis RA, Duffy S, Avery VM, Quinn RJ (2009) Antimalarial benzylisoquinoline alkaloid from the rainforest tree *Doryphora sassafras*. J Nat Prod 72:1541
216. Bick IRC, Douglas GK, Taylor WI (1969) Alkaloids of *Dryadodaphne novoguineensis*. J Chem Soc C 1627
217. Carroll AR, Davis RA, Forster PI, Guymer GP, Quinn RJ (2001) A benzylisoquinoline alkaloid from *Doryphora sassafras*. J Nat Prod 64:1572
218. Bick IRC, Douglas GK (1965) The alkaloids of *Atherosperma moschatum* Labill. II. Non-phenolic alkaloids. Aust J Chem 18:1997
219. Challinor VL, Smith DM, De Voss JJ (2011) Steroidal saponins isolated from an Australian yam *Dioscorea* sp. Aust J Chem 64:545
220. Zraunig A, Pacher T, Brecker L, Greger H (2014) Phenylbenzofuran-type stilbenoids from *Stemona* species. Phytochem Lett 9:33
221. Nguyen TPA, Kim JS, Kim JH (2013) Molecular phylogenetic relationships and implications for the circumscription of Colchicaceae (Liliales). Bot J Linn Soc 172:255
222. Johns SR, Kowala C, Lamberton JA, Sioumis AA, Wunderlich JA (1968) "Homoerythrina" alkaloids from *Schelhammera pedunculata* F. Muell. (family Liliaceae). J Chem Soc D 1102
223. Johns SR, Lamberton JA, Sioumis AA (1969) Alkaloids of *Schelhammera pedunculata* (Liliaceae). III. The structures of schelhammericine and alkaloids A, B, E, G, H, J, and K. Aust J Chem 22:2219
224. Battersby AR, Bradbury RB, Herbert RB, Munro MHG, Ramage R (1967) Structure and synthesis of homoaporphines: a new group of 1-phenethylisoquinoline alkaloids. J Chem Soc D:450
225. Fridrichsons J, MacKay MF, Mathieson AM (1968) The absolute molecular structures of kreysiginine and of morphine. Tetrahedron Lett 24:2887
226. Battersby AR, McDonald E, Munro MHG, Ramage R (1967) Homoaporphine systems and related dienones: isolation, structure, and synthesis. J Chem Soc D:934
227. Cox SD, Jayasinghe KC, Markham JL (2005) Antioxidant activity in Australian native sarsaparilla (*Smilax glyciphylla*). J Ethnopharmacol 101:162
228. Huang AC, Wilde A, Ebmeyer J, Skouroumounis GK, Taylor DK (2013) Examination of the phenolic profile and antioxidant activity of the leaves of the Australian native plant *Smilax glyciphylla*. J Nat Prod 76:1930

229. Sarker SD, Dinan LN, Girault J-P, Lafont R, Waterman PG (1996) Punisterone [(20R,24S)-25-deoxy-11α,20,24-trihydroxyecdysone]: a new phytoecdysteroid from *Blandfordia punicea*. J Nat Prod 59:789
230. Bloor SJ (2001) Deep blue anthocyanins from blue *Dianella* berries. Phytochemistry 58:923
231. Jensen BS, Christensen SB, Jäger AK, Rønsted N (2011) Amaryllidaceae alkaloids from the Australasian tribe Calostemmateae with acetylcholinesterase inhibitory activity. Biochem Syst Ecol 39:153
232. Chuah CH, Yong HS, Goh SH (1997) Methyl eugenol, a fruit-fly attractant, from the browning leaves of *Proiphys amboinensis* (Amaryllidaceae). Biochem Syst Ecol 25:391
233. Jewers K, Burbage MB, Blunden G, Griffin WJ (1974) Brisbagenin and brisbenone, two new spirostanes from *Cordyline* species. Steroids 24:203
234. Jewers K, Manchanda AH, Dougan J, Nagler MJ, Blunden G, Griffin WJ (1974) Cordylagenin, a new steroidal sapogenin diol from *Cordyline cannifolia* and *C. stricta*. Tetrahedron Lett 15:1475
235. Griffin WJ, Blunden G, Jewers K, Burbage MB, Nagler MJ (1976) Steroidal sapogenins from *Cordyline cannifolia* leaves. Phytochemistry 15:1271
236. Blunden G, Jaffer JA, Jewers K, Griffin WJ (2004) Steroidal sapogenins from leaves of *Cordyline* species. J Nat Prod 44:441
237. Tran TD, Olsson MA, Choudhury MA, McMillan DJ, Cullen JK, Parsons PG, Bernhardt PV, Reddell PW, Ogbourne SM (2019) Antibacterial 5*α*-spirostane saponins from the fruit of *Cordyline manners-suttoniae*. J Nat Prod 82:2809
238. Yang M-H, Blunden G, Patel A, Crabb TA, Brain K, Griffin WJ (1990) Two furostane sapogenins from *Cordyline rubra*. Phytochemistry 29:1332
239. Yang M-H, Blunden G, Patel A, Crabb TA, Griffin WJ (1989) Rubragenin, chenogenin and wallogenin, steroidal sapogenins from *Cordyline rubra*. Phytochemistry 28:3171
240. Mimaki Y, Takaashi Y, Kuroda M, Sashida Y (1997) Steroidal glucosides from leaves of *Cordyline stricta*. Phytochemistry 45:1229
241. Blunden G, Jaffer JA, Jewers K, Griffin WJ (1981) Crabbogenin, 1*β*-hydroxycrabbogenin, strictagenin and pompeygenin, four new steroidal sapogenins from *Cordyline stricta* leaves. Tetrahedron 37:2911
242. Mimaki Y, Kuroda M, Takaashi Y, Sashida Y (1998) Steroidal saponins from the leaves of *Cordyline stricta*. Phytochemistry 47:79
243. Allan RD, Correll RL, Wells RJ (1969) A new class of quinones from certain members of the family Cyperaceae. Tetrahedron Lett 10:4669
244. Kassem ME, Shoela S, Marzouk MM, Sleem AA (2012) A sulphated flavone glycoside from *Livistona australis* and its antioxidant and cytotoxic activity. Nat Prod Res 26:1381
245. El-Desouky SK, Kassem MES, Al-Fifi ZIA, El-Deen AMG (2009) A new pyranone derivative from the leaves of *Livistona australis*. Nat Prod Commun 4:499
246. Singab AN, Elnaggar D, Elgindi MR, Abouelala MBE (2015) Biological studies of isolated triterpenoids and phenolic compounds identified from *Wodyetia bifurcata* family Arecaceae. J Pharmacogn Phytochem 3:67
247. Wohlmuth H, Deseo MA, Brushett DJ, Thompson DR, MacFarlane G, Stevenson LM, Leach DN (2010) Diarylheptanoid from *Pleuranthodium racemigerum* with in vitro prostaglandin E(2) inhibitory and cytotoxic activity. J Nat Prod 73:743
248. Raju R, Singh A, Gunawardena D, Reddell P, Münch G (2019) Diarylheptanoids with anti-inflammatory activity from the rhizomes of *Pleuranthodium racemigerum* (Zingiberaceae). Phytochem Lett 30:10
249. He ZH, Gilli C, Yue GG, Lau CB, Greger H, Brecker L, Ge W, But PP (2012) Anti-angiogenic effects and mechanisms of zerumin A from *Alpinia caerulea*. Food Chem 132:201
250. Worberg A, Quandt D, Barniske AM, Löhne C, Hilu KW, Borsch T (2007) Phylogeny of basal eudicots: insights from non-coding and rapidly evolving DNA. Org Divers Evol 7:55
251. Hart NK, Johns SR, Lamberton JA, Suares H (1972) Alkaloids of *Pachygone pubescens* (Menispermaceae). Aust J Chem 25:2289

252. Anet FAL, Hughes GK, Ritchie E (1950) The alkaloids of *Pleogyne cunninghamii*. Aust J Chem 3:346
253. Sioumis AA, Vashist VN (1972) New bisbenzylisoquinoline alkaloids from *Pycnarrhena australiana* (Menispermaceae). Aust J Chem 25:2251
254. Abouchacra ML, Leboeuf M, Guinaudeau H, Cavé A, Cabalion P (1987) The bisbenzylisoquinoline alkaloids of *Pycnarrhena ozantha*. J Nat Prod 50:375
255. Lian L, Peng HW, Ortiz RDC, Jabbour F, Gao TG, Erst AS, Chen ZD, Wang W (2023) Phylogeny and biogeography of Tiliacoreae (Menispermaceae), a tribe restricted to tropical rainforests. Ann Bot 131:685
256. Carroll AR, Arumugan T, Redburn J, Ngo A, Guymer GP, Forster PI, Quinn RJ (2010) Hasubanan alkaloids with δ-opioid binding affinity from the aerial parts of *Stephania japonica*. J Nat Prod 73:988
257. Blanchfield JT, Sands DP, Kennard CH, Byriel KA, Kitching W (2003) Characterisation of alkaloids from some Australian *Stephania* (Menispermaceae) species. Phytochemistry 63:711
258. Bartley JP, Baker LT, Carvalho CF (1994) Alkaloids of *Stephania bancroftii*. Phytochemistry 36:1327
259. Hughes GK, Kaiser FP, Matheson N, Ritchie E (1953) The constituents of *Legnephora moorei* F. Muell. Aust J Chem 6:90
260. Hungerford NL, Sands DPA, Kitching W (1998) Isolation and structure of some constituents of the Australian medicinal plant *Tinospora smilacina* ('snakevine'). Aust J Chem 51:1103
261. Southwell IA, Tucker DJ (1993) Protoanemonin in Australian *Clematis*. Phytochemistry 33:1099
262. Lamont BB, He TH, Milne LA, Cowling RM (2024) Linked continents, overland migration and differential survival explain abundance of Proteaceae in Australia. Perspect Pl Ecol Evol Syst 62
263. Bick IRC, Gillard JW, Leow HM, Preston NW (1979) Alkaloids of *Agastachys odorata*. Aust J Chem 32:2071
264. Bick IRC, Gillard JW, Leow H (1979) Alkaloids of *Bellendena montana*. Aust J Chem 32:1827
265. Bick IRC, Gillard JW, Leow H (1979) Alkaloids of *Darlingia darlingiana*. Aust J Chem 32:2523
266. Katavic PL, Butler MS, Quinn RJ, Forster PI, Guymer GP (1999) Tropane alkaloids from *Darlingia darlingiana*. Phytochemistry 52:529
267. Bick IRC, Gillard JW, Leow H (1979) Alkaloids of *Darlingia ferruginea*. Aust J Chem 32:2537
268. Yang F (2017) Chemotaxonomy study of plants from the family Proteaceae based on its natural product profile (alkaloid). PhD Thesis, Griffith University, Southport, New South Wales, p 297
269. Butler MS, Katavic PL, Davis RA, Forster PI, Guymer GP, Quinn RJ (2000) 10-Hydroxydarlingine, a new tropane alkaloid from the Australian proteaceous plant *Triunia erythrocarpa*. J Nat Prod 63:688
270. Yang F, Zhao H, Carroll AR (2017) Tropane alkaloids from the Australian plant *Triunia montana* (Proteaceae). Tetrahedron Lett 58:736
271. Bick IRC, Gillard JW, Leow H-M (1986) Isolation of tri-(α-methylene-γ-butyrolactonyl)amine from *Bellendena montana*. Phytochemistry 25:972
272. Kohnen-Johannsen KL, Kayser O (2019) Tropane alkaloids: chemistry, pharmacology, biosynthesis and production. Molecules 24:796
273. Lounasmaa M, Pusset J, Sévenet T (1980) Benzyl-2 tropanes et γ-pyranotropanes, alcaloïdes de *Knightia strobilina*. Phytochemistry 19:949
274. Daly JW (2005) Nicotinic agonists, antagonists, and modulators from natural sources. Cell Mol Neurobiol 25:513
275. Foreman DB (1995) Flora of Australia Volume 16—Elaeagnaceae, Proteaceae 1. Australian Biological Resources Study/CSIRO Publishing, Melbourne, Victoria
276. Toxic Tales Tropical Topics (2003) Wet tropics management authority. https://www.wettropics.gov.au/tropical-topics

277. Majewski M, Lazny R (2002) Synthesis of tropane alkaloids via enantioselective deprotonation of tropinone. J Org Chem 60:5825
278. Cirigottis K, Cleaver LEA (1974) Chemical studies of the Proteaceae. VII. An examination of the woods of 17 species for resorcinol derivatives. Aust J Chem 27:345
279. Ritchie E, Taylor WC, Vautin STK (1965) Chemical studies of the Proteaceae. I. *Grevillea robusta* A. Cunn. and *Orites excelsa* R. Br. Aust J Chem 18:2015
280. Ridley DD, Ritchie E, Taylor WC (1968) Chemical studies of the Proteaceae. II. Some further constituents of *Grevillea robusta* A. Cunn.: experiments on the synthesis of 5-*n*-tridecylresorcinol (grevillol) and related substances. Aust J Chem 21:2979
281. Cannon JR, Chow PW, Fuller MW, Hamilton BH, Metcalf BW, Power AJ (1973) Phenolic constituents of *Grevillea robusta* (Proteaceae). The structure of robustol, a novel macrocyclic phenol. Aust J Chem 26:2257
282. Yamashita Y, Matsunami K, Otsuka H, Shinzato T, Takeda Y (2008) Grevillosides A-F: glucosides of 5-alkylresorcinol derivatives from leaves of *Grevillea robusta*. Phytochemistry 69:2749
283. Yamashita Y, Matsunami K, Otsuka H, Shinzato T, Takeda Y (2010) 5-Alkylresorcinol glucosides from the leaves of *Grevillea robusta* Allan Cunningham. J Nat Med 64:474
284. Chuang TH, Chan HH, Wu TS, Li CF (2011) Chemical constituents and biological studies of the leaves of *Grevillea robusta*. Molecules 16:9331
285. Varma RS, Manju M, Parthasarathy MR (1976) New phenolic constituents of *Grevillea robusta* wood. Phytochemistry 15:1418
286. Ahmed AS, Nakamura N, Meselhy MR, Makhboul MA, El-Emary N, Hattori M (2000) Phenolic constituents from *Grevillea robusta*. Phytochemistry 53:149
287. Chuang TH, Wu PL (2007) Cytotoxic 5-alkylresorcinol metabolites from the leaves of *Grevillea robusta*. J Nat Prod 70:319
288. Tully J, Woodruff CM (2022) Allergic contact dermatitis from *Grevillea* masquerading as *Toxicodendron* dermatitis. JAAD Case Rep 23:168
289. Manju M, Varma RS, Parthasarathy MR (1977) New arbutin derivatives from leaves of *Grevillea robusta* and *Hakea saligna*. Phytochemistry 16:793
290. Yamashita-Higuchi Y, Sugimoto S, Matsunami K, Otsuka H (2012) Reinvestigation of structures of robustasides B and C, and isolation of (*E*)-2,5-dihydroxycinnamic acid esters of arbutin and glucose from the leaves of *Grevillea robusta*. Chem Pharm Bull 60:1347
291. Yamashita-Higuchi Y, Sugimoto S, Matsunami K, Otsuka H, Nakai T (2014) Grevillosides J–Q, arbutin derivatives from the leaves of *Grevillea robusta* and their melanogenesis inhibitory activity. Chem Pharm Bull 62:364
292. Deans BJ, Kilah NL, Jordan GJ, Bissember AC, Smith JA (2018) Arbutin derivatives isolated from ancient Proteaceae: potential phytochemical markers present in *Bellendena*, *Cenarrhenes*, and *Persoonia* genera. J Nat Prod 81:1241
293. Mock J, Murphy ST, Ritchie E, Taylor WC (1973) Chemical studies of the Proteaceae. VI. Two naphthoquinones from *Stenocarpus salignus*. Aust J Chem 26:1121
294. Moir M, Thomson RH (1973) Naphthaquinones in *Lomatia* species. Phytochemistry 12:1351
295. Bick IRC, Bremner JB, Gillard JW (1971) Methyl (*p*-hydroxybenzoyl) acetate and an alkaloid, bellendine, from *Bellendena montana*. Phytochemistry 10:475
296. Deng JZ, Starck SR, Sun DA, Sabat M, Hecht SM (2000) A new 7,8-euphadien-type triterpenoid from *Brackenridgea nitida* and *Bleasdalea bleasdalei* that inhibits DNA polymerase β. J Nat Prod 63:1356
297. Jaroszewski JW, Jensen PS, Cornett C, Byberg JR (1988) Occurrence of lotaustralin in *Berberidopsis beckleri* and its relation to the chemical evolution of Flacourtiaceae. Biochem Syst Ecol 16:23
298. Wang D, Murtaza M, Wood SA, Mellick GD, Miao WG, Guymer GP, Forster PI, Feng Y, Quinn RJ (2020) A grand challenge. 3. Unbiased phenotypic function of metabolites from Australia plants *Gloriosa superba* and *Alangium villosum* against Parkinson's disease. J Nat Prod 83:1440

299. Burbage MB, Jewers K, Eade RA, Harper P, Simes JJH (1970) Deoxyemmolactone, a new A(1),28-bisnor-lupane triterpene. J Chem Soc D 814
300. Nascimento Rocha MED, Figueiredo MR, Coelho Kaplan MA, Durst T, Arnason JT (2015) Chemotaxonomy of the Ericales. Biochem Syst Ecol 61:441
301. Khong PW, Lewis KG (1977) New chemical constituents of *Planchonia careya*. Aust J Chem 30:1311
302. McRae JM, Yang Q, Crawford RJ, Palombo EA (2008) Antibacterial compounds from *Planchonia careya* leaf extracts. J Ethnopharmacol 116:554
303. Mills C, Carroll AR, Quinn RJ (2005) Acutangulosides A-F, monodesmosidic saponins from the bark of *Barringtonia acutangula*. J Nat Prod 68:311
304. Quinn R, Mills C (Griffith University, Jarlmadangah Buru Aboriginal Corporation) (2005) Novel analgesic compounds, extracts containing same and methods of preparation. International patent number PCT/AU2004001660
305. Mcrae JM, Yang Q, Crawford RJ, Palombo EA (2008) Acylated flavonoid tetraglycoside from *Planchonia careya* leaves. Phytochem Lett 1:99
306. Cooke RG, Dowd H, Webb LJ (1952) Naphthoquinones from *Diospyros hebecarpa*. Nature 169:974
307. Baigent DR, Lewis KG (1978) Chemical constituents of *Tapeinosperma pseudojambosa*. Aust J Chem 31:1375
308. Courtney JL, Simes JJH, Stern W (1965) Extractives of the Ternstroemiaceae. I. The occurrence of betulic acid in *Ternstroemia cherryi* Bail. and *T. merrilliana* Kobuski. Aust J Chem 18:591
309. Singh A, Cullen JK, Bruce ZC, Reddell P, Munch G, Raju R (2020) Ternstroenols A-E: Undescribed pentacyclic triterpenoids from the Australian rainforest plant *Ternstroemia cherryi*. Phytochemistry 176:112426
310. Raju R, Mathew S, Reddell P, Munch G (2023) Ternstroenol F: a new pentacyclic triterpenoid saponin isolated from the Australian rainforest plant *Ternstroemia cherryi*. Nat Prod Res 37:2421
311. Damtoft S, Jensen SR (1995) Cyclohexyl butenolide glucosides from Epacridaceae. Phytochemistry 40:157
312. Beaulieu JM, Tank DC, Donoghue MJ (2013) A Southern Hemisphere origin for campanulid angiosperms, with traces of the break-up of Gondwana. BMC Evol Biol 13:80
313. Wall ME, Wani MC, Meyer BN, Taylor H (1987) Plant antitumor agents, 26. Isolation, structure, and antitumor activity of alkaloids from *Anopterus glandulosus*. J Nat Prod 50:1152
314. Levrier C, Sadowski MC, Nelson CC, Davis RA (2015) Cytotoxic C_{20} diterpenoid alkaloids from the Australian endemic rainforest plant *Anopterus macleayanus*. J Nat Prod 78:2908
315. Levrier C, Sadowski MC, Rockstroh A, Gabrielli B, Kavallaris M, Lehman M, Davis RA, Nelson CC (2017) 6α-Acetoxyanopterine: a novel structure class of mitotic inhibitor disrupting microtubule dynamics in prostate cancer cells. Mol Cancer Ther 16:3
316. Denne WA, Johnes SR, Lamberton JA, Mathieson AM, Suares H (1972) The absolute structure of anopterine. Tetrahedron Lett 13:2727
317. Hart N, Johns S, Lamberton J, Suares H, Willing R (1976) New alkaloids of the *ent*-kaurene type from *Anopterus* species (Escalloniaceae). I. The structure and reactions of anopterine. Aust J Chem 29:1295
318. Johns SR, Lamberton JA, Suares H, Willing RI (1985) New alkaloids of the *ent*-kaurene type from *Anopterus* species. III. A 2D-NMR study of anopterine and the determination of structure of six minor alkaloids by NMR. Aust J Chem 38:1091
319. Hart N, Johns S, Lamberton J, Suares H, Willing R (1976) New alkaloids of the *ent*-kaurene type from *Anopterus* species. II. The structure of the minor alkaloids. Aust J Chem 29:1319
320. Backer C, Jenett-Siems K, Siems K, Wurster M, Bodtke A, Lindequist U (2014) Cytotoxic saponins from the seeds of *Pittosporum angustifolium*. Z Naturforsch C J Biosci 69:191
321. Bäcker C, Jenett-Siems K, Siems K, Wurster M, Bodtke A, Niedermeyer THJ, Lindequist U (2014) New mono- and bisdesmosidic triterpene glycosides from *Pittosporum angustifolium* Lodd. Z Naturforsch B 69:1026

322. Backer C, Jenett-Siems K, Siems K, Wurster M, Bodtke A, Chamseddin C, Crusemann M, Lindequist U (2013) Triterpene glycosides from the leaves of *Pittosporum angustifolium*. Planta Med 79:1461
323. Bäcker C, Jenett-Siems K, Bodtke A, Lindequist U (2014) Polyphenolic compounds from the leaves of *Pittosporum angustifolium*. Biochem Syst Ecol 55:101
324. Wang Q, Grkovic T, Font J, Bonham S, Pouwer RH, Bailey CG, Moran AM, Ryan RM, Rasko JE, Jormakka M, Quinn RJ, Holst J (2014) Monoterpene glycoside ESK246 from *Pittosporum* targets LAT3 amino acid transport and prostate cancer cell growth. ACS Chem Biol 9:1369
325. Grkovic T, Pouwer RH, Wang Q, Guymer GP, Holst J, Quinn RJ (2015) LAT transport inhibitors from *Pittosporum venulosum* identified by NMR fingerprint analysis. J Nat Prod 78:1215
326. Miller RE, Tuck KL (2013) The rare cyanogen proteacin, and dhurrin, from foliage of *Polyscias australiana*, a tropical Araliaceae. Phytochemistry 93:210
327. Buchanan MS, Carroll AR, Edser A, Parisot J, Addepalli R, Quinn RJ (2005) Tyrosine kinase inhibitors from the rainforest tree *Polyscias murrayi*. Phytochemistry 66:481
328. Wanas AS, Matsunami K, Otsuka H, Desoukey SY, Fouad MA, Kamel MS (2010) Triterpene glycosides and glucosyl esters, and a triterpene from the leaves of *Schefflera actinophylla*. Chem Pharm Bull 58:1596
329. Calviño CI, Teruel FE, Downie SR (2015) The role of the Southern Hemisphere in the evolutionary history of Apiaceae, a mostly north temperate plant family. J Biogeogr 43:398
330. Johns SR, Lamberton JA, Suares H (1985) 3,4,5,6,7,8-Hexahydro-1,7-benzodiazecin-2(1*H*)-one from extracts of *Mackinlaya* species. Aust J Chem 38:1007
331. Hart NK, Johns SR, Lamberton JA (1971) The identification of a minor alkaloid of *Mackinlaya macrosciadia* (Araliaceae) as deoxyvasicinone. Aust J Chem 24:223
332. Fitzgerald JS, Johns SR, Lamberton JA, Redcliffe AH (1966) 6,7,8,9-Tetrahydropyridoquinazolines, a new class of alkaloids from *Mackinlaya* species (Araliaceae). Aust J Chem 19:151
333. Smissen RD (2003) Asterales (Sunflower). Encyclopedia of life sciences. Wiley-Blackwell, Hoboken, NJ, USA
334. Bottomley W, White DE (1950) The chemistry of Western Australian plants. III. Ursolic acid from *Anthotroche* species. Aust J Chem 3:516
335. Zdero C, Bohlmann F, King RM, Robinson H (1987) Sesquiterpene lactones from *Bedfordia arborescens*. Phytochemistry 26:1207
336. Jakupovic J, Pathak VP, Bohlmann F, King RM, Robinson H (1987) Obliquin derivatives and other constituents from Australian *Helichrysum* species. Phytochemistry 26:803
337. Lassak EV, Pinhey JT (1968) The constituents of some *Helichrysum* species (family Compositae). Aust J Chem 21:1927
338. Lyddiard D, Ozga T, Greatrex BW (2020) Diterpene and flavonoid constituents of the newly identified Australian species *Olearia fulgens*. Phytochem Lett 39:19
339. White EP (1969) Alkaloids of some herbaceous *Senecio* species in New Zealand. NZ J Sci 12:165
340. Olmstead RG (2013) Phylogeny and biogeography in Solanaceae, Verbenaceae and Bignoniaceae: a comparison of continental and intercontinental diversification patterns. Bot J Linn Soc 171:80
341. Olmstead RG, Palmer JD (1997) Implications for the phylogeny, classification, and biogeography of *Solanum* from cpDNA restriction site variation. Syst Bot 22:19
342. Srivastava G, Mehrotra RC, Dilcher DL (2018) Paleocene Ipomoea (Convolvulaceae) from India with implications for an East Gondwana origin of Convolvulaceae. Proc Natl Acad Sci USA 115:6028
343. Dorling PR, Colegate SM, Allen JG, Nickels R, Mitchell AA, Main DC, Madin B (2004) Calystegines isolated from *Ipomoea* spp. possibly associated with an ataxia syndrome in cattle in north Western Australia. In: Acamovic, T, Stewart CS, Pennycott TW (eds) Poisonous plants and related toxins. CABI Digital Publishing, p 140

344. Ott SC, Jenett-Siems K, Siems K, Muller F, Hilker M, Eich E (2013) Ipvelutine, 7β-acetoxy-2α-(tigloyloxy)tropane, an unusual tropane alkaloid from *Ipomoea velutina* R. Br. (Convolvulaceae). Sci Pharm 81:543
345. Lee ST, Gardner DR, Cook D (2017) Identification of indole diterpenes in *Ipomoea asarifolia* and *Ipomoea muelleri*, plants tremorgenic to livestock. J Agric Food Chem 65:5266
346. Der Marderosian A, Cho E, Chao JM (1974) The isolation and identification of the ergoline alkaloids from *Ipomoea muelleri*. Planta Med 25:6
347. Quach QN, Clay K, Lee ST, Gardner DR, Cook D (2023) Phylogenetic patterns of bioactive secondary metabolites produced by fungal endosymbionts in morning glories (Ipomoeeae, Convolvulaceae). New Phytol 238:1351
348. Ishimaru K, Shimomura K (1989) 7β-Hydroxyhyoscyamine from *Duboisia myoporoides-D. leichhardtii* hybrid and *Hyoscyamus albus*. Phytochemistry 28:3507
349. Gritsanapan W, Griffin WJ (1991) Alkaloid variation within *Duboisia myoporoides*. Phytochemistry 30:2667
350. Shukla YN, Thakur RS (1992) Tropane alkaloids from *Duboisia myoporoides*. Phytochemistry 31:4389
351. Foley P (2006) *Duboisia myoporoides*: the medical career of a native Australian plant. Hist Rec Aust Sci 17:31
352. Mortimer P (1958) The structure of isopelletierine from *Duboisia myoporoides* R.Br. Aust J Chem 11:82
353. Shukla YN, Thakur RS (1984) 8-Hydroxydotriacontan-30-one and other constituents from *Duboisia myoporoides*. Phytochemistry 23:1516
354. Bird GJ, Collins DJ, Eastwood FW, Gatehouse BMKC, Jozsa AJ, Swan JM (1976) New steroidal alkaloids from *Solanum callium*. Tetrahedron Lett 17:3653
355. Bird GJ, Collins DJ, Eastwood FW, Swan JM (1979) Soladunalinidine, a new dibasic analogue of tomatidine extracted from *Solanum dunalianum*. Aust J Chem 32:611
356. Ripperger HSK, Symon DE (1984) Solanum alkaloids. Part 113: soladulcidine from *Solanum shanesii*. Pharmazie 39:125
357. Burden RS, Loeffler RST, Rowell PM, Bailey JA, Kemp MS (1986) Cyclodebneyol, a fungitoxic sesquiterpene from Tnv infected *Nicotiana debneyi*. Phytochemistry 25:1607
358. Culvenor CCJ, Smith LW (1967) The alkaloids of *Cynoglossum australe* R. Br. and C. amabile Stapf & Drummond. Aust J Chem 20:2499
359. Antonelli A, Clarkson JJ, Kainulainen K, Maurin O, Brewer GE, Davis AP, Epitawalage N, Goyder DJ, Livshultz T, Persson C, Pokorny L, Straub SCK, Struwe L, Zuntini AR, Forest F, Baker WJ (2021) Settling a family feud: a high-level phylogenomic framework for the Gentianales based on 353 nuclear genes and partial plastomes. Am J Bot 108:1143
360. Bremer B, Eriksson T (2009) Time tree of Rubiaceae: phylogeny and dating the family, subfamilies, and tribes. Int J Plant Sci 170:766
361. Johns SR, Lamberton JA, Occolowitz JL (1967) Antirhine, a new indole alkaloid from *Antirhea putaminosa* (F. Muell.) Bail. (family Rubiaceae). Aust J Chem 20:1463
362. Beecham AF, Hart NK, Johns SR, Lamberton JA (1968) The stereochemistry of oxindole alkaloids: uncarines A, B (formosanine), C (pteropodine), D (speciophylline), E (isopteropodine), and F. Aust J Chem 21:491
363. Yang B, Feng YJ, Vu H, McCormick B, Rowley J, Pedro L, Crowther GJ, Van Voorhis WC, Forster PI, Quinn RJ (2016) Bioaffinity mass spectrometry screening. J Biomol Screen 21:194
364. Parry KP, Smith GF (1978) Quadrigemines-A and -B, two minor alkaloids of *Hodgkinsonia frutescens* F. Muell. J Chem Soc, Perkin Trans 1:1671
365. Deans BJ, Bissember AC, Smith JA (2016) Practical isolation of asperuloside from *Coprosma quadrifida* via rapid pressurised hot water extraction. Aust J Chem 69:1219
366. Fridrichsons J, Mackay MF, Mcl. Mathieson A (1974) The absolute molecular structure of hodgkinsine. Tetrahedron 30:85
367. Anet FaL, Hughes GK, Ritchie E (1953) Alkaloids of Australian *Strychnos* species. I. Constituents of *S. lucida* R.Br. and *S. psilosperma* F.Muell. Aust J Chem 6:58

368. Fishbein M, Livshultz T, Straub SCK, Simoes AO, Boutte J, McDonnell A, Foote A (2018) Evolution on the backbone: Apocynaceae phylogenomics and new perspectives on growth forms, flowers, and fruits. Am J Bot 105:495
369. Bitencourt C, Nürk NM, Rapini A, Fishbein M, Simoes AO, Middleton DJ, Meve U, Endress ME, Liede-Schumann S (2021) Evolution of dispersal, habit, and pollination in Africa pushed Apocynaceae diversification after the Eocene-Oligocene climate transition. Front Ecol Evol 9
370. Carroll AR, Hyde E, Smith J, Quinn RJ, Guymer G, Forster PI (2005) Actinophyllic acid, a potent indole alkaloid inhibitor of the coupled enzyme assay carboxypeptidase u/hippuricase from the leaves of *Alstonia actinophylla* (Apocynaceae). J Org Chem 70:1096
371. Allam K, Beutler JA, Le Quesne PW (2004) 14-Ketoalstonidine and other alkaloidal constituents of the stem bark of *Alstonia constricta*. J Nat Prod 50:623
372. Crow WD, Hancox NC, Johns SR, Lamberton JA (1970) New alkaloids of *Alstonia constricta*. Aust J Chem 23:2489
373. Carroll AR, Addepalli R, Fechner G, Smith J, Guymer GP, Forster PI, Quinn RJ (2008) Alkaloids from the Australian rainforest tree *Ochrosia moorei*. J Nat Prod 71:1063
374. Elderfield RC, Wythe SL (1954) Alstonia alkaloids. IV. The structure of alstoniline. J Org Chem 19:683
375. Johns SR, Lamberton JA, Moore BP, Sioumis AA (1975) Alkaloids of *Ochrosia poweri* Bail. The stereochemistry of poweridine and the identity of elliptamine and reserpiline. Aust J Chem 28:1627
376. Labib RM, Zulfiqar F, Ibrahim MA, Balachandran P, Zhang J, Ross SA (2019) FOXO signal activating alkaloids isolated from *Ochrosia elliptica* leaf cultivated in Egypt. Med Chem Res 28:1628
377. Ahond A, Fernandez H, Julia-Moore M, Poupat C, Sánchez V, Potier P, Kan SK, Sévenet T (1981) Contribution à l'étude des Ochrosiinées: alcaloïdes de *Ochrosia moorei*. J Nat Prod 44:193
378. Batchily F, Mehri H, Plat M. Sévenet T (1986) Alkaloids from *Neisosperma kilneri* (F.V. Mueller) Fosberg et Sachet. Ann Pharm Fr 43:513
379. Cook JM, Le Quesne PW (1975) 11-Methoxyakuammicine from *Alstonia muelleriana*. J Org Chem 40:1367
380. Linde HH (1970) Alkaloids from *Melodinus australis* (F. MUELLER) PIERRE (Apocynaceae) and notice on the structure of refractidine and pyrifoline. Pharm Acta Helv 45:248
381. Crow WD, Michael M (1955) The alkaloids of *Kopsia longiflora* Merrill. I. Isolation of the alkaloids. Aust J Chem 8:129
382. Menzies JRW, Paterson SJ, Duwiejua M, Corbett AD (1998) Opioid activity of alkaloids extracted from *Picralima nitida* (fam. Apocynaceae). Eur J Pharmacol 350:101
383. Cook JM, Le Quesne PW (1971) Structure of alstonisidine, a novel dimeric indole alkaloid from *Alstonia muelleriana*. J Org Chem 36:582
384. Burke DE, Cook JM, Le Quesne PW (1972) Biomimetic synthesis and structure of the bisindole alkaloid alstonisidine. J Chem Soc, Chem Commun:697
385. Burke DE, Cook GA, Cook JM, Haller KG, Lazar HA, Le Quesne PW (1973) Further alkaloids of *Alstonia muelleriana*. Phytochemistry 12:1467
386. Hart N, Johns S, Lamberton J (1972) Tertiary alkaloids of *Alstonia spectabilis* and *Alstonia glabriflora* (Apocynaceae). Aust J Chem 25:2739
387. Hu WL, Hesse M (1988) The alkaloids of *Melodinus acutiflorus*. Rhyzicine N-oxide. Planta Med 54:235
388. Chen AH, Liu QL, Ma YL, Jiang ZH, Tang JY, Liu YP, Chen GY, Fu YH, Xu W (2017) A new monoterpenoid indole alkaloid from *Ochrosia elliptica*. Nat Prod Res 31:1490
389. Knox JR, Slobbe J (1975) Indole alkaloids from *Ervatamia orientalis*. I. Isolation of alkaloids and structural identification of two dimers. Aust J Chem 28:1813
390. Morita H, Haseo A, Eko Nugroho A, Hirasawa Y, Kaneda T, Shirota O, Rahman A, Kusumawati I, Cholies Zaini N, Morita H (2015) A new indole alkaloid from *Voacanga grandifolia*. Heterocycles 90:601

391. Douglas B, Kirkpatrick JL, Moore BP, Weisbach JA (1964) Alkaloids of *Ochrosia poweri* Bailey. II. The 2-acylindole stem-bark bases. Aust J Chem 17:246
392. Xu W, Fu Y-H, Chen AH, Liu Y-P, Wang Z-X, Ma Y-L, Jiang Z-H, Lai L, Guo R-R, Long J-T, Lin S-X (2017) Structurally diverse indole alkaloids from *Ochrosia elliptica*. Heterocycles 94:743
393. Linde VHHA (1965) 198. Die Alkaloide aus *Melodinus australis* (F. Mueller) Pierre (Apocynaceae). Helv Chim Acta 48:1822
394. Elderfield RC, Gilman RE (1972) Alkaloids of *Alstonia muelleriana*. Phytochemistry 11:339
395. Cook JM, Le Quesne PW (1971) Macralstonine from *Alstonia muelleriana*. Phytochemistry 10:437
396. Hirasawa Y, Arai H, Rahman A, Kusumawati I, Zaini NC, Shirota O, Morita H (2013) Voacalgines A-E, new indole alkaloids from *Voacanga grandifolia*. Tetrahedron 69:10869
397. Hu W-L Zhu J-P, Akiinl U, Prewo R, Hesse M (1987) Revision of the structures of rhazicine and rhazimine, two alkaloids from *Melodinus acutiflorus*. Phytochemistry 26:2625
398. El-Shiekh RA, Al-Mahdy DA, Hifnawy MS, Tzanova T, Evain-Bana E, Philippot S, Bagrel D, Abdelsattar EA (2017) Chemical and biological investigation of *Ochrosia elliptica* Labill. cultivated in Egypt. Rec Nat Prod 11:552
399. Goodwin S, Smith AF, Horning EC (1959) Alkaloids of *Ochrosia elliptica* Labill. J Am Chem Soc 81:1903
400. O'Sullivan EC, Miller CM, Deane FM, McCarthy FO (2013) Emerging targets in the bioactivity of ellipticines and derivatives. In: Atta-ur-Rahman (ed) Studies in natural products chemistry, vol 39. Elsevier, Amsterdam, p 189
401. Chen AH, Liu Q-L, Ma Y-L, Jiang Z-H, Tang J-Y, Liu Y-P, Chen G-Y, Xu W, Fu Y-H (2018) Pterocarpans from the stems and leaves of *Ochrosia elliptica*. Chem Nat Compd 54:553
402. Anstee JR, Arthur HR, Beckwith AL, Dougall DK, Jefferies PR, Michael M, Watkins JC, White DE (1952) 776. The chemistry of Western Australian plants. Part VI. The occurrence of betulic, oleanolic, and ursolic acids. J Chem Soc:4065
403. Hettiarachchi DS, Locher C, Longmore RB (2011) Antibacterial compounds from the root of the indigenous Australian medicinal plant *Carissa lanceolata* R.Br. Nat Prod Res 25:1388
404. Lindsay EA, Berry Y, Jamie JF, Bremner JB (2000) Antibacterial compounds from *Carissa lanceolata* R.Br. Phytochemistry 55:403
405. Cable J, Coombe RG, Watson TR (1964) The cardiac glycosides of *Cerbera* species. I. The isolation of cerbertin and cerbertatin. Aust J Chem 17:1423
406. Edgar JA, Culvenor CCJ (1975) Pyrrolizidine alkaloids in *Parsonsia* species (family Apocynaceae) which attract Danaid butterflies. Experientia 31:393
407. Orr AG, Trigo JR, Witte L, Hartmann T (1996) Sequestration of pyrrolizidine alkaloids by larvae of *Tellervo zoilus* (Lepidoptera: Ithomiinae) and their role in the chemical protection of adults against the spider *Nephila maculata* (Araneidae). Chemoecology 7:68
408. Baas WJ, Van Berkel IEM, Versluis C, Heerma W, Kreyenbroek MN (1992) Ring-A fissioned 3,4-seco-3-*nor*-triterpene-2-aldehydes and related pentacyclic triterpenoids from the leaf wax of *Hoya australis*. Phytochemistry 31:2073
409. Duff AG, Gellert E, Rudzats R (1973) Flavescin: a new 1-ketopolyhydroxypregnene from *Marsdenia flavescens*. Phytochemistry 12:2943
410. Summons RE, Ellis J, Gellert E (1972) Steroidal alkaloids of *Marsdenia rostrata*. Phytochemistry 11:3335
411. Gellert E, Govindachari TR, Lakshmikantham MV, Ragade IS, Rudzats R, Viswanathan N (1962) 189. The alkaloids of *Tylophora crebriflora*: structure and synthesis of tylocrebrine, a new phenanthroindolizidine alkaloid. J Chem Soc 1008
412. Rao KV (1970) Alkaloids of *Tylophora*. II. Structural studies. J Pharm Sci 59:1608
413. Birch AJ, Massy-Westropp R, Wright SE (1953) Natural derivatives of furan. I. Ngaione. Aust J Chem 6:385
414. Blackburne ID, Park RJ, Sutherland MD (1971) Terpenoid chemistry. XVIII. Myodesmone and isomyodesmone, toxic furanoid ketones from *Myoporum deserti* and *M. acuminatum*. Aust J Chem 24:995

415. Blackburne ID, Park RJ, Sutherland MD (1972) Terpenoid chemistry. XX. Myoporone and dehydromyoporone, toxic furanoid ketones from *Myoporum* and *Eremophila* species. Aust J Chem 25:1787
416. Sutherland MD, Rodwell JL (1989) Terpenoid chemistry. XXVIII. Furanosesquiterpene β-ketols from *Myoporum betcheanum*, *M. deserti*, *M. montanum* and other Myoporaceae. Aust J Chem 42:1995
417. Hart NK, Johns SR, Lamberton JA (1969) Alkaloids of *Jasminum* species (family Oleaceae). II. Isolation of a new monoterpenoid alkaloid and other constituents. Aust J Chem 22:1283
418. Hart NK, Johns SR, Lamberton JA (1968) A new monoterpenoid alkaloid from *Jasminum* species and from *Ligustrum novoguineense* (family Oleaceae). Aust J Chem 21:1321
419. Tan YP, Houston SD, Modhiran N, Savchenko AI, Boyle GM, Young PR, Watterson D, Williams CM (2019) Stachyonic aacid: a dengue virus inhibitor from *Basilicum polystachyon*. Chemistry 25:5664
420. Tan YP, Xue Y, Savchenko AI, Houston SD, Modhiran N, McMillan CLD, Boyle GM, Bernhardt PV, Young PR, Watterson D, Williams CM (2019) Basimarols A, B, and C, highly oxygenated pimarane diterpenoids from *Basilicum polystachyon*. J Nat Prod 82:2828
421. Miller RE, McConville MJ, Woodrow IE (2006) Cyanogenic glycosides from the rare Australian endemic rainforest tree *Clerodendrum grayi* (Lamiaceae). Phytochemistry 67:43
422. Rasikari HL, Leach DN, Waterman PG, Spooner-Hart RN, Basta AH, Banbury LK, Winter KM, Forster PI (2005) Cytotoxic clerodane diterpenes from *Glossocarya calcicola*. Phytochemistry 66:2844
423. Eade RA, Harper P, Simes JJH (1974) The triterpenes of *Faradaya splendida* (Verbenaceae). Aust J Chem 27:2289
424. Rüedi P, Eugster CH (1978) Diterpenoide Drüsenfarbstoffe aus Labiaten: 6 neue p-Chinomethane aus *Plectranthus parviflorus* WILLD. Helv Chim Acta 61:709
425. Dinan L, Whiting P, Scott AJ (1998) Taxonomic distribution of phytoecdysteroids in seeds of members of the Chenopodiaceae. Biochem Syst Ecol 26:553
426. Buchanan MS, Carroll AR, Edser A, Sykes M, Fechner GA, Forster PI, Guymer GP, Quinn RJ (2008) Lysianadioic acid, a carboxypeptidase B inhibitor from *Lysiana subfalcata*. Bioorg Med Chem Lett 18:1495
427. Maurin O, Anest A, Bellot S, Biffin E, Brewer G, Charles-Dominique T, Cowan RS, Dodsworth S, Epitawalage N, Gallego B, Giaretta A, Goldenberg R, Goncalves DJP, Graham S, Hoch P, Mazine F, Low YW, Mcginnie C, Michelangeli FA, Morris S, Penneys DS, Perez Escobar OA, Pillon Y, Pokorny L, Shimizu G, Staggemeier VG, Thornhill AH, Tomlinson KW, Turner IM, Vasconcelos T, Wilson PG, Zuntini AR, Baker WJ, Forest F, Lucas E (2021) A nuclear phylogenomic study of the angiosperm order Myrtales, exploring the potential and limitations of the universal Angiosperms353 probe set. Am J Bot 108:1087
428. Thornhill AH, Ho SY, Kulheim C, Crisp MD (2015) Interpreting the modern distribution of Myrtaceae using a dated molecular phylogeny. Mol Phylogenet Evol 93:29
429. Brown GK, Udovicic F, Ladiges PY (2001) Molecular phylogeny and biogeography of *Melaleuca*, *Callistemon* and related genera (Myrtaceae). Aust Syst Bot 14:565
430. Craven LA, Edwards RD, Cowley KJ (2014) New combinations and names in *Melaleuca* (Myrtaceae). Taxon 63:663
431. Tian LW, Feng Y, Tran TD, Shimizu Y, Pfeifer T, Forster PI, Quinn RJ (2015) Tyrosyl-DNA phosphodiesterase I inhibitors from the Australian plant *Macropteranthes leichhardtii*. J Nat Prod 78:1756
432. Collins DJ, Pilotti CA, Wallis AFA (1992) Triterpene acids from some Papua New Guinea *Terminalia* species. Phytochemistry 31:881
433. Oelrichs PB, Pearce CM, Zhu J, Filippich LJ (1994) Isolation and structure determination of terminalin A toxic condensed tannin from *Terminalia oblongata*. Nat Toxins 2:144
434. Doig AJ, Williams DH, Oelrichs PB, Baczynskyj L (1990) Isolation and structure elucidation of punicalagin, a toxic hydrolysable tannin, from *Terminalia oblongata*. J Chem Soc, Perkin Trans 1:2317

435. Salem MZM, El-Hefny M, Nasser RA, Ali HM, El-Shanhorey NA, Elansary HO (2017) Medicinal and biological values of *Callistemon viminalis* extracts: history, current situation and prospects. Asian Pac J Trop Med 10:229
436. Ritchie E, Snape D, Taylor WC (1961) Chemical studies of the Myrtaceae. III. Triterpenoids from the wood of *Tristania conferta* R. Br. Aust J Chem 14:471
437. Lee CK, Chang MH (1999) Four new triterpenes from the heartwood of *Melaleuca leucadendron.* J Nat Prod 62:1003
438. Lee CK (1998) Ursane triterpenoids from leaves of *Melaleuca leucadendron.* Phytochemistry 49:1119
439. Deng JZ, Starck SR, Hecht SM (1999) DNA polymerase β inhibitors from *Baeckea gunniana.* J Nat Prod 62:1624
440. Setzer WN, Rozmus GF, Setzer MC, Schmidt JM, Vogler B, Reeb S, Jackes BR, Irvine AK (2006) Bioactive principles in the bark of *Pilidiostigma tropicum.* J Mol Model 12:703
441. Setzer WN, Setzer MC, Bates RB, Jackes BR (2000) Biologically active triterpenoids of *Syncarpia glomulifera* bark extract from Paluma, North Queensland, Australia. Planta Med 66:176
442. Su Q, Dalal S, Goetz M, Cassera MB, Kingston DGI (2016) Antiplasmodial phloroglucinol derivatives from *Syncarpia glomulifera.* Bioorg Med Chem 24:2544
443. Hongcheng W, Fujimotot Y (1993) Triterpene esters from *Eucalyptus tereticornis.* Phytochemistry 33:151
444. Potts KT, Roy SK (1965) Triterpenoid constituents of *Backhousia angustifolia* F. Muell. Aust J Chem 18:767
445. Mustafa K, Perry NB, Weavers RT (2003) 2-Hydroxyflavanones from *Leptospermum polygalifolium* subsp. *polygalifolium* equilibrating sets of hemiacetal isomers. Phytochemistry 64:1285
446. Zhang J, Liu F, Jin Q, Li X, Zhan Q, Chen M, Wang S, Wu Z, Ye W, Wang L (2024) Discovery of unusual phloroglucinol-triterpenoid adducts from *Leptospermum scoparium* and *Xanthostemon chrysanthus* by building blocks-based molecular networking. Chin Chem Lett 35:108881
447. Al-Sayed E, Hamid HA, Abu El Einin HM (2014) Molluscicidal and antischistosomal activities of methanol extracts and isolated compounds from *Eucalyptus globulus* and *Melaleuca styphelioides.* Pharm Biol 52:698
448. Brophy JJ, Goldsack RJ, Fookes CJR, Forster PI (1995) Leaf oils of the genus *Backhousia* (Myrtaceae). J Essent Oil Res 7:237
449. Li C, Liu H, Zhao L, Zhang W, Qiu S, Yang X, Tan H (2017) Antibacterial neolignans from the leaves of *Melaleuca bracteata.* Fitoterapia 120:171
450. Mathew S, Zhang K, Zhou X, Munch G, Bodkin F, Li F, Raju R (2023) Myrtinols A-F: new anti-inflammatory peltogynoid flavonoid derivatives from the leaves of Australian indigenous plant *Backhousia myrtifolia.* Molecules 28:2160
451. Ritmejeryte E, Ryan RYM, Byatt BJ, Peck Y, Yeshi K, Daly NL, Zhao G, Crayn D, Loukas A, Pyne SG, Ruscher R, Wangchuk P (2022) Anti-inflammatory properties of novel galloyl glucosides isolated from the Australian tropical plant *Uromyrtus metrosideros.* Chem Biol Interact 368:110124
452. Gohar AA, Maatooq GT, Gadara SR, Aboelmaaty WS (2013) One new pyrroline compound from *Callistemon viminalis* (Sol. ex Gaertner) G. Don ex Loudon. Nat Prod Res 27:1179
453. Qin X-J, Liu H, Yu Q, Yan H, Tang J-F, An L-K, Khan A, Chen Q-R, Hao X-J, Liu H-Y (2017) Acylphloroglucinol derivatives from the twigs and leaves of *Callistemon salignus.* Tetrahedron 73:1803
454. Liu HX, Chen YC, Liu Y, Zhang WM, Wu JW, Tan HB, Qiu SX (2016) Acylphloroglucinols from the leaves of *Callistemon viminalis.* Fitoterapia 114:40
455. Aboutabi EA, Moharram F (1999) Melacoside A, a novel monoterpene glucoside from *Melaleuca quinquenervia* (Cav.) S.T. Blake. Al-Azhar J Pharm Sci 23:110
456. Brophy JJ, Fookes CJR, House APN (1992) 2,4,6-Trimethoxytoluene from the genus *Stockwellia.* Phytochemistry 31:324

457. Liu F, Lu WJ, Li NP, Liu JW, He J, Ye WC, Wang L (2018) Four new cinnamoyl-phloroglucinols from the leaves of *Xanthostemon chrysanthus*. Fitoterapia 128:93
458. Liu F, Chen MH, Hu YX, Zhan Q, Zhang JY, Zhang ZY, Wang SS, Ye WC, Wang L (2023) Xanchryones I–N, six unusual phloroglucinol-amino acid hybrids from the leaves of *Xanthostemon chrysanthus*. Chem Biodivers 20:e202300234
459. Liu F, Tian HY, Huang XL, Wang WJ, Li NP, He J, Ye WC, Wang L (2019) Xanthchrysones A–C: rearranged phenylpropanoyl-phloroglucinol dimers with unusual skeletons from *Xanthostemon chrysanthus*. J Org Chem 84:15355
460. Zhan Q, Wu YY, Liu F, Li NP, Zhou X, Wang CQ, Wu Y, Zhao W, Ye WC, Wang L (2022) (+)- and (–)-Xanthostones A–D: four pairs of enantiomeric cinnamoyl-β-triketone derivatives from *Xanthostemon chrysanthus*. Chem Biodivers 19:e202200356
461. Hyodo S, Etoh H, Yamashita N, Sakata K, Ina K (2014) Structure of resinosides from *Eucalyptus resinifera* as repellents against the blue mussel, *Mytilus edulis*. Biosci Biotechnol Biochem 56:138
462. Gu JH, Liu JS, Lin JH, Liu F, Wu ZL, Zheng YR, Ye WC, Wang L (2023) Five new phenylpropanoyl phloroglucinol derivatives from *Leptospermum scoparium*. Chem Biodivers 20:e202201111
463. Naz T, Packer J, Yin P, Brophy JJ, Wohlmuth H, Renshaw DE, Smith J, Elders YC, Vemulpad SR, Jamie JF (2016) Bioactivity and chemical characterisation of *Lophostemon suaveolens*—an endemic Australian Aboriginal traditional medicinal plant. Nat Prod Res 30:693
464. Venigalla M, Roberts TL, Raju R, Mrad M, Bodkin F, Kopp K, Doyle K, Munch G (2021) Identification of tetragocarbone C and sideroxylin as the most potent anti-inflammatory components of *Syncarpia glomulifera*. Fitoterapia 150:104843
465. Liu F, Wu Y, Li NP, Liu JW, Wang L, Ye WC (2020) Chiral isolation and absolute configuration of (+)- and (–)-xanchryones F and G from *Xanthostemon chrysanthus*. Chem Biodivers 17:e1900683
466. Takasaki M, Konoshima T, Kozuka M, Haruna M, Ito K, Shingu T (1994) Structures of euglobals-G1,-G2,-G3,-G4, and -G5 from *Eucalyptus grandis*. Chem Pharm Bull 42:2591
467. Singh IP, Hayakawa R, Etoh H, Takasaki M, Konoshima T (1997) Grandinal, a new phloroglucinol dimer from *Eucalyptus grandis*. Biosci Biotechnol Biochem 61:921
468. Singh IP, Umehara K, Asai T, Etoh H, Takasaki M, Konoshima T (1998) Phloroglucinol-monoterpene adducts from *Eucalyptus grandis*. Phytochemistry 47:1157
469. Umehara K, Singh IP, Etoh H, Takasaki M, Konoshima T (1998) Five phloroglucinol-monoterpene adducts from *Eucalyptus grandis*. Phytochemistry 49:1699
470. Li JS, Liu YN, Li JY, Lei C, Hou AJ (2023) Acylphloroglucinol-monoterpene meroterpenoids from *Eucalyptus tereticornis* and their inhibitory activity against ATP citrate lyase. Phytochemistry 207:113565
471. Liu H, Feng M-Y, Yu Q, Yan H, Zeng Y, Qin X-J, He L, Liu H-Y (2018) Formyl phloroglucinol meroterpenoids from *Eucalyptus tereticornis* and their bioactivities. Tetrahedron 74:1540
472. Shou Q, Smith JE, Mon H, Brkljača Z, Smith A-S, Smith DM, Griesser HJ, Wohlmuth H (2014) Rhodomyrtals A-D, four unusual phloroglucinol-sesquiterpene adducts from *Rhodomyrtus psidioides*. RSC Adv 4:13514
473. Baxter JR, Holland DC, Gavranich B, Nicolle D, Hayton JB, Avery VM, Carroll AR (2023) NMR fingerprints of formyl phloroglucinol meroterpenoids and their application to the investigation of *Eucalyptus gittinsii* subsp. *gittinsii*. J Nat Prod 86:1317
474. Baxter JR (2022) Chemical investigations of the secondary metabolites of the Australian eucalypts: natural products and their chemotaxonomic implications. PhD thesis, Griffith University, Southport, New South Wales, p 988
475. Pal Singh I, Bharate SB (2006) Phloroglucinol compounds of natural origin. Nat Prod Rep 23:558
476. Qin XJ, Shu T, Yu Q, Yan H, Ni W, An LK, Li PP, Zhi YE, Khan A, Liu HY (2017) Cytotoxic acylphloroglucinol derivatives from *Callistemon salignus*. Nat Prod Bioprospect 7:315
477. Liu HX, Chen K, Liu Y, Li C, Wu JW, Xu ZF, Tan HB, Qiu SX (2016) Callviminols A-E, new terpenoid-conjugated phloroglucinols from the leaves of *Callistemon viminalis*. Fitoterapia 115:142

478. Wu L, Zhang Y-L, Wang X-B, Zhang Y-M, Yang M-H, Luo J, Kong L-Y (2017) Viminalins A-O: diverse [3+2] hybrids of acylphloroglucinol and α-phellandrene from the fruits of *Callistemon viminalis*. Tetrahedron 73:1105
479. Xie X, Wu L, Cui Z, Yang M, Yin Y, Luo J, Kong L (2019) Melaleucadines A and B: two rare benzylic phloroglucinol-terpene hybrids from *Melaleuca leucadendron*. Tetrahedron Lett 60:1011
480. Gu JH, Wang WJ, Chen JZ, Liu JS, Li NP, Cheng MJ, Hu LJ, Li CC, Ye WC, Wang L (2020) Leptosperols A and B, two cinnamoylphloroglucinol-sesquiterpenoid hybrids from *Leptospermum scoparium*: structural elucidation and biomimetic synthesis. Org Lett 22:1796
481. Dethe DH, Nirpal AK (2019) Bio-inspired enantioselective total syntheses of (–)-viminalins A, B, H, I, and N and structural reassignment of (–)-viminalin M. Org Biomol Chem 17:7507
482. Xiang YQ, Liu HX, Zhao LY, Xu ZF, Tan HB, Qiu SX (2017) Callistemenonone A, a novel dearomatic dibenzofuran-type acylphloroglucinol with antimicrobial activity from *Callistemon viminalis*. Sci Rep 7:2363
483. Wu JW, Li BL, Tang C, Ke CQ, Zhu NL, Qiu SX, Ye Y (2019) Callistemonols A and B, potent antimicrobial acylphloroglucinol derivatives with unusual carbon skeletons from *Callistemon viminalis*. J Nat Prod 82:1917
484. Carroll AR (2016) Structure revision of pilidiostigmin from the leaves of *Pilidiostigma glabrum*. Tetrahedron Lett 57:281
485. Shou Q, Sun H, Banbury LK, Lambley EH, Renshaw DE, Griesinger C, Griesser HJ, Wohlmuth H (2013) Pilidiostigmin, a novel bioactive dimeric acylphloroglucinol derivative isolated from *Pilidiostigma glabrum*. Tetrahedron Lett 54:1853
486. Shou Q, Banbury LK, Renshaw DE, Lambley EH, Mon H, Macfarlane GA, Griesser HJ, Heinrich MM, Wohlmuth H (2012) Biologically active dibenzofurans from *Pilidiostigma glabrum*, an endemic Australian Myrtaceae. J Nat Prod 75:1612
487. Igboechi CA, Parfitt RT, Rowan MG (1984) Two dibenzofuran derivatives from fruits of *Rhodomyrtus macrocarpa*. Phytochemistry 23:1139
488. Raju R, Singh A, Gunawardena D, Reddell P, Munch G (2020) Mulgravanols A and B, rare oxidized xanthenes and a new phloroglucinol isolated from the Australian rainforest plant *Waterhousea mulgraveana* (Myrtaceae). Fitoterapia 143:104595
489. Bloor SJ (1992) Antiviral phloroglucinols from New Zealand *Kunzea* species. J Nat Prod 55:43
490. Killeen DP, Larsen L, Dayan FE, Gordon KC, Perry NB, Van Klink JW (2016) Nortriketones: antimicrobial trimethylated acylphloroglucinols from Mānuka (*Leptospermum scoparium*). J Nat Prod 79:564
491. Van Klink JW, Brophy JJ, Perry NB, Weavers RT (1999) β-Triketones from Myrtaceae: isoleptospermone from *Leptospermum scoparium* and papuanone from *Corymbia dallachiana*. J Nat Prod 62:487
492. Mathew S, Zhou X, Munch G, Bodkin F, Wallis M, Li F, Raju R (2022) Tristaenone A: a new anti-inflammatory compound isolated from the Australian indigenous plant *Tristaniopsis laurina*. Molecules 27:6592
493. Brophy JJ, Clarkson JR, Fookes CJR (1989) Angustifolenone, a ketone from *Backhousia angustifolia*. Phytochemistry 28:1259
494. Birch AJ, Elliott P (1956) β-Triketones. III. Xanthostemone. Aust J Chem 9:238
495. Crow WD, Nicholls W, Sterns M (1971) Root inhibitors in *Eucalyptus grandis*: naturally occurring derivatives of the 2,3-dioxabicyclo[4.4.0]decane system. Tetrahedron Lett 12:1353
496. Brophy JJ, Craig DC, Goldsack RJ, Fookes CJR (1994) Hillyl acetate, a keto-acetate from the leaf steam volatiles of *Syncarpia hillii*. Phytochemistry 37:1645
497. Senadeera SPD, Lucantoni L, Duffy S, Avery VM, Carroll AR (2018) Antiplasmodial β-triketone-flavanone hybrids from the flowers of the Australian tree *Corymbia torelliana*. J Nat Prod 81:1588
498. Badalamenti E, Cusimano D, La Mantia T, Pasta S, Romano S, Troia A, Ilardi V (2018) The ongoing naturalisation of *Eucalyptus* spp. in the Mediterranean Basin: new threats to native species and habitats. Aust For 81:239

499. Wu L, Wang XB, Li RJ, Zhang YL, Yang MH, Luo J, Kong LY (2016) Callistiviminenes A-O: diverse adducts of β-triketone and sesqui- or monoterpene from the fruits of *Callistemon viminalis*. Phytochemistry 131:140
500. Senadeera SPD, Robertson LP, Duffy S, Wang Y, Avery VM, Carroll AR (2018) β-Triketone-monoterpene hybrids from the flowers of the Australian tree *Corymbia intermedia*. J Nat Prod 81:2455
501. Khambay BPS, Beddie DG, Hooper AM, Simmonds MSJ (2003) Isolation, characterisation and synthesis of an insecticidal tetramethyltetrahydrochromenedione-spiro-bicyclo[3.1.1]cycloheptane from two species of Myrtaceae. Tetrahedron 59:7131
502. Khambay BPS, Beddie DG, Hooper AM, Simmonds MSJ, Green PWC (1999) New insecticidal tetradecahydroxanthenediones from *Callistemon viminalis*. J Nat Prod 62:1666
503. Wu L, Luo J, Zhang Y, Zhu M, Wang X, Luo J, Yang M, Yu B, Yao H, Dai Y, Guo Q, Chen Y, Sun H, Kong L (2015) Isolation and biomimetic synthesis of (±)-calliviminones A and B, two novel Diels-Alder adducts, from *Callistemon viminalis*. Tetrahedron Lett 56:229
504. Wu L, Luo J, Wang X-B, Li R-J, Zhang Y-L, Kong L-Y (2015) Calliviminones C-H: six new hetero- and carbon-Diels–Alder adducts with unusual skeletons from the fruits of *Callistemon viminalis*. RSC Adv 5:93900
505. Wu L, Zhang Y, Wang X, Liu R, Yang M, Kong L, Luo J (2017) Acylphloroglucinols from the fruits of *Callistemon viminalis*. Phytochem Lett 20:61
506. Lee C-K (1999) Leucadenone A-D, the novel class flavanone from the leaves of *Melaleuca leucadendron* L. Tetrahedron Lett 40:7255
507. Carroll AR, Avery VM, Duffy S, Forster PI, Guymer GP (2013) Watsonianone A-C, anti-plasmodial β-triketones from the Australian tree, *Corymbia watsoniana*. Org Biomol Chem 11:453
508. Carroll AR, Lamb J, Moni R, Guymer GP, Forster PI, Quinn RJ (2008) Myrtucommulones F-I, phloroglucinols with thyrotropin-releasing hormone receptor-2 binding affinity from the seeds of *Corymbia scabrida*. J Nat Prod 71:1564
509. Davis DG, Bax A (1985) Separation of chemical exchange and cross-relaxation effects in two-dimensional NMR spectroscopy. J Magn Reson 64:533
510. Rotstein A, Lifshitz A, Kashman Y (1974) Isolation and antibacterial activity of acylphloroglucinols from *Myrtus communis*. Antimicrob Agents Chemother 6:539
511. Zhao Y, Zhang R, Jiang KW, Qi J, Hu Y, Guo J, Zhu R, Zhang T, Egan AN, Yi TS, Huang CH, Ma H (2021) Nuclear phylotranscriptomics and phylogenomics support numerous polyploidization events and hypotheses for the evolution of rhizobial nitrogen-fixing symbiosis in Fabaceae. Mol Plant 14:748
512. Eade RA, Simes JJH, Stevenson B (1963) Extractives of Australian timbers. IV. Castanogenin (medicagenic acid) and bayogenin, $C_{30}H_{48}O_5$, from *Castanospermum australe* Cunn. et Fras. Aust J Chem 16:900
513. Ahmad VU, Ahmed W, Usmanghani K (1992) Triterpenoid saponins from leaves of *Castanospermum australe*. Phytochemistry 31:2805
514. Ahmed W, Usmanghani K, Ahmad I, Ahmad VU, Miyase T (1994) Isolation and characterization of saponins from *Castanospermum australe* Cunn. et Fraser. Chem Pharm Bull 42:314
515. Zhou BN, Johnson RK, Mattern MR, Fisher PW, Kingston DG (2001) The first naturally occurring Tie2 kinase inhibitor. Org Lett 3:4047
516. Iranshahi M, Vu H, Pham N, Zencak D, Forster P, Quinn RJ (2012) Cytotoxic evaluation of alkaloids and isoflavonoids from the Australian tree *Erythrina vespertilio*. Planta Med 78:730
517. Eade RA, Salasoo I, Simes JJH (1966) Extractives of Australian timbers VII. Bayin. Aust J Chem 19:1717
518. Berry RC, Eade RA, Simes JJH (1977) Extractives of Australian timbers. XVII. The isolation, structure and synthesis of koparin (7,2′,3′-trihydroxy-4′-methoxyisoflavone). Aust J Chem 30:1827
519. Iinuma M, Ohyama M, Tanaka T, Hegarty MP, Egarty EE (1993) Isoflavonoids in roots of *Sophora fraseri*. Phytochemistry 34:1654

520. Van Heerden FR, Brandt EV, Roux DG (1979) Structure and synthesis of a derivative of fasciculiferin, a novel peltogynoid from *Acacia fasciculifera*. Tetrahedron Lett 20:4507
521. Clark-Lewis JW, Dainis I (1967) Flavan derivatives. XIX. Teracacidin and isoteracacidin from *Acacia obtusifolia* and *Acacia maidenii* heartwoods: phenolic hydroxylation patterns of heartwood flavonoids characteristic of sections and subsections of the genus *Acacia*. Aust J Chem 20:2191
522. Foo LY (1986) A novel pyrogallol A-ring proanthocyanidin dimer from *Acacia melanoxylon*. J Chem Soc, Chem Commun:236
523. Foo LY (1989) Isolation of [4-O-4]-linked biflavanoids from *Acacia melanoxylon*: first examples of a new class of single ether-linked proanthocyanidin dimers. J Chem Soc, Chem Commun:1505
524. Al-Sayed E, Abdel-Daim MM (2018) Analgesic and anti-inflammatory activities of epicatechin gallate from *Bauhinia hookeri*. Drug Dev Res 79:157
525. Imperato F (1982) A chalcone glycoside from *Acacia dealbata*. Phytochemistry 21:480
526. Hohenschutz LD, Bell EA, Jewess PJ, Leworthy DP, Pryce RJ, Arnold E, Clardy J (1981) Castanospermine, a 1,6,7,8-tetrahydroxyoctahydroindolizine alkaloid, from seeds of *Castanospermum australe*. Phytochemistry 20:811
527. Nash RJ, Fellows LE, Plant AC, Fleet GWJ, Derome AE, Baird PD, Hegarty MP, Scofield AM (1988) Isolation from and X-ray crystal structure of 3,8-diepialexine, (1*R*,2*R*,3*S*,7*S*,8*R*)-3-hydroxymethyl-1,2,7-trihydroxypyrrolizidine [(2*S*,3*R*,4*R*,5*S*,6*R*)-2-hydroxymethyl-1-azabicyclo[3.3.0]octan-3,4,6-triol]. Tetrahedron 44:5959
528. Molyneux RJ, Tropea JE, Elbein AD (1990) 7-Deoxy-6-*epi*-castanospermine, a trihydroxyindolizidine alkaloid glycosidase inhibitor from *Castanospermum australe*. J Nat Prod 53:609
529. Molyneux RJ, Pan YT, Tropea JE, Benson M, Kaushal GP, Elbein AD (1991) 6,7-Diepicastanospermine, a tetrahydroxyindolizidine alkaloid inhibitor of amyloglucosidase. Biochemistry 30:9981
530. Molyneux RJ, Benson M, Wong RY, Tropea JE, Elbein AD (1988) Australine, a novel pyrrolizidine alkaloid glucosidase inhibitor from *Castanospermum australe*. J Nat Prod 51:1198
531. Harris CM, Harris TM, Molyneux RJ, Tropea JE, Elbein AD (1989) 1-Epiastraline, a new pyrrolizidine alkaloid from *Castanospermum australe*. Tetrahedron Lett 30:5685
532. Nash RJ, Fellows LE, Dring JV, Fleet GWJ, Girdhar A, Ramsden NG, Peach JM, Hegarty MP, Scofield AM (1990) Two alexines (3-hydroxymethyl-1,2,7-trihydroxypyrrolizidines) from *Castanospermum australe*. Phytochemistry 29:111
533. Jones L, Hollinshead J, Fleet GWJ, Thompson AL, Watkin DJ, Gal ZA, Jenkinson SF, Kato A, Nash RJ (2010) Isolation of the pyrrolizidine alkaloid 1-epialexine from *Castanospermum australe*. Phytochem Lett 3:133
534. Tuechler A, Ferrier Å, Cosgrove R (2016) Transforming the inedible to the edible: an analysis of the nutritional returns from Aboriginal nut processing in Queensland's Wet Tropics. Aust Archaeol 79:26
535. Fitzgerald J, Johns SR, Lamberton JA, Redcliffe AH, Sioumis AA, Suares H (1972) Alkaloids of *Hovea longipes* (Leguminosae): the structure of a hypotensive alkaloid (–)-hoveine. Anales de Quím 68:737
536. Hart NK, Johns SR, Lamberton JA, MacKay MF, Mathieson AM, Satzke L (1972) The absolute molecular structure of podopetaline. Tetrahedron Lett 13:5333
537. Cheng PT, Mclean S, Misra R, Nyburg SC (1976) X-ray structure analysis of a further alkaloid from *Podopetalum ormondii*. Tetrahedron Lett 17:4245
538. Misra R, Wong-Ng W, Cheng P-T, McLean S, Nyburg SC (1980) Crystal structure of two quasi-racemates of (–)-podopetaline and (–)-ormosanine isolated from *Podopetalum ormondii*; the absolute configuration of (–)-ormosanine. J Chem Soc, Chem Commun:659
539. Griffin WJ, Phippard JH, Culvenor CCJ, Loder JW, Nearn R (1971) Alkaloids of the leaves of *Erythrophleum chlorostachys*. Phytochemistry 10:2793

540. Loder JW, Culvenor CCJ, Nearn RH, Russell GB, Stanton DW (1972) Isolation of norcassamidide and authentic norcassamidine from *Erythrophleum chlorostachys*. Structural revision of the alkaloids previously known as norcassamidine, norcassamine, norethythrosuamine and dehydro-norerythrosuamine. Tetrahedron Lett 13:5069
541. Loder JW, Culvenor CCJ, Nearn RH, Russell GB, Stanton DW (1974) Tumour inhibitory plants. New alkaloids from the bark of *Erythrophleum chlorostachys* (Leguminosae). Aust J Chem 27:179
542. Loder JW, Nearn RH (1975) Structure of norerythrostachaldine, a cytotoxic alkaloid from *Erythrophleum chlorostachys* (Leguminosae). Aust J Chem 28:651
543. Falkiner MJ, Faux AF, Loder JW, Nearn RH (1975) Isolation of unstable alkaloids from *Erythrophleum chlorostachys* (Leguminosae). Aust J Chem 28:645
544. Benn M, McEwan D, Pass MA, Majak W (1992) Three nitropropanoyl esters of glucose from *Indigofera linnaei*. Phytochemistry 31:2393
545. Pollitt S, Hegarty MP, Pass MA (1999) Analysis of the amino acid indospicine in biological samples by high performance liquid chromatography. Nat Toxins 7:233
546. Fitzgerald JS, Sioumis AA (1965) Alkaloids of the Australian Leguminosae. V. The occurrence of methylated tryptamines in *Acacia maidenii* F. Muell. Aust J Chem 18:433
547. Folk RA, Siniscalchi CM, Doby J, Kates HR, Manchester SR, Soltis PS, Soltis DE, Guralnick RP, Belitz M (2024) Spatial phylogenetics of Fagales: investigating drivers of temperate forest distributions. J Biogeogr 51:1518
548. Steane DA, Wilson KL, Hill RS (2003) Using matK sequence data to unravel the phylogeny of Casuarinaceae. Mol Phylogenet Evol 28:47
549. Takahashi H, Iuchi M, Fujita Y, Minami H, Fukuyama Y (1999) Coumaroyl triterpenes from *Casuarina equisetifolia*. Phytochemistry 51:543
550. Wang P, Wang H, Cai C, Wang H, Kong F, Yuan J, Dai H, Li L, Mei W (2020) Three new compounds from the litters of *Casuarina equisetifolia*. Phytochem Lett 35:58
551. Guise GB, Ritchie E, Taylor WC (1962) Further constituents of *Alphitonia* species. Aust J Chem 15:314
552. Branch G, Burgess D, Dunstan P, Foo L, Green G, Mack JG, Ritchie E, Taylor W (1972) Constituents of *Alphitonia* species. III. Alphitexolide, a new triterpene, and other extractives. Aust J Chem 25:2209
553. Raju R, Gunawardena D, Ahktar MA, Low M, Reddell P, Munch G (2016) Anti-inflammatory chemical profiling of the Australian rainforest tree *Alphitonia petriei* (Rhamnaceae). Molecules 21:1521
554. Eade RA, Rossler LP, Simes HV, Simes JJH (1965) Extractives of Australian timbers. VI. Ebelin lactone. Aust J Chem 18:1451
555. Eade RA, Ellis J, Harper P, Simes JJH (1973) Extractives of Australian timbers. XIII. Jingullic acid, a triterpene of the lupane series containing a C28 $\rightarrow$ 29 lactone group. Aust J Chem 26:831
556. Cooke RG, Johnson BL (1963) Colouring matters of Australian plants. XI. Quinones from *Ventilago viminalis*. Aust J Chem 16:695
557. Huang X, Deng T, Moore MJ, Wang H, Li Z, Lin N, Yusupov Z, Tojibaev KS, Wang Y, Sun H (2019) Tropical Asian Origin, boreotropical migration and long-distance dispersal in nettles (Urticeae, Urticaceae). Mol Phylogenet Evol 137:190
558. Leung CTW, Williams DH, Barna JCJ, Foti S, Oelrichs PB (1986) Structural studies on the peptide moroidin from *Laportea moroides*. Tetrahedron 42:3333
559. Gilding EK, Jami S, Deuis JR, Israel MR, Harvey PJ, Poth AG, Rehm FBH, Stow JL, Robinson SD, Yap K, Brown DL, Hamilton BR, Andersson D, Craik DJ, Vetter I, Durek T (2020) Neurotoxic peptides from the venom of the giant Australian stinging tree. Sci Adv 6:8828
560. Jami S, Deuis JR, Klasfauseweh T, Cheng X, Kurdyukov S, Chung F, Okorokov AL, Li S, Zhang J, Cristofori-Armstrong B, Israel MR, Ju RJ, Robinson SD, Zhao P, Ragnarsson L, Andersson A, Tran P, Schendel V, McMahon KL, Tran HNT, Chin YK, Zhu Y, Liu J, Crawford T, Purushothamvasan S, Habib AM, Andersson DA, Rash LD, Wood JN, Zhao J, Stehbens SJ, Mobli M, Leffler A, Jiang D, Cox JJ, Waxman SG, Dib-Hajj SD, Neely GG, Durek T,

Vetter I (2023) Pain-causing stinging nettle toxins target TMEM233 to modulate Na(V)1.7 function. Nat Commun 14:2442

561. Hart NK, Johns SR, Lamberton JA (1968) 3,4-Dimethoxy-ω-(2′-piperidyl)acetophenone, a new alkaloid from *Boehmeria platyphylla* Don. (family Urticaceae). Aust J Chem 21:1397
562. Hart NK, Johns SR, Lamberton JA (1968) Minor alkaloids of *Boehmeria platyphylla* Don. (family Urticaceae). II. Isolation of cryptopleurine and a new *seco*-phenanthroquinolizidine alkaloid. Aust J Chem 21:2579
563. Farnsworth NR, Hart NK, Johns SR, Lamberton JA, Messmer W (1969) Alkaloids of *Boehmeria cylindrica* (family Urticaceae): identification of a cytotoxic agent, highly active against Eagle's 9KB carcinoma of the nasopharynx in cell culture, as crytopleurine. Aust J Chem 22:1805
564. Russel JH (1963) Alkaloids of *Ficus* species. Occurrence of indolizidine alkaloids in *Ficus septica.* Sci Nat 50:443
565. Galbraith MN, Miller CJ, Rawson JWL, Ritchie E, Shannon JS, Taylor WC (1965) Moretenol and other triterpenes from *Ficus macrophylla* Desf. Aust J Chem 18:226
566. Johns SR, Russel JH (1965) Ficine, a novel flavonoidal alkaloid from *Ficus pantoniana.* Tetrahedron Lett 1987
567. Carroll AR, Davis RA, Addepalli R, Fechner GA, Guymer GP, Forster PI, Quinn RJ (2009) Cytotoxic agarofurans from the seeds of the Australian rainforest vine *Celastrus subspicata.* Phytochem Lett 2:163
568. Wibowo M, Wang Q, Holst J, White JM, Hofmann A, Davis RA (2017) Celastrofurans A-G: dihydro-β-agarofurans from the Australian rainforest vine *Celastrus subspicata* and their inhibitory effect on leucine transport in prostate cancer cells. J Nat Prod 80:1918
569. Wibowo M, Wang Q, Holst J, White JM, Hofmann A, Davis RA (2018) Dihydro-β-agarofurans from the roots of the Australian endemic rainforest tree *Maytenus bilocularis* act as leucine transport inhibitors. Phytochemistry 148:71
570. Wibowo M, Levrier C, Sadowski MC, Nelson CC, Wang Q, Holst J, Healy PC, Hofmann A, Davis RA (2016) Bioactive dihydro-β-agarofuran sesquiterpenoids from the Australian rainforest plant *Maytenus bilocularis.* J Nat Prod 79:1445
571. Wibowo M, Forster PI, Guymer GP, Hofmann A, Davis RA (2019) Using UHPLC-MS profiling for the discovery of new dihydro-β-agarofurans from Australian Celastraceae plant extracts. Molecules 24:859
572. Levrier C, Sadowski MC, Nelson CC, Healy PC, Davis RA (2015) Denhaminols A-H, dihydro-β-agarofurans from the endemic Australian rainforest plant *Denhamia celastroides.* J Nat Prod 78:111
573. Gordon S, Wibowo M, Wang Q, Holst J, Davis RA (2018) Dihydro-beta-agarofurans from the Australian rainforest plant *Denhamia celastroides* that inhibit leucine transport in prostate cancer cells. Magn Reson Chem 57:101
574. Wibowo M, Wang Q, Holst J, White JM, Hofmann A, Davis RA (2016) Dihydro-β-agarofurans from the Australian endemic rainforest plant *Denhamia pittosporoides* inhibit leucine transport in prostate cancer cells. Asian J Org Chem 5:1461
575. Koutentis PA, Jesus MCD, Wapling J, Kiem E, Stockdale T, Whittaker JJ, Clegg JK, Leach D, Church T, Collins R, Leach G, Hewitt D, Voss JJD, Blanchfield JT (2022) Phytochemical composition of *Denhamia obscura* (A. Rich.) Meisn. ex Walp. root bark, seeds and leaves. Arkivoc 2022:123
576. Butler MS, Towerzey L, Pham NB, Hyde E, Wadi SK, Guymer GP, Quinn RJ (2014) Cardenolide glycosides from *Elaeodendron australe* var. *integrifolium.* Phytochemistry 98:160
577. Wagner H, Flitsch K (2007) Neue Monocyclische Lactone aus *Siphonodon australe.* Planta Med 43:245
578. Breck GD, Stout GH (1969) *Calophyllum* products. V. New 4-phenylcoumarin from *Calophyllum australianum.* J Org Chem 34:4203
579. Bhanu S, Scheinmann F, Jefferson A (1975) Xanthones from the heartwood of *Calophyllum ramiflorum.* Phytochemistry 14:298

580. Tokuoka T (2007) Molecular phylogenetic analysis of Euphorbiaceae sensu stricto based on plastid and nuclear DNA sequences and ovule and seed character evolution. J Plant Res 120:511
581. Cullen JK, Boyle GM, Yap PY, Elmlinger S, Simmons JL, Broit N, Johns J, Ferguson B, Maslovskaya LA, Savchenko AI, Mirzayans PM, Porzelle A, Bernhardt PV, Gordon VA, Reddell PW, Pagani A, Appendino G, Parsons PG, Williams CM (2021) Activation of PKC supports the anticancer activity of tigilanol tiglate and related epoxytiglianes. Sci Rep 11:207
582. Panizza BJ, De Souza P, Cooper A, Roohullah A, Karapetis CS, Lickliter JD (2019) Phase I dose-escalation study to determine the safety, tolerability, preliminary efficacy and pharmacokinetics of an intratumoral injection of tigilanol tiglate (EBC-46). EBioMedicine 50:433
583. Erickson KL, Beutler JA, Cardellina II JH, McMahon JB, Newman DJ, Boyd MR (1995) A novel phorbol ester from *Excoecaria agallocha*. J Nat Prod 58:769
584. Maslovskaya LA, Savchenko AI, Gordon VA, Reddell PW, Pierce CJ, Parsons PG, Williams CM (2015) EBC-316, 325–327, and 345: new pimarane diterpenes from *Croton insularis* found in the Australian rainforest. Aust J Chem 68:652
585. Maslovskaya LA, Savchenko AI, Gordon VA, Reddell PW, Pierce CJ, Parsons PG, Williams CM (2011) Isolation and confirmation of the proposed cleistanthol biogenetic link from *Croton insularis*. Org Lett 13:1032
586. Maslovskaya LA, Savchenko AI, Pierce CJ, Gordon VA, Reddell PW, Parsons PG, Williams CM (2014) Unprecedented 1,14-seco-crotofolanes from *Croton insularis*: oxidative cleavage of crotofolin C by a putative homo-Baeyer-Villiger rearrangement. Chemistry 20:14226
587. Maslovskaya LA, Savchenko AI, Gordon VA, Reddell PW, Pierce CJ, Parsons PG, Williams CM (2017) The first casbane hydroperoxides EBC-304 and EBC-320 from the Australian rainforest. Chemistry 23:537
588. Maslovskaya LA, Savchenko AI, Pierce CJ, Boyle GM, Gordon VA, Reddell PW, Parsons PG, Williams CM (2019) New casbanes and the first *trans*-cyclopropane *seco*-casbane from the Australian rainforest plant *Croton insularis*. Chemistry 25:1525
589. Maslovskaya LA, Savchenko AI, Krenske EH, Pierce CJ, Gordon VA, Reddell PW, Parsons PG, Williams CM (2014) EBC-219: a new diterpene skeleton, crotinsulidane, from the Australian rainforest containing a bridgehead double bond. Angew Chem Int Ed Engl 53:7006
590. Maslovskaya LA, Savchenko AI, Krenske EH, Chow S, Gordon VA, Reddell PW, Pierce CJ, Parsons PG, Boyle GM, Williams CM (2020) EBC-342: a novel tetrahydrofuran moiety containing casbane from the Australian rainforest. Eur J Org Chem 2020:1042
591. Maslovskaya LA, Savchenko AI, Gordon VA, Reddell PW, Pierce CJ, Parsons PG, Williams CM (2016) EBC-318 and 339: bicyclo[10.2.1]alkanes from *Croton insularis*. RSC Adv 6:25110
592. Maslovskaya LA, Savchenko AI, Gordon VA, Reddell PW, Pierce CJ, Parsons PG, Williams CM (2016) *seco*-Casbanes from the Australian rainforest: ECD predictions key for determining remote absolute configuration. Eur J Org Chem 2016:1673
593. Maslovskaya LA, Savchenko AI, Krenske EH, Gordon VA, Reddell PW, Pierce CJ, Parsons PG, Williams CM (2014) *Croton insularis* introduces the *seco*-casbane class with EBC-329 and the first casbane endoperoxide EBC-324. Chem Commun 50:12315
594. Maslovskaya LA, Savchenko AI, Gordon VA, Reddell PW, Pierce CJ, Boyle GM, Parsons PG, Williams CM (2019) New halimanes from the Australian rainforest plant *Croton insularis*. Eur J Org Chem 2019:1058
595. Fujita E, Node M, Nishide K, Sai M, Fuji K, McPhail AT, Lamberton JA (1980) Structures of croverin (X-ray analysis) and dihydrocroverin, two new diterpene lactones from *Croton verreauxii* Baill. J Chem Soc, Chem Commun:920
596. Maslovskaya LA, Savchenko AI, Krenske EH, Chow S, Holt T, Gordon VA, Reddell PW, Pierce CJ, Parsons PG, Boyle GM, Kutateladze AG, Williams CM (2020) EBC-232 and 323: a structural conundrum necessitating unification of five in silico prediction and elucidation methods. Chemistry 26:11862

597. Hatt HH, Redcliff AH (1961) Kamlolenic acid from the seed fats of *Mallotus discolor* and *M. claoxyloides*. Aust J Chem 14:321
598. Kite GC, Fellows LE, Lees DC, Kitchen D, Monteith GB (1991) Alkaloidal glycosidase inhibitors in nocturnal and diurnal Uraniine moths and their respective foodplant genera, *Endospermum* and *Omphalea*. Biochem Syst Ecol 19:441
599. Doan TP, Park EJ, Cho HM, Ryu B, Lee BW, Thuong PT, Oh WK (2021) Rugonidines A–F, diastereomeric 1,6-dioxa-7,9-diazaspiro[4.5]dec-7-en-8-amines from the leaves of *Alchornea rugosa*. J Nat Prod 84:3055
600. Grimsson F, Graham SA, Coiro M, Jacobs BF, Xafis A, Neumann FH, Scott L, Sakala J, Currano ED, Zetter R (2019) Origin and divergence of Afro-Indian Picrodendraceae: linking pollen morphology, dispersal modes, fossil records, molecular dating and paleogeography. Grana 58:227
601. Grace MH, Jin Y, Wilson GR, Coates RM (2006) Structures, biogenetic relationships, and cytotoxicity of pimarane-derived diterpenes from *Petalostigma pubescens*. Phytochemistry 67:1708
602. Zhang J, Nguyen DT, Pierens GK, Cock I, Feng Y (2022) Austrobuxusin N, a new picrotoxane terpenoid glycoside, from the Australian endemic plant *Austrobuxus swainii* (Beuzev. & C.T. White) Airy Shaw. Nat Prod Res 36:5199
603. Yin S, Sykes ML, Davis RA, Shelper T, Avery VM, Camp D, Quinn RJ (2010) New galloylated flavanonols from the Australian plant *Glochidion sumatranum*. Planta Med 76:1877
604. Johns SR, Lamberton JA, Sioumis AA, Suares H (1974) The alkaloids of *Poranthera corymbosa* (Euphorbiaceae). Aust J Chem 27:2025
605. White D (2019) Biogeography, diversification, and domestication in the coca family (Erythroxylaceae). Ph.D. thesis, University of Illinois at Chicago, Chicago, IL, USA, p 144
606. Johns SR, Lamberton JA (1967) Meteloidine from *Erythroxylum australe* F. Muell. Aust J Chem 20:1301
607. Griffin WJ (1978) A phytochemical investigation of *Erythroxylon australe* F. Muell. Aust J Chem 31:1161
608. El-Imam YMA, Evans WC, Grout RJ (1988) Alkaloids of *Erythroxylum cuneatum*, *E. ecarinatum* and *E. australe*. Phytochemistry 27:2181
609. Ansell SM, Pegel KH, Taylor DAH (1993) Diterpenes from the timber of *Erythroxylum australe*. Phytochemistry 32:937
610. Fitzgerald JS (1965) (+)-Hygroline, the major alkaloid of *Carallia brachiata* (Rhizophoraceae). Aust J Chem 18:589
611. Shoppee CW, Howden MEH, Johnston GAR (1962) 93. Constituents of the bark of *Balanops australiana* F. Muell. J Chem Soc:498
612. Setzer WN, Setzer MC, Peppers RL, McFerrin MB, Meehan EJ, Chen L, Bates RB, Nakkiew P, Jackes BR (2000) Triterpenoid constituents in the bark of *Balanops australiana*. Aust J Chem 53:809
613. Pokharel SK, Setzer MS, Greenleaf SE, Dosoky NS, Jackes BR, Setzer WN (2017) Bioactive dammarane triterpenoids from the bark of *Drypetes acuminata* from Paluma, North Queensland, Australia. Rec Nat Prod 11:1
614. Webber BL, Miller RE (2008) Gynocardin from *Baileyoxylon lanceolatum* and a revision of cyanogenic glycosides in Achariaceae. Biochem Syst Ecol 36:545
615. Webber BL, Woodrow IE (2009) Chemical and physical plant defence across multiple ontogenetic stages in a tropical rain forest understorey tree. J Ecol 97:761
616. Mosaddik MA, Forster PI, Booth R, Waterman PG (2007) Clerodane diterpenes from the stems of *Casearia grewiifolia* var. *gelonioides* (Flacourtiaceae/Salicaceae sensu lato). Biochem Syst Ecol 35:631
617. Mosaddik A, Forster PI, Booth R, Waterman PG (2006) New clerodane and halimane diterpenes from the leaves and woody stems of *Casearia grayi* (Flacourtiaceae/Salicaceae). Nat Prod Commun 1:441
618. Mosaddik MA, Forster PI, Booth R, Waterman PG (2007) Phenolic glycosides from some Australian species of Flacourtiaceae (Salicaceae sensu lato). Biochem Syst Ecol 35:166

619. Mosaddik A, Forster PI, Waterman PG (2007) Three new 3-benzylbenzofuran-2-one derivatives from *Homalium brachybotrys* (Flacourtiaceae/Salicaceae s. l.). Nat Prod Res 21:1191
620. Bruun T, Jefferies PR, Lang W, Berndt W (1954) *epi*-Friedelinol and derivatives, a re-investigation. Acta Chem Scand 8:1948
621. Meksuriyen D, Nanayakkara DNP, Phoebe CH, Cordell GA (1986) Two triterpenes from *Davidsonia pruriens*. Phytochemistry 25:1685
622. Crayn DM, Rossetto M, Maynard DJ (2006) Molecular phylogeny and dating reveals an Oligo-Miocene radiation of dry-adapted shrubs (former Tremandraceae) from rainforest tree progenitors (Elaeocarpaceae) in Australia. Am J Bot 93:1328
623. Quirion JC, Kan C, Bick IRC, Husson HP (1987) Novel dimeric indole alkaloids from *Aristotelia australasica*. Structural determination and synthesis. J Org Chem 52:4527
624. Kan-Fan C, Quirion JC, Bick IRC, Husson HP (1988) Nouveaux alcaloides indoliques derives de l'aristoteline extraits d'*Aristotelia australasica*. Tetrahedron 44:1651
625. Quirion JC, Kan-Fan C, Bick IRC, Husson HP (1988) Aristolasol and aristolasene: indole alkaloids from *Aristotelia australasica*. Phytochemistry 27:3337
626. Quirion JC, Kan C, Husson HP, Bick IRC (1990) Alcaloïdes indoliques isoles d'*Aristotelia australasica*. J Nat Prod 53:713
627. Quirion JC, Husson HP, Kan C, Laprevote O, Chiaroni A, Riche C, Burkard S, Borschberg HJ, Bick IRC (1992) A revision of the structures of 17-*epi*-aristoteline and aristolasicone, the first two examples of inverted indole alkaloids. J Org Chem 57:5848
628. Bick IRC, Bremner JB, Preston NW, Calder IC (1971) Peduncularine, an indole–pyrrolizidine alkaloid. J Chem Soc D:1155
629. Ros HP, Kyburz R, Preston NW, Gallagher RT, Bick IRC, Hesse M (1979) The structure of the alkaloid peduncularine. Communication no. 172 on organic natural products. Helv Chim Acta 62:481
630. Kyburz R, Schöpp E, Hesse M, Bick IRC (1979) Sorellin und Hobartin, zwei neue Indolalkaloide aus *Aristotelia peduncularis* (LABILL.) HOOK. F. 173. Mitteilung über organische Naturstoffe. Helv Chim Acta 62:2539
631. Hai MA, Preston NW, Kyburz R, Schöpp E, Bick IRC, Hesse M (1980) Aristoserratin, ein neues Indolalkaloid aus *Aristotelia serrata* W.R.B. OLIVER und *A. peduncularis* (LABILL.) HOOK. F. 178. Mitteilung über organische Naturstoffe. Helv Chim Acta 63:2130
632. Kyburz R, Schöpp E, Bick IRC, Hesse M (1981) Tasmanin, ein neues Indolalkaloid aus *Aristotelia peduncularis* (LABILL.) HOOK. F. 181. Mitteilung über organische Naturstoffe. Helv Chim Acta 64:2555
633. Kyburz R, Schöpp E, Hesse M (1984) Peduncularistin, Triabunnin und Aristolarin, drei neue Indolalkaloide aus *Aristotelia peduncularis* (LABILL.) HOOK. F., 191. Mitteilung über organische Naturstoffe. Helv Chim Acta 67:804
634. Bick IRC, Hai MA, Preston NW (1985) Isopeduncularine. Tetrahedron 41:3127
635. Quirion JC, Husson HP, Kan C, Bick IRC (1993) Structure of aristoaristone, a new dimeric indole alkaloid from *Aristotelia australasica*. Nat Prod Lett 2:41
636. Arias HR, Ortells MO, Feuerbach D, Burgos V, Paz C (2019) Alkaloids purified from *Aristotelia chilensis* inhibit the human $\alpha3\beta4$ nicotinic acetylcholine receptor with higher potencies compared with the human $\alpha4\beta2$ and $\alpha7$ subtypes. J Nat Prod 82:1953
637. Johns SR, Lamberton JA, Sioumis AA (1971) New tropane alkaloids, (+)-(3*R*,6*R*)-3α-acetoxy-6β-hydroxytropane and (+)-2α-benzoyloxy-3β-hydroxynortropane, from *Peripentadenia mearsii* (Euphorbiaceae). Aust J Chem 24:2399
638. Lamberton JA, Geewananda YA, Gunawardana P, Bick IRC (1983) Alkaloids of *Peripentadenia mearsii*. I. Isolation, structural determination, and synthesis of peripentadenine. J Nat Prod 46:235
639. Robertson GB, Tooptakong U, Lamberton JA, Geewananda YA, Gunawardana P, Bick IRC (1984) Mearsine, a unique isoquinuclidine alkaloid: X-ray crystal structure of mearsine picrate. Tetrahedron Lett 25:2695

640. Bick IRC, Gunawardana YAGP, Lamberton JA (1985) Alkaloids of *Peripentadenia mearsii*. II. Tetrahedron 41:5627
641. Katavic PL, Venables DA, Guymer GP, Forster PI, Carroll AR (2007) Alkaloids with human δ-opioid receptor binding affinity from the Australian rainforest tree *Peripentadenia mearsii*. J Nat Prod 70:1946
642. Carroll AR, Arumugan G, Quinn RJ, Redburn J, Guymer G, Grimshaw P (2005) Grandisine A and B, novel indolizidine alkaloids with human δ-opioid receptor binding affinity from the leaves of the Australian rainforest tree *Elaeocarpus grandis*. J Org Chem 70:1889
643. Katavic PL, Venables DA, Forster PI, Guymer G, Carroll AR (2006) Grandisines C-G, indolizidine alkaloids from the Australian rainforest tree *Elaeocarpus grandis*. J Nat Prod 69:1295
644. Johns SR, Lamberton JA, Sioumis AA (1968) Elaeocarpidine, a new indole alkaloid from *Elaeocarpus archboldianus* A.C. Sm. Chem Commun:410
645. Hong W, Zhang Y, Yang J, Xia MY, Luo JF, Li XN, Wang YH, Wang JS (2019) Alkaloids from the branches and leaves of *Elaeocarpus angustifolius*. J Nat Prod 82:3221
646. Johns SR, Lamberton JA, Sioumis AA, Suares H, Willing RI (1971) *Elaeocarpus* alkaloids. IV. The alkaloids of *Elaeocarpus sphaericus*. Aust J Chem 24:1679
647. Singh A, Münch G, Reddell P, Radzieta M, Jensen SO, Raju R (2018) A new anti-inflammatory phenolic monosaccharide from the Australian native rainforest plant *Elaeocarpus eumundi*. Nat Prod Commun 13:731
648. Singh SB, Pettit GR (1990) Antineoplastic agents, 195. Isolation and structure of aceratioside from *Aceratium megalospermum*. J Nat Prod 53:1187
649. Miller RE, Stewart M, Capon RJ, Woodrow IE (2006) A galloylated cyanogenic glycoside from the Australian endemic rainforest tree *Elaeocarpus sericopetalus* (Elaeocarpaceae). Phytochemistry 67:1365
650. Buerki S, Callmander MW, Acevedo-Rodriguez P, Lowry PP II, Munzinger J, Bailey P, Maurin O, Brewer GE, Epitawalage N, Baker WJ, Forest F (2021) An updated infra-familial classification of Sapindaceae based on targeted enrichment data. Am J Bot 108:1234
651. Simpson BS, Claudie DJ, Gerber JP, Pyke SM, Wang J, McKinnon RA, Semple SJ (2011) In vivo activity of benzoyl ester clerodane diterpenoid derivatives from *Dodonaea polyandra*. J Nat Prod 74:650
652. Simpson BS, Claudie DJ, Smith NM, McKinnon RA, Semple SJ (2012) Rare, seven-membered cyclic ether labdane diterpenoid from *Dodonaea polyandra*. Phytochemistry 84:141
653. Khong PW, Lewis KG (1976) Chemical constituents of *Harpullia pendula*. Aust J Chem 29:1351
654. Cherry RF, Khong PW, Lewis KG (1977) Chemical constituents of *Harpullia pendula*. II. Further constituents of the bark and leaves. Aust J Chem 30:1397
655. Ghaly NS, Nabil M, Grace MH, Melek FR (2017) Pendulaosides A and B. Two acylated triterpenoid saponins from *Harpullia pendula* seed extract. Phytochem Lett 21:278
656. Sun D-A, Deng J-Z, Starck SR, Hecht SM (1999) Mispyric acid, a new monocyclic triterpenoid with a novel skeleton from *Mischocarpus pyriformis* that inhibits DNA polymerase β. J Am Chem Soc 121:6120
657. Abdelkader MSA, Rateb ME, Mohamed GA, Jaspars M (2016) Harpulliasides A and B: two new benzeneacetic acid derivatives from *Harpullia pendula*. Phytochem Lett 15:131
658. Simpson BS, Claudie DJ, Smith NM, Gerber JP, McKinnon RA, Semple SJ (2011) Flavonoids from the leaves and stems of *Dodonaea polyandra*: a Northern Kaanju medicinal plant. Phytochemistry 72:1883
659. Payne SE, Flematti GR, Reeder A, Kotze AC, Durmic Z, Vercoe PE (2018) Procyanidin A2 in the Australian plant *Alectryon oleifolius* has anthelmintic activity against equine cyathostomins in vitro. Vet Parasitol 249:63
660. Eaton AL, Rakotondraibe LH, Brodie PJ, Goetz M, Kingston DG (2015) Antiproliferative trihydroxyalkylcyclohexenones from *Pleiogynium timoriense*. J Nat Prod 78:1752

661. Oelrichs PB, MacLeod JK, Seawright AA, Ng JC (1997) Isolation and characterisation of urushiol components from the Australian native cashew (*Semecarpus australiensis*). Nat Toxins 5:96
662. Netzel M, Netzel G, Tian Q, Schwartz S, Konczak I (2007) Native Australian fruits—a novel source of antioxidants for food. Innov Food Sci Emerg Technol 8:339
663. Jadulco RC, Koch M, Van Wagoner RM, Pond C, Gideon OG, Matainaho T, Piskaut P, Barrows LR (2011) New antimycobacterial triterpenoids from *Rhus taitensis*. Planta Med 77:1651
664. Noro JC, Barrows LR, Gideon OG, Ireland CM, Koch M, Matainaho T, Piskaut P, Pond CD, Bugni TS (2008) Tetrahdroxysqualene from *Rhus taitensis* shows antimycobacterial activity against *Mycobacterium tuberculosis*. J Nat Prod 71:1623
665. Carman RM, Cowley DE (1964) Canaric acid—a 3,4-secotriterpene acid from *Canarium muelleri*. Tetrahedron Lett 5:627
666. Joyce EM, Appelhans MS, Buerki S, Cheek M, De Vos JM, Pirani JR, Zuntini AR, Bachelier JB, Bayly MJ, Callmander MW, Devecchi MF, Pell SK, Groppo M, Lowry PP II, Mitchell J, Siniscalchi CM, Munzinger J, Orel HK, Pannell CM, Nauheimer L, Sauquet H, Weeks A, Muellner-Riehl AN, Leitch IJ, Maurin O, Forest F, Nargar K, Thiele KR, Baker WJ, Crayn DM (2023) Phylogenomic analyses of Sapindales support new family relationships, rapid Mid-Cretaceous Hothouse diversification, and heterogeneous histories of gene duplication. Front Plant Sci 14:1063174
667. Gibbons S, Craven L, Dunlop C, Gray AI, Hartley TG, Waterman PG (1997) The secondary metabolites of aff. *Samadera* SAC-2825: an Australian Simaroubaceae with unusual chemistry. Phytochemistry 44:1109
668. Waterman PG (2007) The current status of chemical systematics. Phytochemistry 68:2896
669. Appelhans MS, Bayly MJ, Heslewood MM, Groppo M, Verboom GA, Forster PI, Kallunki JA, Duretto MF (2021) A new subfamily classification of the Citrus family (Rutaceae) based on six nuclear and plastid markers. Taxon 70:1035
670. Yamashita M, Saito Y, Rahim A, Fukuyoshi S, Miyake K, Goto M, Nakagawa-Goto K (2020) Novel furoquinolinones from an Indonesian plant, *Lunasia amara*. Tetrahedon Lett 61:151861
671. Subehan TN, Kadota S, Tezuka Y (2011) Cytochrome P450 2D6 inhibitory constituents of *Lunasia amara*. Phytochem Lett 4:30
672. Storer R, Young DW (1972) The structure of halfordamine. Tetrahedron Lett 13:1555
673. Banbury LK, Shou Q, Renshaw DE, Lambley EH, Griesser HJ, Mon H, Wohlmuth H (2015) Compounds from *Geijera parviflora* with prostaglandin E2 inhibitory activity may explain its traditional use for pain relief. J Ethnopharmacol 163:251
674. Agier C, Bury M, Aquette J, Hocquemiller R, Waterman PG (2007) New pseudan (2-alkyl-4(1*H*)-quinolinone) alkaloids from *Boronia ternata* var. *elongata* and *Boronia alulata* (Rutaceae). Nat Prod Res 21:698
675. Lamberton JA (1966) 1,2-Dimethyl-4-quinolone and xanthevodine, minor alkaloids from the leaves of *Acronychia baueri* Schott. Aust J Chem 19:1995
676. Bosson JA, Rasmussen M, Ritchie E, Robertson AV, Taylor WC (1963) The chemical constituents of Australian *Flindersia* species. XVII. The structure of ifflaiamine. Aust J Chem 16:480
677. Dacunha EVL, Armstrong JA, Gray AI, Hockless DCR, Waterman PG, White AH (1993) 3-Monoterpenyl 2,4-dioxygenated quinoline alkaloids from the aerial parts of *Eriostemon australasius* subsp. *banksii* (Rutaceae). Aust J Chem 46:1507
678. Brown RFC, Hobbs JJ, Hughes GK, Ritchie E (1954) The chemical constituents of Australian *Flindersia* species. VI. The structure and chemisty of flindersine. Aust J Chem 7:348
679. Coleman MA, Burchill L, Sumby CJ, George JH (2019) Biomimetic synthesis enables the structure revision of furoerioaustralasine. Org Lett 21:8776
680. Sultana N, Waterman PG (2001) 3-Monoterpenyl-2,4-dioxygenated quinoline alkaloids from the aerial parts of *Halfordia kendack*. Phytochemistry 58:329
681. Johnstone R, Price JR, Todd AR (1958) Alkaloids of the Australian Rutaceae: *Lunasia quercifolia*. I. 7-Methoxy-1-methyl-2-4-quinolone. Aust J Chem 11:562

682. Hart NK, Johns SR, Lamberton JA, Price JR (1968) Alkaloids of the Australian Rutaceae: *Lunasia quercifolia*. IV. Identification of a minor constituent as 5-hydroxy-1-methyl-2-phenyl-4-quinolone and the preparation of an angular isomer of (–)-lunine. Aust J Chem 21:1389
683. Wattanapiromsakul C, Forster PI, Waterman PG (2003) Alkaloids and limonoids from *Bouchardatia neurococca*: systematic significance. Phytochemistry 64:609
684. Halstead CW, Forster PI, Waterman PG (2006) Novel metabolites from the stem bark of *Brombya* sp. nova (Gap Creek) (Rutaceae). Nat Prod Commun 1:351
685. White CT, Ritchie E, Taylor WC, Vautin STK (1961) The chemical constituents of Australian *Flindersia* species. XV. The constituents of *Flindersia acuminata* C. T. White. Aust J Chem 14:469
686. Galbraith MN, Ritchie E, Taylor WC (1960) The chemical constituents of Australian *Flindersia* species. XIII. The constituents of *Flindersia bennettiana* F. Muell. Aust J Chem 13:427
687. Cannon JR, Hughes GK, Ritchie E (1952) The chemical constituents of Australian *Flindersia* species. IV. The constituents of *Flindersia bourjotiana* F. Muell. Aust J Chem 5:420
688. Anet FAL, Gilham PT, Gow P, Hughes GK, Ritchie E (1952) The chemical constituents of Australian *Flindersia* species. III. The alkaloids of *Flindersia collina* Bail. Aust J Chem 5:412
689. Breen GJW, Ritchie E, Taylor WC (1962) The chemical constituents of Australian *Flindersia* species. XVI. The constituents of the wood of *Flindersia laevicarpa* C. T. White & Francis. Aust J Chem 15:819
690. Bowden BF, Cleaver L, Ndalut PK, Ritchie E, Taylor WC (1975) The chemical constituents of Australian *Flindersia* species. XX. An examination of *F. pimenteliana*. Aust J Chem 28:1393
691. Hollis AF, Prager RH, Ritchie E, Taylor WC (1961) The chemical constituents of Australian *Flindersia* species. XIV. The constituents of *Flindersia pubescens* Bail. and *F. schottiana* F. Muell. Aust J Chem 14:100
692. Ritchie E, Taylor WC, Willcocks DV (1960) The chemical constituents of Australian *Flindersia* species. XII. The constituents of *Flindersia xanthoxyla* Domin. Aust J Chem 13:426
693. Johns SR, Lamberton JA (1966) Alkaloids of *Geijera salicifolia* Schott. (family Rutaceae): the identification of platydesmine and platydesmine acetate. Aust J Chem 19:1991
694. Crow WD, Hodgkin JH (1968) Alkaloids of the Australian Rutaceae: *Halfordia scleroxyla* and *Halfordia kendack*. IV. Co-occurrence of oxazole and quinoline alkaloids. Aust J Chem 21:3075
695. Price JR (1959) Alkaloids of the Australian Rutaceae: *Lunasia quercifolia*. II. The nature of lunasine. Aust J Chem 12:458
696. Hart NK, Price JR (1966) Alkaloids of the Australian Rutaceae: *Lunasia quercifolia*. III. Isolation of (–)-*O*-methylluninium salts. Aust J Chem 19:2185
697. Lamberton BJA, Price JR (1953) Alkaloids of the Australian Rutaceae: *Medicosma cunninghamii* Hook. f. Aust J Chem 6:173
698. Blanchi E, Culvenor CCJ, Lamberton JA (1968) Minor alkaloids of *Medicosma cunninghamii* Hook. f. Aust J Chem 21:2357
699. Habtemariam S, Waterman PG, Hartley TG (1996) Fareanine and fareanol from leaves of *Medicosma fareana*. Phytochemistry 43:291
700. Hughes GK, Neill KG (1949) Alkaloids of the Australian Rutaceae: *Evodia xanthoxyloides* I. Evoxanthine. Aust J Chem 2:429
701. Lahey FN, McCamish M, McEwan T (1969) The alkaloids of *Acronychia haplophylla*. Aust J Chem 22:447
702. Lahey FN, Thomas WC (1949) Alkaloids of the Australian Rutaceae: *Acronychia baueri*. I. The isolation of the alkaloids. Aust J Chem 2:423
703. Lamberton JA, Price JR (1953) Alkaloids of the Australian Rutaceae: *Acronychia baueri* Schott. IV. Alkaloids present in the leaves. Aust J Chem 6:66
704. Halstead CW, Forster PI, Waterman PG (2006) Alkaloids from the stem bark of an Australian population of *Zanthoxylum ovalifolium*. Nat Prod Res 20:940

705. Johns BSR, Lamberton JA, Sioumis AA (1968) Furoquinoline alkaloids from *Evodia elleryana* F. Muell. The structure of evellerine. Aust J Chem 21:1897
706. Robertson LP, Makwana V, Voser TM, Holland DC, Carroll AR (2021) Leptanoine D, a new quinoline alkaloid from the Australian tree *Pitaviaster haplophyllus* (Rutaceae). Aust J Chem 74:173
707. Ahsan M, Gray AI, Leach G, Waterman PG (1993) Quinolone and acridone alkaloids from *Boronia lanceolata*. Phytochemistry 33:1507
708. Auzi AA, Hartley TG, Waigh RD, Waterman PG (1996) Acridone alkaloids from *Bosistoa transversa*. Phytochemistry 42:235
709. Price JR (1949) Alkaloids of the Australian Rutaceae: *Melicope fareana*. IV. Some reactions of 1-methyl-4-quinolone-3-carboxylic acid, a degradation product of the alkaloids. Aust J Chem 2:272
710. Quader A, Gray AI, Hartley TG, Waterman PG (1991) 8-Prenylcoumarins: typical secondary metabolites of the Rutaceae from *Pentaceras australis*. Biochem Syst Ecol 19:91
711. Funayama S, Cordell GA (1984) Chemistry of acronycine IV. Minor constituents of acronine and the phytochemistry of the genus *Acronychia*. J Nat Prod 47:285
712. Sultana N, Hartley TG, Waterman PG (1999) Two novel prenylated flavones from the aerial parts of *Melicope micrococca*. Phytochemistry 50:1249
713. Shou Q, Banbury LK, Maccarone AT, Renshaw DE, Mon H, Griesser S, Griesser HJ, Blanksby SJ, Smith JE, Wohlmuth H (2014) Antibacterial anthranilic acid derivatives from *Geijera parviflora*. Fitoterapia 93:62
714. Crow WD, Hodgkin JH (1964) Alkaloids of the Australian Rutaceae: *Halfordia scleroxyla*. II. Isolation and structure of the alkaloids. Aust J Chem 17:119
715. Robertson LP, Duffy S, Wang Y, Wang D, Avery VM, Carroll AR (2017) Pimentelamines A-C, indole alkaloids isolated from the leaves of the Australian tree *Flindersia pimenteliana*. J Nat Prod 80:3211
716. Robertson LP, Lucantoni L, Avery VM, Carroll AR (2020) Antiplasmodial bis-indole alkaloids from the bark of *Flindersia pimenteliana*. Planta Med 86:19
717. Fernandez LS, Buchanan MS, Carroll AR, Feng YJ, Quinn RJ, Avery VM (2009) Flinderoles A-C: antimalarial bis-indole alkaloids from *Flindersia* species. Org Lett 11:329
718. Pousset JL, Cavé A, Chiaroni A, Riche C (1977) A novel bis-indole alkaloid. X-ray crystal structure determination of borreverine and its rearrangement product on diacetylation. J Chem Soc, Chem Commun:261
719. Picker K, Ritchie E, Taylor WC (1976) The chemical constituents of Australian *Flindersia* species. XXI. An examination of the bark and the leaves of *F. laevicarpa*. Aust J Chem 29:2023
720. Haynes HF, Nelson ER, Price JR (1952) Alkaloids of the Australian Rutaceae: *Pentaceras australis* Hook. f. I. Isolation of the alkaloids and identification of canthin-6-one. Aust J Chem 5:387
721. Auzi AA, Gray AI, Hartley TG, Waigh RD, Waterman PG (1994) A xanthine alkaloid from the leaves of *Bosistoa floydii*. Phytochemistry 36:535
722. Deyun K, Gray AI, Hartley TG, Waterman PG (1996) Alkaloids from an Australian accession of *Zanthoxylum nitidum* (Rutaceae). Biochem Syst Ecol 24:87
723. Cannon JR, Hughes GK, Ritchie E, Taylor WC (1953) The chemicial consitituents of Australian *Zanthoxylum* species. II. The constituents of *Z. brachyacanthum* F. Muell., *Z. veneficum* F.M. Bail., and *Z. suberosum* C.T. White. Aust J Chem 6:86
724. Robertson LP, Hall CR, Forster PI, Carroll AR (2018) Alkaloid diversity in the leaves of Australian *Flindersia* (Rutaceae) species driven by adaptation to aridity. Phytochemistry 152:71
725. Ahsan M, Gray AI, Leach G, Waterman PG (1994) Novel angular pyranocoumarins from *Boronia lanceolata*. Phytochemistry 36:777
726. Moreira AS, Mathias L, Braz-Filho R, Schripsema J, Vieira IJ (2002) Two new diprenylated coumarins from *Flindersia brayleyana*. Nat Prod Lett 16:291
727. Anet FAL, Blanks FR, Hughes GK (1949) The chemical constituents of Australian *Flindersia species*. I. Collinin, 7-geranoxy-8-methoxycoumarin. Aust J Chem 2:127

728. Lahey FN, MacLeod JK (1967) The coumarins of *Geijera parviflora* Lindl. Aust J Chem 20:1943
729. Dreyer DL, Lee A (1972) Extractives of *Geijera parviflora*. Phytochemistry 11:763
730. Shou Q, Banbury LK, Renshaw DE, Smith JE, He X, Dowell A, Griesser HJ, Heinrich M, Wohlmuth H (2013) Parvifloranines A and B, two 11-carbon alkaloids from *Geijera parviflora*. J Nat Prod 76:1384
731. Lahey FN, Wluka DJ (1955) Geijerin: a new coumarin from the bark of *Geijera salicifolia* Schott. Aust J Chem 8:125
732. Ritchie E, Taylor WC, Young LM (1968) Some extractives from *Geijera salicifolia* Schott. Aust J Chem 21:1381
733. Latip J, Hartley TG, Waterman PG (1999) Lignans and coumarins metabolites from *Melicope hayesii*. Phytochemistry 51:107
734. O'Donnell F, Ramachandran VN, Smyth TJ, Smyth WF, Brooks P (2009) An investigation of bioactive phytochemicals in the leaves of *Melicope vitiflora* by electrospray ionisation ion trap mass spectrometry. Anal Chim Acta 634:115
735. Raju R, Singh A, Reddell P, Münch G (2018) Anti-inflammatory activity of prenyl and geranyloxy furanocoumarins from *Citrus garrawayi* (Rutaceae). Phytochem Lett 27:197
736. Lamberton JA, Price JR, Redcliffe AH (1967) Micromelin, a new coumarin from *Micromelum minutum* (Forst. f.) Seem (family Rutaceae). Aust J Chem 20:973
737. Bissoué AN, Muyard F, Bévalot F, Hartley TG, Tillequin F, Vaquette J, Waterman PG (1997) A coumarin from aerial parts of *Geijera paniculata*. Phytochemistry 46:383
738. Latif Z, Hartley TG, Rice MJ, Waigh RD, Waterman PG (1998) Isobutylamides and coumarins from *Melicope melanophloia*. Biochem Syst Ecol 26:467
739. Nakayama M, Fujimoto S, Fukui K, MacLeod JK (1971) The isolation, structure, and synthesis of halkendin. Aust J Chem 24:209
740. MacLeod JK (1970) Halfordinin, a 3,4,5-oxygenated furanocoumarin with a novel α,α-dimethylallyl ether substituent. Tetrahedron Lett 11:1319
741. Sultana N, Skelton BW, Waterman PG, White AH (2000) 3,5-Dimethoxy-3-(1,1-dimethylprop-2-enyl)-3,4-dihydropsoralen-4-one: an unusual furanocoumarin derivative from *Halfordia kendack*. Phytochemistry 55:675
742. Lahey FN, Hutchins R (1961) Acrovestone—a new phenolic ketone. In: Proceedings of the symposium on phytochemistry. University of Hong Kong, p 121
743. Quader A, Armstrong JA, Gray AI, Hartley TG, Waterman PG (1991) Chemosystematics of *Acradenia* and general significance of acetophenones in the Rutaceae. Biochem Syst Ecol 19:171
744. Baldwin ME, Bick IRC, Komzak AA, Price JR (1961) Some ketones from *Acradenia franklinii*. Tetrahedron 16:206
745. Raju R, Mathew S, Singh A, Reddell P, Munch G (2022) Acronyols A and B, new anti-inflammatory prenylated phloroglucinols from the fruits of *Acronychia crassipetala*. Nat Prod Res 36:4364
746. Tran TD, Olsson MA, MacMillan DJ, Cullen JK, Parsons PG, Reddell PW, Ogbourne SM (2020) Potent antibacterial prenylated acetophenones from the Australian endemic plant *Acronychia crassipetala*. Antibiotics 9:8
747. Robertson LP, Lucantoni L, Duffy S, Avery VM, Carroll AR (2019) Acrotrione: an oxidized xanthene from the roots of *Acronychia pubescens*. J Nat Prod 82:1019
748. Parsons IC, Gray AI, Hartley TG, Waterman PG (1994) Acetophenones and coumarins from stem bark and leaves of *Melicope stipitata*. Phytochemistry 37:565
749. Kirby KD, Sutherland MD (1956) Some naturally occurring dimethylchromenes II. Alloevodione. Aust J Chem 9:411
750. Ritchie E, Taylor WC, Vautin TK (1965) The constituents of *Melicope broadbentiana* F.M. Bail. The structures of melibentin. melicopol, and methylmelicopol. Aust J Chem 18:2021
751. Chan JA, Shultis EA, Carr SA, Debrosse CW, Eggleston DS, Francis TA, Hyland LJ, Johnson WP, Killmer LB (1989) Novel phloroglucinols from the plant *Melicope sessiliflora* (Rutaceae). J Org Chem 54:2098

752. Auzi AA, Forster PI, Hartley TG, Waigh RD, Waterman PG (1997) Phloroglucinol derivatives from leaves of *Bosistoa pentacocca*. Phytochemistry 45:1673
753. Fazldeen H, Hegarty MP, Lahey FN (1978) Muellitol, a novel trialkenyl inositol from *Evodiella muelleri*. Phytochemistry 17:1609
754. Parsons IC, Gray AI, Waterman PG, Hartley TG (1993) New triterpenes and flavonoids from the leaves of *Bosistoa brassii*. J Nat Prod 56:46
755. Reisch J, Hussain RA, Mester I (1984) Flavonoids from *Flindersia australis*. Phytochemistry 23:2114
756. Reisch J, Wickramasinghe A, Kumar V (1989) Two phenylpropanoids from *Flindersia australis* stem wood. Phytochemistry 28:3242
757. Prager RH, Thredgold HM (1966) Some neutral constituents of *Acronychia baueri*. Aust J Chem 19:451
758. Johns SR, Lamberton JA, Price JR (1967) (±)-*N*-Benzoyl[2-hydroxy-2-(4′-methoxyphenyl)]ethylamine from *Clausena brevistyla* Oliver (family Rutaceae). Aust J Chem 20:2795
759. Latif Z, Hartley TG, Rice MJ, Waigh RD, Waterman PG (1998) Novel and insecticidal isobutylamides from *Dinosperma erythrococca*. J Nat Prod 61:614
760. Parsons IC, Gray AI, Hartley TG, Skelton BW, Waterman PG, White AH (1992) Brombyin-I [(*rel*)- 4*β*-acetyl-3-(1,3-benzodioxol-5-yl)-1,4,4a*β*,5,6,7,8,8a*α*-octahydronaphthalen-1*α*-ol] and brombyin-II [(*rel*)-4*β*-acetyl-3*α*-(1,3-benzodioxol-5-yl)-3,4,4a*β*,5,6,7,8,8a*β*-octahydronaphthalene]: a novel type of natural product isolated from the stem bark of *Brombya platynema*. F. Muell. J Chem Soc, Perkin Trans 1:645
761. Parsons IC, Gray AI, Hartley TG, Waterman PG (1993) Brombyins from the stem bark of *Brombya platynema*. Phytochemistry 33:479
762. Hambley TW, Rideout JA, Taylor WC (1990) The chemical constituents of Australian *Flindersia species*. XXIV. 1,3-Diphenylpropanes from the wood of *Flindersia brassii*. Aust J Chem 43:1327
763. Nimgirawath S, Ritchie E, Taylor WC (1977) The chemical constituents of Australian *Flindersia* species. XXII. Some extractives of *F. brassii*. Aust J Chem 30:451
764. Breen GJW, Ritchie E, Sidwell WTL, Taylor WC (1966) The chemical constituents of Australian *Flindersia* species. XIX. Triterpenoids from the leaves of *F. bourjotiana* F. Muell. Aust J Chem 19:455
765. Abdul Q, Gray AI, Waterman PG, Lavaud C, Massiot G, Sadier IH (1991) Phebaloparvi-lactone: a new protolimonoid from the leaves of *Phebulium souamulosum* ssp. *parvifolium*. Tetrahedron 47:3611
766. Bosson JA, Galbraith MN, Ritchie E, Taylor WC (1963) The chemical constituents of Australian *Flindersia* species. XVIII. The structure of ifflaionic acid. Aust J Chem 16:491
767. Breen GW, Picker K, Raston CL, Skelton BW, Taylor WC, White AH (1984) Chemical constituents of the *Flindersia*. XXIII. Determination of the structure of flindercarpin-2 as methyl (S)-(acetoxy){[3′*S*-(3*α*,3a*α*,4*α*,5*β*,6*α*,6a*α*,7*α*,10a*α*,11a*α*)]-5′,6′-diacetoxy-7′-(furan-3′-yl)-2′,2′,3a′,6a′, 11a′-pentamethyl-12′-methylene-9′-oxo-2′,3′,3a′,4′,5′,6′,6a′,9′,10′,11a′-decahydro-7′*H*-4′, 10a′-methano-10a′*H*-furo[2,3-b]pyrano[3,4-g]oxocin-3′-yl)acetate. Aust J Chem 37:1461
768. Jones RVJ, Sutherland MD (1968) Terpenoid chemistry. XV. 1,5-Dimethylcyclodeca-1,5,7-triene, the precursor of geijerene in *Geijera parviflora* (Lindley). Aust J Chem 21:2255
769. Bates RB, Onore MJ, Paknikar SK, Steelink C, Blanchard EP (1967) D- and L-filifolones, monoterpenoid cyclobutanones with the bicyclo[3,2,0]heptane ring system. Chem Commun:1037
770. Batey IL, Hellyer RO, Pinhey JT (1971) The structure of squamulosone, a new sesquiterpene ketone from *Phebalium squamulosum*. Aust J Chem 24:2173
771. Kongkathip N, Kongkathip B (2009) Constituents and bioactivities of *Clausena excavata*. Heterocycles 79:121
772. Khandokar L, Bari MS, Seidel V, Haque MA (2021) Ethnomedicinal uses, phytochemistry, pharmacological activities and toxicological profile of *Glycosmis pentaphylla* (Retz.) DC.: a review. J Ethnopharmacol 278:114313

773. Yohanes R, Harneti D, Supratman U, Fajriah S, Rudiana T (2023) Phytochemistry and biological activities of *Murraya* species. Molecules 28:5901
774. Michael JP (2017) Acridone alkaloids. In: Knölker H-J (ed) The alkaloids: chemistry and biology, vol 78. Elsevier, Amsterdam, p 1
775. Szewczyk A, Peczek F (2023) Furoquinoline alkaloids: insights into chemistry, occurrence, and biological properties. Int J Mol Sci 24:12811
776. Yadav TT, Murahari M, Peters GJ, Yc M (2022) A comprehensive review on acridone based derivatives as future anti-cancer agents and their structure activity relationships. Eur J Med Chem 239:114527
777. Negi JS, Bisht VK, Bhandari AK, Bharti MK, Sundriyal RC (2011) Chemical and pharmacological aspects of *Toona* (Meliaceae). Res J Phytochem 5:14
778. Hieu TT, Chung NT, Dung VC, Duc DX (2023) Chemical composition and bioactivities of *Melia azedarach* (Meliaceae): a comprehensive review. Curr Org Chem 26:2160
779. Muellner-Riehl AN, Rojas-Andrés BM (2022) Biogeography of neotropical Meliaceae: geological connections, fossil and molecular evidence revisited. Braz J Bot 45:527
780. Mulholland DA, Nair JJ (1994) Triterpenoids from *Dysoxylum pettigrewianum*. Phytochemistry 37:1409
781. Mulholland DA, Taylor DAH (1992) Limonoids from Australian members of the Meliaceae. Phytochemistry 31:4163
782. Mulholland DA, Kotsos M, Mahomed HA, Taylor DAH (1998) Triterpenoids from *Owenia cepiodora*. Phytochemistry 49:2457
783. Mulholland DA, Monkhe TV (1993) Two glabretal-type triterpenoids from the heartwood of *Aglaia ferruginaea*. Phytochemistry 34:579
784. Mohamad K, Sévenet T, Dumontet V, Pais M, Vantri M, Hadi H, Awang K, Martin M (1999) Dammarane triterpenes and pregnane steroids from *Aglaia lawii* and *A. tomentosa*. Phytochemistry 51:1031
785. Mulholland DA, Naidoo N (1998) A revision of the structure of ferrugin from *Aglaia ferruginaea*. Phytochemistry 47:1163
786. Dean FM, Monkhe TV, Mulholland DA, Taylor DAH (1993) An isoflavanoid from *Aglaia ferruginaea*, an Australian member of the Meliaceae. Phytochemistry 34:1537
787. Seger C, Pacher T, Greger H, Saifah E, Hofer O (2002) Aglairubine: structure revision of a chemotaxonomically interesting bisamide in *Aglaia* (Meliaceae). Monatsh Chem 133:97
788. Mulholland DA, Iourine S, Taylor DAH (1998) Sesquiterpenoids from *Dysoxylum schiffneri*. Phytochemistry 47:1421
789. Tyler MI, Howden ME (1985) Antineoplastic and piscicidal 1-alkyldaphnane orthoesters from *Pimelea* species. J Nat Prod 48:440
790. Johns SR, Lamberton JA, Loder JW, Redcliffe AH, Sioumis AA (1969) Alkaloids of *Argyrodendron peralatum* (Sterculiaceae): identification of Nα-cinnamoylhistamine. Aust J Chem 22:1309
791. Cook IF, Knox JR (1975) Flavonoids from *Cochlospermum gillivraei*. Phytochemistry 14:2510
792. Hendra R, Keller PA (2016) Flowers in Australia: phytochemical studies on the Illawarra flame tree and Alstonville. Aust J Chem 69:925
793. Doskotch RW, Ray AB, Beal JL (1971) Codonocarpine, a new lunaria-type alkaloid from *Codonocarpus australis* A. Cunn. J Chem Soc D:300
794. Ji W, Osbourn A, Liu Z (2024) Understanding metabolic diversification in plants: branchpoints in the evolution of specialized metabolism. Phil Trans R Soc Lond B Biol Sci 379:20230359
795. Kohonen T (1982) Self-organized formation of topologically correct feature maps. Biol Cybern 43:59
796. Sander T, Freyss J, Von Korff M, Rufener C (2015) DataWarrior: an open-source program for chemistry aware data visualization and analysis. J Chem Inf Model 55:460
797. Kim HW, Wang M, Leber CA, Nothias LF, Reher R, Kang KB, van der Hooft JJJ, Dorrestein PC, Gerwick WH, Cottrell GW (2021) NPClassifier: a deep neural network-based structural classification tool for natural products. J Nat Prod 84:2795

798. Zhang Y, Deng T, Sun L, Landis JB, Moore MJ, Wang H, Wang Y, Hao X, Chen J, Li S, Xu M, Puno PT, Raven PH, Sun H (2021) Phylogenetic patterns suggest frequent multiple origins of secondary metabolites across the seed-plant 'tree of life'. Natl Sci Rev 8:nwaa105
799. Thon FM, Muller C, Wittmann MJ (2024) The evolution of chemodiversity in plants—from verbal to quantitative models. Ecol Lett 27:e14365
800. Achenbach H, Renner C, Waibel R (1995) Constituents of tropical medicinal plants, LXIX. The hexalobines, diprenylated indoles from *Hexalobus crispiflorus* and *Hexalobus monopetalus*. Liebigs Ann Chem:1327
801. Izuogu NB, Bello OE, Bello OM (2020) A review on *Borreria verticillata*: a potential bionematicide, channeling its significant antimicrobial activity against root-knot nematodes. Heliyon 6:e05322
802. Wulff P, Carlé JS, Christophersen C (1981) Marine alkaloids. Part 4. A formamide, flustrabromine, from the marine bryozoan *Flustra foliacea*. J Chem Soc, Perkin Trans 1:2895
803. Suzuki S, Umezawa T (2007) Biosynthesis of lignans and norlignans. J Wood Sci 53:273
804. Juan C, Martinez V, Yoshida M, Gottlieb OR (1981) ω-Ethyl, ω-ethenyl and ω-ethynyl-α-alkylidene-γ-lactones from *Clinostemon mahuba*. Phytochemistry 20:459
805. Cao Y, Xuan B, Peng B, Li C, Chai X, Tu P (2015) The genus Lindera: a source of structurally diverse molecules having pharmacological significance. Phytochem Rev 15:869
806. Liaw CC, Liou JR, Wu TY, Chang FR, Wu YC (2016). Acetogenins from Annonaceae. In: Kinghorn AD, Falk, H, Gibbons, Kobayashi J (eds) Progress in the chemistry of organic natural products, vol 101. Springer, Cham, Switzerland, p 113
807. Guo D, Wang H, Zhang S, Lan T (2022) The type III polyketide synthase supergene family in plants: complex evolutionary history and functional divergence. Plant J 112:414
808. Naake T, Maeda HA, Proost S, Tohge T, Fernie AR (2021) Kingdom-wide analysis of the evolution of the plant type III polyketide synthase superfamily. Plant Physiol 185:857
809. Miyanaga A, Horinouchi S (2009) Enzymatic synthesis of bis-5-alkylresorcinols by resorcinol-producing type III polyketide synthases. J Antibiot 62:371
810. Joshi BS, Viswanathan N, Gawad DH, Balakrishnan V, Von Philipsborn W (2004) Piperaceae alkaloids: part IV. Structure and synthesis of cyclostachine A, cyclostachine B and cyclopiperstachine. Helv Chim Acta 58:2295
811. Miller JS (1987) Host-plant relationships in the Papilionidae (Lepidoptera): parallel cladogenesis or colonization? Cladistics 3:105
812. Kamanna K, Kamath A (2024) Prenylation of natural products: an overview. In: Zhan X, Jabbari A (ed) Modifications in biomacromolecules, biochemistry. IntechOpen, London
813. Shen Y, Liang WJ, Shi YN, Kennelly EJ, Zhao DK (2020) Structural diversity, bioactivities, and biosynthesis of natural diterpenoid alkaloids. Nat Prod Rep 37:763
814. Hernandez-Garcia J, Briones-Moreno A, Blazquez MA (2021) Origin and evolution of gibberellin signaling and metabolism in plants. Semin Cell Dev Biol 109:46
815. Sorokina M, Merseburger P, Rajan K, Yirik MA, Steinbeck C (2021) COCONUT online: collection of open natural products database. J Cheminform 13:2

Anthony (Tony) Carroll has a bachelor's degree in chemistry and botany and studied the alkaloid and lignan chemistry of Australian rainforest plants (B.Sc. (Hons) and Ph.D.) under the supervision of Prof. Walter C. Taylor at the University of Sydney (1985–1989). Marine natural products research became a major focus after postdoctoral fellowships at the University of Hawaii with Prof. Paul Scheuer and at James Cook University, Australia with Drs. John Coll and Bruce Bowden. A 15-year period as Head of Natural Products Chemistry for the AstraZeneca/Griffith University drug discovery project (1993–2008) allowed him to combine his interests in marine and plant natural products to discover new treatments of human diseases. Over this period, he also managed the collection program of plant and marine organisms that ultimately generated a library of over 40,000 samples. He also expanded his interests to include high-throughput purification and structure determination techniques and cheminformatics. Since 2008, he has held a faculty position at Griffith University, Gold Coast, Queensland, Australia where he is currently a Professor. He has supervised over 35 doctoral candidates and published more than 250 peer-reviewed journal articles. Over the last nine years, he has been the lead author of the highly cited annual review on Marine Natural Products published in *Natural Product Reports*. The challenges associated with identifying unique molecules in complex mixtures has been a major focus of his current research and he continues to develop NMR spectroscopic methods to tackle this challenging problem. He has had a keen interest in rainforests from a young age and uses science, photography, and painting to explore these rich ecosystems.

FSC
www.fsc.org
MIX
Papier | Fördert gute Waldnutzung
FSC® C083411

Kolibri 360 GmbH
Ferdinand-Jühlke-Straße 7
99095 Erfurt, Deutschland
produktsicherheit@kolibri360.de